THE CITY

A GEOGRAPHY OF URBAN PLACES

URBAN GEOGRAPHY

A GEOGRAPHY OF URBAN PLACES

Edited by

ROBERT G. PUTNAM, FRANK J. TAYLOR AND
PHILIP G. KETTLE

LONDON AND NEW YORK

First published in 1970

This edition published in 2007
Routledge
2 Park Square, Milton Park, Abingdon, Oxon, OX14 4RN

Routledge is an imprint of Taylor & Francis Group, an informa business

Transferred to Digital Printing 2009

The publishers have made every effort to contact authors and copyright holders of the works reprinted in the *The City* series. This has not been possible in every case, however, and we would welcome correspondence from those individuals or organisations we have been unable to trace.

These reprints are taken from original copies of each book. In many cases the condition of these originals is not perfect. The publisher has gone to great lengths to ensure the quality of these reprints, but wishes to point out that certain characteristics of the original copies will, of necessity, be apparent in reprints thereof.

British Library Cataloguing in Publication Data
A CIP catalogue record for this book
is available from the British Library

A Geography of Urban Places
ISBN10: 0-415-41934-4 (subset)
ISBN10: 0-415-41318-4 (set)

ISBN13: 978-0-415-41790-7 (volume)
ISBN13: 978-0-415-41934-5 (subset)
ISBN13: 978-0-415-41318-3 (set)

ISBN10: 0-415-41790-2 (hbk)
ISBN10: 0-415-48957-1 (pbk)

ISBN13: 978-0-415-41790-7 (hbk)
ISBN13: 978-0-415-48957-7 (pbk)

Routledge Library Editions: The City

A Geography of Urban Places

Selected Readings

Edited by

Robert G. Putnam

Frank J. Taylor

Phillip G. Kettle

Methuen

Toronto London Sydney Wellington

Library of Congress Catalog Card Number 77-116479
ISBN 0-458-90630-1 (hc)
ISBN 0-458-90620-4 (pb)

Design by Carl Brett
Illustrations by Gus Fantuz
Printed and Bound in Canada by
McCorquodale & Blades Printers Limited

74 73 72 71 70 1 2 3 4 5 6

Preface

e growth in the number of people who live cities and the diffusion of urban life to ery part of the habitable world is one of outstanding characteristics of twentieth-tury life. In Canada, nearly 75 per cent of people reside in urban centres as defined the census. Mississaga,* the great Southern tario conurbation, focussed around the stern end of Lake Ontario, best illustrates situation. In Mississaga there are five cen-s metropolitan areas with populations of re than 100,000 each, and four other ma- cities, each with populations exceeding ,000. Mississaga now contains more than 500,000 people, 92 percent of whom live an urban environment. Growth has been pid. Mississaga in 1951 contained about 100,000 people; between 1951 and 1966 it perienced a population increase of 75 per nt. Today the average population density is out 635 per square mile. In the densely pulated areas, such as the City of Toronto, e population density exceeds 19,000 people r square mile.

How does a geographer evaluate and assess ban places such as Mississaga? What tools, techniques and theories are useful in understanding the geography of urban places? It is the intent of this book to provide a selection of readings that will help the geography student who is beginning to study the urban environment to realize the concepts and analytical techniques possible. The articles have been chosen to provide a wide range of ideas and levels of understanding about urban geography. Many of these ideas and concepts have been previously published only in professional journals and would not be readily available to students. These articles have not been modified from their original content. In all cases the original footnotes and references have been retained.

The book is divided into three major sections. The first is concerned with the definition of urban geography, the origin location characteristics of cities. In the second section, emphasis is placed upon the economic base of cities. How cities function, the role of the Central Business District and the intricate relationships that exist between cities and their hinterlands are all part of the economic base. The third section concentrates upon the effects of urbanization and examines closely the urban environment, and the ways in which this environment can be improved.

*For a discussion of the use of *Mississaga* see P. F. Putnam and R. G. Putnam, *Canada: A Regional Analysis* (Toronto, J. M. Dent and Sons, 1970), Chapter *111*.

Contents

Contributors

:croft, Eric

rmerly Director of the Community Planning Association of Canada. Now
·ector of the Urban and Regional Studies Program, University of
:stern Ontario, London, Ontario.

ımenfeld, Hans

rmerly Assistant Director of Planning for the Metropolitan Toronto Planning
ard. Now Planning Consultant, Toronto, Ontario.

al, Frederick W.

partment of Geography, The Queen's University of Belfast, Northern Ireland.

rchert, John R.

ɔfessor of Geography, University of Minnesota, Minneapolis, Minnesota.

ıwson, Marion

sources for the Future, Inc., Washington, D.C.

ton, Leonard K.

ntributor to Landscape.

stein, Bart J.

ınager, Retail Real Estate, The B. F. Goodrich Co., Akron, Ohio.

ıuntain, G. F.

rmerly Director of Planning for the City of Vancouver.

ıffney, Mason

:sources for the Future, Inc., Washington, D.C.

ırner, Barry J.

ofessor, Geographical Institute, University of Aarhus, Aarhus, Denmark.

:tis, Arthur

ıtgers, The State University, Brunswick, New Jersey.

Getis, Judith

Rutgers, The State University, Brunswick, New Jersey.

Goodwin, William

Associate Professor of Geography, University of Wisconsin, Milwaukee, Wisconsin.

Harris, Chauncy D.

Chairman of the Department of Geography, University of Chicago, Chicago, Illinois.

Horwood, Edgar M.

Professor of Urban Geography, University of Washington, Seattle, Washington.

Jacobs, Jane

Author of Downtown is for People, The Exploding Metropolis, The Death and Life of Great American Cities.

Kelley, Eugene J.

Director, Division of Business Administration of Clark University, Worcester, Massachusetts.

Kiang, Ying-Cheng
Professor of Geography, University of Chicago, Chicago, Illinois.

Kinsel, John
Contributor to Community Planning Review.

Kolosova, Yu A.

Moscow University, Moscow, U.S.S.R.

MacNair, Malcolm D.

The Office of the Urban Renewal Coordinator of the City of Seattle, Seattle, Washington.

Mayer, Harold M.

Professor of Geography, Kent State University, Kent, Ohio.

Murphey, Rhoads

Director, Center for Chinese Studies, University of Michigan, Ann Arbor, Michigan.

Nelson, Howard J.

Chairman of the Department of Geography, University of California, Los Angeles, California.

inemann, Martin W.

ɔfessor of Earth Science, Northern Illinois University, DeKalb, Illinois.

:hardson, N. H.

ıff Consultant, City of Toronto Planning Board, Toronto, Ontario.

oney, John F., Jr.

sociate Professor of Geography, Southern Illinois University, Carbondale, Illinois.

elt, Jacob

ofessor of Geography, University of Toronto, Toronto, Ontario.

ıfford, Howard A., Jr.

ofessor, Head of the Department of Geography, University of Cincinnati,
ncinnati, Ohio.

ınford, John H.

ntributor to Landscape.

ɔkes, Charles J.

ıarles Anderson Dana Professor, Chairman, Department of Economics,
ıiversity of Bridgeport, Bridgeport, Connecticut.

ınner, Ogden

me-Life Books, New York, New York.

ıorsell, James

ɔntributor to Ontario Geography.

llman, Edward L.

ofessor of Geography, University of Washington, Seattle, Washington.

ance, James E., Jr.

ofessor of Geography, University of California, Berkeley, California.

oorhees, Alan M.

'estgate Research Park, MacLean, Virginia.

'ilson, L. H.

ɔrmerly Chief Architect and Planning Officer for New Town of Cumbernauld.
ow Planning Consultant, Glasgow, Scotland.

city Origin
nd
ocation

Section

What is urban geography, and how do geographers study the urban landscape? Garner, in *Aspects and Trends of Urban Geography* suggests that urban geographers are concerned with towns as items in the general fabric of settlement and in the interaction that exists between these items. Towns can also be studied in terms of site, situation, layout and build. Internal structure, city interaction and city growth are all important. In *Cities and Urban Geography,* Mayer raises the problems associated with defining the area occupied by a city, for in most cases the urban area involved exceeds the official corporate limits. He also considers what is meant by the term *metropolis,* and points out some of the difficulties urban geographers have in dealing with the census metropolitan areas. The concept of megalopolis is also examined with the intention of showing how it has evolved since the time of the Greeks.

City Origin and Location

"The city is as old as civilization." Murphey, in *Historical and Comparative Urban Studies,* views the origins of cities as centres of exchange. Cities have possessed this important function right through history, from the early beginnings in Egypt, Mesopotamia and Asia, to the present day. It is his opinion that, as commercial and manufacturing functions evolve with the changes in technology, cities everywhere are becoming more like one another. Cities in North America, however, have not the long history of development of those in Eurasia. Most are less than two centuries old, and have evolved rapidly in response to the changing technology of the industrial revolution. As such, the origins and developments are traced in *American Metropolitan Evolution* by Borchert. He suggests that four epochs of development have occurred and that these reflect changes in t[he] technology of transportation and industri[al] energy. The pattern of evolution that he su[g]gests can be applied directly to Canada, a[nd] with some modifications is applicable els[e]where.

City location is important. Ullman, in h[is] classic paper *A Theory for the Location [of] Cities,* suggests that urban places are not sca[t]tered illogically over the surface of the ear[th] but that orderly spacing does occur. His com[]ments on the settlement theory proposed b[y] Christaller and others have been influenti[al]. More insight into Christaller's urban hiera[r]chy system is found in the paper by Arth[ur] and Judith Getis. Here the k = 3 network [is] explained, and some of the limitations to th[e] theory are explored.

City Structure

The structure and form of cities are far fro[m] static. They evolve through time. In *Tec[h]nology and Urban Form,* Frederick Bo[] makes the strong point that the city is cha[r]acterized by flows — flows of people, com[]modities and information. The way the[se] flows take place shapes the city, for they re[p]resent the interaction between various par[ts] of the city. Transportation and constructio[n] technology have profound influences on th[e] city because they are part of the "space a[d]justing" mechanism. He then shows the ro[le] of technology on urban form during thr[ee] technological periods: the pedestrian city; th[e] steam engine and wheel-track city; and th[e] flexible city.

However, the structure of cities involv[es] not only the build-up of urban uses within th[e] city, but also the territorial expansion to are[as] that have become urbanized beyond the ci[ty] limits. Voorhees, in *Urban Growth Chara[c]teristics,* illustrates how urban growth can b[e] died. An interesting viewpoint of city wth is provided by Kolosova of Moscow iversity in *The Territorial Expansion of nerican Cities and Their Population owth.* He concentrates on the role that an-xation has played in the growth of Ameri-a cities.

Since it was published in 1945, *The Nature Cities* by Harris and Ullman has been con-ered by many geographers to be a classic its type. It reviews the reasons why cities m, what functions cities require, and how internal structure of cities is arranged tention is focussed on three generalizations arrangement — concentric zone, sector, d multiple nuclei. These generalizations are panded somewhat by Nelson's *The Form d Structure of Cities: Urban Growth Pat-ns.* Nelson also emphasizes the major ele-nts in the urban structure of cities. One of se elements, which is very significant for y growth, is manufacturing. Reinemann, in *e Pattern and Distribution of Manufactur-ing in the Chicago Area,* carefully studies the dispersal of this land use between 1945 and 1960. While his conclusions regarding the outward movement of manufacturing to sub-urban areas are significant, the methods of analysis employed are also of considerable importance.

However, city structure also involves man. Within cities people live, work, and play in close proximity to one another. Many North American cities also benefit from the wide range of backgrounds possessed by their resi-dents. One obvious characteristic is that of income, which has been mentioned earlier in the study by Harris and Ullman. Another is that of the ethnic origin of the people. Kiang, in *The Distribution of Ethnic Groups in Chi-cago, 1960,* outlines a procedure for studying the distribution of ethnic groups which could be applied to most cities in the world. He also documents the mobility of ethnic groups with-in a city, and shows the intra-urban migration pattern that has evolved.

1

Aspects and Trends of Urban Geography

Barry J. Garner

Perhaps the most general, although surely the least satisfactory, definition of geography is, "Geography is what geographers do." Accompanying the continuing trend toward urbanization in the world, an increasingly large part of contemporary geographical literature comprises studies of various aspects of the urban environment. Geographers are doing the city. But what are they doing in the city? What problems do they study? What are their concepts about the city? How do they, as distinct from sociologists, for example, approach the study of urban areas? Can the findings be useful in helping to improve the social and economic health of cities, which, it has been suggested, represents one of the gravest challenges in the twentieth century? This paper will answer some of these questions — perhaps not as fully as they deserve, but as adequately as space will allow.

It is often said that geographers are "Jacks-of-all-trades" since they seem to study a little of everything, much of which borders on other fields of study. One has only to glance at the table of contents of a geography textbook to see how true this is. In the same way, the subject matter of urban geography is a rich and varied potpourri. Moreover, there are many different approaches to the geographical study of towns, although perhaps two are most common. First, the town can be considered as forming a discrete phenomenon in the general fabric of settlement. Concepts and generalizations may be formed regarding their distribution, size, function and growth. Areas served by urban places may be delimited and the spatial interaction between places may be studied. Second, the town may be studied in terms of its layout and build, which express its origin, growth, and function. Concepts and generalizations may be related to the character and intensity of land use within the urban area and to the spatial interaction between its constituent parts.

Perhaps the best way to understand the topics studied and the concepts used in urban geography is to imagine a town and view it through the eyes of an urban geographer and the studies he might undertake there. The first question he might ask is, "Where is it?" It has a *site*, the actual ground on which it stands. Why did it grow up here and not somewhere else? It also has a location, both

Reprinted from *Journal of Geography* (May, 1966), pp. 206-11 by permission.

absolute and, more important, a relative ation, a *situation* with respect to other phy- al and human things around about it. To- y, many of the early site factors have ised to be important. Similarly, relative ations have changed with improved com- inications. The communications have en- led distances to be covered more cheaply d in less time than ever before and have ulted in the intensification of interaction tween places. Site and situation characteris- s are thus constantly changing through ie. The nature of these changes, the pro- sses underlying them, and the effects they ve on the changing spatial relationships tween towns constitute one set of problems urban geography.

The study of the town itself may be from aggregative or an elemental viewpoint; at is, the geographer may consider the wn as a discrete whole, as a collection of ements which distinguish it from others, or may pay specific attention to some or all e constituent elements within it as separate rts that go to make up the whole. In both ses the studies may be cross-sectional in at they present the situation at a given point time, or they may be time-oriented with nphasis on tracing the evolution of the pre- nt pattern. In fact, most studies contain ements of both of these approaches.

Most aggregative studies are concerned th form and function. In answer to ques- ons about form or shape, concepts from ology have proved useful and various mea- res of shape have been attempted by ge- graphers. In distinguishing towns by the nctions they perform, analysis is usually in rms of the dominant activities revealed in nsus materials on employment and/or oc- pations, in terms of the presence or absence functions difficult to quantify, such as gov- nment or universities, and in terms of their roles, for example, as service centers or ports. The end product of these studies is the formulation of generic (type) classifications of towns, not only as ends in themselves, but also as a way of ordering information conveniently for the purpose of further analysis and investigation.

STUDIES OF INTERNAL CITY STRUCTURE

The town itself occupies an area of appreciable size, and activities are separated from one another within it. It has an internal structure which can most easily be expressed in terms of the differences in character and intensity of land uses at various locations. When the geographer recognizes residential areas, shopping districts, or industrial zones, he is also identifying different functions and forms, all of which give the basis for the recognition of uniform regions — areas which are homogeneous in terms of specified characteristics — within towns. He might say that it is the description of the nature of these urban regions, their disposition and their social interdependence, that constitutes a geographical analysis of the internal structure of the town.

Residential Land Use

A geographer may very well want to consider only one of the major functions or land uses within the town in isolation from all the others. Many studies are of this nature. For example, attention might be focused on residential land use, and questions pertaining to the patterns and the principles underlying those patterns would most surely be asked, for there is considerable internal variation in residential structure. The older, inner areas of the town are usually ones of high density, with old and multi-story buildings, whereas the outer and progressively newer parts are lower in density. He can view this as a surface of differing intensity and character of use and apply concepts of density gradients, decreasing outward from the core of the town, for purposes of analysis. Identifying the shapes of these gradients and studying the changes in them over time and in different-sized urban areas are increasingly important research topics.

Business and Commercial Structure

The commercial structure of towns has received much attention in recent investigations. The complex business structure of the town can be disaggregated into various component parts, such as ribbon development along arterials, into clusters or centers of activities of varying size and functional composition, and into specialized concentrations of similar types of function, such as printing or medical districts. A geographer would want to understand these patterns and would rely in part on the concepts of the *threshold size* (the minimum amount of support necessary for a business to survive in the economic landscape) and the *range of a good* (which defines the area from which this support can be obtained), for these help determine the spacing of business activities. He could use concepts of agglomeration of economic activity and of differences in shopping habits in the analysis of the location, functional composition, and interaction among activities.

He would most certainly recognize that part of the business structure consists of a hierarchy of vari-sized shopping centers — both newer, planned ones and older, unplanned nucleations — ranging from the smallest street corner cluster to the most complex concentration in the central business district, or the downtown area. Each level or center in the hierarchy can be characterized by th numbers and types of activities it contain The central business district alone furnish ample study topics, ranging from its delimit tion to an analysis of its internal structure an the interconnections existing within it. H would identify a *core*, comprising a numb of highly interrelated functions, and a su rounding *frame* of loosely connected and les intensive land uses, and trace the connection between them.

Industrial Structure

Differences exist also in the nature and distri bution of industries within and betwee towns. Some industrial types are highly con centrated in the inner, older parts of the cit while others are located in isolation towar the periphery. Clusters of similar types occu in some areas while others consist of appar ently disparate kinds. Indeed, towns them selves are in many instances characterized b dominance of a single kind of industry or b the diversification of types. In the analysi of patterns of industrial location the geogra pher could use concepts of *scale economic* — which relate quantity of goods pro duced directly to costs of production — t understand the localization and grouping o industries within the town. He could also us the *economic base concept* (the idea tha most cities exist primarily as centers of em ployment opportunities) to identify thos activities which are *basic* to the town's exist ence, in that they bring money into the tow from outside by trading with other areas, an those which are *nonbasic*, or city serving, i that they serve the people living within th area of the town itself. He would relate th town's growth and economic well-being t the basic industrial structure, since town cannot exist without trade with other areas.

'ERACTION WITHIN THE CITY

have asserted that activities occupy dif-
:nt locations within the city. For example,
:e of residence is separated by varying dis-
:es from place of work for most people.
kages and movements between the various
ts must therefore take place. People and
ds must overcome distance in order to
sfy their demands, and in so doing they
: rise to a variety of patterns of movement
l flows, the most pronounced one being, of
rse, the daily commuting patterns. These
:ractions are reflected in the amount,
:ction, time, and character of movements
ween the various functional areas. Con-
uently, the study of urban transportation
become a major topic in urban geogra-
/. The town, in fact, comprises a complex
tem of overlapping functional areas and
nodes and foci about which human activity
organized. Concepts of *intervening oppor-
ity,* of least-effort behavior resulting in
imizing distances traveled, and the prin-
les relating trip generation to the character
l intensity of land use afford partial ex-
nations of these patterns.

Movement and focality are intimately con-
:ted with the concept of *accessibility*. Since
ivities have different needs from the view-
nt of accessibility, and since some loca-
ns are more accessible than others to the
:ious parts of the urban area and its func-
ns, there is competition for the use of land,
d this competition affects its value. The
nomic concepts relating land values to dif-
ences in accessibility are fundamental to
: analysis of the distribution and intensity
land uses within the urban area. Urban
d can be represented as a rent surface,
ich, in general, resembles a contour map of
ill in that values and rents are highest in
: core of the town, where accessibility is
presumably greatest, and decrease by differing amounts in various directions to the periphery. Regularities exist in the relationship betwen this surface and the distribution of urban functions.

THE DYNAMIC NATURE OF CITIES

So far we have considered the town as static. In reality, however, its internal structure is constantly changing. New buildings replace old ones, and land uses are changed as functions move from one location to another or disappear altogether. Patterns are in a state of flux as adjustments are made to changing conditions. Centripetal forces of various kinds attracts activities to the centers of towns while centrifugal forces result in decentralization and dispersal. At the same time, areas of one kind of land use are invaded by other uses which eventually take over, resulting in change in function, in intensity of use, and in form.

An urban geographer would therefore be interested in the processes of residential decentralization, in the expansion of the suburbs at the town's periphery. He would note that the older, inner areas of the town, no longer meeting the requirements of the original users, decline and are taken over by alien uses. He would study slums and trace the evolution of uses at any given location by means of the concept of *sequent occupance*. He would study the changes in business structure as the quality of the market deteriorates and changes occur in the technology of retailing. He would note other changes — the growth of out-of-town shopping centers and the associated reductions in sales volumes and vacancy rates in the downtown area, the emergence of commercial blight as areas run down, and the trend toward industrial decentralization as older sites with their congestion, scarcity of land for expansion, and high costs become increasingly less suitable for most kinds of modern industry. He would want to know the effect of such changes as these on the patterns of movement within the city. In fact, he is interested in defining the relationships that exist and in identifying the processes underlying the changing internal structure of the town.

Towns do not exist only to serve the people living within their bounds; they are also intimately connected with the areas surrounding them. Another major group of study topics is thus concerned with the interaction between towns and the areas that comprise their hinterlands. The city has a sphere of influence, a trade area, in the same way that a magnet has a field of influence. People go to towns to purchase goods or attend a game and, through its function as a collecting and distributing center, the town serves the surrounding area in a variety of ways. The urban geographer would want to know what these relationships are. He would delimit the trade area, and analyze its character and size.

The extent of trade areas is directly related to the proximity and functional structure of other settlements. He would look at the size and spacing of towns in a region, asking questions about the degree of clustering or dispersal of settlement. He would find the ideas related to *central place theory** useful in understanding patterns. He would be concerned with the principles underlying distributions of towns and the evolution of patterns of town-spacing over time.

METHODS OF STUDY

Although the foregoing does not exhaust the concepts associated with urban geograph a word about methods and techniques is order at this juncture. Much geograph representation is purely descriptive in cha acter. The identification of patterns, the description, and the generation of classific tions are the traditional and necessary pr requisites for geographical study. Howeve much of the recent work appears to b markedly different from this, although o closer inspection it will be found that th differences are mainly in the way research undertaken rather than in the kinds of prob lems studied. Greater emphasis is bein placed on the use of mathematics as a analytical techinque, and there is increase concern with the general rather than th specific. The subject is changing from a idiographic one, in which the emphasis wa on the intensive study of individual cases, the nomothetic approach, in which there is search for general laws and spatial regular ties.

We are really saying that the subject is b coming more scientific, with greater emphas on the explanation and prediction of geogra phic phenomena. For, although descriptio may be a logical starting point, it is inad quate by itself, since it can only lead to a re arrangement of the facts without addin principles from which to strike out anew Moreover, once we emphasize explanatio we become engaged in the search for theorie It is in this connection that another interest ing change has taken place in the approac to urban geography, namely, the increasin use that is being made of models.

The usual concept of a model is that of miniature; for example, a model railroad se This really is an analogue to the real thing It is, in fact, not an exact replica but a like ness, since only the most salient features hav

* See A. Getis and J. Getis, "Christaller's Central Place Theory," on p. 68.

n represented. Models are just that — tractions or simplifications of the real rld in which the irrelevant material is dis- ded to lay bare the bones of what are con- ered to be the simplest and most significant ects of the problem under investigation.

Although models can be of various kinds, se most widely used in urban geography mathematical ones, in which words are laced by mathematical symbols, and con- of a set of mathematical assertions from ich consequences can be derived by logical thematical argument. For example, in dying the field of influence of a town, we w the analogy to the magnetic field. Phy- ists tell us that if two magnets are placed r each other, the boundary of their mag- ic fields is directly related to their size strength, and inversely related to the dis- ce separating them. In the same way, the ndary between the trade areas of towns d the interaction between towns is also ated to their sizes and to the distances ween them. Translating this to symbols, can express the same thing simply as P_b/D where P_a and P_b stand for the size of two towns and D represents the spacing. ing this very simple model, we can esti- te the extent of trade area of say place *b*, the amount of interaction expected be- een towns *a* and *b*.

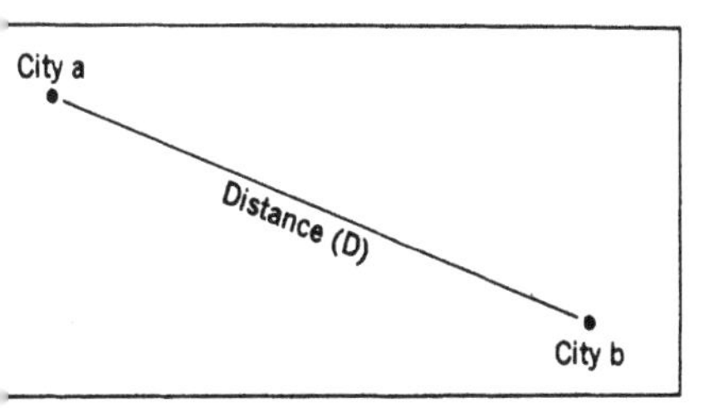

More complicated models have been devel- ed recently to help in the study of journey- work and journey-to-shop patterns. They have been applied to the analysis and explanation of the distribution of land uses within the city; for example, to residential land use and to commercial structure. Much effort is now being devoted to the development of models which can simulate entire urban areas to aid in the formulation of planning policies.

But the model gives us an artificial situation which must be compared to the findings obtained from actual field observation. There is, as a result, a problem of appraisal. It is in the testing of theoretical results derived from the use of a model that we find an added departure from the traditional way of doing things. The traditional way was largely to use "eyeball" methods of direct visual comparison — methods which are both highly subjective and generally inaccurate. In order to gain greater objectivity and reliability, more and more urban geographers are turning to the use of statistical methods of testing hypotheses or in evaluating results.

Mathematical methods are also being used now to describe phenomena (in the description of patterns, for example, or in the formulation of classifications) and in the identification and measure of relationships or correlations between phenomena. Recent work has therefore tended to become more quantitative with (a) greater problem orientation, (b) hypothesis testing using mathematics, and (c) emphasis on the building of models as an aid in the formulation of theory.

Once explanation is available, predictions may become logical. It is here that urban geography appears to be able to make an even greater contribution to urban planning than it has in the past. Continued search for the understanding of the spatial structure of the city can help in making decisions about its future. It is the application of the findings of geographical studies to this purpose that many urban geographers consider to be their most worthwhile goal. They are contributing in a large way to the understanding of the urban world and the eventual betterment of this environment for an increasingly urbanized society.

Further Reading

Dickinson, Robert E., *City and Region: Geographical Interpretation* (London, Routled and Kegan Paul, Ltd., 1964).
Mayer, Harold and Kohn, Clyde eds., *Readin* in Urban Geography (Chicago, University Chicago Press, 1959).

2 Cities and Urban Geography

Harold M. Maye

One of the most rapidly developing fields of geography during the generation which followed World War II has been that of urban studies. Currently, a very high proportion of the research effort and interests is in the phenomena of cities and metropolitan areas. In this, geography has shared with other social sciences many common concepts an methods, but has contributed a distinctiv set of viewpoints and a unique focus: th which is primarily concerned with the organ zation of man's use of space and re sources in the development and functionin of urban settlements. Research in urba

Reprinted from *Journal of Geography* (January, 1969), pp. 6-19 by permission.

ɔgraphy has contributed to man's under-nding of the complex of interrelated urban ənomena. In addition, the application of ɔgraphic concepts and viewpoints has ɔduced substantial practical application in y and metropolitan planning.[1] In this paper ne of these geographic concepts and con-ɔutions to urban studies will be outlined.

ity" and "Urban"

ıat is a city? What is meant by "urban"? though there are many common definitions these terms, the geographer seeks a defini-n which is unambiguous, definite, and ar. If we are to compare one city with other, or a metropolitan area or urban ag-moration with another, or if we wish to mpare the extent and character of urban pulation in one country or region with that another, we need some criteria which can applied in the measurement of these enomena. There have been many attempts geographers and others to develop defini-ns that can be used in comparative mea-rement, but none is completely satisfactory r all such purposes. The difficulty lies in the ture of the phenomena themselves, for, in ite of some common characteristics, cities d urbanization, as spatial phenomena and ocess, exhibit many interregional and inter-tional differences, rendering comparative studies of cities in different areas and at different times extremely difficult; yet the characteristics which all cities share render the attempts to compare them not entirely futile. There is a practical aspect to these attempts, as well as an academic: by learning of the successes and failures to solve urban problems in given cities, the probability of success or failure of similar solutions may be evaluated in terms of the extent to which the respective cities resemble, or do not resemble, each other with respect to the relevant characteristics. Solutions, for example, to the problems of housing in Swedish cities may not be entirely successful if applied to cities of the northeastern United States, because the political, cultural, racial, and economic backgrounds of the respective regions and countries differ in several important respects.

Nevertheless, cities throughout the world exhibit certain common characteristics which we identify as "urban," and which therefore help us to define the cities.

One of the definitions is administrative and legal. In most nations, and in each state in the United States, there is a set of criteria by which certain areas, called "cities" are designated by the sovereign power — the nation or state — to carry out certain functions and responsibilities, in return for which certain privileges are granted. In most of the British commonwealth countries, a City is such by royal decree, and the designation bears little, if any, relation to the population size of the place. Thus, the City of London, for example, embraces an area which constitutes only a very small portion of what in the United States would be considered a part of the central business district of the metropolitan complex. British usage of the term "City" thus is a purely historical and legal one; what Americans conceive of as a city is embraced in the British term "Town," while

A general review of the main directions and trends in urban geography is contained in Philip M. Hauser and Leo F. Schnore, eds., *The Study of Urbanization* (New York, John Wiley & Sons, Inc., 1965), especially pp. 81-114, which include an extensive bibliography. Recent applications to planning are presented in Leo F. Schnore and Henry Fagin, eds., *Urban Research and Policy Planning* (Beverly Hills, Calif., Sage Publications, Inc., 1967), especially Chapter 8, "Urban Geography and City and Metropolitan Planning" by Harold M. Mayer, pp. 221-38.

the Town in America has different meanings in different parts of the country. British "town planning" has the same meaning as the American term "city planning," while some British geographers and planners refer to the urban landscape as a "townscape."

In most American states, a city is a municipality — a governmental unit embracing a proscribed area with definitely-located boundaries, and with a minimum population which, if reached, entitles the area to apply for city status. Smaller urban units of government, with smaller populations, or larger ones which choose not to apply for city status, may incorporate in most states as municipalities with lesser status, such as villages, towns, or boroughs, the various designations differing from state to state. In many states, municipalities may be further differentiated by "class," depending upon population size, each class being granted certain rights and obligations. Depending upon the amount of "home rule" permitted municipalities in each of the various states, state enabling legislation delegates authority over certain matters to municipalities in varying degrees, depending upon the class. Thus, for example, in Illinois the state does not pass legislation applying to any particular city, but it designates "cities of the first class," the only such city being Chicago.

Rarely do the boundaries of the municipalities correspond to the actual boundaries of the respective built-up areas; cities are virtually always "over-bounded" or "under-bounded," more usually the latter, depending upon whether, in the first instance, substantial areas of land which are not "urban" in character are included within the municipal boundaries, or, in the second instance, there is "overspill" of urban character beyond the municipal boundaries. Areas can, and frequently are, added to municipalities by annexation, and most cities have expanded their extent many times as the population grows with the urbanization of land beyond the older boundaries. Thus, in 1854 Philadelphia city and county were consolidated, expanding the extent of the city severalfold, but there have been no annexations to Philadelphia since then. Other major American cities which once expanded by consolidating with their counties include St. Louis and San Francisco, and, more recently, Nashville, Tennessee, and, for certain functions, Miami (Dade County), Florida. Such consolidation is rare, however. More commonly, cities add territory by piecemeal annexation, usually in order to provide municipal services and controls to areas beyond the previous borders, or to take in commercial or industrial areas, existing or prospective, which would furnish an enhanced tax base while at the same time not requiring substantial expenditures for extension of municipal services, such as schools, which account for more costs in most urban areas than do any other municipal functions. Factories and shops do not produce children, but they do produce substantial shares of the municipal tax revenues.

Except in the Southwestern United States, where cities such as Dallas, Houston, El Paso, Phoenix, and Tucson have more than doubled their municipal areas by annexation in recent years, it has become increasingly difficult for most cities to expand their administrative boundaries by this means. One reason is that many people, business establishments, and industries have located, and continue to locate, outside of the areas of the central cities, but still within convenient reach of them, in order to escape some of the obligations of internal location, such as supporting the many services which central cities must supply for the respective metropolitan areas, and also to avoid the pressures of con-ion, pollution, and social pathology ciated with high central densities of ulation, traffic, and activity, including the land values which reflect the mutual conience of such high densities. In other ds, a locational decision within an urban plex reflects the respective weights which, each instance, are given to central conience but high density on the one hand, peripheral amenity and low density on other; centripetal, or centralizing forces, centrifugal, or decentralizing forces, are balance, with the fulcrum at different locas for each function or establishment at the e the locational decision is made. Since re is a limit to the number and intensity of ctions which can occupy any given area, ond which "scale diseconomies" such as gestion and high land costs result, urban wth tends to take place at the outer edges the built-up area, the result being periral expansion.[2]

As expansion takes place on the edges of previously built-up urban area, it is not g before the need for services and facilities ich are provided by local governments is nifest. Schools, paved streets, utilities, fire police protection, and other services must provided, and they cannot be provided ciently if the population density is too low, f effective municipal government does not er the area. Thus, there is a lower limit, hreshold, under which "urban" conditions do not constitute characteristics of an area, and which is necessary for effective provision of many needed facilities and services.

This brings us to another form of definition of "city" and of "urban," based upon a minimum density.

As we drive through an urban complex, or inspect an air photo of one, we can readily identify variations in density of buildings, streets, and both working and residential population. In most "western" cities, the density drops off very sharply from the core or center, with the decline in density decreasing at a decreasing rate as we move outward; this typical pattern is described mathematically as a negative exponential curve.[3] At some distance from the city center, the density is so low that it can no longer be considered typically urban; here the urban area ends. We can draw contours of equal density around the city cores, and one of these contours can be used to describe the location of the outer limit of the urbanized area. For this purpose, we can use residential population — although with respect to this measure there tends to be a "crater" in the middle where intensive non-residential land uses tend to outbid the residential — or we can use some ratio of building volume or percent of land covered by structures, or,

A classic statement of this principle in the geographic literature is that of Charles C. Colby, "Centrifugal and Centripetal Forces in Urban Geography," *Annals of the Association of American Geographers*, Vol. XXIII, no. 1 (March, 1933), pp. 1-20. In the literature of urban planning, an excellent statement of the process is: Hans Blumenfeld, "The Tidal Wave of Metropolitan Expansion," *Journal of the American Institute of Planners*, Vol. XX, no. 1 (Winter, 1954), pp. 3-14.

[3] There is extensive literature describing and documenting the negative exponential pattern of urban density. For example, Colin Clark, "Urban Population Densities," *Journal of the Royal Statistical Society*, Ser. A, Vol. CXIV (1951), pp. 490-96; Richard F. Muth, "The Spatial Structure of the Housing Market," *Papers and Proceedings of the Regional Science Association*, Vol. VII (1961), pp. 207-20; C. A. Doxiadis *et al.*, "Densities of Human Settlements," *Ekistics*, Vol. XXII, no. 128 (July, 1966), pp. 77-101; A critical review of some of the literature is Bruce E. Newling, "Urban Population Densities and Intra-urban Growth," *Geographical Review*, Vol. LIV, no. 3 (July, 1964), pp. 440-42.

perhaps, traffic volume, or land costs. In nearly every city, these various density measures are highly correlated; any one greatly influences all or most of the others, and the patterns which result are very similar.

The concept of the density threshold has been used by governmental agencies in defining, for example, the "urbanized areas" surrounding each central city with at least 50,000 population in 1950, and again, with some slight changes in measurement criteria, in 1960. Thus, it was found that, in the United States, there is a high correlation between the contours representing the outer limits of a typical "urban" density of streets (whether rectilinear or not), a housing density of about 500 units per square mile, or a residential population of roughly 2,000 per square mile. Lower than these limits indicates nonurban status, although, as mentioned previously, the administrative municipal limits rarely if ever coincide with these density contours.

The pre-industrial city, whether in the past or at present in some non-industrialized parts of the world, was perhaps more easily defined in physical and functional terms than the modern city. Before the advent of strong national governments, the city-state was common; in the middle ages the feudal estate was somewhat analogous. In each instance, the settlement was more or less economically self-contained, and it had to provide for its own defense. It was a center of government, of economic activity, and of defense. Not uncommonly, it was walled; within the walls and gates was urban, outside was not. With growth of population, not all citizens could locate within the walls, so "sub-urbs" developed outside, often performing trade functions as well as serving as residential locations for the population which could not be accommodated inside.

With industrialization and the advent modern transportation, and especially t development of large national areas, the c concept of a dichotomy between "urban" a "nonurban" or rural has been breached. Fi uratively, as well as literally, the walls ha come down, and the distinctions betwe urban and rural are losing much of the former significance. Until the past fe decades, urban development had to ta place along lines of mass transportation people and goods, and peripheral expansi of urban areas, along good transportatio was characterized by clear-cut distinctio between the areas which were urbanized ar those which were not. Transportation ma the difference; it not only permitted urba expansion, but it also permitted extreme high concentrations of activities and peop to develop where the routes converged; th was a cause-and-effect relationship reci rocally. With the ubiquity of the automobi truck, and modern highways, accessibility good transportation is still important in t process of development of the urban patter but it takes different forms. Many industri no longer need to be located on navigab waterways, or alongside railroads to recei and ship goods; inter-model transportatio with break-of-bulk at intermediate transf points or "interfaces," or, of increasing in portance, the elimination of much goo handling at interfaces by development of co tainers interchangeable between highwa railroad, water, and even air carriers, ha made locations much more footloose for mo industries than was formerly the case. Sim larly, shopping locations and residenti development no longer need be entirely de pendent upon large-scale mass transport tion, so that many more locations a available for development. Densities, in th western world, are declining almost ever ere, and the cities are spreading out at an r-increasing rate, but with no decrease in essibility. The urbanized areas, therefore, expanding at a rate even more rapid than rapidly expanding urban population. In United States, it is anticipated that if pre- it trends continue, the amount of land oc- pied by urban settlement will double every enty years.[4] The resulting urban "sprawl" resents a challenge to governments at all els, from national to local, to find ways of ding and controlling urban growth in order maximize opportunities for all potential d uses to locate properly, and in the right ounts, for maximum efficiency and to pro- le for the maximum amenity and aesthetic isfactions.[5]

The processes of guiding and controlling patterns of urban land use have been the bject of much experimentation in recent cades, with the expansion of the concern government with urban problems, the velopment of comprehensive city and tropolitan planning, and the adoption of rious forms of regulation such as subdivi- n controls and zoning. None of these have en completely successful, and the problem ll remains unsolved of how to provide a sirable degree of guidance and control of ban land development while at the same ne retaining the maximum possible scope the interplay of market forces in the tradi- nal American free enterprise system. The idance and control of patterns of urban owth, however, are enormously com- cated by the proliferation of local governmental units and the fragmentation of the functional geographic urban areas into many local governmental jurisdictions. There are currently over 18,000 local governments in the United States — cities, towns, villages, and boroughs — most of them in the 230-odd metropolitan areas which are discussed later, in addition to special-purpose districts in at least equal number: school districts, park districts, sanitary districts, forest preserve districts, drainage districts, flood control districts, mosquito abatement districts, transit authorities, port authorities, housing authorities, and others.[6] Only school districts, among these, have decreased in number; for in the case of school districts it has been recognized that a certain minimum population and economic support, beyond reach of very small governmental units, is required in order to finance and operate adequate facilities with adequately qualified personnel. The study of the spatial aspects of inter-governmental relations and organization within urban areas is a field in which a few geographers have recently begun to produce significant studies, and it presents major opportunities for research and application in the future.

Thus, the problem of defining "urban" and "city" in areal or spatial terms is greatly complicated by the present patterns of government, with fragmentation of urban units in a geographic sense. On the other hand, these local governmental units, even though rarely coinciding with functional units in most geographic senses, are significant, for they represent organizational units which, in themselves, represent internal homogeneity with regard to certain policies, practices, and political and fiscal situations which are geographically significant, in spite of the fact that they

Marion Clawson, R. Burnell Held, and Charles H. Stoddard, *Land for the Future* (Baltimore, The Johns Hopkins Press, 1960), especially pp. 109-14.

Jean Gottmann and Robert A. Harper, eds., *Metropolis on the Move; Geographers Look at Urban Sprawl* (New York, John Wiley & Sons, Inc., 1967).

6 *The Municipal Yearbook, 1967* (Chicago, The International City Managers' Association, 1967), pp. 11-17.

make more difficult the recognition of larger geographical units of which they are a part.

Metropolis

Geographers have long recognized that the concept of "city" cannot be exclusively concerned with the administrative or jurisdictional municipal unit, and that most cities, in a functional sense, transcend political boundaries. A major professional concern of some geographers has been to set up and apply criteria for delimiting and describing the functional city, as distinguished from the administrative city. In this process, the practical problems of collating statistics for various areas, with comparability of criteria and measurements among such urban areas from place to place and time to time, has been, and is, a major concern. The problem is complicated by the fact that, in most urban areas, a decreasing proportion of the population and economic activity is located in the central city, and in most central cities both are suffering an absolute decline. In a rapidly increasing proportion of the metropolitan areas, the suburbs and unincorporated fringe areas now contain larger residential populations than do the respective central cities, while at the same time commercial and industrial activities are increasing both relatively and absolutely in the suburban areas. These trends have been enormously accelerated by the ubiquity of highway transportation, with its flexibility, producing a much greater freedom of choice among locations than was ever before possible.

The metropolitan area, in some ways, constitutes the geographic city. These areas consist of one or more "central cities," together with suburban and fringe areas which are related functionally to the central city. The Federal agencies define Standard Metropolitan Statistical Areas, of which there are a proximately 230, as consisting of a centr city, or cities, with minimum population 50,000, together with counties having fun tional inter-relations, as measured mainly b community patterns, with the county co taining the central city. Thus, the concept the labor-market area, or "commuter shed, is basic to the defining and delimitation of a urban unit. Such a unit contains the territo and population which is essentially depen ent upon the central city for many service In spatial terms, the component portions metropolitan areas are mutually inter-relate and for many purposes the metropolitan are constitutes a more suitable unit for geogr phic study than does the central city alone.

A metropolitan area, then, is a geographi functional city, and for many purposes ma be regarded as the basic areal urban unit.

Yet the concept of the metropolitan are too, like that of the administrative or munic pal city, has certain serious shortcomings.

One set of shortcomings in defining metropolitan areas arises from the use of the statistical areal units. In the United States, th county is the basic areal unit for which statistics are compiled and tabulated; metropolita areas are defined as consisting of one or mor counties. In many instances, the urbanize area within a county may be located onl in a small portion of the county, the res of which has very little population or deve opment, yet the entire county is include within the metropolitan area. Thus, the un statistical area, the county, may be far fro homogeneous with respect to the essenti urban characteristics. This is a typical ex ample of the use for geographic purposes o what has been called "modifiable units, or areas which, variously delimited an used in differing combinations, produc

fering area definitions.[7] For example, the n Bernardino-Riverside-Ontario, Califor-, Standard Metropolitan Statistical Area, largest in the United States, consists of er 27,000 square miles, an area consider-ly larger than all of southern New England, cause San Bernardino County, whose pulation is nearly entirely confined to the thwestern corner, the rest being virtually inhabited desert, must, by definition, be in-ded in its entirety. Similarly, St. Louis unty, Minnesota, a part of the Duluth-perior Standard Metropolitan Statistical ea, includes much wilderness and extends about 150 miles northward from Lake perior to the Canadian border.

By no stretch of the imgaination could the lderness of the Mohave Desert or the rthern Minnesota arrowhead be considered urban or metropolitan in inherent charac-istics, yet the rule that whole counties must included within metropolitan areas if any rt of them is metropolitan, produces such omolous situations.

Recently, with the increased federal par-ipation in the financing of many programs urban areas, including highways, mass nsit, open space land acquisition, urban ewal, and sewer and water facilities among ers, the pressures to provide comprehen-e metropolitan planning have increased. ice 1965, comprehensive plans prepared metropolitan regional agencies, have been quired as prerequisites for local assistance by the federal government, which has at the same time provided matching funds to set up and operate such metropolitan planning agencies. These agencies, however, are not substitutes for metropolitan government, which, in most parts of the United States, the electorate is unwilling, as yet, to accept. The lack of such acceptance springs largely from the pervasive trend of people and commercial and industrial establishments to move away from the central cities toward the suburbs and outer portions of the metropolitan areas. The vacuum thus created in the inner portions of cities tends to be filled by the recent in-migrants who, at least in many northern cities, tend to be members of minority groups, predominantly Negro. These in-migrants, having been long disadvantaged educationally and culturally, are least able to command the better and newer housing available on the outskirts of the respective urban areas, which, in turn, expand as the result of the peripheral growth of the city. Thus there tends to be a series of conflicts of interest between the inner city, or the central city, on the one hand, and the "middle class" areas on the edge of the city and in the suburbs beyond. This is commonly reflected in political differences, with the central cities tending to reflect the more liberal attitudes of the minorities as the proportion of central city population which they represent increases. It is not at all unexpected that Cleveland, Gary, and Washington have Negro mayors; in earlier generations many of the city governments were predominantly reflections of the succession of ethnic and national in-migration patterns. On the other hand, the suburbs tend, on the whole, to reflect the attitudes of those who have "arrived" economically and socially, and who resist the assimilation of the more recently urbanized

The problems of selecting suitable areal units are discussed in Otis Dudley Duncan, Ray P. Cuzzort, and Beverly Duncan, *Statistical Geography* (Glencoe, The Free Press, 1961), and by David Griff, "Regions, Models and Classes," *Models in Geography* by Richard J. Chorley and Peter Haggett (London, Methuen & Co., Ltd., 1967), pp. 461-509. The latter contains an extensive bibliography on the problem of geographically significant statistical units.

in-migrant populations. Zoning, for example, as an instrument of land-use control, has frequently been misused as an instrument of economic segregation, and hence ethnic segregation in many suburban areas, by providing minimum lot sizes and density standards which only the more affluent could attain, thereby excluding people whose economic status tended to the lower.[8]

These city-suburban differences, of course, are reflected in resistance to metropolitan government and the continuation of proliferating local governments within metropolitan areas. In order to overcome some of the consequent inefficiencies and provide for the multiplicity of services and facilities which are the responsibilities of local governments, many devices short of comprehensive metropolitan government are being developed. One such device is the special-purpose government, in many instances functioning in a geographic area which does not coincide with any of the areas of local government, but dealing with problems of metropolitan scope beyond the ability of local governments. Among the early noteworthy examples are the Metropolitan Sanitary District of Greater Chicago and the Port of New York Authority, the latter representing an early noteworthy example of a government set up across state boundaries by interstate compact. Special-purpose governments, however, have certain shortcomings; among them are the lack of coincidence of boundaries with local governments and often with each other, the tendency to add to the proliferation of governmental units, and, in some instances, the interposition of an additional governmental unit without direct control by th electorate. Many of the governmental an administrative problems of cities and metr politan areas arise from an essentially geo graphic concern, namely, the determinatio of the nature and extent of those function which are best dealt with locally and thos which are of extra-local regional concern For example, local access streets should b the responsibility of the local community, bu regional expressways must be planned, buil and operated by a regional or larger unit c government. Similarly, local parks and re gional parks must be differentiated. School constitute major problems in this regard; th different standards of inner-city and suburba schools have been adequately documentec and few people are unaware of the problem in this regard.[9] To the geographer, problem such as these constitute interesting and im portant challenges: to what extent should th planning, building, and operation of facilitie and services be localized, and to what exten and in what manner — and especially withi what area — should these facilities and serv ices be the concern of the entire metropolita region, or of major portions of it? How an by what criteria can determinations be mad of the degree and areal extent of involve ment of each level of government appropri ate to the problem? How can local autonom be preserved for matters of local concern an at the same time assure that regional an metropolitan concerns can be reconciled wit the local ones?

Short of metropolitan government, and i addition to regional special-purpose govern ments, many official and unofficial device are being tried to coordinate urban function

[8] Richard F. Babcock, *The Zoning Game* (Madison, The University of Wisconsin Press, 1966); Sidney M. Wilhelm, *Urban Zoning and Land-Use Theory* (New York, The Free Press of Glencoe, 1962).

[9] One of the most widely known presentations c the disparities between city and suburban schoo is James B. Conant, *Slums and Suburbs* (Ne York, McGraw-Hill Book Co., 1961).

order that metropolitan areas may overcome the deficiencies of intergovernmental fragmentation and function as urban units. In the past two decades, many metropolitan areas have set up comprehensive transportation studies as prerequisite to planning of highway and transit systems; since 1965 these operations have been required, as previously mentioned, by the federal government. But it was soon realized that projections of traffic loads were inadequate, and that circulation patterns merely reflect land-use patterns, because it is the land uses which generate traffic. Many of the transportation study agencies thus evolved into comprehensive regional and metropolitan planning agencies, concerned not only with transportation and land-use patterns, but also with economic and social planning within the respective areas. The professional planners themselves, through such organizations as the American Institute of Planners, have formally recognized the broadening of their responsibilities and the scope of their profession, to include social and economic planning along with the physical. Geographers, likewise, are increasingly concerned with the spatial aspects of social and economic conditions and problems, many of which concentrate heavily in urban areas.

To coordinate planning and its effectuation, then, a variety of new organizational forms is evolving, and these constitute, in their spatial aspects, appropriate subjects for geographic investigation. Metropolitan comprehensive government is an ultimate solution in some instances, but, short of that, comprehensive regional and metropolitan planning agencies, special-purpose regional governments, informal and formal councils of government, and intergovernmental arrangements for purchasing or leasing of facilities and services, as well as pooling arrangements are all effective in various situations. The analysis and planning of governmental and inter-governmental arrangements in their spatial aspects within urban areas constitute a set of challenges of great prospective opportunity to the geographer; very likely these problems will constitute a new frontier of geography in the near future.

Megalopolis: Inter-Metropolitan Coalescence

Just as the traditional concept of the city as a unit of spatial organization of urban areas has largely been superseded by the concept of the metropolitan area, so the metropolitan area is itself inadequate as a descriptive device for contemporary and prospective patterns of urbanization. With respect to many functions, the metropolitan areas are not discrete functional areal units; a more comprehensive and larger unit has had to be developed to describe groups of metropolitan areas with overlapping functions.

Modern transportation is largely responsible for the growth of super-metropolitan conurbations. Transportation and land use are but two sides of the same coin, one static, the other dynamic. Urban land is useful primarily because of its accessibility to other land. There is a complementarity[10] among land uses which gives rise to cities, and, at the local

[10] The principle of complementarity is effectively presented by Edward L. Ullman, "The Role of Transportation and the Bases for Interaction," *Man's Role in Changing the Face of the Earth*, edited by William L. Thomas, Jr. (Chicago, The University of Chicago Press, 1956), pp. 862-80. Complementarity is also related to the gravity analog in spatial interaction, the subject of many articles in the literature of both geography and planning. For example, Gunnar Olsson, *Distance and Human Interaction: A Review and Bibliography* (Philadelphia, Regional Science Research Institute, 1965).

scale, there is a complementarity among the component functional areas and establishments within cities and metropolitan regions. Transportation is one of two forms of mutual accessibility among land uses and establishments; it overcomes the friction of distance, but at a cost. The other form of accessibility is mutual proximity, which, at the cost of high-value land and high density, reduces or eliminates the necessity for providing transportation. Thus, the dichotomy of transport costs and land costs is reflected in the balance between centralizing and decentralizing forces, previously mentioned.[11] Geographers, land economists, sociologists, and others have evolved many idealized forms of spatial patterns to describe urban structure and growth, but all of them are based upon a balance between the two forms of accessibility: mutual proximity on the one hand and transportation on the other. Transportation — measured by traffic flows — depends, in turn, upon complementarity in which each area served produces specialized goods and/or services which are utilized elsewhere. In cities, the result is a functional differentiation of local areas; among cities, modern transportation and communication makes possible specialization, so that each city in a group or set of cities — such as a conurbation — does not need to perform all of the urban functions in equal degree, providing that interchange of people and services, as well goods, can take place among the areas cities within the group.[12] A common form this in the modern city is the separation working and residential areas, giving rise the daily commuting between them.

As transportation improves — such with the development of modern expre highways — the area accessible within given time-cost limit from any given poi increases. This produces a greater flexibili

[11] The literature on relations between transportation and land use is enormous. A few examples: Lowdon Wingo, Jr., *Transportation and Urban Land* (Washington, Resources for the Future, Inc., 1961); Walter G. Hansen, "How Accessibility Shapes Land Use," *Journal of the American Institute of Planners*, Vol. XXV, no. 2 (May, 1959), pp. 73-76; numerous reports of the Highway Research Board (Washington), and the proliferating reports of the many comprehensive metropolitan transportation-land use studies, such as the *Final Report, Chicago Area Transportation Study* (Chicago, Chicago Area Transportation Study, 1959-61), 3 vols.

[12] Functional specialization among cities is subject of a vast literature on "economic bas Among the more useful items are Ralph Pfouts, *The Techniques of Economic Bc Analysis* (West Trenton, N.J., Chandler-Da Publishing Co., 1960), selected readings; Char M. Tiebout, *The Community Economic Bc Study* (New York, Committee for Econon Development, 1962); John W. Alexander, "T Basic–Nonbasic Concept of Urban Econon Functions," *Economic Geography*, Vol. XX no. 3 (July, 1954), pp. 246-61; Walter Isa "Interregional and Regional Input–Output Tec niques," Chapter 8 of his *Methods of Regior Analysis* (Cambridge, Technology Press a New York, John Wiley & Sons, Inc., 1960), p 309-74; Edward L. Ullman and Michael Dac "The Minimum Requirements Approach to t Urban Economic Base," *Proceedings of the IG Symposium in Urban Geography Lund 19* Lund Studies in Geography Ser. B. Hum Geography, no. 24, Lund, 1962), pp. 121-4 The differences among cities with respect their economic functions also may serve as basis for functional classification of cities, subject of a proliferating literature, examples which are Chauncy D. Harris, "A Function Classification of Cities in the United State *Geographical Review*, Vol. XXXIII, no. (January, 1943), pp. 86-99; Howard J. Nelso "A Service Classification of American Citie *Economic Geography*, Vol. XXXI, no. 3 (Jul 1955), pp. 189-210; Richard L. Forstall, "Ec nomic Classification of Places over 10,000, 196 1963," *The Municipal Yearbook 1967* (Chicag The International City Managers' Associatio 1967), pp. 30-65. Also R. H. T. Smith, "Meth and Purpose in Functional Town Classification *Annals of the Association of American Geo raphers*, Vol. LV, no. 3 (September, 1965 pp. 539-48.

movement. Whereas a relatively short time ɔ, cities were highly concentrated, with lial transportation patterns focusing upon : core of the city, now the patterns of ffic flow and accessibility tend to be more fuse and the densities of development ver, while the character of the urban func-ns within the cores is also changing. With pect to many activities, the central cores of cities are rapidly becoming less important. the same time, densities are becoming less, d is being absorbed by urbanization at an elerating rate, and new nodes or foci of essibility, such as regional shopping cen-s and industrial parks, are developing with luced dependence upon access to the old y cores.

But even with modern transportation, ıal access cannot be provided to all areas, d the form and shape of the urban pattern still, no less than formerly, controlled in ge measure by the pattern of transporta-n routes. Multiple nuclei, rather than a gle dominant core, constitute the pattern of modern city. Furthermore, the city does t spread with equal density or equal idity in all directions; those areas with st transportation access tend to develop re rapidly, and generally with higher den-es than those less well provided with trans-rtation access. Formerly, cities tended to velop a star-shaped pattern, with higher ısities and earlier development along lial railroads, rapid transit lines, or trolley es. More recently, with the ubiquity of hways, automobiles, and trucks, the inter-ial areas between the older radial prongs growth tended to fill in, generally, how-er, at somewhat lower densities than the ler radial prongs. The outlines of the anized areas thus tended to depart from star-shaped patterns characteristic of a v decades ago and to assume somewhat more circular patterns. More recently, however, with the advent of the express highway, the tendency has been for the development of axial "corridors" along such routes of maximum accessibility, with higher density of development closest to such routes, and especially to the intersections of such routes. Indeed, the development of such "corridors of high accessibility" has been a major physical element in the recently-published comprehensive plans of a number of cities and metropolitan areas. Serving as the axes along which such corridors are expected to develop are efficient transportation routes, including various combinations of railroads, high-speed rail rapid transit lines, and express highways.[13]

Such corridors, however, are not confined to within the limits of individual metropolitan areas. With efficient high-speed transportation of people and goods, the expanding tentacles of urbanization of cities and metropolitan areas formerly separated by many miles tend to coalesce. The pressures for lower densities and consequent increasing amounts of urbanized land tend to expand along the lines of least resistance: the most efficient transportation routes. As the tentacles along these routes connecting adjacent cities and metropolitan areas extend toward each other, they eventually form more-or-less continuous chains of urban development.[14] Cross com-

[13] National Capital Regional Planning Council, "The Regional Development Guide 1966-2000, Washington, D.C.," (Washington, Government Printing Office, 1966); *Comprehensive Plan of Chicago* (Chicago, Department of Development and Planning, 1966); *The Comprehensive General Plan for the Development of the Northeastern Illinois* (Chicago, Northeastern Illinois Planning Commission, 1968).

[14] An interesting study of one such multi-nodal urban agglomoration is F. Stuart Chapin, Jr., and Shirley F. Weiss, eds., *Urban Growth Dynamics in a Regional Cluster of Cities* (New York, John Wiley & Sons, Inc., 1962).

muting, and functional specialization among the several or many cities forming such chains or clusters gives rise to common labor-market and housing-market areas; super-cities, transcending metropolitan area boundaries, thus tend to develop.

To this concept of the multi-nodal inter-metropolitan super-city, whether arranged as a cluster, or lineally along a corridor, the term Megalopolis has been applied.[15] The term is by no means a new one; it is derived from the Greek "mother city," and was first used a half-century ago by Patrick Geddes to describe an urbanized region.[16] Popularized as a generic term by Lewis Mumford,[17] it was adopted as a proper noun by Jean Gottmann[18] to describe the area between Boston and Washington, which, in fact, had developed even in colonial times as a chain of cities at the head of navigation for ocean-going vessels along the "fall line" marking the boundary between the Piedmont hard-rock area and the Atlantic coastal plain. When effective land transportation developed to replace coastal water transportation in the area during the late nineteenth century, the cities were linked by heavily-trafficked main-line railroads, and later by express highways and by shuttle airlines, all of which carry the heaviest volumes of inter-city traffic in North America. With these efficient transportation links, the urbanization spread along the axis, an the adjacent tentacles coalesced, eventuall merging the older nodes into a lineal city.[19]

Many other "megalopolises," consisting clusters or lineal arrangements of cities ar recognized. Cross-commuting among citie fifty or more miles apart is now quite com mon, and a new form of urban structure i emerging. Meanwhile, the multi-nodal cluste type of super-city has its prototype i southern California, while the Boston-Wash ington axis constitutes the classic example c a lineal inter-city corridor. The federal gov ernment, in fact, now officially recognizes th new urban corridor form, for the so-calle "Northeast Corridor" is the scene of a com prehensive series of experiments in high speed intercity transportation.[20] Actually, thi program springs directly from the work c Jean Gottmann, whom Senator Claiborn Pel the author of the federal legislation, credit for the idea of the corridor.[21] Other corrido like megalopoli exist in the area betwee Cleveland and Pittsburgh, Dallas and Fo Worth, Chicago and Milwaukee, Portlan Oregon and Vancouver, British Columbi and in many other places, each of which i characterized by increased accessibility alon a major axis of transportation. At a large scale, the area between the Midwest and th Atlantic seaboard, long identified by geogra phers as the core urban-industrial area of th

15 Jean Gottman, "Megalopolis, or the Urbanization of the Northeastern Seaboard," *Economic Geography*, Vol. XXXIII, no. 3 (July, 1957), pp. 189-200.

16 Patrick Geddes, *Cities in Evolution* (London, 1915; reprinted London and New York, Oxford University Press, 1949).

17 Lewis Mumford, *The Culture of Cities* (New York, Harcourt, Brace and Company, 1938), and *The City in History* (New York, Harcourt, Brace & World, Inc., 1961).

18 Jean Gottmann, *Megalopolis* (New York, The Twentieth Century Fund, 1961).

19 *The Region's Growth: A Report of the Secon Regional Plan* (New York, Regional Plan Assc ciation, 1967).

20 *Highway Travel in Megalopolis* (New Yor Wilbur Smith and Associates, 1963).

21 Robert A. Nelson, "A Quick Rundown on th Northeast Corridor Transportation Project," *Report on the 1966 Conference on Mass Trans portation*, University of Chicago Center fo Continuing Education, March 4, 1966 (Cleve land, Brotherhood of Railroad Trainmen, 1967) pp. 117-22.

ited States,[22] has recently been called by a nner the "Great Lakes Megalopolis,"[23] in ich the east coast Megalopolis is at one d, the Chicago-Milwaukee conurbation at other, and the Cleveland-Pittsburgh galopolis in the center. Thus, just as ge-raphers have identified a hierarchy of ntral places," ranging from the crossroads nlet to the world metropolis,[24] so there is erging a concept of a hierarchy of conurba-ns, corridors, or megalopoli, constituting framework of the pattern of urban devel-ment and urban functional organization.

e Nature of Urban Functions

Lewis Mumford has identified cities as ntainers," since they contain the ag-meration of people and facilities which rform certain functions identified as ur-n.[25] What are the urban functions, in the formance of which the structural and tial forms of organization discussed above ve evolved?

The urban functions are predominantly, indicated above, those which depend upon accessibility, whether by mutual proximity or transportation, or, most commonly, a combination of the two. Thus, in a sense, all basic urban functions are "central place" functions. Most urban geographers and economists differentiate functions as economic and non-economic, although even the essentially non-economic functions, such as education, religious shrines, and residence, have important economic aspects. The economic functions include, in one sense, the "central place" administrative and control functions, such as government, as well as the administration of business organizations. Many people differentiate economic functions into three or four categories. The primary functions are those which result in the exploitation or extraction of resources of the site, such as mining, forestry, and agriculture, whether or not the resources are renewable. Secondary functions are those involving the physical handling and processing of goods, including manufacturing, packaging, storage, and transportation. Tertiary functions involve transfer of title or ownership, and include retailing and wholesaling. Some writers include record-keeping among the tertiary functions; others designate them in a separate category as quaternary functions.[26]

The primary functions are not predominantly urban in character, although some give rise directly to urban settlements, such as fishing communities, mining camps, and others, but, invariably, other urban functions arise wherever numbers of people gather in

This "core region," the American Manufacturing Belt, is identified in virtually all of the standard geographic texts on North America. For example: C. Langdon White, Edwin J. Foscue, and Tom L. McKnight, *Regional Geography of Anglo-America* (3rd ed., Englewood Cliffs, N.J., Prentice-Hall, Inc., 1964), pp. 32-66.

C. A. Doxiadis, "The Great Lakes Megalopolis," *Ekistics*, Vol. XXII, no. 128 (July, 1966), pp. 14-31.

The concept of the central-place hierarchy is one of the most intensively explored in the recent American literature of urban geography, and has attracted much world-wide attention. A comprehensive bibliography on the subject is Brian J. L. Berry and Allen Pred, *Central Place Studies: A Bibliography of Theory and Applications*, 2nd ed. (Philadelphia, Regional Science Research Institute, 1965). This bibliography contains annotations on over 1,000 items.

Mumford, *op. cit.*

26 William Goodwin, "The Management Center in the United States," *Geographical Review*, Vol. LX, no. 1 (January, 1965), pp. 1-16; William Applebaum, "Formation and Location of National Trade Associations in the United States," *The Professional Geographer*, Vol. XX, no. 1 (January, 1968), pp. 1-4; Sidney M. Robbins and Nestor F. Terleckyj, *Money Metropolis* (Cambridge, Harvard University Press, 1960).

mutual proximity. The locational patterns of the secondary functions depend upon relative access to raw materials, components, energy resources, manpower, markets, and site availability in varying proportions, and there is extensive literature on the factors of industrial location.[27] The tertiary and quarternary functions tend to locate at points of maximum accessibility to their respective service areas or "hinterlands," which are, in turn, "nested" in a hierarchical arrangement, giving rise to a hierarchy of central places or urban nodes within which the functions are concentrated.[28] A variation of the urban hierarchy in some areas is the "primate city" arrangement, in which one city dominates an extensive area, and several of the lower orders of urban places, between the dominant city on the one hand and the smaller places on the other, are absent. At the top of the hierarchy is the "world city," whose service area for some functions may be intercontinental.[29]

Cities are also centers for the orgination and diffusion of both material and nonmaterial culture.[30] By providing concentrations of people, interaction and the development of new ideas and concepts is facilitated in cities; the larger the city the more likely is it to be seat of new ideas and inventions, and the more attractive it becomes for the leaders in all fields. Thus, the gravitative pull of the city depends upon its mass, or population size and diversity, and the larger the city the more likely is it to attract additional population. This, in turn, creates more interaction and reinforces the role of the city as a center from which innovations diffuse outward. The increased diversity of the city, in turn, makes it more attractive, for maximum freedom of choice is available in the larger urban agglomerations. In larger metropolitan and megalopolitan agglomerations the greatest variety of economic, social, and cultural opportunities exist, as well as the greatest choice of housing accommodation and residential environments. Thus, there is a rural-urban migration, and a differential rate of growth, other factors being equal, favoring the larger urban concentrations over the smaller ones. The increased availability of the automobile, which releases urbanization from dependence upon a limited number of areas especially favored by mass transportation, is spreading the accessibility to the big-city advantages, homogenizing the culture, reducing the formerly sharp differences between urban and non-urban areas, and constituting new and abundant opportunities for geographers to investigate the changing relationships, in urban settings and beyond, between man and his culture on the one hand and the emerging spatial patterns and environmental challenges and opportunities on the other.

[27] Benjamin H. Stevens and Carolyn A. Brackett, *Industrial Location: A Review and Annotated Bibliography of Theoretical, Empirical and Case Studies* (Philadelphia, Regional Science Research Institute, 1967).

[28] A very useful recent statement is Brian J. L. Berry, *Geography of Market Centers and Retail Distribution* (Englewood Cliffs, N.J., Prentice-Hall, Inc., 1967).

[29] Peter Hall, *The World Cities* (New York, McGraw-Hill Book Co., 1966).

istorical and Comparative rban Studies

3

oads Murphey

e city is as old as civilization. The two rds come from the same root, which originally meant city, suggesting an interrelation ich was especially close in the early his- cal period, but which is still relevant. *ilization* is generally equated with a liter- tradition preserved by organized groups, societies, living in permanently fixed sters rather than in small, scattered units. ese groups are also characterized by some *ision of labor* or task specialization, which ludes a variety of *secondary* and *tertiary* ading, services, manufacturing) rather n merely *primary* (agriculture, fishing, nting, and gathering) activities. Such a cription also fits the city. Historically the t cities began to appear with the emer- ce of a division of labor. It seems reason- y certain that this development first took ce somewhere in the hill country bordering Tigris-Euphrates Valley, where the wild estors of wheat and barley were native, l where, well before the appearance of es, goats, pigs, dogs, and cattle had been nesticated and permanent field agriculture l begun to produce consistent surpluses. ly when this stage was reached could a division of labor take place so that some members of the population could engage in non-food-producing activities and still be fed from surpluses produced by the remaining farmers.

Cities As Centers Of Exchange

There were obvious advantages in conducting most of these non-agricultural functions in a concentrated center — a city where trade goods, raw materials, and food could most conveniently be assembled and from which a set of goods and services could be made available to surrounding rural areas or to other cities as trading partners. The first cities profited in these ways from *economies of scale* (lower costs resulting from size and concentration of the enterprise) and demonstrated the universal function of all cities at every time and place — as *centers of exchange*. Some aspects of this function, as in nearly all cities now, may not have been strictly economic, for example, political administration, religious and ritual services, or the wider social and cultural rewards present in a large community, which is also more

rinted from *Journal of Geography* (May, 1966), pp. 212-19 by permission.

diverse than a farmers' village. From the time of its origin the city has been the chief engine of economic and institutional growth and change through specialization based on exchange.

But in order to perform every aspect of its function as a center of exchange, the city depended, as it still does, on good access to and from the areas served, the city's *hinterland*. Any city implies and is based on *spatial interaction,* or relations between places. It cannot exist as an isolated unit, but only as it can receive from outside the city, often from great distances, the materials and food which it needs, and can provide in return a set of goods and services to consumers in its hinterland. It must therefore be concerned with maximizing *transferability* (the means of moving goods from one point to another at bearable cost) by building artificial routes of easy movement, such as roads, and/or by selecting a site and a situation which make access to and from the city easy. *Site* refers to the actual ground on which a city is built, *situation* to the wider pattern of spatial interactions with other places. Situation is thus by definition a relative term rather than an absolute one. It describes the relation of a given place to other places in simple distance, or, more importantly, in cost or effort. It can, and usually does, change over time, especially as accessibility and spatial interaction are affected by changes in transport and transferability.

In this broad sense all cities are *central places*. They are nodes or nuclei of their respective hinterlands, performing in one concentrated and easily accessible (or central) place specialized functions on behalf of the wider area which they serve. A final essential basis of exchange, the city's universal function, is *complementarity* — the actual or potential relationship between places which possess different sets of goods or ser- vices and which therefore have a basis f mutually profitable exchange. No two plac are alike, and each place tends to have i own set of *comparative advantages* for t production of different sets of goods. Cor plementarity results in trade, however, on if transferability between the places co cerned is great enough to overcome the fri tion of distance at bearable cost, and if *intervening opportunity* lies between the where either potential trade partner cou satisfy its needs with less effort. These a universal conditions for and characteristics cities in every area and at every historic period.

Early Cities Of Mesopotamia And Egypt

One of the results of historical or compar tive urban studies is their demonstration urban universals rather than differences. I about 4000 B.C., city building in the Midd East seems to have spread from its probab earlier hearth in the hill country of Syri Palestine, and Iraq down onto the flood pla of the Tigris-Euphrates, and perhaps equal early to the lower Nile Valley in Egypt. The are both potentially productive areas whe exchange was relatively easy and transfe ability great. The Nile and Tigris-Euphrat are both exotic rivers, rising in well-water highlands and flowing across a desert reach the sea. Their anual floods deposit highly fertile alluvium, especially in t deltas, and in addition provided semi-aut matic irrigation. It was possible on this bas to produce large and reliable food surplus and thus to enable a division of labor as we as to create a prosperous hinterland for citi to serve. But the advantages of both areas f the growth of large and numerous cities f the first time were at least as important derived from the relative ease with whic goods could be moved and therefore cente ɜxchange could function. Both areas were ɜl and largely or entirely treeless, so that vement overland was relatively easy. The es which arose, however, made more im- 'tant use of the rivers themselves as easy ıtes of spatial interaction where transfer- lity was maximized. City sites were ripar- (on the river banks) and their situational 'antages meant that they could assemble water transport and at low cost, large and ky shipments of delta-grown grain, dates, l other food commodities, and could also ng in heavy materials like stone or metals m great distances in addition to trade ɔds, since they could use sea as well as ɜr routes. Goods and services could be dis- uted from these cities in the same way. ansport costs tend to vary inversely with capacity of the carriers, especially for ıvy or bulky goods. For many millennia ɜr the first cities arose and until the devel- nent of the railway, large cities could not maintained except on navigable water- ys.

The early cities of Mesopotamia and Egypt o depended, as all cities do, on comple- ntarity as a basis for exchange. The funda- ntal complementary relationship is be- ɜen the city itself and the non-urban agri- tural or primary-producing areas in its nterland. But even in the ancient period lower Nile and Tigris-Euphrates were void of stone, wood, or metal, and these nmodities had to be imported to the cities, ich were thus involved in complementary change (finished goods or services in ex- ange for raw materials) with a variety of ces by river and sea routes. Cedars came Egypt by sea from Lebanon, stone in great antities from quarries along the upper le, metals and ores from many distant ırces. In the early cities of Mesopotamia, ɜn nails and simple tools were originally de from sun-dried or kilin-fired clay until exchange had been established with distant sources of ores and metals as well as wood and stone. There was thus a relatively rapid expansion of a commercial network, or field of spatial interaction, accompanying the rise of the first cities.

Urbanization In Asia

Connections of this sort were established with complementary areas as distant as India (to which the Mesopotamian urban model spread by at least 2500 B.C.). Cities of the Indus plain, notably Harappa and Mohenjo Daro, became trade partners with Mesopota- mian cities in a mutually profitable exchange.

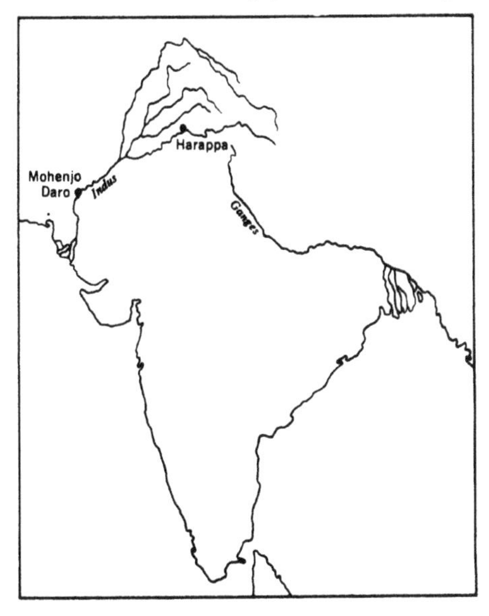

There was little or no exchange between these early Indian cities and Egypt, since urban Mesopotamia presented an intervening op- portunity. Similarly trade and cultural ex- change between Mesopotamia and China was apparently both late and weak, since the Indus cities intervened spatially and since transferability between Mesopotamia and India was much greater by sea routes along the shores of the Persian Gulf and Arabian

Sea. The flood plain of the Yellow River in north China was probably in historical terms the third major center of urbanization to develop (by about 2000 B.C.). It and the Indus plain shared the same set of physical characteristics as had earlier favored the rise of Mesopotamian and Egyptian cities — exotic river systems flowing across an arid but level and largely treeless plain where fertile alluvial soils and water for irrigation were combined with high transferability. Within each of these Asian areas urbanization spread first along the major rivers and their tributaries and distributaries, as avenues of easy access to and from productive agricultural areas. Asian urbanization is still heavily water-oriented.

With the rise of the first territorial empires in Asia, however, the dominant urban function became increasingly one of political administration and cultural synthesis. Trade did, of course, continue to grow in and from urban bases, but most traditional Asian cities were only secondarily involved in trade or manufacturing in terms of employment or investment, and were more importantly concerned with managing the administrative machinery of large bureaucratic states, including government trade monopolies. As with many modern political capitals, they also played an important cultural and ceremonial role as centers of the literate, intellectual, and artistic tradition in each area, and invested a large share of urban resources in monument building for both political and religious rather than strictly economic goals. City sites and situations had still, nevertheless, to maximize access, which is as important for an administrative city dealing in services as for a commercial city dealing in goods.

Urbanization In The West

From the beginning of the urban tradition of the West, however, the city was pr dominantly a commercial center. By the fir millennium B. C., the dynamic centers of u banization and expanding trade had for variety of reasons shifted from Mesopotami and the Nile into the Mediterranean Basin. small enclosed sea, generally free of storm and with a highly indented coastline, th Mediterranean was one of the earliest heartl for the growth of navigation and ocean trans port. Transferability within the basin wa high, once the elementary techniques of nav gation had been mastered. It is to Phoenician Cretan, Greek, and Roman forms that th principal direct roots of modern Wester urbanization extend, rather than to Luxor Memphis, Ur, or Babylon. Tyre, Knossos Athens, Syracuse (and Rome, to a degree were centers of trade on a greatly increase scale and spatial spread. While it is true tha

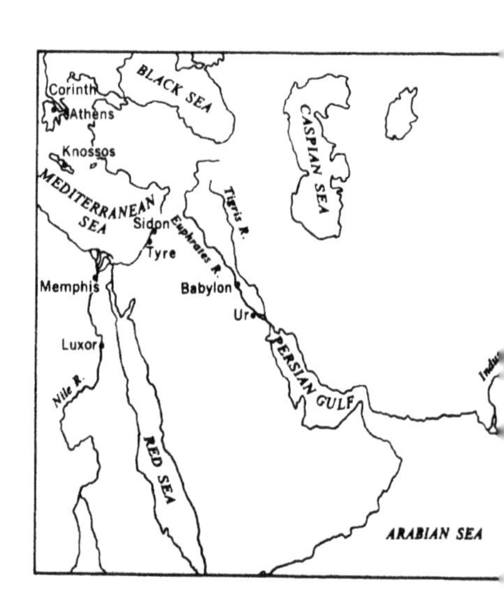

ancient urban Mesopotamia and Egypt show striking similarities to modern urbanism, confronted many of the same problems, and solved (or failed to solve) them in many o the same ways, the modern parallel with the pre-classical and classical Mediterranean is

ser still, as part of a continuous Western
ban tradition which stretches at least from
lon and Carthage to Corinth, Rome, Lon-
n, New York, and San Francisco. The
at majority of Western cities has been and
nains of dominantly commercial rather
n administrative or symbolic, although, as
traditional Asia (where the proportions
re reversed), most cities have performed
th functions. The *polis*, or city-state of
ssical Greece, and the colonies which
eeks founded elsewhere along the Mediter-
nean shores as their trade by sea expanded,
re primarily commercial centers involved
widely-extended spatial interaction pat-
ns stretching from the Black Sea coasts
d the Crimea to the Pillars of Hercules,
ere the Mediterranean joins the Atlantic.
ese sorts of cities, of which the Greek
velopments were prototypes, have been
elled *heterogenetic* — influenced by and
olved with a variety of interactions with a
at range of distant and different places —
opposed to the *orthogenetic* cities of the
at Asian empires, which were proportion-
ly more shaped by and involved in inter-
ions within culturally homogeneous hinterds and which functioned as urban pin-
cles of the several Asian "Great Tradi-
ns."

man Cities

th the rise of the Roman Empire, the
minantly commercial nature of the Greek
lis was overlaid by a new and wider set of
ministrative functions. The city of Rome,
led by its dual function in both trade and
ministration and by its central position
thin the Mediterranean for assembling and
tributing food and other goods by sea, was
obably the first city in history to reach or
proach a population of one million. Its size
s a reflection of the enormous extent and
productivity of the hinterland it served. People and goods were transported by sea and by an impressive network of paved roads, all of which proverbially led to Rome and greatly augmented transferability and centrality. Cities founded or expanded by the Romans throughout their empire in Europe, North Africa, and the Middle East bore the imperial stamp of central authority. Many began as military camps or garrison towns, grew to provincial or regional capitals, and acquired some commercial functions as well. They were usually walled and in most instances carefully laid out on a gridiron plan, with major avenues leading from each of the gates at the four points of the compass. In this they resembled the administrative cities of the contemporary Chinese empire as its spatial extent also expanded and was marked by the establishment of walled regional centers of control laid out spatially on a uniform imperial plan.

The collapse of Roman control in the West was followed by a period of several centuries in which both the number and the size of cities shrank. Roman roads degenerated, transferability was everywhere lessened by brigandage and civil disorder, and commercial production was similarly affected. The basis for urban growth was thus sharply reduced as compared with the period of the *Pax Romana*, when exchange and transferability were maximized. The city of Rome itself may have shrunk to as few as 5,000 inhabitants at its lowest point, a dramatic illustration of what happens to a city when it loses its hinterland. No political or administrative functions remained, and economic functions were reduced to those of a tiny local exchange center for the immediate environs of the town. Elsewhere in the former empire urbanism was also in retreat. Many of the cities which survived the Dark Ages did so because they had become the seats of bishops (cathedral towns) or grew up in the shelter or even within the walls of the fortified castle of a feudal lord. As in Italy itself, lowland sites of easy access which had attracted cities under the *Pax Romana* were often abandoned, and nearly all cities sought the protection of elevated or defensible sites where accessibility was sacrificed for security.

Cities In The Middle Ages

Venice is the most notable example of a city which arose during the chaos following the Roman collapse as a protected refuge — in the marshes near the seaward edge of the Po River delta on the Adriatic — and was able to combine defensibility with high transferability by sea. With the combined advantages of its site and its situation at the northernmost extension of the Mediterranean, where it could serve the north European market and could assemble goods by sea from the Levant and the Orient, Venice became the biggest city in medieval Europe. Elsewhere in the West exchange was to a large extent provided by periodic fairs, as an adjustment to conditions of low transferability and limited commercial production. Fairs in effect brought the market to the hinterland on a rotating basis in the same way and for the same reasons that periodic fairs or markets still operate in areas of low transferability and restricted commercial production, such as large parts of North Africa or rural China.

Toward the close of the Middle Ages, and for a variety of reasons, larger and better-ordered political units began to grow, security and transferability began to increase, roads reappeared, barrier forests were removed, trade began to revive, and cities once more increased in numbers and size. The age of the great discoveries and the rapid improvements in shipping and navigation from the 15th century on meant a further and enormous widening of the limits of spatial interaction by se routes which soon encompassed the entir world. This, incidentally, meant the declin of Venice, since first its old rival Genoa an then the new port cities along the northwes coast of Europe now stood as a series c intervening opportunities between Venic and the new sea routes to world markets which no longer flowed eastward through th Mediterranean but westward and southwar in the Atlantic to the New World and to Asi around Africa. The Asian tropics, in partic ular, offered the basis for a strong comple mentarity with Europe as producers of spice and a growing variety of other goods whic Europe demanded but could not produce For the same reasons of complementarity an high transferability by sea, a large and profit able trade in sugar arose with the West Indies

Commercial Centers And The Coming Of The Industrial Revolution

Booming urban commercial centers grew i northwest Europe to manage the expande volume of trade, with the greatest advantage accruing to those cities with the most appro priate situations — ports in the Low Coun tries near the mouths of the Rhine whicl offered easy access to the variety of con tinental markets, and on the other side of th Channel, the city of London, well placed a the head of the Thames estuary to serve as major *entrepôt* (center of assembly and dis tribution of goods by water) for the Euro pean market as a whole. This same urba model — the commercial city dominated b merchants — spread across the Atlantic an was reproduced in Boston, New York Philadelphia, Baltimore, and other Nortl American centers. It later spread to Australia New Zealand and to parts of Latin Americ ere commercial production of primary ds for export became prominent and re- red large urban centers of exchange and nmercial services, such as Sydney, Mel- rne, Buenos Aires, or Montevideo, whose oments of grain and meat helped to feed mushrooming urban populations of ope and North America.

The coming of the industrial revolution forced the growth of most of the earlier- ablished commercial cities, for they were ters of investment capital and of cheap ess for the assembly of raw materials and distribution of finished goods. But there also a new growth of manufacturing es close to sources of the bulky materials ch industrialization now required in enor- us amounts and which therefore meant atly increased sensitivity to transport costs. nchester, Birmingham, Sheffield, Essen, sseldorf, Pittsburgh, Breslau, Magnit- rsk, Jamshedpur, Anshan, Yawata, and er virtually new cities arose in association n local deposits of coal or ore. The spread railways did, however, make a greater ree of urban concentration possible. The road and other innovations, such as the mship, and ultimately the truck and the eline, so heightened transferability at low t that even the older commercial centers, from raw material sources, could bring in at manufacturing required and could also g in food to maintain urban populations considerable size. One result of the rapid ease in the size of ocean carriers was that ny of the harbors adequate for an earlier could no longer accommodate them. *Out- ts* had to be developed, farther down the er or estuary on which the city lay, or har- s had to be artificially deepened or en- ged on the coast. Thus, for example, Le vre came to serve as the outport for is, Southampton and Gravesend for London, Cuxhaven for Hamburg, and Kobe for Osaka.

As one consequence of the expansion of trade and the growth of a worldwide trade network, exchange between Europe and the Afro-Asian area greatly increased. In Asia there had in the past been few predominantly commercial cities, and in particular few coastal ports. The cities of the Great Tradition had been inland centers of administration. European traders and later colonial merchants had therefore to establish, or to expand from small indigenous nuclei, a whole set of new port cities to handle the new trade and to service the expanding commercial hinterlands from which they drew their export goods. From Karachi, Bombay, Madras, Colombo, and Calcutta through the major ports of Southeast Asia to Hong Kong, Shanghai, Tientsin, and Yokohama there arose a series of similarly organized cities which were either founded by Western traders where none had existed before or were largely developed by them (in Japan, by what were clearly Western methods). Not merely their coastal sites and their situations, which maximized access by sea, but their physical appearance and their institutions, which were designed to generate and safeguard the accumulation of commercial capital, the increase of trade, and the security of the private entrepreneur and his goods, were replicas of the urban models already developed by post-Renaissance Europe in the merchant cities of the modern West. With the coming of national independence in the wake of the second World War, many of these Western-developed colonial cities have also become political capitals. In every country of Asia (if Tokyo may be regarded as a city developed on Western lines, if not in Western hands), the colonial or semi-colonial (e.g., Bangkok) port became well before independence a strong or even over-whelming *primate* city, i.e., a city which is at least twice as populous as its nearest domestic rival. In the several small countries of Southeast Asia, in particular, the ex-colonial primate city almost monopolizes commercial and industrial as well as political functions for the entire state, and thus has no rational rivals as a national capital despite its alien origins. Urban development in parts of Africa during the past century has followed parallel lines. Colonial port cities were built by Europeans to handle the export of primary products and became much larger than the indigenous inland urban centers. Similarly, many of these ports, originally colonial, have become primate cities and, with national independence, political capitals.

The Growing Similarity Of Cities

With the creation of a global commercial network, the spread of industrialization, and the technological revolution in transport and transferability, cities everywhere are becoming more like one another. The urban differences which once distinguished various cultural and economic areas are lessening. The degree of urbanization, or the proportion of the total population living in cities, tends strongly to vary with the level of commercialization of the economy. Hence, for example, Australia, Uruguay, Germany, or Japan are much more highly urbanized than China, India, Mexico, or the Congo. But all cities are increasingly involved in spatial interaction with each other and with other parts of the world. This interaction involves what is more and more the same set of cor mercial and industrial functions depende on the same kinds of techniques and co fronting the same sorts of problems. Tl growing spatial spread of *conurbations* (ci clusters and expanding metropolitan areas) appearing in Asia and Africa and Lat America as in Europe and North Americ The foreshadowings of *megalopolis* (literall giant city, a growing together over a vast extended urban area of what were orginal widely separated cities) are apparent not on in southern England or between Boston, Ma sachusetts, and Richmond, Virginia, but b tween Tokyo and Osaka-Kobe, along tl Hooghly River, and along the Rio de la Plat The growth of the city since urban-base civilization first developed some 6000 yea ago is still continuing and the city still fun tions as the center of exchange which increa ingly unites all areas of the world.

Further Reading

1 Chelde, V. G., *What Happened in History*, Re Ed. (Baltimore, Penguin Books, 1965).

2 Dickinson, R. E., *The West European Ci* (London, Routledge and Kegan Paul, Lt 1951).

3 Mumford, L., *The City in History* (New Yor Harcourt, Brace, & World, 1961).

4 Murphey, Rhoads, "The City as a Centre Change: Western Europe and China," *Anna of the Assoc. of Am. Geographers* (1954), p 349-62.

5 Pirenne, Henri, *Medieval Cities* (Princeton, N. Princeton University Press, 1925).

6 Ullman, E. L., "The Role of Transport ar the Bases for Interaction," in *Man's Role Changing the Face of the Earth*, ed. W. Thomas (Chicago, University of Chicago Pres 1956), pp. 862-80.

4

.merican Metropolitan volution

ohn R. Borchert

e landscapes of any American city reflect ıntless decisions and actions from the time settlement to the present. The results are parent not only in differences in land use t in the kaleidoscopic variety of building ades, street patterns, and lot sizes. Early ions precluded or frustrated many later ational decisions. The metropolitan phy-al plant has accumulated through various torical epochs, and clearly those epochs re distinguished from one another by dif-ent ideas and technologies. Increasingly, proportion to its size and age, the metrop-s is becoming a complicated puzzle of erogeneous and anachronistic features.

The evolutionary nature of the metropoli- anatomy is, of course, widely recognized, d this fact is reflected in a wealth of studies the historical-geographical development of lividual cities and of the anachronistic acies that make up much of the urban ysical plant. Yet research on systems of ies, cities as central places, cities and trans-rtation networks, internal spatial structure cities, and rank-size distributions has lacked a general historical context.[1] A structured urban history of the country or of its major regions, which would help to bring order to the mixture of historical-locational forces that generate the urban landscape,[2] has not yet appeared. Meanwhile, future populations are projected, and increasingly massive clearance and redevelopment proceed in the old central areas of metropolises without a fully developed theory of metropolitan

[1] See the excellent summary by Brian J. L. Berry, "Research Frontiers in Urban Geography," in *The Study of Urbanization* ed. by Philip M. Hauser and Leo F. Schnore (New York, John Wiley and Sons, 1965), pp. 403-30. For further comment see Fred Lukermann "Empirical Expressions of Nodality and Hierarchy in a Circulation Manifold," *East Lakes Geographer*, Vol. 2 (1966), pp. 17-44. A notable exception is the work of Allan Pred, "Industrialization, Initial Advantage, and American Metropolitan Growth", *Geogr. Rev.*, Vol. 55 (1965), pp. 158-85.

[2] This is the thesis of a detailed review of the literature of United States urban history by Charles N. Glaab, "The Historian and the American City: A Bibliographic Survey," in *The Study of Urbanization* (see footnote 1 above), pp. 53-80.

growth and form.[3] Stages of urban economic growth and social evolution have been postulated. Wilbur Thompson has observed that these are "highly impressionistic generalizations" and "leave much too strong a feeling of the inevitability of growth and development." And he has asked, "What are some of the dampening and restraining forces that surely must exist? We see all about us evidence of local economic stagnation and decay and even demise."[4]

Central questions relate to the factors that have influenced the location of relative growth and decline, the relationship of anachronistic regions within cities to the evolution of the national pattern of urban growth, the threads that run consistently through the evolutionary process, and the nature of future change as suggested by experience to date. These questions can be illuminated by examining the evolution of the present pattern of standard metropolitan statistical areas through a series of historical epochs, from the first census, in 1790, to the most recent, in 1960.

Major Innovations and Epochs

Most American metropolitan areas, throughout much of their history, have functioned chiefly as collectors, processors, and distributors of raw materials and goods. Consequently, it might be expected that changes in their growth rates would have been partic-

[3] William Alonso, "The Historic and the Structural Theories of Urban Form: Their Implications for Urban Renewal," *Land Economics*, Vol. 40 (1964), pp. 227-31.

[4] Wilbur R. Thompson, "Urban Economic Growth and Development in a National System of Cities," in *The Study of Urbanization* (see footnote 1 above), pp. 431-90; reference on pp. 438-39. See also his "A Preface to Urban Economics" (Baltimore, 1965), pp. 12-17.

ularly sensitive to changes in (1) the si and resource base of the hinterland and (2 the technology of transport and industri energy for the processing of primary r sources. These two sets of variables are inte related. The technology partly defines th resource base, and the transportation tech nology, in particular, strongly affects the siz and therefore the resources, of a city's hinte land. There is, of course, no implication th the technological changes have been inde pendent variables or basic causes of growt The presumptions are, rather, that within th given framework of values and institution they were stimulated by the economic grow and geographical expansion of the nation an that they in turn not only further stimulate growth but also helped to differentiate geographically.[5]

Among many possibilities, this paper em phasizes three relatively brief periods sinc the 1790 census in which major innovation appeared in the technology of transport an industrial energy.

THE INNOVATIONS

Steamboat and "Iron Horse"

The first of the innovations was the use o the steam engine in water and land trans portation. The census year selected is 183 To be sure, the early steamboats in Americ preceded that date, but the real buildup o steamboat tonnage on the Ohio-Mississipp Missouri system begin in the 1830's (Figur 1),[6] and the main period of increase in th

[5] For an excellent summary of the operation contingent, interrelated variables in the grow process see Robert W. Fogel, *Railroads ar American Economic Growth* (Baltimore, 1964 pp. 234-37 (section on "Implications for th Theory of Economic Growth").

[6] John W. Oliver, *History of American Tec nology* (New York, 1956), pp. 192-93 and 20

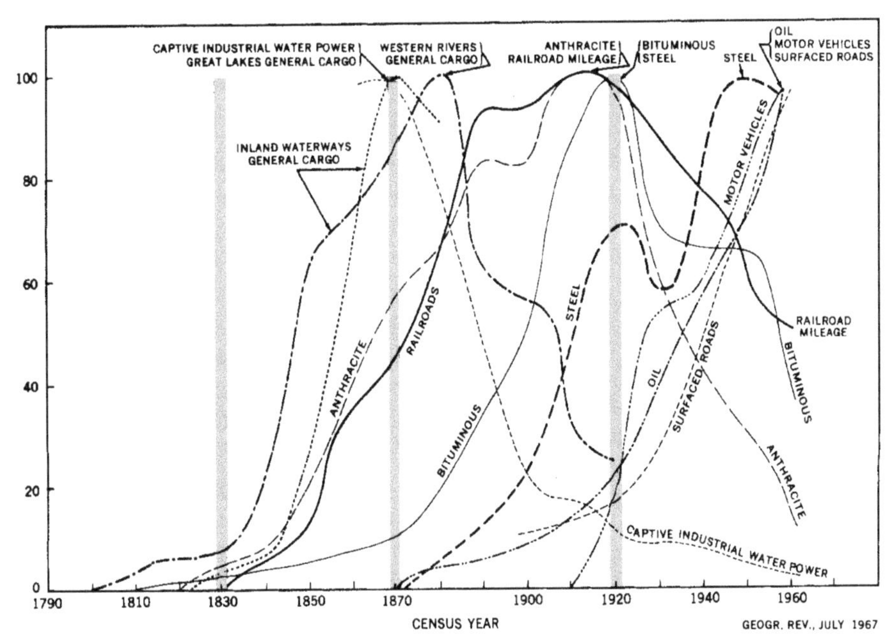

gure 1. Rise and decline of ten indicators of the technology of transport and industrial energy. Peak ues of past years concentrate around 1870 and 1920. Sources: "Historical Statistics of the United States," . 416-417, 427-429, 446, and 458; *Statistical Abstract of the United States 1965,* pp. 561, 569, 718, 729, d 811 (see text footnote for both); and "A Compendium of the Ninth Census" (U.S. Bureau of the nsus, 1872), p. 706.

nnage of general-cargo vessels on the Great kes also began in the 1830's and 1840's. il mileage, likewise, grew rapidly after tial development in 1829. By the end of the cade "the major mechanical features of the nerican locomotive were established," box-rs had been introduced, regular mail routes re in operation on the railroads, and the st transatlantic steamer had arrived in New ork.[7]

The introduction of steam power created ajor transportation corridors on the western ers and the Great Lakes and resulted in largement of the hinterlands of ports on th the inland waterways and the Atlantic. made possible the development of a na-nal transportation system through the integration of these major waterways and regional rail webs. These changes favored the growth of ports with relatively large harbors and proximity to important resource concentrations. Simultaneously, however, they hurt the economy of nearby smaller ports.

Steam power was also applied in manufacturing, but its impact was apparently more localized because of the impracticality of long hauls of coal or other bulk commodities with the comparatively light equipment and iron rails of the time. As a result, local waterpower sites continued to influence industrial location. By 1870 waterwheels were still providing roughly half of the inanimate energy for manufacturing, especially in the major manufacturing region. About half of the entire inanimate power for industry was in the five

Ibid., pp. 184-85, 189, and 202.

states of Massachusetts, Connecticut, New York, Pennsylvania, and Ohio. Oliver observes that steam "was not universally used in cotton mills until the railroads were sufficiently developed to transport coal cheaply."[8] That ability came generally in the 1870's.

Steel Rails and Electric Power

The second major innovation was the appearance of abundant, and hence low-priced, steel. The census year chosen for this is 1870. The preceding decade had seen the first commercial output of Bessemer steel in America, and by the mid-1870's American steel products were breaking into the world market (Figure 1).

A number of related events, each with geographical ramifications of great importance, occurred in dramatic sequence in the decade of the 1870's. Steel rails replaced iron on both newly built and existing lines. Heavier equipment and more powerful locomotives permitted increased speed and the long haul of bulk goods. Rail gauge and freight-car parts were standardized[9] (there had been eleven gauges among the northern systems in 1860), so that interline exchange and coast-to-coast shipment were possible. Refrigerated cars made their entry, ushering in a new er of regional specialization in agriculture an centralization of the packing industry a major rail nodes. Other ramifications favore industrial, hence urban, centralization. Th practical length of coal haul was extende and the cost reduced. The effect was to ope vast central Appalachian bituminous deposit and to facilitate the movement of coal to th great ports whose growth had been launche four decades earlier. The greater availabilit of coal was soon supplemented by the avai ability of central-station electric power, whic followed in the 1880's.

For the first time massive forces were ar rayed favoring market orientation of industr and the metropolitanization of America. A the same time there were negative impacts The long rail haul spelled the doom of mos passenger traffic and cargo movement on th inland waterways, especially the rivers. Sma river ports were destined to become virtua museums. It is noteworthy that general-carg shipping capacity on the western river peaked not on the eve of the Civil War but i the 1870's; thereafter it fell precipitously fo half a century (Figure 1). The easier avail ability of coal and central-station electricit doomed the small waterpower sites. Mos small industrial cities retained their functior many were rail nodes large and importan enough to continue to grow with the nationa economy. But for subsequent decisions th decentralizing factor of many small water power sites had yielded to the centralizin force of the metropolitan rail centers, thei giant markets, and their superior acces sibility.[10]

[8] *Ibid.*, p. 160.

[9] The urgency of rebuilding existing lines with steel rails at standard gauge in the 1870's is related in Robert J. Casey and W. A. S. Douglas, *The Lackawanna Story* (New York, 1951), pp. 92-94. The impact of the introduction of the long haul on the geography of an existing system is described in Louis Jackson, *A Brief History of the Chicago, Milwaukee, and St. Paul Railway* (1900), pp. 6-8. See also Harlan W. Gilmore, *Transportation and the Growth of Cities* (Glencoe, Ill., 1953), p. 51. For a full account of the beginning of the steel era in the United States and for further references see Oliver, *op. cit.* (see footnote 6 above), pp. 319-425, especially pp. 416-25.

[10] Allan Pred has developed at length the fa that the period from the Civil War to Worl War I saw the major growth of large cities a industrial centers, *op. cit.* (see footnote above), pp. 161-62.

ernal-Combustion Engine and Shift to vices

e third major innovation, and probably least debatable, was the introduction of internal-combustion engine in transporta- a and related technology. The census date sen is 1920. To be sure, the automobile l entered the American scene in the 0's, but motor-vehicle registration was gnificant before 1910, and road surfacing l petroleum production began their steep nb in the 1920's (Figure 1). The need for ational system of highways was recognized 1916 with the first federal aid for road struction.

The impact of the internal-combustion ine on the geography of American cities ds little review. But some of the most found changes affecting the city occurred griculture.[11] True, the new technology put farmer in an automobile and thus en- raged the centralization of urban growth the larger, diversified centers in all the mercial farming regions. But also, by ting the farmer on a tractor, it multiplied land area he could work alone, initiated a olution in family farm size, and sped the anization of much of rural America. In ition, air passenger transport helped to ourage centralization of the national busi- s management function in a few cities,[12] the auto stimulated the decentralization nost metropolitan functions. The internal- bustion engine had a profound and happy act on the growth prospects of cities in the oil fields, but the opposite effect on cities in the coal fields and at railroad division points.

Another change, of overriding importance, coincided with the beginning of the auto-air age. Throughout the nation's history the primary (agriculture, forests, fisheries, mining) and secondary (manufacturing) sectors of the economy had dominated the employment picture. Their share of total employment had been gradually diminishing, but in 1920 they still accounted for 56 percent. Since 1920, however, the share has been less than half and has been falling rapidly (Figure 2). The trend is, of course, a reflection of the combined technological advances that have been leading the nation gradually toward an era of automation.

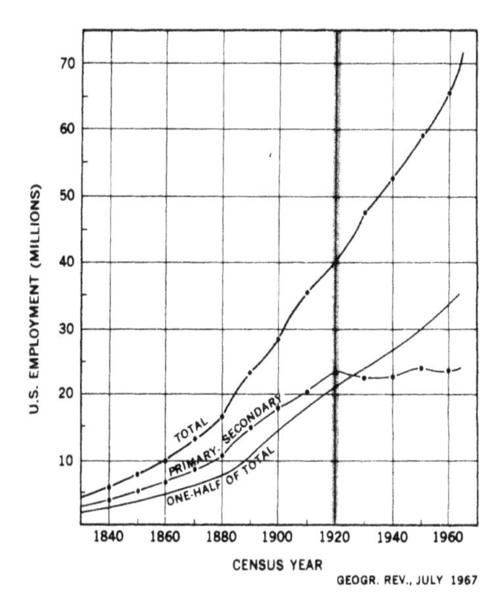

Figure 2. The changing relationship of employment in primary and secondary industries to total employment in the United States. Source: "Historical Statistics of the United States," p. 74; *Statistical Abstract of the United States 1965*, p. 220 (see text footnote 24 for both).

ee Chauncy D. Harris, "Agricultural Production in the United States, The Past Fifty Years nd the Next," *Geogr. Rev.*, Vol. 47 (1957), pp. 75-93.

ee William Goodwin, "The Management enter in the United States," *Geogr. Rev.*, Vol. 5 (1965), pp. 1-16.

When employment was mainly in resource and processing industries, it was fair to look on cities mainly as assemblers and processors of the nation's resources. It was appropriate to assume that changes in the technology of transport and industrial energy would be crucial for the growth or decline of cities. In the auto-air age, when primary and secondary employment occupies only a decreasing minority of the labor force, such technological changes are of declining importance in the life and death of cities. Two new factors have come to the fore. One is the increase in service employment. With a fast-growing majority of new jobs since 1920 in the least mechanized and least automated part of our economy — the personal and professional services — the most likely locations for new employment growth have been the places where there were already large concentrations of people to be served. Hence in the auto-air age, even more than in the preceding epoch, growth breeds growth. The second factor is the large and growing amount of leisure time available. As Ullman pointed out some years ago, this has led to the great importance of amenities as an urban location factor, both for commuting workers and for retired people.[13] It had also led to an increase in the time available for, and, presumably, the need of, formal education. Hence educational centers as well as high-amenity locations have been blessed by the fruits of changing technology in this present age.

THE EPOCHS

In short, four epochs in American history can be identified that have been characterized by changes in technology crucial in the location of urban growth and development: (1) Sa Wagon, 1790-1830; (2) Iron Horse, 183 1870; (3) Steel Rail, 1870-1920; (4) Aut Air-Amenity, 1920-.[14]

Although emphasis here is on factors a fecting the differential growth of Americ cities as entities, these periods are diffe entiated also by internal features of urban g ography and morphology.[15] The railroad h many impacts on the structure and locati of industrial and central business distric and these districts changed with the entry steel and the long haul. The coming of ste

[13] Edward L. Ullman, "Amenities as a Factor in Regional Growth," *Geogr. Rev.*, Vol. 44 (1954), pp. 119-32.

[14] This differentiation is somewhat related to E Lampard's formulation of three critical peric in the regional economic development of United States up to 1910: (1) a period initial resource exploitation in the histo eastern base region, from colonial ti through the Civil War; (2) a period of "exte sion of accessibility" from the eastern ba region to the rest of the country associated w an enlargement of the resource hinterland of t eastern base, from the Civil War to Wo War I; and (3) the era of "nationalization" the economy, utilizing and improving on a vir ally fully developed transportation system sin World War I. See Eric E. Lampard, "Regior Economic Development, 1870-1950", in *Regio Resources, and Economic Growth* by Harv S. Perloff, Edgar S. Dunn, Jr., Eric E. La pard, and Richard F. Muth (Baltimore, 196(pp. 107-292 (Part 3). See also Constan McLaughlin Green, *American Cities in t Growth of the Nation* (New York, 1957), Cha 10, and the works cited by Sjoberg in co nection with his discussion of the Technologic School (Gideon Sjoberg, "Theory and Resear in Urban Sociology" in *The Study of Urbaniz tion* (see footnote 1 above), pp. 157-89, refe ence on pp. 170-71).

[15] For an extensive discussion of urban evoluti in relation to transportation technology a development in specific metropolitan areas s James E. Vance, Jr., "Labor-shed, Employme Field, and Dynamic Analysis in Urban Geo raphy," *Econ. Geogr.*, Vol. 36 (1960), pp. 18 220, and his "Geography and Urban Evoluti in the San Francisco Bay Area" (Berkele 1964).

l electric power made possible the sky-
aper and rapid transit. The auto and coin-
ent developments in electronics need
elaboration. Each innovation brought
jor changes in land-use patterns, densities,
sizes, nodality of the central business dis-
t, and other intraurban variables.

Reservations and qualifications apply, of
urse, to this fourfold historical division.
e periods are not homogeneous. Even if
y should stand up as useful divisions for
description and study of American metro-
itan evolution, they contain many subdivi-
ns, which vary from one region to another.

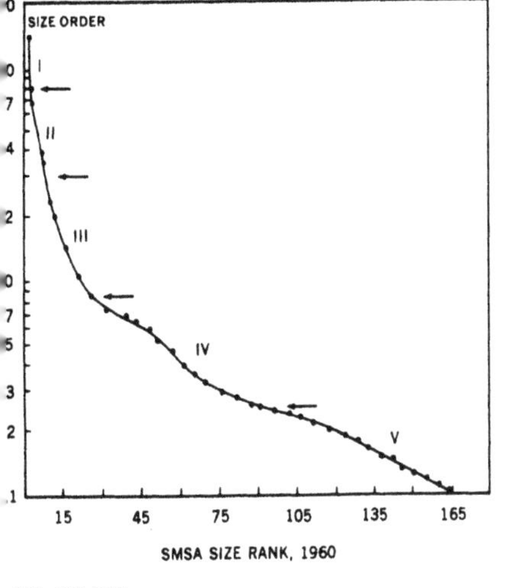

ure 3. Population rank-size distribution and -order limits for SMSA's in 1960. Except for the e largest metropolitan areas, dots represent y every fifth place in the sequence. The basis this graph and for subsequent maps is a list of 1960 SMSA's, with populations for the same nties or groups of counties for 1790, 1830, 1870, 0, and 1960. The 212 census SMSA's on the have been reduced to 178 by dropping areas h fewer than 80,000 inhabitants in 1960 and nbining certain others, mainly in New England several large multicentered complexes. Where SA's on the list differ in definition from those he census, the differences are specified. Copies the list are available from the author.

Furthermore, the boundaries between the epochs, although characterized by the near simultaneity of important innovations, are nevertheless complex transition periods. Some of the features of transition constitute little epochs of their own; examples might be the canal epoch (*ca.* 1810's to 1840's) and the electric interurban railway epoch (*ca.* 1900's to 1930's). Oliver[16] observes, "The canal was at best a temporary and an inadequate answer to the need for inland transportation." In a sense, it represented an attempt to adapt the technology of water transportation to the quickly growing need for tapping inland resources as the frontier advanced. Likewise, the rash of interurban electric rail lines that appeared about the turn of the present century may be viewed as a "temporary and inadequate" attempt to adapt rail-transport technology to the growing need for a flexible, rapid linkage between farm, small town, and city as the populations of large regions became commercialized and urban-oriented.[17]

Finally, throughout virtually all the first three epochs the settled area of the United States was expanding westward. The rate, timing, and direction of advance of the settlement frontier were in many ways quite independent of the major technological innovations that opened each of these epochs. On the other hand, the westward expansion helped to press the need for these innovations. More important from the viewpoint of this paper, cities were needed and built as

[16] Oliver, *op. cit.*, p. 180.

[17] See Mildred M. Walmsley, "The Bygone Electric Interurban Railway System," *Professional Geographer*, Vol. XVII, no. 3 (1965), pp. 1-6; and the detailed maps and descriptive data in G. W. Hilton and J. F. Due, *The Electric Interurban Railways in America* (Stanford, Calif., 1960).

new lands were opened. Hence the land pioneered during each of these epochs constitutes a region within which all city sites were chosen, and subsequent investments made, under a particular sequence of technological considerations.

Metropolitan Size Classes

In order to compare sizes and growth rates during the four historical epochs postulated here, American cities were divided into fiv population size categories. First, the 21 standard metropolitan statistical areas of th 1960 census were reduced to 178 by com bining some and dropping those unde 80,000 population. The 178 were the ordered by size. A smoothed curve joinin them in rank-size distribution is shown i Figure 3. A change in slope is noticeable a four points — at populations of abou 250,000, 820,000, 3,000,000 and 8,000,000 Above each of these critical points lies group of cities whose growth at some perio in their history has been accelerated as com pared with the places below the critical poin These division points break the 1960 SMSA into five groups, which may be labeled as fo lows: first order, more than 8,000,000 (Ne York); second order, 2,300,000 to 8,000, 000; third order, 820,000 to 2,300,000 fourth order, 250,000 to 820,000; and fift order, less than 250,000. Although the divi sions are somewhat arbitrary, they seem t identify significantly different groups of citie The New York metropolis is, of course, in class by itself no matter how the SMSA's ar divided. The second and third groups appea as clusters on the graph that portrays Jerom Pickard's population and functional siz orders (Figure 4).[18] The fifth group seems t match the primary wholesale-retail categor identified, by the analysis of business func tions, at the top level below Minneapolis-St Paul in a regional hierarchy of trade center in the northern Midwest and Great Plains.[19]

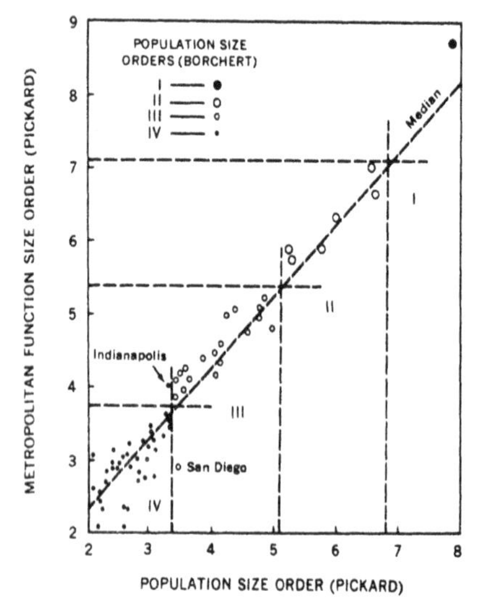

Figure 4. The 26 SMSA's defined as I, II, or III order in Figure 3 shown in their positions on Pickard's combined population and metropolitan-function scales. With only two exceptions, places that fall within the top three population size orders in Figure 3 also form discrete groups on the basis of metropolitan functions. "Metropolitan functions" in this case include measures of bank deposits, wholesale trade, Sunday newspaper circulation, federal-government employment, and manufacturing employment. The exceptions are San Diego, a third-order city based on population but weak in its metropolitan functions, and Indianapolis, a fourth-order city in Figure 3 (1960 SMSA definition of the Indianapolis SMSA) but with exceptional metropolitan strength. Graph modified from Pickard, Metropolitanization of the United States (see text footnote 18 for reference), Figure 21.

[18] Jerome P. Pickard, "Metropolitanization of th United States," *Urban Land Inst. Researc Monograph*, Vol. 2 (Washington, D.C., 1959 Fig. 21, p. 67.

[19] John R. Borchert and Russell B. Adams, "Trad Centers and Trade Areas of the Upper Mic west," *Upper Midwest Economic Study Urba Rept. No. 3* (Minneapolis, 1963), pp. 36-39 an Figs. 1 (p. 4), 8 (p. 25), and 9 (p. 27).

For the earlier census years two definitions d to be formulated before the procedure uld be applied. First, it was necessary to fine "SMSA" for those years. In the 1960 1sus an SMSA by definition had to contain, effect, a central city of at least 50,000 pulation. For the present study the mini-m size of the central city was reduced to commensurate with the smaller total pulation of the United States in the earlier 1sus years. Hence the minimum-sized 1tral city for an "emerging SMSA" in 1920 s 29,500; in 1870, 11,100; in 1830, 3600; d in 1790, 1100. The population of the unty was used at each point in the time ies. Where a county was split during the ies, appropriate adjustments were made. t each SMSA or emerging SMSA that ap-ars at any point in the time series is always ined by the same county or counties. To sure, this is only one way of achieving a asure of consistency in dealing with a blem for which there is no entirely satis-tory solution at present.[20] Few of these unties in the early epochs were "metropoli-" in any modern sense. But they included forty to fifty largest places in 1790 and 30; and the definition identified, then as w, the nation's principal population sters.

Second, it was necessary to define for the lier census years the limits of the five size lers defined above for 1960. Again, each the size-order limits was reduced to be nmensurate with the smaller national pulation in the earlier years. These values, wever, were then adjusted upward to account for the smaller proportion of a given "metropolitan" county covered by the smaller central city of earlier times.[21] The result was the set of limits shown in Table I.

Thus five size orders and four historical epochs were established. By comparing the numbers of newcomers, dropouts, and shifts in size order through the series of epochs it is possible to observe the evolution of the modern array of metropolitan areas. It is also possible to observe the impact of several major changes in technology and the expansion of resource base and metropolitan hinterlands that accompanied the westward movement. Figure 5 shows all the places included in these five size orders during at least one of the four historical epochs.

ee Karl Gustav Grytzell, "The Demarcation of Comparable City Areas by Means of Population Density," *Lund Studies in Geography*, Ser. B, Human Geography, no. 25 (1963), especially pp. 5-9.

21 For the years before 1960 each size-order threshold, T_y, was first defined by the relationship $T_y = T_0 (P_y/P_0)$ where T_0 is the threshold population in 1960, P_0 is the United States population in 1960, and P_y is the United States population in the earlier year. For thresholds under 100,000 "SMSA" population a further adjustment was made, using the relationship $T_r = T_y (F_y/F_0)$, where T_y is the threshold defined for the earlier year by the initial adjustment, T_r is the readjusted value, F_0 is the percentage of United States population that was rural in 1960, and F_y is the percentage of United States population that was rural in the earlier year. Where the country population exceeded 100,000 it was assumed that rural (especially farm) population within the "SMSA" was negligible, and no second adjustment was made.

The net effect of this definition and procedure is to overstate the populations of urban areas in the earlier periods, especially before 1920 and especially in areas with populations under 100,000. Because of the second adjustment of threshold values, the size orders are roughly comparable throughout the series. Lukermann (see footnote 1 above) approximated urban-area populations for the eastern and central United States 1790-1890. Comparison of his rank sizes and geographical patterns with those in this paper shows no significant discrepancies resulting from the different definitions.

TABLE I

Limits of Size Orders for SMSA's in 1960 and Corresponding Areas in Earlier Years*

	Population Threshold (thousands)				
Size Order	*1790*	*1830*	*1870*	*1920*	*1960*
First	180	530	1,300	4,750	8,000
Second	90	160	400	1,480	2,300
Third	40	90	130	470	820
Fourth	15	35	750	150	250
Fifth	5	15	30	60	80
Central-city minimum	1.1	3.6	11.1	29.5	50

* The fifth order is truncated by the lower size limits for an SMSA; a number of urban areas not lar enough to be defined as SMSA's are fifth-order centers.

Evolution of the Pattern[22]

Sail-Wagon Epoch, 1790-1830

At the time of the 1790 census almost all the major urban population clusters were ports on Atlantic bays or estuaries, or on the navigable reaches of the Connecticut, Hudson, Delaware, and Savannah Rivers, or on the Chesapeake Bay system (Figure 6). Among the centers of third or high order, only Worcester, Massachusetts, was not port, and only Worcester and Pittsburgh we not on the Atlantic waterway. Lower-orde centers also were mainly Atlantic port though they included inland centers of agr cultural trade and local industry. There wa no primate city or national metropolis; th Boston, New York, and Philadelphia area were of about equal size. Lukermann has ob served that the entire family of Atlantic por was characterized by small hinterlands an a primary orientation toward the sea an Europe and could in fact be considered pa of the West European urban system.[23]

Change was modest during the epoc (Figure 7). Virtually all places that rose i size order — that is, grew faster than the na tional growth rate — were in areas of west ward expansion and accompanying develop ment of new resources: the drift-filled valley and drift-capped plateaus of western Ne York, the Ontario plain, the Great Valley the Bluegrass, and the Nashville Basin. Othe important resources lay within these ne agricultural regions or adjacent to them – notably waterpower, the timber of th northern Appalachians and the Adirondack

22 This section is an attempt not to write history but to interpret briefly Figures 6 through 11 in the light of readily accessible secondary materials and historical census data. The regional framework and regional economic changes discussed are elaborated in Perloff and others (see footnote 14 above); in Ralph H. Brown, *Historical Geography of the United States* (New York, 1948); and in several standard regional geographies, notably J. Russell Smith, *North America* (New York, 1925); George J. Miller and Almon E. Parkins, *Geography of North America* (New York and London, 1928); and Harold Hull McCarty *The Geographic Basis of American Economic Life* (New York and London 1940). On the other hand, the detailed history and geography of specific cities within these regions, over time, are dispersed through myriad local studies. Important references appear in Harold M. Mayer, "Urban Geography" in *American Geography: Inventory & Prospect*, edited by Preston E. James and Clarence F. Jones (Syracuse, N.Y., 1954), pp. 142-66; and in Glaab, *op. cit.*, pp. 73-80.

23 Lukermann, *op. cit.*

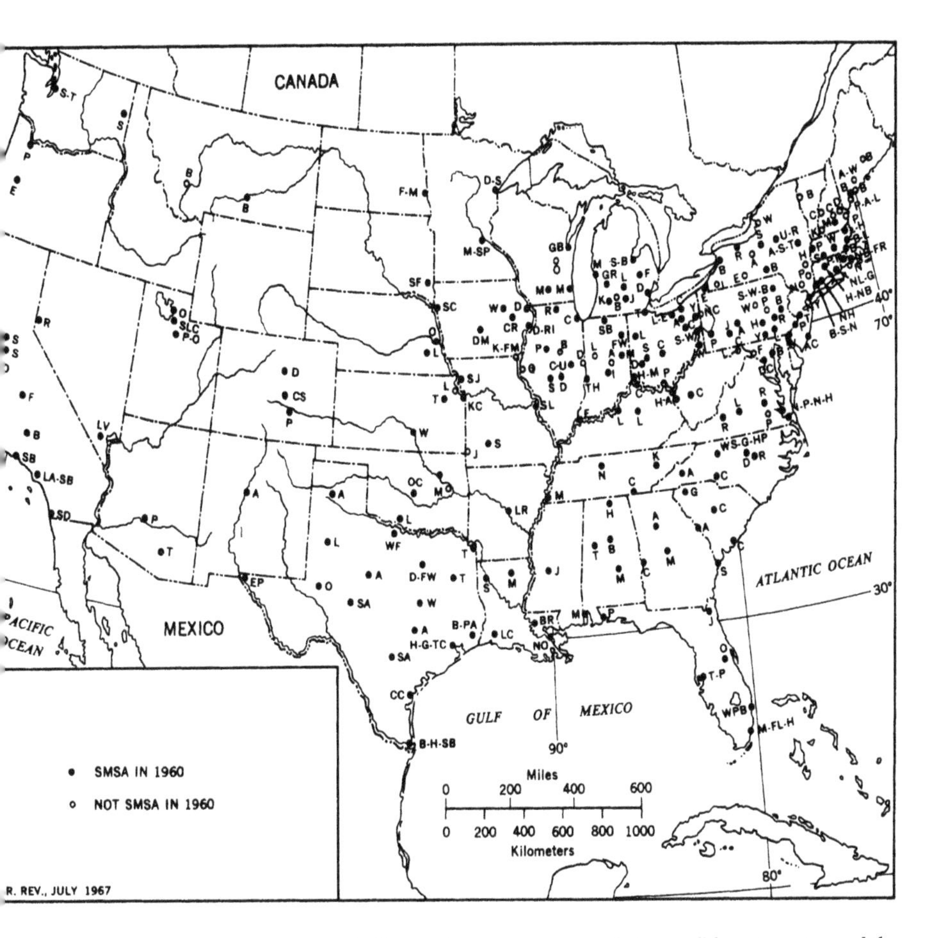

ıre 5. Places in the five size orders of SMSA's or "emerging metropolitan areas" in one or more of the ɜus years 1790, 1830, 1870, 1920 or 1960. Names of the central cities are indicated by initials.

the anthracite of northeastern Pennsyl- ıia. The "boom" cities, those which rose or more ranks, were mainly along the in- d waterways that penetrated the new ɜtern lands — the Erie Canal, the lower eat Lakes, and the Ohio River system. ceptions were the Great Valley cities near anthracite fields.

As agricultural settlement expanded in stern New York there was a relative cline in growth rate, and a drop in size ler, at a number of small ports and inland nters serving agricultural areas in eastern New York and New England. Meanwhile, although the struggle for deeper hinterlands had begun, the absence of any major change in the technology of land transport permitted most Atlantic ports to retain essentially the same functions through most of the epoch and to register neither relative increase nor relative decrease in size order up to 1830.

Iron Horse Epoch, 1830-1870

At the beginning of this epoch all emerging metropolitan areas of third or higher order

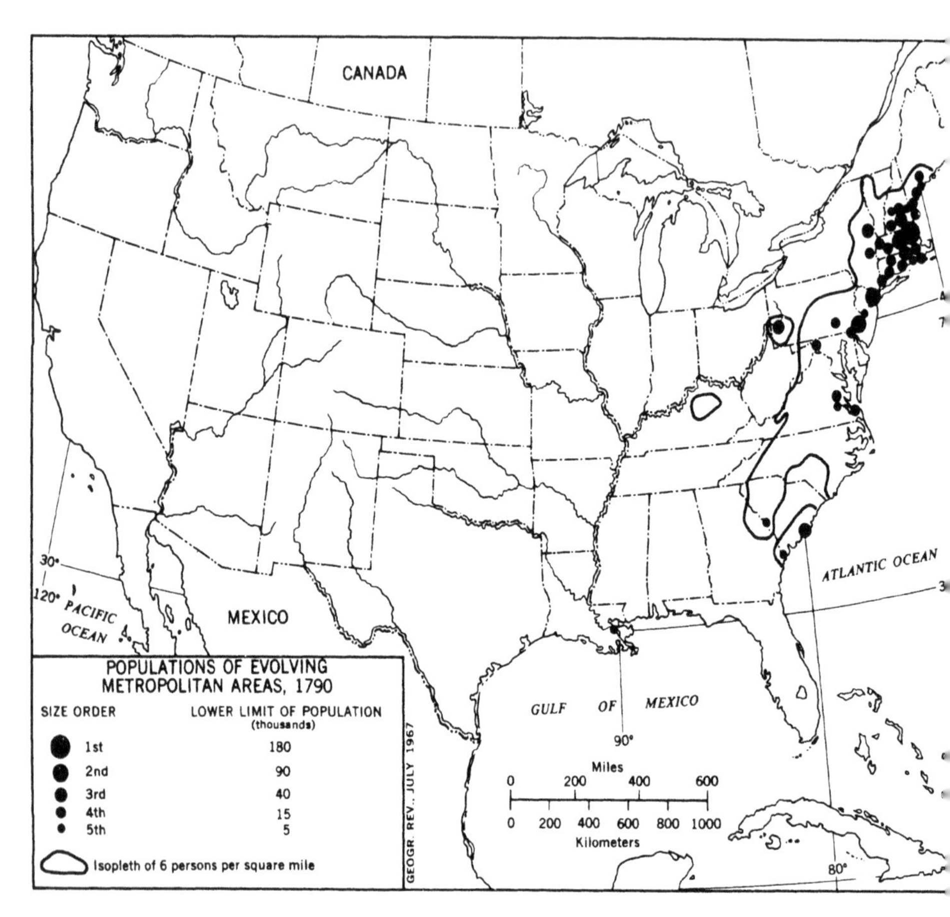

Figure 6. Distribution of major towns and neighboring county populations by size order, 1790. Source of the population data for Figures 5 - 8, "A Compendium of the Ninth Census" (1872). Population-densit isopleths generalized from Clifford L. Lord and Elizabeth H. Lord; Historical Atlas of the United States (New York, 1953), p. 46.

except Pittsburgh were east of the Appalachians or in western New York. The area of continuous settlement was spreading westward toward the Mississippi. But commercialization of the newly exploited land resources still awaited an effective network of transportation lines and cities.

The coming of the railroad brought drastic changes. A series of regional rail networks developed. The larger networks converged at critical port locations on the inland waterways that penetrated the vast agricultur land resource of the Interior Plains. Sma networks or individual lines focused on th smaller ports. The emergence of these "gre ports" of the Midwest accounted for most the boom cities of the epoch (Figure 8) ar laid the metropolitan base for an importa part of the market-oriented industrial grow of the Steel-Rail Epoch. Limitations of tec nology made the rail networks general tributary to the water-transport system; the

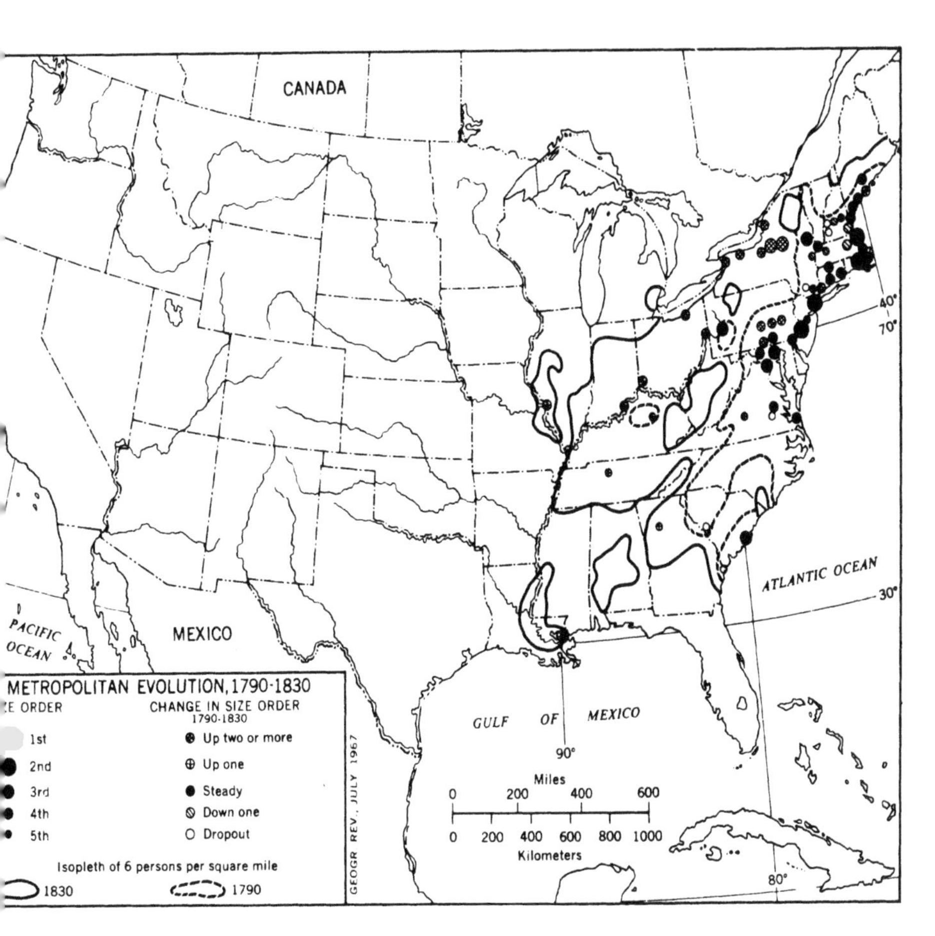

ure 7. Changes in the order, major towns and neighboring counties, 1790-1830. Population-density pleths generalized from Lord and Lord, *op. cit.* (see Figure 6), pp. 46 and 49.

re built outward from the major ports or to se ports from principal neighboring con- ntrations of farmland, mineral, or timber ources. In their initial effect they were refore complementary to the waterways as g-haul general-freight carriers.

In the older settled areas of the North er important changes in the urban pat- n emerged. Boom centers appeared in the thracite fields, and Pittsburgh advanced to cond order, where it would remain until the d of the Steel-Rail Epoch. These changes lected the accelerated demand for coal that came with the development of the rail-road, the wider industrial application of steam, and accompanying changes in the iron industry. On the Atlantic seaboard New York became a first-order center; it had been about the same size as Philadelphia at the beginning of the epoch, it was twice as large at the close.

Thus the urban pattern of the United States was revolutionized by the development of a national system of transportation, albeit a crude one. The epoch saw not only the emergence of a first-order center but also the

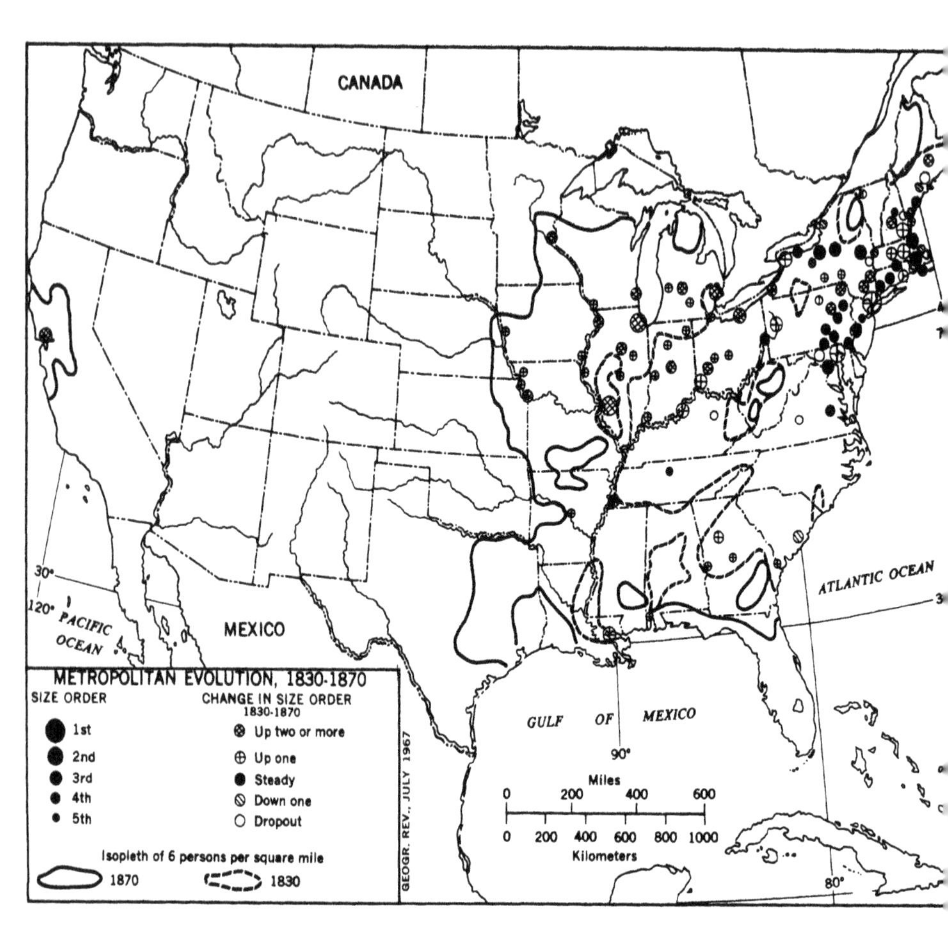

Figure 8. Changes in size order, major towns and neighboring counties, 1830-1870. Population-density isopleths generalized from Lord and Lord, *op. cit.*, pp. 49 and 104.

greatest increase, both relative and absolute, in the number of second- and third-order centers in the nation's history (Table II).

The map for this epoch (Figure 8) also reflects the aftermath of the Civil War and, probably more important, the slow rate of investment in urban, industrial, and transportation facilities in the South in the preceding decades. New Orleans was an exception. It was a critical point in the national transportation system that comprised the northern regional rail networks and the inland waterways. Before the Civil War it had risen to third order and largest city in the South. Meanwhile Charleston, despite pioneer railroad building into its comparatively static agricultural hinterland, dropped to fourth order and began a prolonged relative decline.

Steel-Rail Epoch, 1870-1920

By 1870 major urban areas had arisen as far west as the Missouri River frontier, and one (Little Rock, Arkansas) had appeared west of the Mississippi in the South. Many resource concentrations remained unexploited

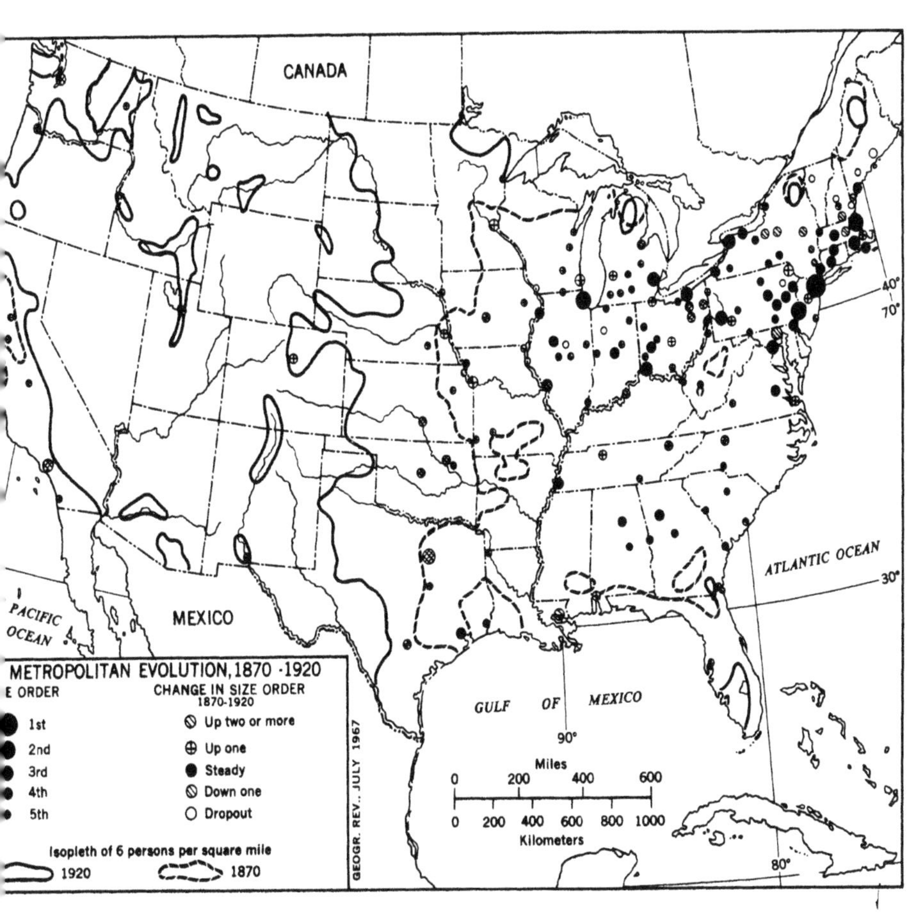

ure 9. Emerging metropolitan areas in the Steel-Rail Epoch. Changes in size order of major cities, luding neighboring county populations, 1870-1920. Population-density isopleths generalized from Lord 1 Lord, *op. cit.*, pp. 104 and 107-108. Sources of population data: "A Compendium of the Ninth Census" 372) and "Abstract of the Fifteenth Census of the United States" (1933).

nding further improvement of the land nsportation system and creation of a net- rk of urban centers in the West and South. the beginning of the following epoch all ese regions had been knit together by a ndardized, nationwide system of rail lines, d the modern pattern of major urban nters was beginning to emerge (Figure 9).

New urban centers reflected the opening commercialization of the remaining im- rtant agricultural land resources of the West — the Texas and Oklahoma prairies, the Colorado piedmont, the Wasatch piedmont, the Central Valley and Southern California, the Puget Sound–Willamette lowland, and the Palouse. They also reflected the exploitation of hitherto isolated major mineral deposits, such as Butte copper, southwest Missouri lead and zinc, and Lake Superior iron ore, and of mineral and timber resources in the mountains adjoining the agricultural oases and valleys. Finally, they

TABLE II

Number of Centers and Total Population in Each Size Order

Size Order	*1790*	*1830*	*1870*	*1920*	*196*
			Number of Centers		
First	0	0	1	1	
Second	3	3	6	4	
Third	8	8	14	16	1
Fourth	20	29	33	51	7
Fifth	8	12	37	75	8
Total	39	52	91	147	17
			Total Population (thousands)		
First	—	—	2,171	8,490	14,76
Second	514	1,120	3,301	10,364	28,82
Third	499	784	3,627	13,918	26,49
Fourth	530	1,812	2,533	12,829	30,47
Fifth	95	300	1,826	6,972	12,64
"SMSA" total	1,638	4,016	13,458	52,573	113,19
U.S. total	3,929	12,866	39,818	105,711	179,32

reflected the advance of the agriculture-timber-mineral frontier into Florida.

In the older settled areas the boom cities were associated mainly with the upward leap in the importance of coal — especially high-grade bituminous — that accompanied the growth of the modern iron and steel industry. A cluster of boom cities emerged on the western Pennsylvania coalfields and in the area between Pittsburgh and Lake Erie; and metropolitan Birmingham appeared on the map in the South. Other, but less spectacular, advances in rank occurred along the Norfolk-Toledo axis as long-haul technology opened the rich bituminous deposits of West Virginia and eastern Kentucky. Still others resulted from industrial growth based on the forest resources along the northern frontier of the agricultural Midwest and on the resources readily available for the hydroelectricity-based industrial development on the Piedmont in the South. Meantime, nearly all the great metropolitan commercial centers of the Midwest and Northeast, while establishin themselves as major industrial cities, retaine their positions or advanced one level in th hierarchy.

Two groups of metropolitan areas ac counted for most of the declines in size orde in this epoch. The largest group comprise the towns along the Ohio–Mississippi–Mis souri and principal tributaries. Smaller cen ters such as Dubuque and Quincy (Illinois dropped out of the "metropolitan" ranks; S Louis, Louisville, and Wheeling fell in th hierarchy, never to recover the relative posi tions they had held during the epoch of th steam packet and the iron horse. A secon group consisted of a number of importan industrial cities at historic waterpower site along the Mohawk, the Merrimack, and th Blackstone and minor ports on the Hudso and the New England coast. None of thes places was to regain the level it had held be fore the epoch of the steel rail and central station electric power.

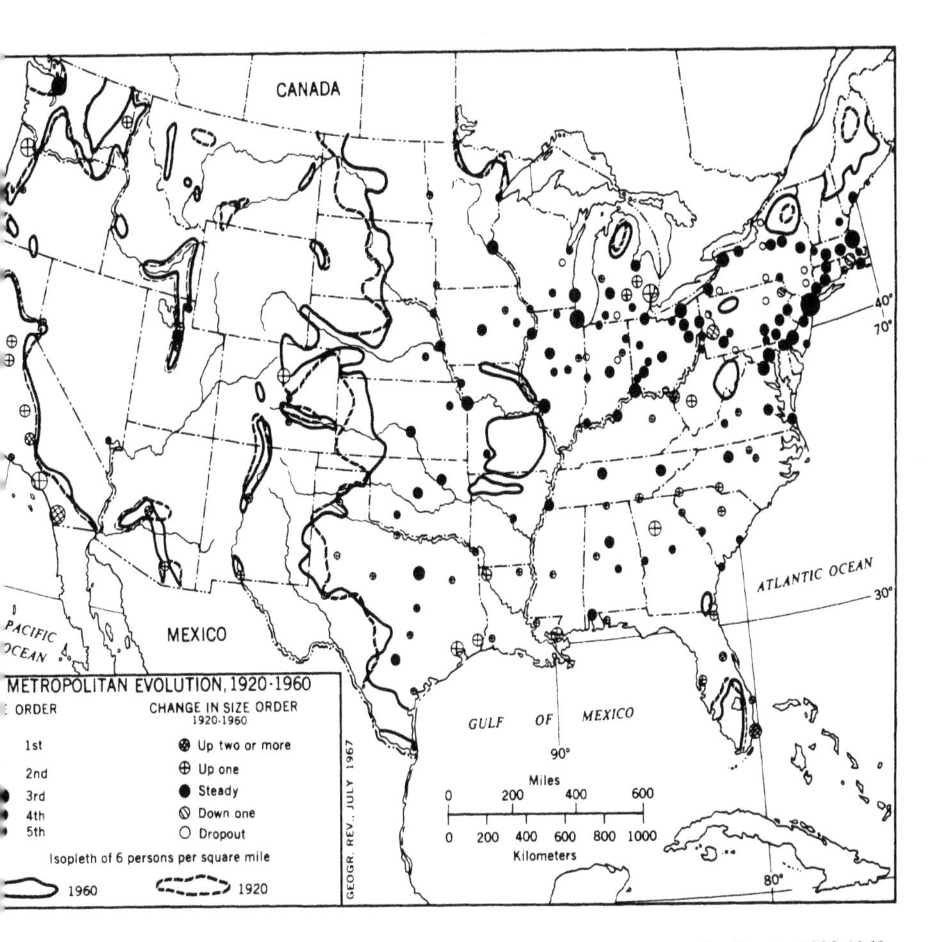

ure 10. Changes in metropolitan-area size and distribution in the Auto-Air-Amenity Epoch, 1920-1960. ulation-data source: *County and City Data Book, 1962,* U.S. Bureau of the Census. Population-density leths have been generalized from "Goode's World Atlas," 12th edit., 1964, p. 58.

The main shifts during the epoch are sum- rized in Table II. The new centralization ndustry in major metropolitan areas was ected in the growth of the five largest ASA's." That group increased its share of national population faster than in any er epoch. The number of third- or higher- ler centers, which had increased greatly in Iron Horse Epoch, was stabilized, but ir share of the nation's people rose sharply. e extension of national accessibility to isolated parts of the South and West augmented the number of lower-order metropolitan centers. The total number of fourth- and fifth-order cities registered its greatest growth in this epoch.

Auto-Air-Amenity Epoch, 1920-

By 1920 the present pattern of settled areas had been established, and subsequent metropolitan changes have taken place within that pattern. Nevertheless, new resources and

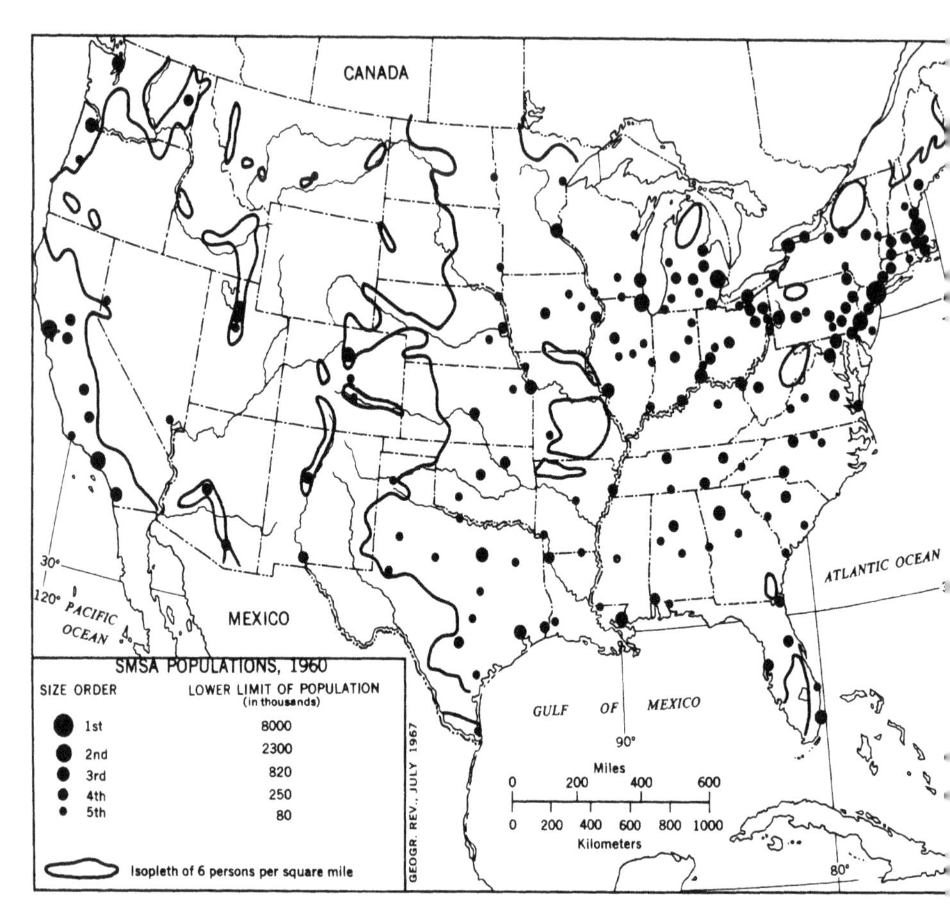

Figure 11. Geographic distribution of SMSA's by size order, 1960. Population-density isopleths have be generalized from "Goode's World Atlas," *op. cit.*, p. 58.

locations have been exploited and old ones abandoned (Figures 10 and 11).

The much smaller labor force required per unit of production in the extractive industries is reflected on the map in the relative decline of Butte and Joplin and, along with the shift from steam to internal combustion, in the decline of coal and railroad centers in the Appalachians and across the Midwest. The shift to internal combustion is, of course, largely behind the outbreak of new or higher-order metropolitan areas in the oil fields from central Kansas to western Texas and th western Gulf Coast and in the concentratic of growth in the SMSA's of southern Mich gan.

Regional and metropolitan dispersa inherent in the shift to auto and truck an in the development of a dense highway ne work, is reflected on Figures 10 and 11 i two principal ways. One is the entry int the metropolitan ranks of numerous "sate lite" cities on the fringes of the historic Manu facturing Belt and within 100 to 150 miles c at metropolitan industrial centers. The er is the effect of suburban dispersal on definition of a metropolitan area. In the palachians, for example, half a dozen MSA's" have dropped out because their tral cities failed to maintain a population al to, or larger than, that of other metro-itan centers. In these cases it is typical to d a central city crowded on a valley floor, ghted by obsolescent buildings, air pollu-n, narrow streets, and rusty rail lines, and osed to flood risk. Population and com-rcial growth have dispersed to the uplands exploit the resources of open space, a ise rural road net, panoramic views, and atively clean air. The result is that the tropolitan area has grown and ceased to "metropolitan" by definition, and the tral city has declined and ceased to be ntral" in fact.

This diffusion of metropolitan structure is o evidence of the increasing importance amenities in determining the growth pat-n of individual cities and regions. The loca-n of the boom centers is further evidence the force of amenity on a national scale. are in Florida, the desert Southwest, and uthern California. The migration to the uthwest and Florida in the Auto-Air-enity Epoch has been as massive as any the earlier westward movement. There was net migration of 11.4 million persons to lifornia, Arizona, and Florida from other ions between 1920 and 1960. This ualed the total population[24] gain — and was probably double the immigration — for the twelve North Central States during the Iron Horse Epoch, 1830-1870.

At the regional scale the increase in importance of educational centers, related in part to the growing importance of amenity, is illustrated by two pairs of Midwestern cities. The coal-rail center of Danville, Illinois, dropped out of the "metropolitan area" group, and the nearby university center of Champaign-Urbana entered it. Among the urban centers of the old eastern Indiana gas belt, the industrial city of Anderson dropped out, and neighboring Muncie, with both industrial and university functions, entered.

Effects of Technological Change

Throughout the evolution of the present pattern of American metropolitan areas two factors, great migrations and major changes in technology, have particularly influenced the location of relative growth and decline. Both factors have repeatedly been given specific geographical expression through their relationship to resource patterns. Major changes in technology have resulted in critically important changes in the evaluation or definition of particular resources on which the growth of certain urban regions had previously been based. Great migrations have sought to exploit resources — ranging from climate or coal to water or zinc — that were either newly appreciated or newly accessible within the national market. Usually, of course, the new appreciation or accessibility had come about, in turn, through some major technological innovation.

Nor can one see the end of these changes in locational advantage due to technological change and migration. Speculate, for instances, on the possible outcome of three

'Historical Statistics of the United States, Colonial Times to 1957" (U.S. Bureau of the Census, Washington, D.C., 1960), pp. 44-45 1920-1960 migration data), p. 13 (1830-1870 opulation changes by states), and pp. 23-30 early birth- and death-rate data); *Statistical Abstract of the United States 1965*, 86th edit., Washington, D.C., 1965), U.S. Bureau of the Census, p. 34.

changes, quite conceivable within the next half-century whose seeds may well be lying in our midst at present. Assume the automation of, say, 80 percent of the office work heavily concentrated in the downtown skyscrapers of major metropolitan centers. Or assume the production of low-cost, mass-produced, single-family dwellings varied in style and superior in structure and maintenance to those now in use. Or assume the introduction and success of a lightweight family vehicle that requires neither steel to build nor oil to power. Clearly, the process of urban growth in an open system is open-ended. To be sure, the rate of change from any point in time is constrained by the existing physical plant and institutions. But there is unlikely to be any "end product" of the process. Each epoch will simply be succeeded by another.

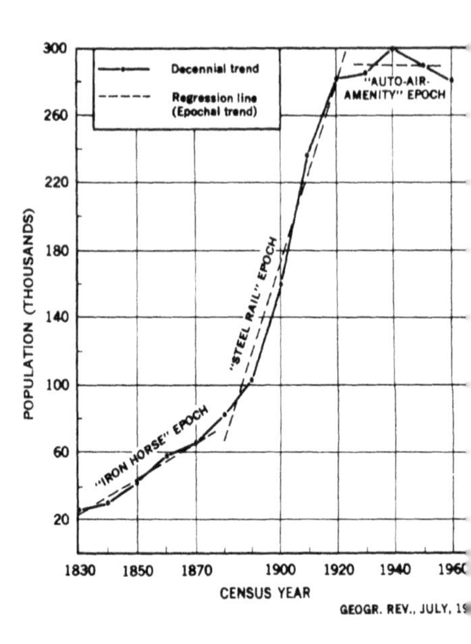

Figure 12. Population of the two counties comprising the Johnstown, Pa., SMSA (1960), by decades, 1830-1960. The depression of the 1930's, with its temporary decrease in mobility of a substantial economically "stranded" population, appears as a relatively short-term, low-amplitude event in the growth history in comparison with either the rise of the steel industry and dominance of steam-powered rail transportation or the advent of the internal-combustion engine.

VARYING PREDICTABILITY OF METROPOLITAN GROWTH

If metropolitan growth tends to be epochal and open-ended this suggests two probable characteristics of its predictability.

On the one hand, during any given epoch, similar conditions of paramount importance are likely to govern the rate and direction of growth over wide regions or types of location. Of course, countless short-term random effects are superimposed on the long-term trend. One might therefore expect the growth trend for a given metropolis during a given epoch to be regressive. That is, short-term spurts or declines will expectably be offset by succeeding short-term trends in the opposite direction. A regression line, fitted to the points representing these frequent ups and downs, describes the long-term trend. Thus the Upper Midwest Economic Study's urban research disclosed that past growth trends (during the automobile epoch) provided by far the most significant independent variable in a multiple regression equation to project 1950-1960 population growth rates of urban areas in its study region.[25]

On the other hand, when some basic component of the nation's society or economy or technology "turns a corner" and a new epoch opens, a new set of overriding and "long term" forces goes into effect. Thereafter, one might expect past growth to cease to be

[25] John R. Borchert and Russell B. Adams, "Projected Urban Growth in the Upper Midwest: 1960-1975," *Upper Midwest Economic Study, Urban Rept. No. 8*, (Minneapolis, 1964), pp. 1-2 and Appendix.

od predictor. At the least, its validity uld have to be reestablished for a new set conditions. The old regression line for a en metropolitan area would not necessarily present the long-term growth trend in the w epoch. Short-term fluctuations would be s likely to regress toward the same line as the preceding epoch (Figure 12).

As the new epoch unfolds, a new pattern "initial advantage" also emerges; for tain advantages are created that could not ve existed before.[26] Business and civic in-tutions must reorganize to meet new chal-ges. This seems to have been done most ectively in the places least tied to natural-source exploitation or secondary produc-n, and with the largest and most diversified iterlands, hence the most important centers circulation and management. Even some the high-order centers have had to make issive adjustments from one epoch to the xt; St. Louis and Pittsburgh are probably outstanding cases so far.

TERNAL DIFFERENTIATION OF ETROPOLITAN AREAS

ring each epoch a new increment of physical development has been added to each metropolitan area. Each increment is eventually differentiated from the adjoining ones not only by the age of its structures but also by their scale, design, use, degree of obsolescence, and, often, site or location. The successive increments form distinctive regions in the internal geography of any metropolitan area, and the regions have certain characteristics in common wherever they appear across the country.

But historically different increments tend to differentiate American cities at least as much as they tend to standardize them (Figure 13), because cities differ profoundly in their epoch of initial settlement and in their periods of boom or decline. For example, it is possible that a Chicago or a Los Angeles metropolitan area will someday be as populous as the present metropolitan area of New York. But both are most unlikely to have a physical structure similar to New York's, even aside from the differences among the natural settings. Chicago's growth so far belongs 47 percent to the Steel-Rail Epoch and 45 percent to the Auto-Air-Amenity Epoch; corresponding percentages for Los Angeles are 15 and 85 (Table III). Hence their historical increments have been markedly different, and their future increments will belong to a different technology from that which has built present-day New York.

If the system is indeed evolving and open-ended, it is patently incorrect to consider either Los Angeles or Chicago illustrative of a stage en route to the development of another New York or, for that matter, to consider any American city to be at any stage in any rigid model of development.

EXPLOITATION OF LAND

America's metropolitan centers grew initially

This suggests a modification of Allan Pred's model of self-generating urban growth during a period of rapid industrialization (Pred, *op. cit.*) Innovation in a particular industry, because of its impact on the evaluation of a particular resource, may result in new or expanded industry at a place other than the one at which the innovation occurred. In that case, the flow of benefits may be diverted to a new location. The observed evolution also suggests that technological change may be considered an integral part of the basic process that generates an urban hierarchy. Although this was recognized by Walter Christaller in the section on "Dynamic Processes" in his *Central Places in Southern Germany* translated by Carlisle W. Baskin (Englewood Cliffs, N.J., 1966), pp. 84-132, technological change has generally been handled as a secondary "modifier" of the basic model.

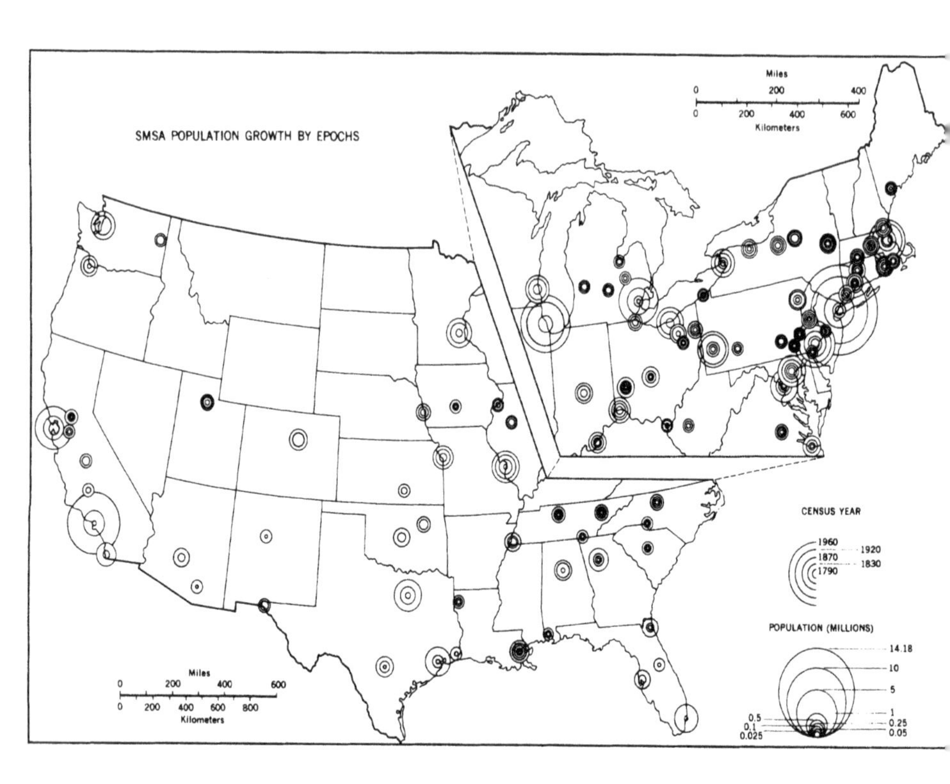

Figure 13. The varied historical layering of American SMSA's in 1960.

on land whose sites and locations were regarded as advantageous, given the contemporary technology and migration pattern. Improvement and use of the land made redevelopment or restoration much more costly, if not impossible. At first the land was "improved;" but as the improvements aged and grew obsolete the land appeared instead to have been "despoiled". New technologies hastened obsolescence and transferred the locational and site advantages to other land. Men moved to this new land and began again the sequence of improvement and abandonment: they abandoned the obsolete buildings, locations, or cities to those who remained behind to adapt and abandon in their turn.

This sequence of land selection in the light of existing technology, development, use, despoliation, and abandonment has cha acterized in varying degree the past utilizatio of timber, soil, water, and mineral resource It appears also to have characterized the us of land for urban purposes. The process ca be visualized at three different scales on th accompanying maps. At the metropolita scale it is illustrated by the partial abandon ment of the central city for outlying area during successive technological epochs; a the regional scale, by the shift of new devel opment from mining towns to universit towns, from railroad centers to recreationa centers; at the national scale, by the shi from older cities in the Northeast to newe cities in the Southwest or Florida. It could b argued that, at any scale, the basic attitud toward the potential urban land resource ha

TABLE III

Percentages of Populations of Selected 1960 SMSA's Attained in Major Historical Epochs

e Order	SMSA[a]	*Wagon-Sail Pre-1830*	*Iron Horse 1830-1870*	*Steel Rail 1870-1920*	*Auto-Air Amenity 1920-1960*
st	New York[b]	3	11	44	42
ond	Philadelphia	9	15	38	38
	Boston[c]	9	18	48	25
	Chicago[b]	0	8	47	45
	Detroit	0	6	29	65
	San Francisco-Oakland and San Jose	0	7	27	66
	Los Angeles	0	0+	15	85
ird	Washington	5	5	19	71
	Pittsburgh	7	10	56	27
	St. Louis	2	21	34	43
	New Orleans	7	18	23	52
	Seattle-Tacoma	0	0+	42	58
	Denver	0	1	35	64
	Dallas and Fort Worth	0	3	30	67
	Miami and Fort Lauderdale	0	0	4	96
urth	Albany-Schenectady-Troy	23	25	23	29
	New Bedford and Fall River	13	12	65	10
	Scranton and Wilkes-Barre-Hazelton	4	20	63	13
	Birmingham	1	1	47	51
	Omaha	0	10	50	40
	Flint	0	7	27	66
	Jacksonville	0	3	22	75
	Phoenix	0	0	14	86
th	Corpus Christi	0	2	8	90
	Altoona	0	21	72	7
	Charleston, S.C.	40	1	9	50
	Lubbock	0	0	7	93
	Las Vegas	0	0	4	96

60 SMSA except where noted to contrary.
andard consolidated area.
orfolk, Suffolk and Middlesex Counties.

n exploitive. There is no general provision "recycling" the resource of developed land en the initial development has become solete.

The result is a gigantic, national "filter wn" process with important geographical l historical dimensions. The nation's new nstruction has been concentrated, in any given epoch, not only in new neighborhoods and new suburbs but also in what have been, for all practical purposes, new cities (Table III). The residue of obsolescent physical plant has also become concentrated, not only in certain districts of most cities but in virtually the entire area of some. Vast big-city cores and nearly the whole of some smaller metropolitan areas are approaching the condition of inhabited ruins, and the residue of old structures continues to expand, thanks to the lagging national rate of replacement. Analysis of available historical housing data indicates that the construction of new dwelling units over the first sixty years of the twentieth century was enough to replace, on the average, only 4 percent of the units standing at the end of each decade.[27] For later generations the legacy of buildings, like the earlier "natural" endowment, has become an exploitable physical resource.

On the one hand, the traditional exploitive development of the land resource for urban purposes is understandable in view of the abundance of undeveloped land, the high costs of acquiring and clearing used land, and the high replacement costs under prevailing conditions.[28] On the other hand, the growing accumulation and low level of maintenance of obsolescent districts and cities suggest the inadequacy of the present approach.

ADAPTABILITY AND CONTROL

Two major problems seem to result from the nature of metropolitan evolution. First, lon term changes in size and physical charact are highly uncertain. Second, the exploitatic of new land and accompanying abandonme of old in successive periods lead to a gigant accumulation of residual structures. Th residue is a drag on both the improvement the general health and welfare and the mark for new, low-cost buildings.

It appears unlikely that the tendencies i herent in this evolutionary process w change significantly. To be sure, the fractic of the total number of metropolitan areas th changed size order diminished during t Steel-Rail and Auto-Air-Amenity Epoch Also, there have been fewer booms (Tab IV). This increasing stabilization was to b expected as the national transportation ne work was completed and improved and nea ly every part of the country raised its lev of participation in the national economy. Nevertheless, metropolitan areas contin to grow at differential rates. There has bee virtually no decrease over the past two epoc in the number of places that advanced or rank in the size order, and only a slight d crease in the number that declined one ran At the beginning of each epoch a maj change in technology registered its impact o the values of existing metropolitan locatio and on the pattern of migration; and som new cities were also established. As long a America remains an open society, there wi surely be unforeseen major changes affectin old cities and new alike — new rounds c initial advantage, reorganization, and adapt tion, new reasons to exploit new land an abandon old.

Given the two problems of uncertain and migration-abandonment, pressure

[27] John R. Borchert, Earl E. Stewart, and Sherman S. Hasbrouck, "Urban Renewal: Needs and Opportunities in the Upper Midwest," *Upper Midwest Economic Study Urban Rept. No. 5* (Minneapolis, 1963), pp. 1-4.

[28] Louis F. Winnick, "Housing and Urban Development: The Private Foundation's Role" (New York, The Ford Foundation, 1965), p. 3, makes the following observations: "No other important consumer good has been as inflation-prone as housing. Over the past seventy years the cost of a unit of housing space has risen twice as fast as other costs." Also, Chauncy D. Harris has pointed out the very small fraction of the land resource required by cities in "The Pressure of Residential-Industrial Land Use," in *Man's Role in Changing the Face of the Earth* edited by William L. Thomas, Jr., (Chicago, 1956), pp. 881-95, reference on p. 889.

[29] Lukermann, *op. cit.*

TABLE IV

Number of SMSA's and Emerging SMSA's Experiencing Shifts in Size Order

ft in Size Order	*1790-1830*	*1830-1870*	*1870-1920*	*1920-1960*
one rank	7	37	66	65
two ranks	11	16	15	6
three ranks	2	4	3	1
four ranks	0	1	0	0
wn one rank	10	14	25	20
wn two ranks	2	2	2	0
ady	26	24	49	103
w entries	19	47	69	48
opouts	6	8	13	17
t increase	13	39	56	31

unting to make the metropolitan settle-nt pattern more adaptable to change. For s purpose two types of development ap-ar to be of great potential importance. One he production, for the full range of urban ctions, of soundly engineered and attrac-ely designed structures that can be em-ced or removed at much lower costs than the past.[30] The other is the improvement of ormation-education systems[31] to increase the extent, accuracy, and currency of knowledge of the changing metropolis. An important consequence of this might be the development of a degree of public objectivity that would permit more rapid adaptation of institutions, notably local government.[32]

On the other hand, one might expect mounting pressure to create new institutions and shift values in order to retard the rate of change and thereby reduce the need for rapid and massive adaptation. For example, in some cases truly comprehensive, long-range planning could lower the permissible rate of technological change throughout an urban or regional system to that of the least tractable, slowest-changing component of the system, in the interest of preserving orderly development.

The mounting pressures for greater adaptability and greater control may often conflict. Where and how to compromise them seems likely to be an important and recurring issue in the future course of American metropoli-

For a concise recent summary of one aspect of this topic see William K. Wittausch, "New Con-cepts for the Housing Industry," *Urban Land,* Vol. XXV, no. 5 (1966), pp. 11-12.

See Edward F. R. Hearle and Raymond J. Mason, *A Data Processing System for State and Local Governments* (Englewood Cliffs, N.J., 1963). Chapter 4 presents a long list of classes of data now generally noncomparable and in-complete in coverage that will inevitably be standardized and automated yet already form a vital body of information about the internal geography and other aspects of the structure of each metropolitan area. See also W. L. Garrison, "Urban Transportation Planning Models in 1975," *Journ. Amer. Inst. of Planners,* Vol. 31, (1965), pp. 156-58. This is an extension of the logic of Garrison's forecast of traffic planning that assumes more rapid adjustment to crises, through improved information systems and de-signs. See also Edward L. Ullman, "The Nature of Cities Reconsidered," *Papers and Proc. Regional Science Assn.,* Vol. 9 (1962), pp. 7-23.

32 See Robert C. Wood, "1400 Governments," with Vladimir V. Almendinger, New York Metropolitan Region Study, Vol. 8 (Cambridge, Mass., 1961), especially the introductory and concluding chapters.

tan evolution. Furthermore, the issue is likely to be debated and resolved on many different grounds, since no one city is the evolutionary prototype for all others.

5

A Theory for the Location of Cities

Edward L. Ullman

I

Periodically in the past century the location and distribution of cities and settlements have been studied. Important contributions have been made by individuals in many disciplines. Partly because of the diversity and uncoordinated nature of the attack and partly because of the complexities and variables involved, a systematic theory has been slow to evolve, in contrast to the advances in the field of industrial location.[1]

The first theoretical statement of modern importance was von Thünen's *Der isolierte Staat*, initially published in 1826, wherein he postulated an entirely uniform land surface

[1] CF. Tord Palander, *Beitragezur Standortstheorie* (Uppsala, Sweden, 1935), or E. M. Hoover, Jr. *Location Theory and the Shoe and Leather Industries* (Cambridge, Mass., 1937).

Reprinted from *American Journal of Sociology*, Vol. XLVI, (May, 1941), pp. 853-64 by permission of the author.

d showed that under ideal conditions a city uld develop in the center of this land area d concentric rings of land use would velop around the central city. In 1841 hl investigated the relation between cities d the natural and cultural environment, ying particular attention to the effect of nsport routes on the location of urban nters.[2] In 1894 Cooley admirably demon-ated the channelizing influence that trans-rtation routes, particularly rail, would have the location and development of trade nters.[3] He also called attention to break in nsportation as a city-builder just as Ratzel d earlier. In 1927 Haig sought to deter-ne why there was such a large concentra-n of population and manufacturing in the gest cities.[4] Since concentration occurs ere assembly of material is cheapest, all siness functions, except extraction and nsportation, ideally should be located in ies where transportation is least costly. ceptions are provided by the processing of rishable goods, as in sugar centrals, and of ge weight-losing commodities, as in elters. Haig's theoretical treatment is of different type from those just cited but ould be included as an excellent example a "concentration" study.

In 1927 Bobeck[5] showed that German geographers since 1899, following Schlüter and others, had concerned themselves largely with the internal geography of cities, with the pattern of land use and forms within the urban limits, in contrast to the problem of location and support of cities. Such preoccupation with internal urban structure has also characterized the recent work of geographers in America and other countries. Bobeck insisted with reason that such studies, valuable though they were, constituted only half the field of urban geography and that there remained unanswered the fundamental geographical question: "What are the causes for the existence, present size, and character of a city?" Since the publication of this article, a number of urban studies in Germany and some in other countries have dealt with such questions as the relations between city and country.[6]

II

A theoretical framework for study of the distribution of settlements is provided by the work of Walter Christaller.[7] The essence of the theory is that a certain amount of productive land supports an urban center. The center exists because essential services must be performed for the surrounding land. Thus the primary factor explaining Chicago is the productivity of the Middle West; location at

J. G. Kohl, *Der Verkehr und die Ansiedlungen der Menschen in ihrer Abhangikeit von der Gestaltung der Erdoberflache*, 2nd. ed., (Leipzig, 1850).

C. H. Cooley, "The Theory of Transportation," *Publications of the American Economic Association*, Vol. IX (May, 1894), pp. 1-148.

R. M. Haig, "Toward an Understanding of the Metropolis: Some Speculations Regarding the Economic Basis of Urban Concentration," *Quarterly Journal of Economics*, Vol. XL (1926), pp. 179-208.

Hans Bobeck, "Grundfragen der Stadt Geographie," *Geographischer Anzeiger*, Vol. XXVIII (1927), pp. 213-24.

6 A section of the International Geographical Congress at Amsterdam in 1938 dealt with "Functional Relations between City and Country." The papers are published in Vol. II of the *Comptes rendus* (Leiden: E. J. Brill, 1938). A recent American study is C. D. Harris, *Salt Lake City: A Regional Capital* (Ph.D. diss., University of Chicago, 1940). Pertinent also is R. E. Dickinson, "The Metropolitan Regions of the United States," *Geographical Review*, Vol. XXIV (1934), pp. 278-91.

7 *Die zentralen Orte in Suddeutschland* (Jena, 1935); also a paper (no title) in *Comptes rendus du Congres international de geographie Amsterdam* (1938), Vol. II, pp. 123-37.

the southern end of Lake Michigan is a secondary factor. If there were no Lake Michigan, the urban population of the Middle West would in all probability be just as large as it is now. Ideally, the city should be in the center of a productive area.[8] The similarity of this concept to von Thünen's original proposition is evident.

Apparently many scholars have approached the scheme in their thinking.[9] Bobeck claims he presented the rudiments of such an explanation in 1927. The work of a number of American rural sociologists shows appreciation for some of Christaller's preliminary assumptions, even though done before or without knowledge of Christaller's work and performed with a different end in view. Galpin's epochal study of trade areas in Walworth County, Wisconsin, published in 1915, was the first contribution. Since then important studies bearing on the problem have been made by others. These studies are confined primarily to smaller trade centers but give a wealth of information[10] o distribution of settlements which indepenc ently substantiates many of Christaller's basi premises.

As a working hypothesis one assumes tha normally the larger the city, the larger it tributary area. Thus there should be cities c varying size ranging from a small hamle performing a few simple functions, such a providing a limited shopping and marke center for a small contiguous area, up to large city with a large tributary area com posed of the service areas of many smalle towns and providing more complex services such as wholesaling, large-scale banking specialized retailing, and the like. Service performed purely for a surrounding area ar termed "central" functions by Christalle

[8] This does not deny the importance of "gateway" centers such as Omaha and Kansas City, cities located between contrasting areas in order to secure exchange benefits. The logical growth of cities at such locations does not destroy the theory to be presented — cf. R. D. McKenzie's excellent discussion in *The Metropolitan Community* (New York, 1933), pp. 4 ff.

[9] Cf. Petrie's statement about ancient Egypt and Mesopotamia: "It has been noticed before how remarkably similar the distances are between the early nome capitals of the Delta (twenty-one miles on an average) and the early cities of Mesopotamia (averaging twenty miles apart). Some physical cause seems to limit the primitive rule in this way. Is it not the limit of central storage of grain, which is the essential form of early capital? Supplies could be centralized up to ten miles away; beyond that the cost of transport made it better worth while to have a nearer centre" — W. M. Flinders Petrie, *Social Life in Ancient Egypt* (London, 1923; reissued, 1932), pp. 3-4.

[10] C. J. Galpin, "Social Anatomy of an Agricu tural Community," University of Wisconsi Agricultural Experiment Station Research Bul 34 (1915), and the restudy by J. H. Kolb an R. A. Polson, "Trends in Town-Country Rela tions, University of Wisconsin Agricultura Experiment Station Research Bull. 117 (1933) B. L. Melvin, "Village Service Agencies of Ne York State, 1925," Cornell University Agricu tural Experiment Station Bull. 493 (1929), an "Rural Population of New York, 1855-192 Cornell University Agricultural Experiment Sta tion Memoir 116 (1928); Dwight Sandersor *The Rural Community* (New York, 1932), esp pp. 488-514, which contains references to man studies by Sanderson and his associates; Carl C. Zimmerman, "Farm Trade Centers in Minne sota, 1905-29," University of Minnesota Agri cultural Experiment Station Bull. 269 (1930) T. Lynn Smith, "Farm Trade Centers in Louisi ana 1905 to 1931," Louisiana State Universit Bull. 234 (1933); Paul H. Landis, "Sout Dakota Agricultural Experiment Station Bul 274 (1932), and "The Growth and Decline c South Dakota Trade Centers, 1901-1933," Bul 279 (1938), and "Washington Farm Trade Cen ters, 1900-1935," State College of Washingto Agricultural Experiment Station Bull. 36 (1938). Other studies are listed in subsequen footnotes.

the settlements performing them "cen-" places. An industry using raw materials orted from outside the local region and pping its products out of the local area uld not constitute a central service.

deally, each central place would have a ular tributary area, as in von Thünen's position, and the city would be in the ter. However, if three or more tangent les are inscribed in an area, unserved ces will exist; the best theoretical shapes hexagons, the closest geometrical figures circles which will completely fill an area gure I).[11]

Christaller has recognized typical-size lements, computed their average popula-, their distance apart, and the size and ulation of their tributary areas in accor-ce with his hexagonal theory as Table I ws. He also states that the number central places follows a norm from gest to smallest in the following order: :6:18:54, etc.[12]

All these figures are computed on the basis South Germany, but Christaller claims m to be typical for most of Germany and stern Europe. The settlements are classi- on the basis of spacing each larger unit in a hexagon of next-order size, so that the distance between similar centers in the table above increases by the $\sqrt{3}$ over the preceding smaller category (in Figure I, e.g., the distance from *A* to *B* is $\sqrt{3}$ times the distance from *A* to *C*). The initial distance figure of 7 km. between the smallest centers is chosen because 4–5 km., approximately the distance one can walk in one hour, appears to be a normal service-area limit for the smallest centers. Thus, in a hexagonal scheme, these centers are about 7 km. apart. Christaller's maps indicate that such centers are spaced close to this norm in South Germany. In the larger categories the norms for distance apart and size of centers appear to be true averages; but variations from the norm are the rule, although wide discrepancies are not common in the eastern portion of South Germany, which is less highly industrialized than the Rhine-Ruhr areas in the west. The number of central places of each rank varies rather widely from the normal order of expectancy.

The theoretical ideal appears to be most nearly approached in poor, thinly settled farm districts — areas which are most nearly self-contained. In some other sections of Germany industrial concentration seems to be a more important explanation, although

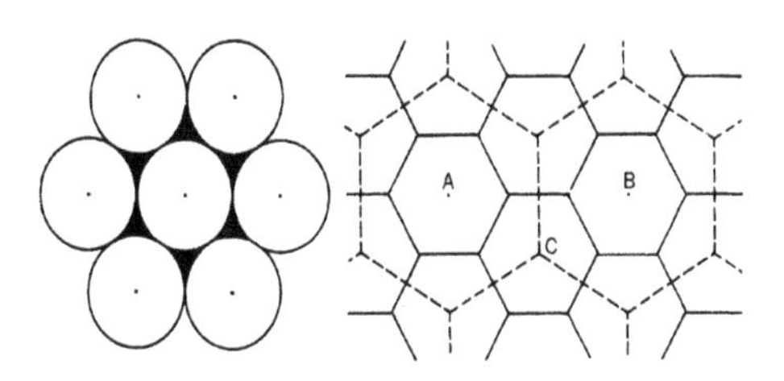

Figure 1. Theoretical shapes of tributary areas. Circles leave unserved spaces, hexagons do not. Small hexagons are service areas for smaller places, large hexagons (dotted lines) represent service areas for next higher-rank central places.

ee August Lösch, "The Nature of the Economic Regions," *Southern Economic Journal*, Vol. V 1938), p. 73. Galpin (*op. cit.*) thought in erms of six tributary-area circles around each enter. See also Kolb and Polson, *op. cit.*, pp. 0-41.

arnes and Robinson present some interesting naps showing the average distance apart of armhouses in the driftless area of the Middle Vest and in southern Ontario. Farmhouses night well be regarded as the smallest settle-nent units in a central-place scheme, although hey might not be in the same numbered se-uence (James A. Barnes and Arthur H. Robin-on, "A New Method for the Representation of Dispersed Rural Population," *Geographical Review*, Vol. XXX (1940), pp. 134-37).

TABLE I

Central Place	*Towns* Distance Apart (Km.)	Population	*Tributary Areas* Size (Sq. Km.)	Populati
Market hamlet (*Marktort*)	7	800	45	2,7
Township center (*Amtsort*)	12	1,500	135	8,1
County seat (*Kreisstadt*)	21	3,500	400	24,0
District city (*Bezirksstadt*)	36	9,000	1,200	75,0
Small state capital (*Gaustadt*)	62	27,000	3,600	225,0
Provincial head city (*Provinzhauptstadt*)	108	90,000	10,800	675,0
Regional capital city (*Landeshauptstadt*)	186	300,000	32,400	2,025,0

elements of the central-place type of distribution are present. Christaller points out that Cologne is really the commercial center for the Ruhr industrial district even though it is outside the Ruhr area. Even in mountain areas centrality is a more important factor than topography in fixing the distribution of settlements. Christaller states that one cannot claim that a certain city is where it is because of a certain river — that would be tantamount to saying that if there were no rivers there would be no cities.

III

Population alone is not a true measure of the central importance of a city; a large mining, industrial, or other specialized-function town might have a small tributary area and exercise few central functions. In addition to population, therefore, Christaller uses an index based on number of telephones in proportion to the average number per thousand inhabitants in South Germany, weighted further by the telephone density of the local subregion. A rich area such as the Palatinate supports more telephones in proportion to population than a poor area in the Bavarian Alps; therefore, the same number of telephones in a Palatinate town would not gi it the same central significance as in the Alp He claims that telephones, since they are use for business, are a reliable index of centralit Such a thesis would not be valid for most the United States, where telephones are common in homes as in commercial and pro fessional quarters.

Some better measures of centrality cou be devised, even if only the number of out-o town telephone calls per town. Better sti would be some measure of actual central se vices performed. It would be tedious an difficult to compute the amount, or percen age, of business in each town drawn fro outside the city, but some short cuts migh be devised. If one knew the average numbe of customers required to support certai specialized functions in various regions, the the excess of these functions over the norm required for the urban population would b an index of centrality.[13] In several states rur sociologists and others have computed th

[13] In Iowa, e.g., almost all towns of more than 45 inhabitants have banks, half of the towns 250-300, and 20 per cent of the towns of 10 150 (according to calculations made by th author from population estimates in *Rand M Nally's Commercial Atlas for 1937*).

ːrage number of certain functions for vns of a given size. With one or two excep-ns only small towns have been analyzed. tail trade has received most attention, but ofessional and other services have also en examined. These studies do not tell us ually what population supports each ser-e, since the services are supported both by vns and by surrounding rural population, t they do provide norms of function pectancy which would be just as useful.[14]

A suggestive indicator of centrality is pro-led by the maps which Dickinson has made per capita wholesale sales of cities in the nited States.[15] On this basis centers are tributed rather evenly in accordance with gional population density. Schlier has com-ted the centrality of cities in Germany on the basis of census returns for "central" oc-cupations.[16] Refinement of some of our census returns is desirable before this can be done entirely satisfactorily in the United States, but the method is probably the most promising in prospect.

Another measure of centrality would be the number of automobiles entering a town, making sure that suburban movements were not included. Figures could be secured if the state-wide highway planning surveys in forty-six states were extended to gather such statistics.

IV

The central-place scheme may be distorted by local factors, primarily industrial concen-tration or main transport routes. Christaller notes that transportation is not an areally operating principle, as the supplying of central goods implies, but is a linearly work-ing factor. In many cases central places are strung at short intervals along an important transport route, and their tributary areas do not approximate the ideal circular or hexagonal shape but are elongated at right angles to the main transport line.[17] In some areas the reverse of this normal expectancy is true. In most of Illinois, maps depicting tri-butary areas show them to be elongated

See particularly the thorough study by B. L. Melvin, "Village Service Agencies, New York State 1925"; C. R. Hoffer, "A Study of Town-Country Relationships, Michigan Agricultural Experiment Station Special Bull. 181 (1928), (data on number of retail stores and professions per town); H. B. Price and C. R. Hoffer, "Ser-vices of Rural Trade Centers in Distribution of Farm Supplies," Minnesota Agricultural Experi-ment Station Bull. 249 (1938); William J. Reilly, "Methods for the Study of Retail Rela-tionships," Bureau of Business Monographs, no. 4, University of Texas Bull. 2944 (1929), p. 26; J. H. Kolb, "Service Institutions of Town and Country," Wisconsin Agricultural Experiment Station Research Bull. 66 (1925) (town size in relation to support of institutions); Smith, *op. cit.*, pp. 32-40; Paul H. Landis, South Dakota Town-Country Trade Relations, 1901-1931," p. 20 (population per business enterprise), and pp. 24-25 (functions per town size); Zimmer-man, *op. cit.*, pp. 16 and 51 ff.

For a criticism of population estimates of unincorporated hamlets used in many of these studies see Glenn T. Trewartha, "The Unincor-porated Hamlet: An Analysis of Data Sources," (paper presented December 28 at Baton Rouge meetings, Association of American Geographers; forthcoming, probably, in March number of *Rural Sociology*, Vol. VI (1941).)

Ibid., pp. 280-81.

16 Otto Schlier, "Die zentralen Orte des Deutschen Reichs," *Zeitschrift der Gesellschaft für Erd-kunde zu Berlin* (1937), pp. 161-70. See also map constructed from Schlier's figures in R. E. Dickinson's valuable article, "The Economic Regions of Germany," *Geographical Review*, Vol. XXVIII (1938), p. 619. For use of census figures in the United States see Harris, *op. cit.*, pp. 3-12.

17 For an illustration of this type of tributary area in the ridge and valley section of east Tennessee see H. V. Miller, "Effects of Reservoir Construc-tion on Local Economic Units," *Economic Geography*, Vol. XV (1939), pp. 242-49.

parallel to the main transport routes, not at right angles to them.[18] The combination of nearly uniform land and competitive railways peculiar to the state results in main railways running nearly parallel and close to one another between major centers.

In highly industrialized areas the central-place scheme is generally so distorted by industrial concentration in response to resources and transportation that it may be said to have little significance as an explanation for urban location and distribution, although some features of a central-place scheme may be present, as in the case of Cologne and the Ruhr (p. 858).

In addition to distortion, the type of scheme prevailing in various regions is susceptible to many influences. Productivity of the soil,[19] type of agriculture and intensity of cultivation, topography, governmental organization, are all obvious modifiers. In the United States, for example, what is the effect on distribution of settlements caused by the sectional layout of the land and the regular size of counties in many states? In parts of Latin America many centers are known as "Sunday towns"; their chief functions appear to be purely social, to act as religiou and recreational centers for holidays — henc the name "Sunday town."[20] Here socia rather than economic services are the primar support of towns, and we should accordingl expect a system of central places with fewe and smaller centers, because fewer function are performed and people can travel farthe more readily than commodities. These under lying differences do not destroy the value c the theory; rather they provide variations c interest to study for themselves and for pur poses of comparison with other regions.

The system of central places is not stati or fixed; rather it is subject to change an development with changing conditions.[21] Im provements in transportation have ha noticeable effects. The provision of goo automobile roads alters buying and market ing practices, appears to make the smalles centers smaller and the larger centers larger and generally alters trade areas.[22] Since goo

18 See, e.g., *Marketing Atlas of the United States* (New York, International Magazine Co., Inc.) or *A Study of Natural Areas of Trade in the United States* (Washington, D.C., U.S. National Recovery Administration, 1935).

19 Cf. the emphasis of Sombart, Adam Smith, and other economists on the necessity of surplus produce of land in order to support cities. Fertile land ordinarily produces more surplus and consequently more urban population, although "the town . . . may not always derive its whole subsistence from the country in its neighborhood. . . ." Adam Smith, *The Wealth of Nations*, "Modern Library" edition, (New York, 1937), p. 357; Werner Sombart, *Der moderne Kapitalismus* (zweite, neugearbeitete Auflage, Munich and Leipzig, 1916), Vol. I, pp. 130-31).

20 For an account of such settlements in Brazil se Pierre Deffontaines, "Rapports fonctionne entre les agglomérations urbaines et rurales: u example en pays de colonisation, le Brésil, *Comptes rendus du Congrès internationale d géographie Amsterdam*, Vol. II (1938), pp 139-44.

21 The effects of booms, droughts, and other fac tors on trade-centre distribution by decades ar brought out in Landis' studies for South Dakot and Washington. Zimmerman and Smith als show the changing character of trade-cente distribution (see footnote 10 of this paper fo references). Melvin calls attention to a "villag population shift lag"; in periods of depresse agriculture villages in New York declined i population approximately a decade after th surrounding rural population had decreased (B L. Melvin, *Rural Population of New York 1855-1925*, p. 120).

22 Most studies indicate that only the very smalles hamlets (under 250 population) and crossroad stores have declined in size or number. Th larger small places have held their own (se

ads are spread more uniformly over the
d than railways, their provision seems to
ake the distribution of centers correspond
ore closely to the normal scheme.[23]

Christaller may be guilty of claiming too
eat an application of his scheme. His cri-
ia for determining typical-size settlements
d their normal number apparently do not
actual frequency counts of settlements in many almost uniform regions as well as some less rigidly deductive norms.[24]

Bobeck in a later article claims that Christaller's proof is unsatisfactory.[25] He states that two-thirds of the population of Germany and England live in cities and that only one-third of these cities in Germany are real central places. The bulk are primarily industrial towns or villages inhabited solely by farmers. He also declares that exceptions in the rest of the world are common, such as the purely rural districts of the Tonkin Delta of Indo-China, cities based on energetic entrepreneurial activity, as some Italian cities, and world commercial ports such as London, Rotterdam, and Singapore. Many of these objections are valid; one wishes that Christaller had better quantitative data and was less vague in places. Bobeck admits, however, that the central-place theory has value and applies in some areas.

The central-place theory probably provides as valid an interpretation of settlement distribution over the land as the concentric-zone theory does for land use within cities. Neither theory is to be thought of as a rigid framework fitting all location facts at a given moment. Some, expecting too much, would jettison the concentric-zone theory; others, realizing that it is an investigative hypothesis of merit, regard it as a useful tool for comparative analysis.

V

Even in the closely articulated national economy of the United States there are strong

Landis for Washington, *op. cit.*, p. 37, and his *South Dakota Town-Country Trade Relations 1901-1931*, pp. 34-36). Zimmerman in 1930 (*op. cit.*, p. 41) notes that crossroads stores are disappearing and are being replaced by small villages. He states further: "It is evident that claims of substantial correlation between the appearance and growth of the larger trading center and the disappearance of the primary center are more or less unfounded. Although there are minor relationships, the main change has been a division of labor between the two types of centers rather than the complete obliteration of the smaller in favor of the larger" (p. 32).

For further evidences of effect of automobile on small centers see R. V. Mitchell, *Trends in Rural Retailing in Illinois 1926 to 1938*, University of Illinois Bureau of Business Research Bull., Ser. 59 (1939), pp. 31 ff., and Sanderson, *op. cit.*, p. 564, as well as other studies cited above.

Smith (*op. cit.*, p. 54) states: "There has been a tendency for centers of various sizes to distribute themselves more uniformly with regard to the area, population, and resources of the state. Or the changes seem to be in the direction of a more efficient pattern of rural organization. This redistribution of centers in conjunction with improved methods of communication and transportation has placed each family in frequent contact with several trade centers. . . ."

In contrast, Melvin (*Rural Population of New York, 1855-1925*, p. 90), writing about New York State before the automobile had had much effect, states: "In 1870 the villages . . . were rather evenly scattered over the entire state where they had been located earlier in response to particular local needs. By 1920, however, the villages had become distributed more along routes of travel and transportation and in the vicinity of cities."

[24] This statement is made on the basis of frequency counts by the author for several midwestern states (cf. also Schlier, *op. cit.*, pp. 165-69, for Germany).

[25] Hans Bobeck, "über einige functionelle Stadttypen und ihre Beziehungen zum Lande," *Comptes rendus du Congrès internationale de géographie Amsterdam*, Vol. II (1938), p. 88.

forces at work to produce a central-place distribution of settlements. It is true that products under our national economy are characteristically shipped from producing areas through local shipping points directly to consuming centers which are often remote. However, the distribution of goods or imports brought into an area is characteristically carried on through brokerage, wholesale and retail channels in central cities.[26] This graduated division of functions supports a central-place framework of settlements. Many nonindustrial regions of relatively uniform land surface have cities distributed so evenly over the land that some sort of central-place theory appears to be the prime explanation.[27] It should be worth while to study this distribution and compare it with other areas.[28] In New England, on the other hand, where cities are primarily industrial center based on distant raw materials and extra regional markets, instead of the land sup

[26] Harris, *op. cit.*, p. 87.

[27] For a confirmation of this see the column diagram on p. 73 of Lösch (*op. cit.*), which shows the minimum distances between towns in Iowa of three different size classes. The maps of trade-center distribution in the works of Zimmerman, Smith, and Landis (cited earlier) also show an even spacing of centers.

[28] The following table gives the average community area for 140 villages in the United States in 1930. In the table notice throughout that (i) the larger the village, the larger its tributary area in each region and (ii) the sparser the rural population density, the larger the village tributary area for each size class (contrast mid-Atlantic with Far West, etc.).

Community Area in Square Miles

Region	Small Villages (250-1,000 Pop.)	Medium Villages (1,000-1,750 Pop.)	Large Villages (1,750-2,500 Pop.)
Mid-Atlantic	43	46	87
South	77	111	146
Middle West	81	113	148
Far West	—	365	223

Although 140 is only a sample of the numbe of villages in the country, the figures are si nificant because the service areas were carefull and uniformly delimited in the field for a villages (E. deS. Brunner and J. D. Kolb, *Rur Social Trends* (New York, 1933), p. 95; se also E. deS. Brunner, G. S. Hughes, and M Patten, *American Agricultural Villages* (Ne York, 1927), chap. ii).

In New York 26 sq. mi. was found to b the average area per village in 1920. Villag refers to any settlement under 2,500 populatio Nearness to cities, type of agriculture, an routes of travel are cited as the three mo important factors influencing density of village Since areas near cities are suburbanized in som cases, as around New York City, the villag density in these districts is correspondingly high Some urban counties with smaller cities (Roch ester, Syracuse, and Niagara Falls) have fe suburbs, and consequently the villages are fa ther apart than in many agricultural countie (B. L. Melvin, *Rural Population of New York 1855-1925*, pp. 88-89; table on p. 89 show number of square miles per village in eac New York county).

In sample areas of New York State the aver age distance from a village of 250 or under t another of the same size or larger is about miles; for the 250-749 class it is 3-5 miles; fo the 750-1,249 class, 5-7 miles (B. L. Melvin *Village Service Agencies, New York, 1925*, p 102; in the table on p. 103 the distance average cited above are shown to be very near th modes).

Kolb makes some interesting suggestions as t the distances between centers. He shows th spacing is closer in central Wisconsin than i Kansas, which is more sparsely settled — J. H Kolb, "Service Relations of Town and Country, Wisconsin Agricultural Experimental Statio Research Bull. 58 (1923), see pp. 7-8 for theo retical graphs.

In Iowa, "the dominant factor determinin the *size* of convenience-goods areas is distance — *Second State Iowa Planning Board Repor* (Des Moines, April, 1935), p. 198. This repor contains fertile suggestions on trade areas fo Iowa towns. Valuable detailed reports on reta trade areas for some Iowa counties have als been made by the same agency.

ting the city the reverse is more nearly
: the city supports the countryside by
viding a market for farm products, and
s infertile rural areas are kept from being
n more deserted than they are now.

The forces making for concentration at
ain places and the inevitable rise of cities
these favored places have been em-
sized by geographers and other scholars.
phenomenal growth of industry and
ld-trade in the last hundred years and the
comitant growth of cities justify this em-
sis but have perhaps unintentionally
sed the intimate connection between a city
its surrounding area partially to be over-
ked. Explanation in terms of concentra-
is most important for industrial districts
does not provide a complete areal theory
distribution of settlements. Furthermore,
re is evidence that "of late . . . that rapid
wth of the larger cities has reflected their
reasing importance as commercial and
vice centers rather than as industrial centers."[29] Some form of the central-place theory should provide the most realistic key to the distribution of settlements where there is no marked concentration — in agricultural areas where explanation has been most difficult in the past. For all areas the system may well furnish a theoretical norm from which deviations may be measured.[30] If might also be an aid in planning the development of new areas. If the theory is kept in mind by workers in academic and planning fields as more studies are made, its validity may be tested and its structure refined in accordance with regional differences.

[29] U.S. National Resources Committee, *Our Cities — Their Role in the National Economy: Report of the Urbanism Committee* (Washington, Government Printing Office, 1937), p. 37.

[30] Some form of the central-place concept might well be used to advantage in interpreting the distribution of outlying business districts in cities. cf. Malcolm J. Proudfoot, "The Selection of a Business Site," *Journal of Land and Public Utility Economics*, Vol. XIV (1938), esp. 373 ff.

6

Christaller's Centra
Place Theor

Arthur Geti
Judith Geti

Much present-day research in urban geography has its roots in the work of Walter Christaller, a German scholar from Bavaria.* He was attempting to find the laws which determine the number, size, and distribution of towns. He was convinced, he wrote, that just as there are economic laws which determine the life of the economy, so are there special economic-geographic laws determining the arrangement of towns.

Since 1933, when his book on central places in southern Germany was published, many writers have praised and criticized, reformulated and expanded parts of Christaller's theory. Today very few accept all aspects of his work, but they realize that it stimulated some of the most advanced scientific work in geography. The following summary of the rudiments of central place theory is included here in order to acquaint the reader with the meaning of some of the terminology prevalent in urban geography today — a terminology introduced into our literature by tho dealing with Christaller's work. Hopeful this will be a reference when reading abo urban geography.

No attempt is made to summarize all Christaller's work. The emphasis is c Christaller's marketing principle or k = network. Other networks were derived whic had their foundation in principles of tran portation and administration.

A Central Place

The chief function or characteristic of town, Christaller said, is to be the center a region. Settlements which are prevalent centers of regions he called *central places.* contrast to these are dispersed places, i.e., a those places which are not centers. The might be areally-bound places (the inhab tants live from their agricultural activities pointly-bound places (the inhabitants mak their living from resources which occur specific locations, such as mining settlement customs places, and so on), or settlemen which are indifferent with regard to the location (monastery settlements). Christalle was concerned with the central places only.

* Walter Christaller, *Die zentralen Orte in Sud-deutschland*, trans. C. Baskin, (Jena, Gustave Fischer, 1933 and Charlottesville, University of Virginia, Bureau of Population and Urban Research, 1954).

Reprinted from *Journal of Geography* (May, 1966), pp. 220-26 by permission.

Some central places are more important n others — their central functions extend er regions in which other central places less importance exist. Christaller devised neans of measuring the centrality of towns their relative importance in regard to the rounding region.

ntral Goods and Services

ods produced at a central place, and the vices offered there, are called *central ods and services*. Dispersed goods and seres, in contrast, are ubiquitous; they are ered and produced everywhere. Further, industry using raw materials imported m outside the local region and shipping its oducts out of the local area would not conute a central service. The goods must be oduced for the surrounding region.

e Range Of A Good

is is the distance the dispersed population willing to travel to buy a good offered at a ntral place. The good has both an upper d a lower limit to its *range*. The *upper limit* he maximum radius of sales beyond which price of the good is too high for it to be d. The upper limit may be either an ideal a real limit.

Ideal limit: the maximum radius results m the increase of price with distance until nsumers will no longer purchase the good. *Real limit:* the radius is determined by the oximity of an alternate center which can er the good at a lower price at a certain tance from the first center. The *lower limit* the range encloses the number of conmers necessary to provide the minimum es volume required for the good to be oduced and distributed profitably from the ntral place. This has been called the *eshold level* of the good. It should be noted that there is no fixed distance between the lower and upper limits of a range; sometimes the distance between the two is small, at other times it is great.

Each good will have its own range, due to the fact that the prices of various goods increase at different rates with increasing distances from the center, and to the fact that different goods have different thresholds. Further, Christaller notes that the range of any one good may be different at each central place and at each point in time.

The Complementary Region

This is the area enclosed about a central place by the range of a good. Christaller assumes that the central place has a monopoly in the supply of the good to its complementary region by virtue of the price at which it can offer the good.

Ideally, each central place would have a circular tributary (market) area, with itself at the center. However, either unserved places would exist, if this were the case, or the circles would overlap, in which case the condition of monopoly would not be fulfilled.

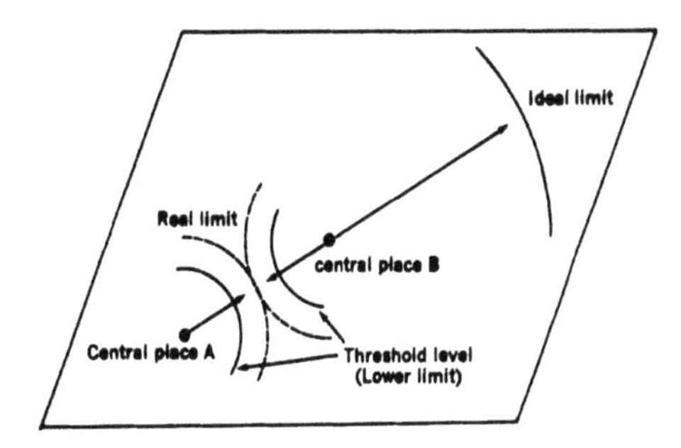

Figure 1. The range of a good, say a radio, offered at both A and B. Since the threshold level is less than the real limit, both central places will produce it. The real limit, halfway between the two central places, shows the trade area for radios for each place. Those living beyond the ideal limit must either do without radios or establish a new central place supplying the good.

Next to circles, hexagons are the most efficient figures both to serve an area (central places and distances traveled will be minimized) and to fill an area completely, as the figure below indicates. Therefore the complementary region of a central place assumes the form of a hexagon.

Using the terms defined above, and a set of assumptions and conditions, Christaller evolved a system of central places. The assumptions, which are listed below, tell us about the kind of landscape on which his system would be erected.

(1) An unbounded plain with soil of equal fertility everywhere and an uneven distribution of resources.

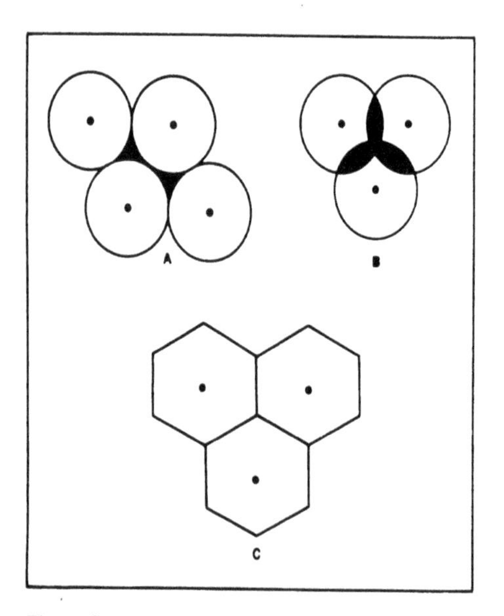

Figure 2. Three arrangements of complementary regions.

A. The unserved areas are shaded.

B. Shaded areas indicate places where the condition of monopoly would not be fulfilled.

C. Hexagons completely fill an area, with no overlap.

(2) An even distribution of population an purchasing power.

(3) A uniform transportation network in a directions, so that all central places c the same type are equally accessible.

(4) A constant range of any one centra good, whatever the central place fro which it is offered.

Given this landscape, we have to kno what the desires of the people are — i.e what constraints will exist on the systen These conditions follow:

(1) A maximum number of demands fo the goods and services should be satis fied.

(2) The incomes of the people offering th goods and services should be maxi mized.

(3) Distances moved by consumers to pur chase the goods and services should b minimized; i.e., goods are purchase from the closest point.

(4) The number of central places should b the minimum possible.

The System Of Central Places: The k = 3 Network

Under the assumptions and conditions state above, Christaller's system of central place may be derived. Let us assume first that ther

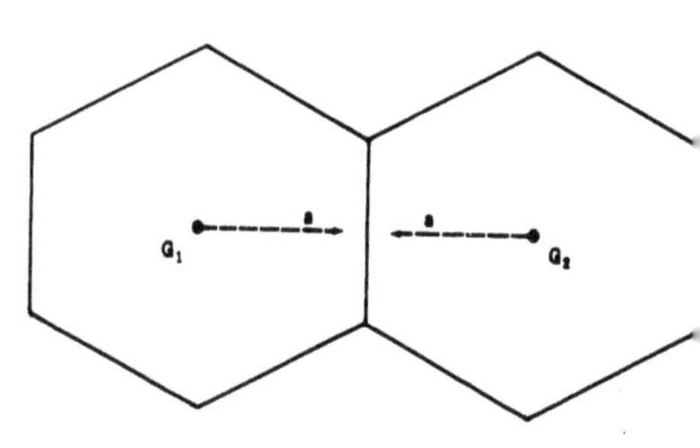

Figure 3. Real range of good 1 is shown by distance a.

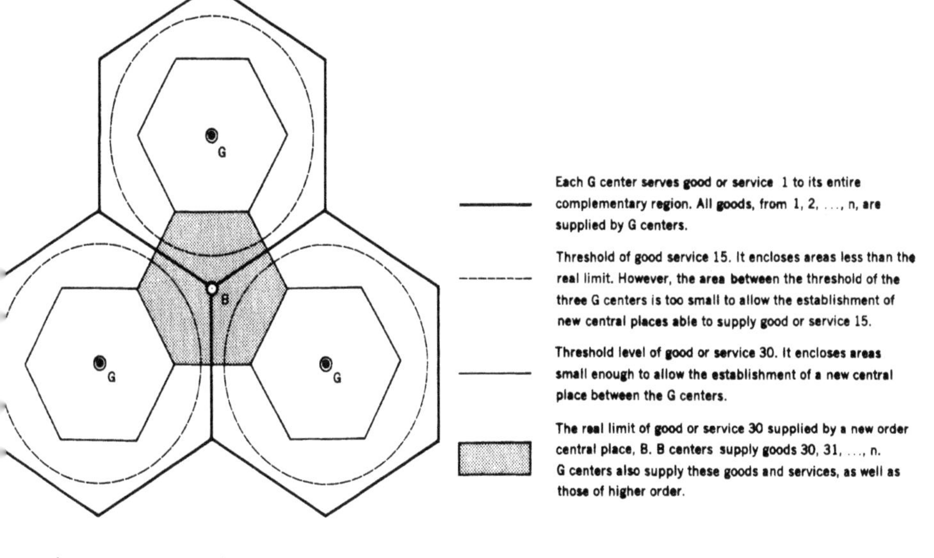
G
B
G
G
Each G center serves good or service 1 to its entire complementary region. All goods, from 1, 2, . . ., n, are supplied by G centers.
Threshold of good service 15. It encloses areas less than the real limit. However, the area between the threshold of the three G centers is too small to allow the establishment of new central places able to supply good or service 15.
Threshold level of good or service 30. It encloses areas small enough to allow the establishment of a new central place between the G centers.
The real limit of good or service 30 supplied by a new order central place, B. B centers supply goods 30, 31, . . ., n. G centers also supply these goods and services, as well as those of higher order.

gure 4.

e two central places, called G centers, ich offer all of the goods and services, from good or service called order 1 with the high- t threshold, to a good or service of order 0, with the lowest threshold. Of necessity, e two G centers have the largest market eas (complementary regions) of all central aces. The "real" range of the highest order od demanded, good of order 1, defines the undary between the two G centers. There- re, the two hexagon-shaped complementary gions have one side in common.

As was noted above, the ranges of the ods decline successively; good of order 2 s a smaller range than that of order 1. As e ranges decline, larger and larger numbers consumers are left between the two G cen- s. With some good, say good of order 30, ere are enough "surplus" consumers over d above the thresholds of the G centers to ow the development of alternate centers. ese are called B centers.

B centers supply goods, 30, 31, . . . , 100 at wer prices than the G centers *in the areas between the threshold ranges of those goods from the G centers.* B centers are located at the maximum economic distance from the G centers — i.e., on the outermost edges of the areas defined about G centers by the real range of good of order 1. In this way, consumer movements are kept to a minimum, and a maximum number of demands are satisfied from a minimum number of centers.

B centers in turn leave progressively larger numbers of surplus demands, and with some good, say good of order 50, these are large enough to permit the existence of a third rank of centers: K centers. K centers provide goods 50, 51, . . . , 100. The existence of four other types of centers is accounted for in the same manner.

Besides its definite spatial pattern two things should be noted about Christaller's system. First, a very rigid class structure has been described. That is, each central place supplies all the goods and services — the *identical* goods and services — that the centers below it provide, plus some additional ones. It is due to this fact that Christaller was able to assume that discrete population levels could be assigned to centers of the same type. Since the population of a town depends upon the number and types of functions it performs, then centers performing similar functions will have similar populations. Further, since no centers not of the same type offer identical goods and services, the population levels will be unique.

Second, the system of central places and their complementary regions is characterized by interdependency. All centers except the smallest have other centers dependent upon them for the supply of certain goods. Thus, B centers have K centers and their complementary regions, and all centers of a lower rank than K, dependent upon them for th supply of goods 30, 31, . . . , 49. In turn, I centers and their complementary regions de pend on G centers for the supply of goods 1 2, . . . , 29. Each complementary region of B place is served with those goods by thre G centers, and Christaller assumes that one third of its trade goes to each. Each G cente thus serves its own region for the supply c goods 1, 2, . . . , 29 as well as one-third o the complementary regions of the six B cen ters. In all, a G center serves three total B type regions. This is called a k = 3 network where k equals the total number of comple mentary regions of next lowest order serve by the central place of next highest orde Likewise, of course, three complete K-typ regions are served with goods 30, 31, . . . , 4 by a B center, and so on.

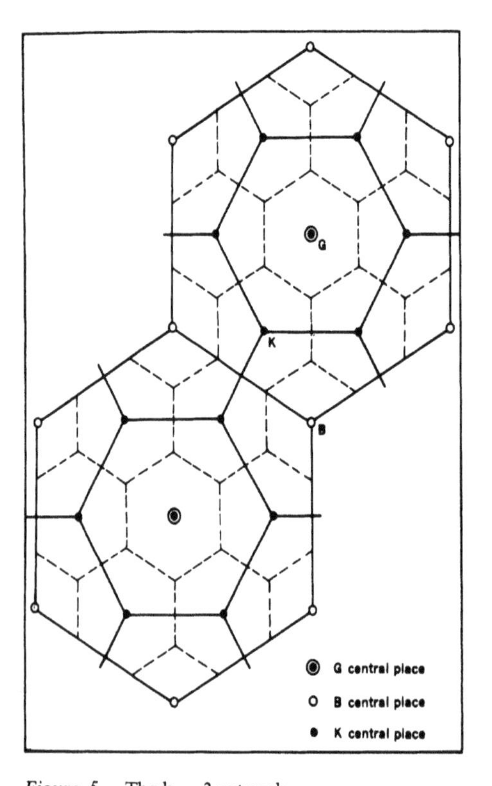

Figure 5. The k = 3 network.

In summary, the chief contribution c Christaller to central place theory is, c course, its basic formulation. More specifi cally, his identification of concepts relevan to the location of cities, his logically-derive system of central places, and the conclusion following from it are to be noted. Further, i should once more be pointed out that ther is much in the book which has not been men tioned here. Christaller recognizes certain de ficiencies of his system and qualifies i whenever he deems necessary. In general, h noted that:

> The strict mathematical scheme developed previously is imperfect in some respects. It is even incorrect in this strictness. The scheme should approximate reality; therefore we should study the factors under whose influence it undergoes change (p. 217).

erry, Brian J. L. and Pred, Allen, *Central Place tudies: A Bibliography of Theory and Applica-ons* (Philadelphia, Pa., Regional Science Re-earch Institute). A supplement through 1964 of entral place studies is also available.

hristaller, Walter, trans. by Carlisle Baskin, *Central Places in Southern Germany, The Pioneer Work in Theoretical Geography* (Englewood Cliffs, N.J., Prentice-Hall, 1966).

Ullman, E. L., "A Theory of Location for Cities," *Am. Jour. of Sociology*, Vol. XLVI, no. 6 (1941), pp. 853-64.

echnology and Urban Form

7

ederick W. Boal

. The "message" of any medium or tech-ogy is the change of scale or pace or pat-1 that it introduces into human affairs. e railway did not introduce movement or nsportation or wheel or road into human society, but it accelerated and enlarged the scale of previous human functions, creating totally new kinds of cities and new kinds of work and leisure. — Marshall McLuhan, *Understanding Media*, p. 24.

rinted from *Journal of Geography*, (April, 1968), pp. 229-36 by permission.

Transportation and construction technology have profound influences on the internal geography of urban areas.[1] A city is characterized by flows — flows of people, commodities, and information. The way these flows take place shapes the city just as, in turn, the form of the developed city influences the ways in which the flows occur. These flows represent interaction between the various parts of the city, and at the present time they occur along a whole series of networks, such as streets, railroad tracks, pipelines, electricity and telephone lines, and even the channels for radio and television transmission.

Urban areas are also characterized by relatively intensive use of ground space. This is, of course, partly dependent on transportation technology, but it is also dependent on the nature of building construction. The taller a building, the greater the potential for intensive use of the piece of ground on which the structure is placed.

The impact of transportation and construction technology on the city is one of "space adjusting"[2] — in part of the adjustment of the spatial relationships between various areas by the development and use of various forms of transportation and communication, and in part the creation of additional space by construction. In terms of "space adjus ing," the city can be viewed at three tec nological periods, which we will call t period of the *Pedestrian City*, the period the *Steam Engine and Wheel-Track City*, ar the period of the *Flexible City*. Viewing t impact of technology on urban form throu time clarifies the nature of the impact ar also helps explain the form of existing citie which in most cases have developed durin more than one technological period. In broader cultural context it is interesting note that the three periods above close parallel the three periods through whic Marshall McLuhan suggests western socie has moved — village society, mechaniz society, and the electric era.[3]

The Pedestrian City

The urban areas of the pre-industrial perio were dominated by movement on foot. Th had a number of consequences. First, the di tances that could be covered in a given tim period were short and in so far as connectio with the center of the town was important th set a strict limit to the radius of town develo ment. Secondly, because movement was pr dominantly on foot, there could be a co siderable number of routes and they could b narrow and quite irregular. The Swiss arch tect Le Corbusier suggested this when h wrote, "The pack-donkey meanders alon meditates a little in his scatter-brained an distracted fashion, he zigzags in order t avoid the larger stones, or to ease the clim or to gain a little shade; he takes the line c least resistance. . . . The Pack Donkey's Wa is responsible for the plan of every continent

[1] For discussion of certain aspects of these influences see John R. Borchert, "American Metropolitan Evolution," *Geographical Review*, Vol. 57 (July, 1967), pp. 301-32; A. Fleisher, "The Influence of Technology on Urban Forms," *Daedalus*, Vol. 90 (1961), pp. 48-60; and James E. Vance, Jr., "Labor-Shed, Employment Field, and Dynamic Analysis in Urban Geography," *Economic Geography*, Vol. 36 (July, 1960), pp. 189-220.

[2] E. A. Ackerman, *Geography as a Fundamental Research Discipline*, Research Paper No. 53 (Chicago, University of Chicago, Department of Geography, 1958), p. 24.

[3] Marshall McLuhan, *Understanding Media: Th Extensions of Man* (New York, New America Library, 1966).

,."[4] E. A. Gutkind, though disagreeing h some of Le Corbusier's ideas, reinforces concept of irregularity to street pattern ich he claims was due to attaching more portance to the houses than to the streets.[5] this sense the streets were almost relict ce left between buildings. In addition, the ole street system was of equal importance, d ease of movement, with the exception of few principal streets, developed along ıtes that joined the town to neighboring tlements. However, this was possible only a pedestrian and animal transport domin-d town. Transportation modes with higher ıce demands and lower directional flexibil- would change all this.

A further feature of interest was the ar-ıgement of the living areas of the various sses. Sjoberg has noted the universality of pattern in which the well-to-do and power- congregated in the city center while the orer folks found what accommodation they ıld round the edge. The upper class needed dy access to the headquarters of the gov-ımental, religious, and educational organi-ions and because "the feudal society's hnology permits relatively little spatial bility,"[6] this ready access could be ob-ned only by physical proximity. Interest-ly, a mid-nineteenth century English oposal by John Silk Buckingham for a opian city to be called Victoria also had a out of well-to-do in the middle and low ome round the edge. This proposal was ıde just too early for recognition to be en to the significance of the railroad.

Le Corbusier, *The City of Tomorrow* (London, The Architectural Press, 1947), pp. 23-24.

E. A. Gutkind, *Urban Development in Central Europe* (New York, Free Press, 1964).

G. Sjoberg, *The Pre-Industrial City* (New York, Free Press, 1960), p. 99.

The urban growth of this pedestrian period was achieved by creating new towns rather than adding area to existing ones. In this way the requirements of the pedestrian scale were not violated.

The Steam Engine and Wheel-Track City

The development by Watt of the steam engine in 1788 and the first tentative efforts of the steam locomotive in 1804 were to have revolutionary effects on urban areas.

The stationary steam engine can be characterized in three ways. First of all, up to a point, the larger the steam engine, the more efficient it was. Secondly, the steam engine burned bulky coal as fuel and was inefficent as a converter of energy. Thirdly, the energy made available by the steam engine could only be transferred to operate machinery by direct shaft and belt. Taken together these characteristics led to the development of large factory complexes with a very close locational association of coal mine, steam engine, and factory. Many factory buildings were developed to six or more stories, because, by so doing, the most efficient connections could be made between steam engine and factory machinery using short axles and belts.

The build-up of large industrial populations would not have been possible without the steam engine when used in long distance transport. However, the railroad had a profound effect, not only on the size of urban population concentrations, but also on the form of the urban areas themselves. The locomotive is capable of delivering great power while on the rails but is helpless and delivers no power off them. Its effectiveness is ribbon-like in form, introducing into an area relatively high mobility where railroad tracks exist. Because of the concomitant change in time-distance relationships where railroads were built, urban areas began to develop most markedly along the tracks. Thus the star-like form of the nineteenth century city began. However, the railroad does not improve accessibility all along its length but only at the stations, and because closely spaced stations would have led to excessive losses of energy in stopping and starting, the stations were spaced. In discussing London, Carter[7] has stated that "a steam train takes a long time to accelerate and draw up, so that for economy of operation the stations had to be fairly widely spaced. The urban units whose growth was stimulated by the railways remained well separated and maintained their identity." In this way main directions of urban growth developed, not, however, in a continuous fashion, but as a series of discrete beads (Figure 1). The developments around the suburban stations themselves were restricted and compact because foot travel was dominant within the suburban node. Later, the electrification of suburban lines enabled stations to be placed closer together, because of decreased energy losses in stopping and starting, and this in time led to fusion of the nodes.

The development of speedy transportation began to change fundamentally the social geography of cities. It was no longer necessary for the upper classes to congregate round the center. They could buy more space and seclusion for themselves on the now not-so-distant periphery. Robbins[8] has noted that "[the railroads] . . . enabled the wealthier and middle classes to live at some distance from the places of their daily work; this helped to create sharp differences of class between the different parts of a city and its suburbs, whic had not existed in the earlier age when 'gooc town houses mostly lay close beside house which were not so good."

While railroad transport in urban area provided only a very coarse network c radial routes, urban street transport wer some way towards providing a more finel articulated system. Like the railroad, it prc vided linear accessibility and only a limite number of points at which one could get o and off. Unlike the railroad, the number c generally radial routes was much greate though still restricted because of minimur use levels below which it was uneconomic to provide public transport. In addition, c course, the stops were much closer togethe As urban street transport developed from th stagecoach to the horse bus and the use c rails to the electrification of traction, so th distance that could be covered in a given tim period was increased and fares per mil reduced, making the street-car a means c "mass" transport. In some cases the streetc lines were constructed through pre-existin built-up areas, but frequently they wer extended out into the rural fringe as a specu lative measure to increase land values an

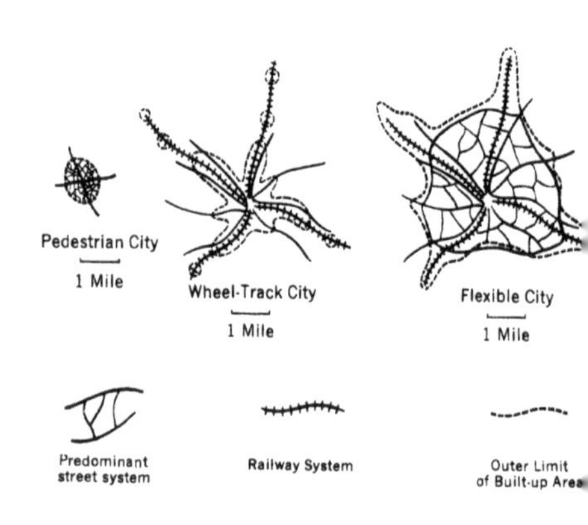

Figure 1. Urban shape and extent and predominant transportation networks.

[7] E. Carter, *The Future of London* (London, Penguin Books, 1962), p. 142.

[8] M. Robbins, *The Railway Age* (London, Penguin Books, 1965), pp. 48-49.

mulate development. The "street-car" burb of older American cities is a typical sult of this. Together, the railroad and eet-car made large areas available for ban development, producing, in both in- nces, a star like form, the "fingers" otruding out along the radial routes. Con- ual extension in this fashion led to the sion of previously separate urban areas re- lting in the conurbations, a term coined by trick Geddes in 1915.

e Flexible City

number of technological innovations dat- g initially from the end of the nineteenth ntury are having very widespread effects on e form of urban areas today. In particular, e widespread use of electricity, the develop- ent of the internal combustion engine, tele- mmunications, and steel frame buildings ust be given consideration.

Electricity as an energy form is unique in e variety of ways it can be applied and ales at which it can be used. In terms of its ect on urban form, the emphasis must be aced on flexibility. Economies of scale can obtained by large scale generation, but un- e the nineteenth century steam engine- ctory combination, the power can be used veral hundred miles from where it is gen- ated. Consequently industrial establish- ents can be located with much greater edom than at any time since the introduc- n of large scale production units. Use of ectrical power has also revolutionized fac- ry layout. Each machine can have its in- vidual power link, overhead belting and les are unnecessary and highly developed rms of internal transportation can be ilized, whether they be conveyor, overhead ane, or small trucks (e.g. forklift). All ese put a premium on horizontal, one floor layouts. The factory has thus changed from a multi-story mill to the one level structure, consequently increasing the demand for land. Locations on the urban periphery, where space is available and per acre land costs lower, become the most desirable.

We have noted previously the influence of electricity in transportation used in the spacing of suburban railroad stations. However, it is the internal combustion engine that has had the greatest effect of all on urban form. The car and the truck have introduced a combination of speed and flexibility previously unknown. Unlike fixed rail transportation, the car and the truck can move on almost any street in an urban area. Movement patterns are more complex, choice of route much wider. This has meant that the various functional parts of the city, whether industrial, residential, commercial, or any other, have to a considerable extent lost their dependence on close association with the railroad or the tramway. Public transport, in the form of the diesel bus, has greater route flexibility than the street-car, though bus routes are still overwhelmingly radial and deviate little from the previous street-car routes. To a large extent, the bus lacks the flexibility of the car, and the layout of bus routes makes it relatively easy to move from the periphery to the city center, but very difficult to move from one suburban area to another. The street-car in the past and the bus today reinforce the locational significance of the city center. The car and the truck are not affected by these route constraints. Lateral movement is relatively easy, though the previously developed predominantly radial road system does produce difficulties.

Car and truck transport opens up large areas for potential development. An all around increase in the effective radius for urban development from 3 to 12 miles increases the area available for development from 29 to 452 square miles. This in itself is an inbuilt mechanism to produce lower densities for urban areas than has been true in the past.

The dispersive effects of the internal combustion engine are also seen in the space demands generated by cars and trucks for movement and storage (parking). The high space demands of the motor vehicles have produced most of the urban planning problems associated with them. The vaunted flexibility of the motor car looks weak if you can't find anywhere to stop — "the pedestrian in the vehicle he cannot shed" as Fisher has put it.[9] The Table contrasts the space requirements of an individual with those of the car he may use.

However, the present day city is not entirely dominated by the motor vehicle. There still remain areas where the pedestrian scale is dominant, as in the central business district and in residential areas planned as "neighborhoods" in some European cities. In fact several recent studies have demonstrated that the horizontal dimensions of the commercial core of cities are internally conditioned by the distances people are willing to walk. These pedestrian dominated areas of the Flexible City have strong affinities with the Pedestrian City in terms of form.

While areal scales have been revolutionized by the use of internal combustion engined vehicles, there are intriguing parallels with the Pedestrian City of the past, in that the whole street pattern is available for movement and a restricted number of radial routes is no longer the predominant influence on urban form (Figure 1). Consequently many of the interstitial areas of the nineteenth century star-shaped city are being filled in while widespread development at low densities in the urban fringe is possible, unless, of course there are legislative restraints. The only partial parallel with the railroad era is the freeway, which functions rather like a railroad in producing ribbon-like facilities for

TABLE

Space demands: One person and one car

Man standing comfortably	1 sq. ft.
Average car space	150 sq. ft.
Office worker	275 sq. ft.
Parked car of office worker	1500 sq. ft.
Space per person/mile	
Walk	3 sq. ft.
By car, assuming 1.5 persons per car	20-70 sq. ft.

* Assumes 1.5 persons per car.

In terms of space needed for movement, the car also contrasts with other forms of mechanical transport. Ritter refers to an estimate that during the peak period, using a nine foot wide traffic lane, 1,200 people per hour could be carried by car, 11,000 per hour using an 80 seater bus, and 40,000 per hour using "rapid track transport."

[9] H. T. Fisher, "Radials and Circumferentials," in T. H. William, *ed., Urban Survival and Traffic* (London, E. and F. N. Spon, 1962), p. 52.

ı-speed movement and critical points of ı accessibility at the interchanges (sta-ıs), though the form of development as-iated with the interchanges will be very erent from the pedestrian dominated ds of the nineteenth century suburban rail-d station.

Telecommunications are also having im-tant effects on urban form and function. lio and television provide universal com-nication within an area, though, in terms public systems, restricted in possible tent and one-directional in flow. The phone influences urban form more signi-ntly in that it already acts as a substitute certain forms of face-to-face communica-ı. Thus information and instructions can v between various parts of an industrial anization even though the parts may be in ely dispersed locations. Central area head ces can keep in contact with factory or ehouse in the periphery. Like the car, the phone appears as a dispersive influence, ugh this may not be entirely true, because person makes a telephone call rather than actual trip, transportation space is saved l higher densities may result. This is also e where people stay at home and watch vision instead of going out to the movie ater. In fact, several aspects of modern hnology are ambiguous in their effects on an form. Overall, however, electricity, the rnal combustion engine, and telecom-nications have generally been conducive to an spread. Equally important, they have ded to introduce greater flexibility into the to which a given plot of land can be put. as been suggested that, in this sense, urban d is being "homogenized."

The technology of building construction, ile adapting to the demands for coverage large ground areas (factories, stadiums, .), has been most noted for the development of the tall building.[10] The key technological breakthrough was the development by William Jenney of the iron frame building (Home Insurance Building, Chicago, 1885). Use of the iron frame meant that the walls ceased to be load bearing structures and became purely weather proofing. Thus thickness of wall and height of building did not need to bear any relationship to each other. However, as efficient functional units, tall buildings were not feasible without some means of rapid vertical transportation. This was provided by the elevator developed by Elisha Otis in 1857 with refinements through to 1889. It was said that Jenny's "sky-scrapers" could be built just as high as Otis' "elevator" would go.

The development of the tall building enabled tremendous intensification of space use to take place. First in the United States, and now generally, central areas of large cities have become dominated by the tall commercial structure. This predominance is essentially a twentieth century constructional response to nineteenth century urban transportation/communication means in that, until recently, the city center was by far the most accessible point in any given urban complex. Under these conditions high land values justify the intensive use of space, or, as in this case, the actual creation of space.

The pressure on space in the centers of the largest cities has led to the development of subterranean transportation networks, which not only increase transportation capacity but also introduce route flexibility partially independent of the overlying street network. Recent proposals under the general head of "Traffic Architecture" further demonstrate

[10] James H. Johnson, "The Geography of the Sky-scraper," *Journal of Geography,* Vol. LV (November, 1956), pp. 379-87.

the intensification of space use in central areas, partly associated with the need to segregate the motor vehicle and the pedestrian.

We have characterized the city as it is presently developing as flexible. It is increasingly flexible in terms of locational choice for various functions. It can be flexible in terms of social patterns. Thus the associations of an individual can be widely dispersed in a large urban area, linked by car and telephone. We can have "community without propinquity"[11] at least for all but the poor, the very young, and the very old. Flexibility is also seen in the multiplicity of different urban forms being proposed for new towns and cities in various parts of the world — dispersed grids, rings, linear cities, high density, low density, and so on. Technology provides this flexibility, but the profusion of proposed forms may suggest absence of adequate research as much as richness of alternatives.

[11] M. M. Webber, "Order in Diversity: Community without Propinquity," in L. Wingo, *ed., Cities and Space* (Baltimore, Johns Hopkins Press, 1963), pp. 23-54.

Conclusion

This paper has examined some aspects of t interaction of technology and urban form it has occurred in the urban areas of the mc highly developed countries. In the unde developed countries, public investment transportation systems is low and priva ownership and use of the car and telepho are generally restricted to a very small pr portion of the total population. In additio rates of population growth of the urban are have been very high. Consequently it wou be erroneous to conclude that urban form the underdeveloped countries is similar that in the developed. The vast low incon shack areas on the edges of cities, such Rio and Calcutta, make this only too clea In addition, because of lack of resources, it unrealistic to assume that the cities of Lat America, Africa, and South and East As represent, at present, stages of urban grow preliminary to the type of growth present occurring in North America and Europe.

rban Growth Characteristics

8

an M. Voorhees

w techniques have been developed to ılyze growth characteristics of urban areas. ese techniques have been applied in ten ıerican cities — Hartford, Baltimore, ıshington, and a group of Iowa cities in- ding Des Moines, Sioux City, Council ıffs, Waterloo, Cedar Rapids, Dubuque, ı Davenport.

Although these communities may not be ly representative of all American cities, the dings show that former concepts of urban velopment are rapidly changing, along with cial, economic, and technological changes, rticularly as they relate to our rising stan- rd of living.

The exact pattern of growth that material- s in a community is the result of many indi- lual decisions — a manufacturer decides ere to locate his next plant, a merchant de- les which of his stores to expand, a govern- nt agency decides where to relocate, a nily decides where to rent or buy a house. e final choice in all such cases depends on the alternatives available at any given ne.

In making such decisions many factors are weighed. For example, among other factors, a family will consider type, size and kind of house, financial cost arrangements, neighbor- hood amenities, distance to stores and schools, accessibility to jobs, and adequacy of governmental service.

The studies undertaken in the ten cities measured the importance people place on these and other factors in making their loca- tional decisions. Though different individuals may weigh the same factors differently, on a group basis the most significant factors can be defined and measured. From the analysis, it was possible to develop mathematical for- mulas to indicate what weight people give to the various factors of location.

Use of formulas

In addition to serving as an analytical device to appraise past growth trends, formulas have other advantages. They can be used to esti- mate the potential impact of public policies or decisions. Thus a formula can be used to help predict a pattern of growth that will be brought about by different urban plans or highway programs, as well as the influence that zoning plans might exert on urban growth. Cities like Hartford, Washington, and Baltimore have been employing such formulas to evaluate alternative schemes of land development.

Usually four or five formulas are developed to reflect aspects of urban growth. For example, in the Hartford area separate formulas were used to analyze manufacturing service and retail employment, as well as population.

To establish past growth trends of manufacturing employment in the Hartford area, nine variables were used in the analysis. The variables considered were highway accessibility, availability of industrial land, tax rate, sewer and water service, rail service, proximity of industrial land to airports, size of existing industrial activities, and promotional aspects.

With highway accessibility, tax rate, and proximity to airports, a quantitative measurement was fairly easy. However, for such things as promotional aspects subjective evaluations were necessary to develop a rating index.

The influence that the several variables had on urban growth was appraised by multiple correlation. In effect, a formula was developed by multiple correlation analysis which would give a growth index for a particular zone. This growth index, when compared with the sum of the indices for all the zones, reflects the amount of the growth that can be expected in a zone anticipating a certain amount of overall growth.[1, 2, 3]

This test indicated that the most importa factors were *available land* and *sewer servic* Next in order of importance were accessibili to airport, highway access, and rail servic The other factors were of relatively minc significance in influencing community growt Because this information can be of conside able value to developers and planners, a sum mary is presented based on the finding derived from these ten cities.

Population Patterns

In reviewing population and residenti growth trends, the studies recognized that builders are to build for the low-cost home market, they must search for low-priced lanc Doubtless, low land costs are probably th most influential factors affecting urba growth. The analysis in these cities reveal that, other things being equal, an area havin low-priced land may develop up to three c four times as fast as it theoretically shoulc on the basis of accessibility and availabilit of land. Evansdale, Iowa, a small incor porated area outside Waterloo, grew at rapid rate during the last decade because c inexpensive land. The pattern change abruptly when FHA withheld mortgages i the town because of sewer problems.

However, as people rise in the income scal they tend to place more emphasis on the typ of area in which their home is located. The ask: Is it an attractive neighborhood? Ar the public services such as schools and polic

[1] Charles F. Barnes, "Integrating Land Use and Traffic Forecasting (Washington, D.C., Highway Research Board, 1961).

[2] Rex H. Wiant, "A Simplified Method for Fore casting Urban Traffic" (Washington, D.C Highway Research Board, 1961).

[3] *Baltimore-Washington Interregional Study* (Ba timore Regional Planning Council & Nationa Capital Regional Planning Council, Novembe 1960).

otection adequate? Does the locality have y claim to distinction as a "prestige area?" eas that meet these tests grow nearly twice fast as those which do not. Notable ex- ples of phenomenal growth of prestige eas are found in Montgomery County, aryland, northwest of Washington, D.C. d in the Towson area, north of Baltimore.

But, as already indicated, there are many ıer elements involved in the rate of growth a residential area. Lack of sewer or water rvice, for example, can hold growth rates to out one-third of normal, everything else ing equal. An example of this is the south- st section of the Washington Metropolitan ea. Otherwise attractive areas where land vners are deferring sale of property for tax vantages, or for capital gains, often have ly half the growth that might be antici- ted. A good example of this is Suffield, nnecticut, a town north of Hartford near e state line. This factor is much more pre- lent than one might expect, particularly in e large metropolitan area.

Nevertheless, one of the prime growth ctors is the availability of land. The extent d intensity of its use, of course, depend on the zoning policy for the area.

Although accessibility of an area is an portant consideration for a family, a family eighs other factors that have been cited. ost sections in a metropolitan area are rved by highways. The difference in acces- bility as between one place and another ay be only a few minutes. Hence it is ap- rent that transportation facilities and ser- ces do not now exert the dominant influence urban growth as formerly. *The automobile s freed the family transportation-wise so at it can afford to weigh other factors in e selection of a home.*

Employment Patterns

In considering the spatial distribution of employment, it was recognized that different types of work activities place varying emphasis on locational factors. Therefore, the various categories of employment were grouped by locational requirements in the studies cited here.

Manufacturing

As in residential selection, the employer in his choice of plant location also has been freed by the automobile. He now knows that he can locate a plant almost any place within an urban region and be assured of the necessary labor force. In short, today accessibility to the labor force is not the important factor to industrialists or other employers as in the past. More often the chief concern is in a specific site requirement such as water, sewer, freeway, port or rail service.

In the Baltimore area most of the largest manufacturing industries have gravitated to the corridor lying to the east and northeast of Baltimore City which has good rail and highway access as well as accessibility to water.

Only a few manufacturing activities are attracted to residential areas. Usually these are light industries looking for "prestige sites," as is particularly true in the area north of Baltimore where a number of smaller industrial plants have located. Overall, however, five-sixths of the industrial expansion that occurred in the Baltimore region was influenced largely by site requirements. Only one-sixth of the expansion was influenced primarily by the residence of population.

The studies also revealed a definite tendency for existing large employment centers to draw new industrial activities to their general vicinity. Similarly, greater growth was evidenced where there were promotional programs for industrial land development.

Certainly the availability of rail, port, and highway played a part in the location of some manufacturing plants. Areas with relatively low residential prestige often provided land to industrial developers at a price they could afford and, as a result, these areas had considerable growth. However, if the parcels of available land in such areas were too small, the growth was retarded.

An interesting sidelight in these studies is that industrialists are more interested in the capacity of the sewer system than in the cost of connecting to it. On the other hand, home builders are more interested in the cost of sewer service.

Retail

Retail employment trends in the study cities changed quite apparently with the changing pattern of population. As the families shifted to the suburbs, the neighborhood retail trade went along. However, retail expansion generally follows the first wave of residential development. Retailers like to have an "insured" market before they move out. The exact location of the store will depend largely upon the accessibilty factors.

The exact location of shopping centers and other commercial activities is largely dictated by street patterns. Most merchants want the point of highest accessibility. In various analyses made in these studies it was quite clear that commercial development will shift with changes in accessibility. For example, some of the older out-lying shopping areas were located at junctions of fairly narrow streets. When new and wider streets were built, new commercial areas developed along the new routes.

With the development of freeways in the area, the new commercial establishments located near the junctions of the freeways. Often these commercial areas were within a short distance of each other. All of these shifts were brought about by changes in the heightened level of transportation service which in effect influenced the relative accessibility of the specific areas.

Shopping center location is also influenced by the accessibility factor. The location and size of existing centers have marked impact upon the location of new centers. Generally a new center is apt to locate within three or four miles of an existing center which does not have an adequate range of merchandise. But the pace of residential development will determine whether the old center is expanded or a new one is developed.

Other Employment

Service activities, such as dry cleaning outlets, hairdressers, banks, etc., have followed a pattern similar to that of retail establishments. They have moved to the suburbs to be closer to the people and industries they serve. Certain types of activity, however, still find the downtown area very attractive. These are organizations that depend upon accessibility to a large labor pool or specialize in certain lines. For example, in Baltimore the Commercial Industrial Trust located in the downtown area mainly because it needed a large pool of office workers. On the other hand, some of the more or less standardized office operations which do not require large labor forces will locate in the suburbs.

Governmental

In studying the location choice pattern for many government activities in the Washing-area, it was clear that a strictly personal ision by the head of the agency was often olved. However, different factors of in- nce were noted. First, accessibility to the or force is important. Thirty per cent of governmental expansion between 1948 1957 occurred in the downtown area.

Certain governmental activities concen- e in outlying suburban zones where other ernmental activities have already been ated. On the basis of the analysis made in study, nine zones in the suburban area eived more than their share of growth.

n addition to increase in the number of leral employees, an increase in employ- nt by local governments took place. Most this growth followed changes in popula- , since a large portion of it was aimed at ving the new people.

n general, the study has shown that many he changes in employment patterns have owed population adjustments. Almost all il and service employment has shifted h population. Many governmental and ce activities have moved out to be closer he people they serve. In effect, about two- ds of the employment follows the migrat- population.

nsportation No Longer Key Factor

most important conclusion to be drawn m these studies is that transportation is not key factor in shaping our cities today. h the universal use of the automobile and development of metropolitan area street tems, the urban dweller has been given al- st unlimited latitude in where to live or ate his business. Because of this, he gives siderable attention to factors other than nsportation in making a final decision.

At first glance, this conclusion seems to be odds with other recent research. For ex- ample, the Bone and Wohl study[4] of the industrial expansion along Route 128 outside Boston has indicated that transportation was a very important factor in this growth. These results were based largely upon attitudes expressed by industrialists who have located along this freeway.

However, a further look at the replies by these industrialists reveals that they also were giving weight to other factors such as the attractiveness of the area and its advertising value. Though many of them did say that "commercial access" was very important to them, the fact remains that they could have obtained practically the same commercial access if they located one or two miles away from the expressway. A few minutes of travel would not have made that much difference.

Certainly the fact that the firm of Cabot, Cabot & Forbes was promoting the area along Route 128 increased its growth potential. In fact, the area along the route was not developing until Cabot, Cabot & Forbes were able to convince the first industrialist to locate along the route. The first venture began the wave of development found there today. The fact that industrialists follow "fads" just as people do was observed in most of the cities that were studied.

This would also hold with regard to the accessibility to the labor force. Perhaps, if overall employment patterns in the Boston area were analyzed, the findings would be similar to those found in the ten cities, mentioned earlier.

Also often cited, when discussing the influence of freeways on urban development, is the great expansion of residential development that occurred near the Gulf Freeway in

[4] A. J. Bone and Martin Wohl, *Massachusetts Route 128 Impact Study*, Highway Research Board Bulletin 227.

the southeast section of Houston.[5] It is true that when this facility was first opened there was a tremendous growth in its vicinity for a period of about five years. (See Figure 1.) However, recent growth trends in the Houston areas have swung to the southwest portion of the city, which is generally considered the prestige area. Furthermore, a new freeway has been built which serves a rather rundown section of Houston. This has had no apparent influence on the growth patterns.

Such illustrations show that the dynamic factors in urban growth must be considered together. When a change in transportation facilities is introduced it may modify the influence of the other factors. Improved accessibility, particularly at the beginning of a new development, if combined with other favorable factors may stimulate a great deal of growth. But, as in Houston, the other factors, like blight, may become dominant, and changes in accessibility are not enough to really stimulate growth.

HOUSTON POPULATION INCREASES 1940-1957

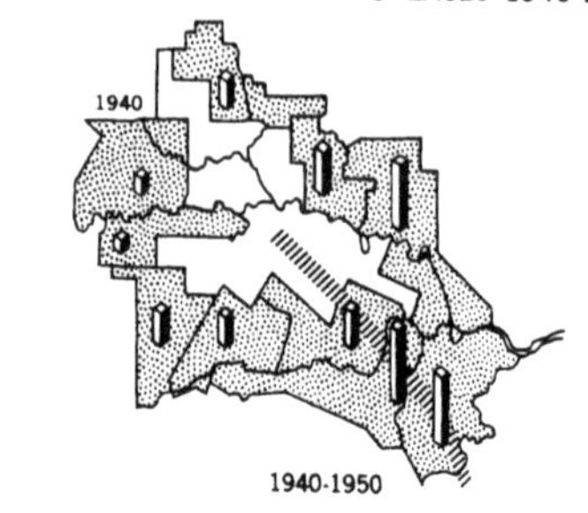

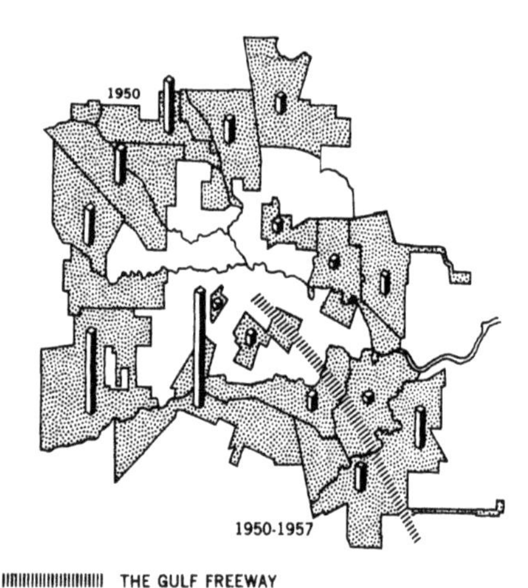

Figure 1.

Decisions, Alternatives, and Time

All of this tends to emphasize the basic point brought out at the beginning of this paper — namely, that urban growth depends not only upon many individual decisions, but on the alternatives offered at the time the decisions are made. If the expansion of large electronic activities in the Boston area had occurred before Route 128 was developed, these industries would have located in some other place. If Route 128 were being built today, it might not attract anywhere near the number of firms it has attracted there because the industries might not see the site advantages in the same light today that they did few years ago. Urban growth is sensitive t many factors modified by time.

In conclusion, this point should be strongl emphasized: it is quite obvious from th analysis of growth patterns in the ten Amer can cities that our standard of living is be coming a major factor in our developmer patterns. The individual, whether hom owner or industrialist, is becoming more an more concerned with other factors beside transportation. This change must be recog nized in urban plans.

[5] *Economic Evaluation of the Gulf Freeway* (City of Houston Department of Traffic and Transportation, July, 1949).

he Territorial Expansion of merican Cities and their opulation Growth

9

ı A. Kolosova

y analysis of United States census data the population growth of cities must not e sight of the fact that one of the reasons growth, and sometimes the principal rea-ı, may be the expansion of the city area l changes in its limits. These changes are ntified as annexations in the census re-·ts. *The Municipal Yearbook* (1963) re-·ts (p. 53) that 152 cities expanded their a in 1945, 382 in 1950, 526 in 1955, and 2 in 1960. (*The Municipal Yearbook* is source for data in expansions of city areas ntercensal periods; each annual issue con-ıs information on cities that expanded ir area in the given year.) The data sug-t that cities are expanding their area with reasing frequency, and annexations are as-ning increasing importance as a factor in pulation growth. Because of this trend, the 60 census data (series "Number of abitants," state volumes) included a parate table (No. 9) showing the popula-n of cities within the annexed territories. e need for a proper assessment of factors ulting in population growth of the cities of United States during the 1950-1960 riod, a time when annexations assumed in-easing importance, prompted us to analyze these data. The principal results of that analy-sis are the subject of the present paper.

Before discussing the effect of annexation on city growth in the United States as a whole and in various regions, a few words should be said about the reasons for, and the essence of, this process. Judging from the paucity of information on the subject in the literature, Americans do not seem to have dealt with this question to any great extent.

Harland Bartholomew, author of *Land Uses in American Cities*, notes that the needs of a growing city for additional territory can be met in a variety of ways: (1) by peri-pheral expansion, (2) by redistribution of existing land uses, (3) by building up of empty lands, and (4) by more intensive use of existing lands and buildings (Harland Bartholomew, *Ispol'zovaniye territorii v amerikanskikh gorodakh*. Moscow: Gosstro-yizdat, 1959 [Russian edition of: *Land Uses in American Cities*. Cambridge: Harvard University Press, 1955]). On the basis of his observations, he reports that the fourth ap-proach is relatively little used in American cities because it is slower and more com-plicated. "More often community growth flows into areas offering the least physical

printed by permission from *Soviet Geography: Review and Translation.* Translated by Theodore Shabad.

or economic resistance to expansion. Thus the predominant type of growth occurs in the form of lateral expansion into surrounding agricultural areas where raw land is converted to urban purposes" (p. 33 in Russian edition; p. 13 in American edition). The first objective of city expansion is usually territory adjoining the city's periphery that is already being used for urban purposes but is still unincorporated.

Annexations involve a complex knot of contradictions. The actual annexation procedure is regulated in different ways in various states. In some states it is simpler, in others more complicated. In some states, Rhode Island for example, the law altogether prohibits changes in municipal boundaries; in eight states, annexations must be approved by state legislatures, and in others a local municipal ordinance is sufficient. Incorporation of a territory into a city may be initiated by residents of that particular territory in the form of a petition, on which the vote is restricted to property owners who reside permanently in the territory. (In 1962, 216 cities annexed new territory of at least one quarter square mile; only in 50 cases did the initiative stem from residents of the annexed territory.) Annexations may also be inspired by businessmen and politicians of the central city, interested in new land for industrial and residential construction. They also seek to control the course of development of territory adjoining the city, thus enhancing its importance. Another factor is the desire to gain additional taxpayers and thus reduce the deficit in the municipal budget so characteristic of many American cities.

The procedure is much simpler if the annexation does not require the approval of the residents of the area. There have been cases where annexation was inevitable because the central city had virtually surrounded a particular area, making i separate existence impractical. One such cas is described in *Western City*, February, 196 relating to North Sacramento, which was ir corporated into Sacramento in 1964. As lat as 1963, North Sacramento was still resistin annexation, but it gave in after it found itse surrounded by Sacramento territory. Simila cases are reported by Raymond Murphy i *The American City*, (New York, 1966 They are Hamtramck and Highland Park i the Detroit SMSA, [standard metropolita statistical area], University Park and High land Park in Dallas, and San Fernando in Lo Angeles.

The annexation procedure may becom more complex and may be frustrated by op position on the part of the authorities of th county in which the particular territory i situated, or of residents of the county. Ac cording to the American literature, th system of representation in local legislature does not reflect in some cases the extensiv shifts in the distribution of population tha have taken place as a result of urbanizatior As a result rural areas sometimes carr weight out of proportion to their populatior And it is their representatives who ma frustrate city expansion plans. Opposition t annexation often is motivated by fear that th city may impose new local taxes and may gai control over land use. (*The Municipal Year book*, 1963, p. 52). The separate incorpora tion of territory is used as a countermeasur against annexation. In cases where unincor porated territory lies near two cities, com petition between the two may also dela annexation.

Although annexation involves friction an aggravates relationships, the data cited earlie suggest that an increasing number of citie are making use of it. The 1960 Census note that cities frequently expanded their territor ough annexation in the 1950s, with some es achieving substantial territorial expan-n. In many cases this significantly affected th absolute and relative population growth.

In our analysis of the 1960 Census, we ited ourselves to cities with a population at least 10,000 in 1960. There were 1899 h cities, of which 1080 (56.8%) had ex-nded their territory since 1950.

The distribution of cities that annexed ritory in the 1950-1960 period is shown Table I for various population classes and ions.

TABLE I

Distribution of cities that annexed territory in 1950-1960 by population classes and by regions

ulation Class (n thousands)	Number of Cities that Expanded Their Territory			
	Total	*North*	*South*	*West*
10-25	627	294	216	117
25-100	363	156	120	87
00-500	80	27	34	19
500-1000	8	3	3	2
ver 1000	2	1	—	1
Total	1080	481	373	226
centage share of cities of 000 or more in the total nber of cities	56.8%	44.5%	75.9%	69.1%

The Census does not contain data on the e of the territory annexed by any particular y. *The Municipal Yearbook, 1963,* reports t the total area of annexations during 51-1960 was 11,941.2 square kilometers. 1962, when 754 cities had annexations, average territorial gain of large cities, ording to the same source, was 1.5 square es, or 3.9 square kilometers, and in the e of small cities 0.5 mi², or 1.3 km², but cities acquired more than 10 mi², or 25.9 ², each.)

The total population of newly annexed ritories was 8.5 million in 1960. The popu-on of these cities increased by 11.8 million in the 1950-1960 period. Consequently, 71.8% of the total growth was the result of annexations. This factor must especially be borne in mind in cases where territorial expansion was the principal source of population growth, for example, the central cities of the 212 standard metropolitan statistical areas. Their combined population increased by 10.7% during the 1950-1960 period, with 9.3% of the increase stemming from annexations. Among the 58 million residents of these cities, 4.9 million lived in newly annexed territory. Annexation played a relatively minor role (only 0.4 million people) in the growth of population of the central cities of the largest SMSA (with more than 3 million). The central cities of the next group of SMSA (1 to 3 million) lost 2.2% of their population in their old territory, but annexations gave them an increase of 7.8%, and in the case of the smaller SMSA (under 1 million), more than two-thirds of the population growth stemmed from the expansion of city limits (U.S. Census of Population, 1960, *Number of Inhabitants, U.S. Summary.* Washington, 1961, p. XXVII).

In 1958 cities of 10,000 or more, the population of the newly annexed territory accounted for 25 to 50% of the city's total 1960 population (78 of these cities were in the South, 47 in the West and 33 in the North.) In 45 cities (21 in the South, 19 in the West and 5 in North), the annexed territory acounted for 50 to 75% of the population, and in six cities (three in the South, two in the West and one in the North) for more than 75%. For a list of cities with the most significant annexations, see U.S. Census of Population, 1960. *Number of Inhabitants, U.S. Summary.* Washington, 1961, pp. XXIV.)

Territorial expansion has been especially intensive in cities of the South and West. There are several reasons for this. The South, which used to lag behind other regions in the level of urbanization, has been accelerating its rate both because of major social-economic shifts in agriculture and the development of industry. Around the large cities of the South and its industrial centers, an urban fringe has arisen and, as time goes on, it becomes the first candidate for inclusion in the city limits. In Texas and some of the other southwestern states, annexation is also furthered by its relative juridical simplicity. There are quite a number of examples of substantial expansion in this region. Thus, 26.8% of the 1960 population of Houston lived in annexed territory, 28% in Dallas, 44.9% in El Paso and 63.7% in Odessa. Oklahoma City repeatedly expanded its city limits, and after its greatest expansion (354.2 km² in 1962) it extended into four counties and now has the largest area of any United States city — 1605.8 km² with 355,000 people (*The Municipal Yearbook,* 1963, p. 55. Oklahoma City and its environs contain 1800 oil wells; oil extraction and refining is the leading economic activity). Previously, Los Angeles had the largest city area, 1177.9 km² with a population of 2,479,000.

Territorial expansion of cities in the We is being furthered by the rapid growth c population in this region (mainly as a resu of migration) and by intensive industri development. Some cities registered virtuall no population gain within their old limit but annexed territory with a population fa exceeding the former population. The ou standing example is Phoenix, Ariz. whic had a population of 107,000 in 1950 an 439,000 in 1960, the difference (332,000 being equivalent to the population of th newly annexed territory. A similar situation found in Tucson, Ariz., and Vallejo, Sa Jose, and San Leandro, Calif.

In the North, such great population growt as a result of annexations is much rarer an occurs mainly in the case of small citie Among the larger cities in which populatio growth through annexation was significar are Milwaukee (17% of the total 196 population), Wichita (33%), and Colum bus, Ohio (16%). The presence of a dens network of incorporated places around large city tends to block annexation attempt as in the case of Cleveland and, to som extent, Chicago.

Despite annexations, some cities seem t be unable to maintain their former popula tion level. Of the 481 cities of the North (wit a population of 10,000 or more) which an nexed territory in the 1950-60 period, 6 had a smaller population in 1960 within th enlarged territory than they had in 195 within the smaller territory. Of these citie 37 were small, 22 medium, and 8 large (in cluding Chicago and Cincinnati). There ar fewer such cities in the South (16) and the are not characteristic of the West (4), wher the tendency toward suburbanization (th outflow of population to the suburbs) is le pronounced.

he Nature of Cities

10

auncy D. Harris

ward L. Ullman

es are the focal points in the occupation utilization of the earth by man. Both a duct of and an influence on surrounding ions, they develop in definite patterns in oonse to economic and social needs.

Cities are also paradoxes. Their rapid wth and large size testify to their superty as a technique for the exploitation of earth, yet by their very success and conuent large size they often provide a poor al environment for man. The problem is to ld the future city in such a manner that advantages of urban concentration can be served for the benefit of man and the disantages minimized.

Each city is unique in detail but resembles ers in function and pattern. What is rned about one helps in studying another. cation types and internal structure are eated so often that broad and suggestive eralizations are valid, especially if limited cities of similar size, function, and regional ting. This paper will be limited to a discusn of two basic aspects of the nature of es — their support and their internal structure. Such important topics as the rise and extent of urbanism, urban sites, culture of cities, social and economic characteristics of the urban population, and critical problems will receive only passing mention.

THE SUPPORT OF CITIES

As one approaches a city and notices its tall buildings rising above the surrounding land and as one continues into the city and observes the crowds of people hurrying to and fro past stores, theaters, banks, and other establishments, one naturally is struck by the contrast with the rural countryside. What supports this phenomenon? What do the people of the city do for a living?

The support of a city depends on the services it performs not for itself but for a tributary area. Many activities serve merely the population of the city itself. Barbers, dry cleaners, shoe repairers, grocerymen, bakers, and movie operators serve others who are engaged in the principal activity of the city, which may be mining, manufacturing, trade, or some other activity.

printed from *Annals of the America Academy of Political and Social Science,* Vol. 242 (1945), pp. 7 by permission of the author and the Academy.

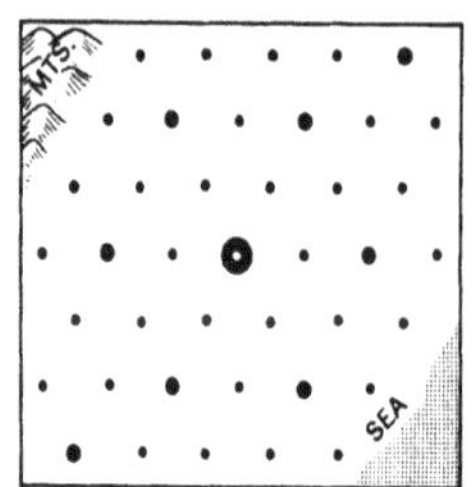

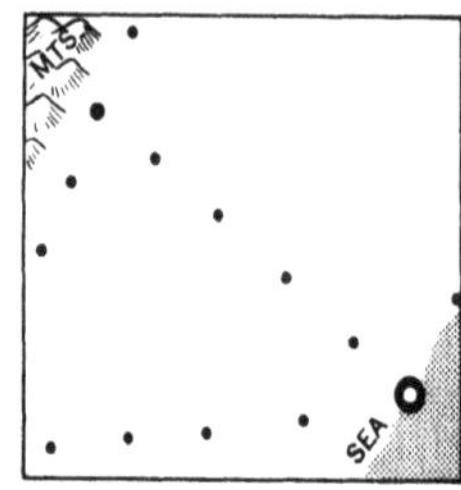

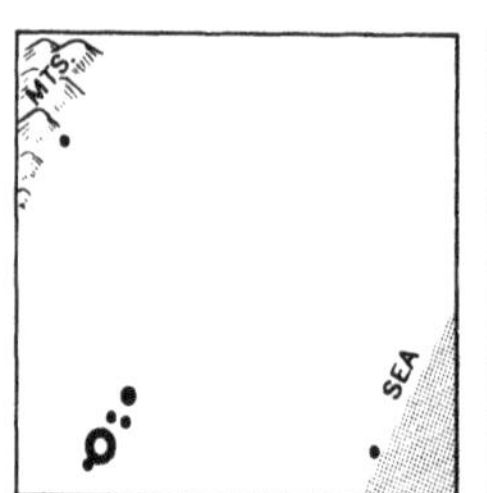

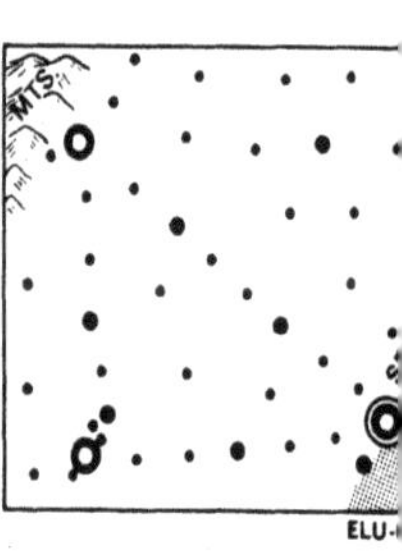

Figure 1. Theoretical distribution of central places. In a homogeneous land, settlements are evenly spaced; largest city in center surrounded by 6 medium-size centers which in turn are surrounded by 6 small centers. Tributary areas are hexagons, the closest geometrical shapes to circles which completely fill area with no unserved spaces.

Figure 2. Transport centers, aligned along railroad or at coast. Large center is port; next largest is railroad junction and engine-changing point where mountain and plain meet. Small centers perform break of bulk principally between rail and roads.

Figure 3. Specialized-function settlements. Large city is manufacturing and mining center surrounded by a cluster of smaller settlements located on a mineral deposit. Small centers on oce and at edge of mountains are resorts.

Figure 4. Theoretical composite grouping. Port becomes the metropolis and, although off center, serves as central place for whole area. Manufacturing-mining and junction centers are ne largest. Railroad alignment of many towns evident. Railroad route in upper left of Figure 2 has been diverted to pass through manufacturin and mining cluster. Distribution of settlements in upper right follows central-place arrangement.

The service by which the city earns its livelihood depends on the nature of the economy and of the hinterland. Cities are small or rare in areas either of primitive, self-sufficient economy or of meager resources. As Adam Smith stated, the land must produce a surplus in order to support cities. This does not mean that all cities must be surrounded by productive land, since strategic location with reference to cheap ocean highways may enable a city to support itself on the specialized surplus of distant lands. Nor does it mean that cities are parasites living off the land. Modern mechanization, transport, and a complex interdependent economy enable much of the economic activity of mankind to be centered in cities. Many of the people engaged even in food production are actually in cities in the manufacture of agricultural machinery.

The support of cities as suppliers of urban services for the earth can be summarized in three categories, each of which presents factor of urban causation:[1]

1. Cities as central places performi comprehensive services for a surroundi area. Such cities tend to be evenly spac throughout productive territory (Figure 1 For the moment this may be considered t "norm" subject to variation primarily in r sponse to the ensuing factors.

2. Transport cities performing break-o bulk and allied services along transpo routes, supported by areas which may b remote in distance but close in connectic because of the city's strategic location c transport channels. Such cities tend to be a ranged in linear patterns along rail lines or coasts (Figure 2).

3. Specialized-function cities performi

[1] For references see Edward Ullman, "A Thec of Location for Cities," *Am. Jour. of Sociolo* Vol. 46, no. 6 (1941), pp. 853-64.

e service such as mining, manufacturing, recreation for large areas, including the neral tributary areas of hosts of other ies. Since the principal localizing factor is en a particular resource such as coal, water wer, or a beach, such cities may occur gly or in clusters (Figure 3).

Most cities represent a combination of the ee factors, the relative importance of each rying from city to city (Figure 4).

ties as central places

ties as central places serve as trade and cial centers for a tributary area. If the land se is homogeneous these centers are uni-mly spaced, as in many parts of the agri-tural Middle West (Figure 1). In areas of even resource distribution, the distribution cities is uneven. The centers are of vary-; sizes, ranging from small hamlets closely aced with one or two stores serving a local butary area, through larger villages, towns, d cities more widely spaced with more ecial services for larger tributary areas, up the great metropolis such as New York Chicago offering many specialized services a large tributary area composed of a ole hierarchy of tributary areas of smaller aces. Such a net of tributary areas and nters forms a pattern somewhat like a fish t spread over a beach, the network regular d symmetrical where the sand is smooth, t warped and distorted where the net is ught in rocks.

The central-place type of city or town is despread throughout the world, partic-arly in nonindustrial regions. In the United ates it is best represented by the numerous ail and wholesale trade centers of the ricultural Middle West, Southwest, and est. Such cities have imposing shopping nters or wholesale districts in proportion to their size; the stores are supported by the trade of the surrounding area. This contrasts with many cities of the industrial East, where the centers are so close together that each has little trade support beyond its own population.

Not only trade but social and religious functions may support central places. In some instances these other functions may be the main support of the town. In parts of Latin America, for example, where there is little trade, settlements are scattered at relatively uniform intervals through the land as social and religious centers. In contrast to most cities, their busiest day is Sunday, when the surrounding populace attend church and engage in holiday recreation, thus giving rise to the name "Sunday town."

Most large central cities and towns are also political centers. The county seat is an example. London and Paris are the political as well as trade centers of their countries. In the United States, however, Washington and many state capitals are specialized political centers. In many of these cases the political capital was initially chosen as a centrally located point in the political area and was deliberately separated from the major urban center.

Cities as transport foci and break-of-bulk points

All cities are dependent on transportation in order to utilize the surplus of the land for their support. This dependence on transportation destroys the symmetry of the central-place arrangement, inasmuch as cities develop at foci or breaks of transportation, and transport routes are distributed unevenly over the land because of relief or other limitations (Figure 2). City organizations recognize the importance of efficient transportation, as witness their constant concern with freight-rate regulation and with the construction of new highways, port facilities, airfields, and the like.

Mere focusing of transport routes does not produce a city, but according to Cooley, if break of bulk occurs, the focus becomes a good place to process goods. Where the form of transport changes, as transferring from water to rail, break of bulk is inevitable. Ports originating merely to transship cargo tend to develop auxiliary services such as repackaging, storing, and sorting. An example of simple break-of-bulk and storage ports is Port Arthur-Fort William, the twin port and wheat-storage cities at the head of Lake Superior; surrounded by unproductive land, they have arisen at the break-of-bulk points on the cheapest route from the wheat-producing Prairie Provinces to the markets of the East. Some ports develop as entrepôts, such as Hong Kong and Copenhagen, supported by transshipment of goods from small to large boats or vice versa. Servicing points or minor changes in transport tend to encourage growth of cities as establishment of division points for changing locomotives on American railroads.

Transport centers can be centrally located places or can serve as gateways between contrasting regions with contrasting needs. Kansas City, Omaha, and Minneapolis-St. Paul serve as gateways to the West as well as central places for productive agricultural regions, and are important wholesale centers. The ports of New Orleans, Mobile, Savannah, Charleston, Norfolk, and others served as traditional gateways to the Cotton Belt with its specialized production. Likewise, northern border metropolises such as Baltimore, Washington, Cincinnati, and Louisville served as gateways to the South, with St. Louis a gateway to the Southwest. In recent years the South has been developing its ow central places, supplanting some of the mono poly once held by the border gateway Atlanta, Memphis, and Dallas are example of the new southern central places and trans port foci.

Changes in transportation are reflected i the pattern of city distribution. Thus the de velopment of railroads resulted in a railroa alignment of cities which still persists. Th rapid growth of automobiles and widesprea development of highways in recent decade however, has changed the trend toward more even distribution of towns. Studies i such diverse localities as New York an Louisiana have shown a shift of centers awa from exclusive alignment along rail routes Airways may reinforce this trend or stimulat still different patterns of distribution for th future city.

Cities as concentration points for specialized services

A specialized city or cluster of cities perform ing a specialized function for a large are may develop at a highly localized resourc (Figure 3). The resort city of Miami, for ex ample, developed in response to a favorabl climate and beach. Scranton, Wilkes-Barre and dozens of nearby towns are specialize coal-mining centers developed on anthracit coal deposits to serve a large segment of th northeastern United States. Pittsburgh and its suburbs and satellites form a nationall significant iron-and-steel manufacturing clus ter favored by good location for the assembl of coal and iron ore and for the sale of stee to industries on the coal fields.

Equally important with physical resource in many cities are the advantages of mass pro duction and ancillary services. Once started a specialized city acts as a nucleus for simi-

or related activities, and functions tend to ramid, whether the city is a seaside resort ch as Miami or Atlantic City, or, more im- rtant, a manufacturing center such as tsburgh or Detroit. Concentration of indus- in a city means that there will be a con- ntration of satellite services and industries supply houses, machine shops, expert nsultants, other industries using local in- strial by-products or waste, still other in- stries making specialized parts for other nts in the city, marketing channels, spe- lized transport facilities, skilled labor, and host of other facilities; either directly or lirectly, these benefit industry and cause o expand in size and numbers in a concen- ted place or district. Local personnel with know-how in a given industry also may cide to start a new plant producing similar like products in the same city. Further- re, the advantages of mass production elf often tend to concentrate production a few large factories and cities. Examples localization of specific manufacturing in- stries are clothing in New York City, fur- ure in Grand Rapids, automobiles in the troit area, pottery in Stoke-on-Trent in gland, and even such a specialty as tennis kets in Pawtucket, Rhode Island.

Such concentration continues until oppos- forces of high labor costs and congestion lance the concentrating forces. Labor costs y be lower in small towns and in indus- ally new districts; thus some factories are ving from the great metropolises to small vns; much of the cotton textile industry s moved from the old industrial areas of w England to the newer areas of the rolinas in the South. The tremendous ncentration of population and structures in ge cities exacts a high cost in the form of ngestion, high land costs, high taxes, and strictive legislation.

Not all industries tend to concentrate in specialized industrial cities; many types of manufacturing partake more of central-place characteristics. These types are those that are tied to the market because the manufacturing process results in an increase in bulk or perishability. Bakeries, ice cream establishments, ice houses, breweries, soft-drink plants, and various types of assembly plants are examples. Even such industries, however, tend to be more developed in the manufacturing belt because the density of population and hence the market is greater there.

The greatest concentration of industrial cities in America is in the manufacturing belt of northeastern United States and contiguous Canada, north of the Ohio and east of the Mississippi. Some factors in this concentration are: large reserves of fuel and power (particularly coal), raw materials such as iron ore via the Great Lakes, cheap ocean transportation on the eastern seaboard, productive agriculture (particularly in the west), early settlement, later immigration concentrated in its cities, and an early start with consequent development of skilled labor, industrial know-how, transportation facilities, and prestige.

The interdependent nature of most of the industries acts as a powerful force to maintain this area as the primary home of industrial cities in the United States. Before the war, the typical industrial city outside the main manufacturing belt had only a single industry of the raw-material type, such as lumber mills, food canneries, or smelters (Longview, Washington; San Jose, California; Anaconda, Montana). Because of the need for producing huge quantities of ships and airplanes for a two-ocean war, however, many cities along the Gulf and Pacific coasts have grown rapidly during recent years as centers of industry.

Applications of the three types of urban support

Although examples can be cited illustrating each of the three types of urban support, most American cities partake in varying proportions of all three types. New York City, for example, as the greatest American port is a break-of-bulk point; as the principal center of wholesaling and retailing it is a central-place type; and as the major American center of manufacturing it is a specialized type. The actual distribution and functional classification of cities in the United States, more complex than the simple sum of the three types (Figure 4), has been mapped and described elsewhere in different terms.[2]

The three basic types therefore should not be considered as a rigid framework excluding all accidental establishment, although even fortuitous development of a city becomes part of the general urban-supporting environment. Nor should the urban setting be regarded as static; cities are constantly changing, and exhibit characteristic lag in adjusting to new conditions.

Ample opportunity exists for use of initiative in strengthening the supporting base of the future city, particularly if account is taken of the basic factors of urban support. Thus a city should examine: (1) its surrounding area to take advantage of changes such as newly discovered resources or crops, (2) its transport in order to adjust properly to new or changed facilities, and (3) its industries in order to benefit from technological advances.

INTERNAL STRUCTURE OF CITIES

Any effective plans for the improvement or rearrangement of the future city must tak account of the present pattern of land us within the city, of the factors which have pro duced this pattern, and of the facilities re quired by activities localized within particu lar districts.

Although the internal pattern of each cit is unique in its particular combination of de tails, most American cities have business, in dustrial, and residential districts, The force underlying the pattern of land use can b appreciated if attention is focused on thre generalizations of arrangement — by con centric zones, sectors, and multiple nuclei.

Concentric zones

According to the concentric-zone theory, th pattern of growth of the city can best be un derstood in terms of five concentric zones (Figure 5).

1. *The central business district.* — This i the focus of commercial, social, and civic lif and of transportation. In it is the downtow retail district with its department store smart shops, office buildings, clubs, bank hotels, theaters, museums, and organizatio headquarters. Encircling the downtown reta district is the wholesale business district.

2. *The zone in transition.* — Encirclin the downtown area is a zone of residenti deterioration. Business and light manufac turing encroach on residential areas char acterized particularly by rooming houses. I this zone are the principal slums, with the submerged regions of poverty, degradatior

[2] Chauncy D. Harris, "A Functional Classification of Cities in the United States," *The Geographical Review*, Vol. 33, no. 1, (January, 1943), pp. 85-99.

[3] Ernest W. Burgess, "The Growth of the City in *The City*, Robert E. Park, Ernest W. Burge and Robert D. Mackenzie, *eds.* (Chicago, Un versity of Chicago Press, 1925), pp. 47-6 Ernest W. Burgess, "Urban Areas," in *Chicag an Experiment in Social Science Research*, T. Smith and Leonard D. White, *eds.* (Chicag University of Chicago Press, 1929), pp. 113-3

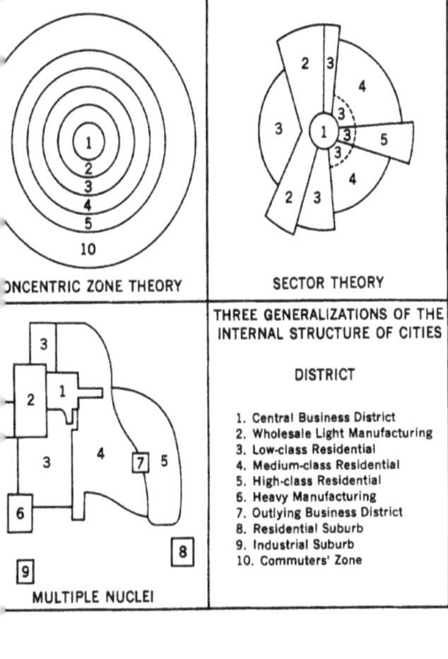

ıre 5. Generalizations of internal structure :ities The concentric-zone theory is a eralization for all cities. The arrangement of the ors in the sector theory varies from city to city. diagram for multiple nuclei represents one sible pattern among innumerable varions.

l disease, and their underworlds of vice. nany American cities it has been inhabited gely by colonies of recent immigrants.

3. *The zone of independent workingmen's nes.* — This is inhabited by industrial rkers who have escaped from the zone in nsition but who desire to live within easy ess of their work. In many American es second-generation immigrants are im-tant segments of the population in this a.

4. *The zone of better residences.* — This nade up of single-family dwellings, of ex-sive "restricted districts," and of high-ss apartment buildings.

5. *The commuters' zone.* — Often beyond city limits in suburban areas or in satellite es, this is a zone of spotty development of high-class residences along lines of rapid travel.

Sectors

The theory of axial development, according to which growth takes place along main transportation routes or along lines of least resistance to form a star-shaped city, is refined by Homer Hoyt in his sector theory, which states that growth along a particular axis of transportation usually consists of similar types of land use[4] (Figure 5). The entire city is considered as a circle and the various areas as sectors radiating out from the center of that circle; similar types of land use originate near the center of the circle and migrate outward toward the periphery. Thus a high-rent residential area in the eastern quadrant of the city would tend to migrate outward, keeping always in the eastern quadrant. A low-quality housing area, if located in the southern quadrant, would tend to extend outward to the very margin of the city in that sector. The migration of high-class residential areas outward along established lines of travel is particularly pronounced on high ground, toward open country, to homes of community leaders, along lines of fastest transportation, and to existing nuclei of buildings or trading centers.

Multiple nuclei

In many cities the land-use pattern is built not around a single center but around several

[4] Homer Hoyt, "City Growth and Mortgage Risk," *Insured Mortgage Portfolio*, Vol. I, nos. 6-10 (December 1936-April 1937), *passim*; U.S. Federal Housing Administration, *The Structure and Growth of Residential Neighborhoods in American Cities*, Homer Hoyt (Washington, Government Printing Office, 1939), *passim*.

discrete nuclei (Figure 5). In some cities these nuclei have existed from the very origins of the city; in others they have developed as the growth of the city stimulated migration and specialization. An example of the first type is Metropolitan London, in which "The City" and Westminster originated as separate points separated by open country, one as the center of finance and commerce, the other as the center of political life. An example of the second type is Chicago, in which heavy industry, at first localized along the Chicago River in the heart of the city, migrated to the Calumet District, where it acted as a nucleus for extensive new urban development.

The initial nucleus of the city may be the retail district in a central-place city, the port or rail facilities in a break-of-bulk city, or the factory, mine, or beach in a specialized-function city.

The rise of separate nuclei and differentiated districts reflects a combination of the following four factors:

1. Certain activities require specialized facilities. The retail district, for example, is attached to the point of greatest intracity accessibility, the port district to suitable water front, manufacturing districts to large blocks of land and water or rail connection, and so on.

2. Certain like activities group together because they profit from cohesion.[5] The clustering of industrial cities has already been noted above under "Cities as concentration points for specialized services." Retail districts benefit from grouping which increases the concentration of potential customers and makes possible comparison shopping. Financial and office-building districts depend upon facility of communication among offic within the district. The Merchandise Mart Chicago is an example of wholesale cluste ing.

3. Certain unlike activities are detriment to each other. The antagonism between fa tory development and high-class residenti development is well known. The heavy co centrations of pedestrians, automobiles, ar streetcars in the retail district are antagonist both to the railroad facilities and the stre loading required in the wholesale district a to the rail facilities and space needed by lar industrial districts, and vice versa.

4. Certain activities are unable to affo the high rents of the most desirable sites. Th factor works in conjunction with the for going. Examples are bulk wholesaling ar storage activities requiring much room, low-class housing unable to afford the luxu of high land with a view.

The number of nuclei which result fro historical development and the operation localization forces varies greatly from city city. The larger the city, the more numero and specialized are the nuclei. The followi districts, however, have developed arou nuclei in most large American cities.

The central business district. — Th district is at the focus of intracity tran portation facilities by sidewalk, private ca bus, streetcar, subway, and elevated. Becau of asymmetrical growth of most large citie it is generally not now in the areal center the city but actually near one edge, as in t case of lake-front, riverside, or even inla cities; examples are Chicago, St. Louis, a Salt Lake City. Because established intern transportation lines converge on it, howeve it is the point of most convenient access fro all parts of the city, and the point of highe land values. The retail district, at the poi of maximum accessibility, is attached to t

[5] Exceptions are service-type establishments, such as some grocery stores, dry cleaners and gasoline stations.

ewalk; only pedestrian or mass-transporta-n movement can concentrate the large mbers of customers necessary to support partment stores, variety stores, and cloth- shops, which are characteristic of the trict. In small cities financial institutions d office buildings are intermingled with ail shops, but in large cities the financial trict is separate, near but not at the point greatest intracity facility. Its point of at-hment is the elevator, which permits three-nensional access among offices, whose st important locational factor is accessibil- to other offices rather than to the y as a whole. Government buildings also commonly near but not in the center of retail district. In most cities a separate utomobile row" has arisen on the edge of central business district, in cheaper rent as along one or more major highways; its achment is to the highway itself.

The wholesale and light-manufacturing trict. — This district is conveniently with- the city but near the focus of extra city nsportation facilities. Wholesale houses, ile deriving some support from the city it-f, serve principally a tributary region ched by railroad and motor truck. They , therefore, concentrated along railroad es, usually adjacent to (but not surround-) the central business district. Many types light manufacturing which do not require cialized buildings are attracted by the ilities of this district or similar districts: od rail and road transportation, available t buildings, and proximity to the markets d labor of the city itself.

The heavy industrial district. — This near the present or former outer edge of city. Heavy industries require large tracts space, often beyond any available in sec-ns already subdivided into blocks and eets. They also require good transportation, either rail or water. With the development of belt lines and switching yards, sites on the edge of the city may have better transportation service than those near the center. In Chicago about a hundred industries are in a belt three miles long, adjacent to the Clearing freight yards on the southwestern edge of the city. Furthermore, the noise of boiler works, the odors of stockyards, the waste disposal problems of smelters and iron and steel mills, the fire hazards of petroleum refineries, and the space and transportation needs which interrupt streets and accessibility — all these favor the growth of heavy industry away from the main center of the large city. The Calumet District of Chicago, the New Jersey marshes near New York City, the Lea marshes near London, and the St. Denis district of Paris are examples of such districts. The stockyards of Chicago, in spite of their odors and size, have been engulfed by urban growth and are now far from the edge of the city. They form a nucleus of heavy industry within the city but not near the center, which has blighted the adjacent residential area, the "back-of-the-yards" district.

The residential district. — In general, high-class districts are likely to be on well-drained, high land and away from nuisances such as noise, odors, smoke, and railroad lines. Low-class districts are likely to arise near factories and railroad districts, wherever located in the city. Because of the obsolescence of structures, the older inner margins of residential districts are fertile fields for invasion by groups unable to pay high rents. Residential neighborhoods have some measure of cohesiveness. Extreme cases are the ethnically segregated groups, which cluster together although including members in many economic groups; Harlem is an example.

Minor nuclei. — These include cultural

centers, parks, outlying business districts, and small industrial centers. A university may form a nucleus for a quasi-independent community; examples are the University of Chicago, the University of California, and Harvard University. Parks and recreation areas occupying former wasteland too rugged or wet for housing may form nuclei for high-class residential areas; examples are Rock Creek Park in Washington and Hyde Park in London. Outlying business districts may in time become major centers. Many small institutions and individual light manufacturing plants, such as bakeries, dispersed throughout the city may never become nuclei of differentiated districts.

Suburb and Satellite. — Suburbs, either residential or industrial, are characteristic of most of the larger American cities.[6] The rise of the automobile and the improvement of certain suburban commuter rail lines in a few of the largest cities have stimulated suburbanization. Satellites differ from suburbs in that they are separated from the central city by many miles and in general have little daily commuting to or from the central city, although economic activities of the satellite are closely geared to those of the central city. Thus Gary may be considered a suburb but Elgin and Joliet are satellites of Chicago.

Appraisal of land-use patterns

Most cities exhibit not only a combination of the three types of urban support, but also aspects of the three generalizations of the land-use pattern. An understanding of both is useful in appraising the future prospects of the whole city and the arrangement of its parts.

As a general picture subject to modification because of topography, transportatio and previous land use, the concentric-zor aspect has merit. It is not a rigid pattern, ina much as growth or arrangement often reflec expansion within sectors or developme around separate nuclei.

The sector aspect has been applied parti ularly to the outward movement of residenti districts. Both the concentric-zone theory an the sector theory emphasize the gener tendency of central residential areas to d cline in value as new construction takes plac on the outer edges; the sector theory is, hov ever, more discriminating in its analysis that movement.

Both the concentric zone, as a general pa tern, and the sector aspect, as applie primarily to residential patterns, assume (a though not explicitly) that there is but single urban core around which land use arranged symmetrically in either concentr or radial patterns. In broad theoretical term such an assumption may be valid, inasmuc as the handicap of distance alone would fav as much concentration as possible in a sma central core. Because of the actual physic impossibility of such concentration and th existence of separating factors, howeve separate nuclei arise. The specific separatir factors are not only high rent in the cor which can be afforded by few activities, b also the natural attachment of certain activ ties to extra-urban transport, space, or oth facilities, and the advantages of the separ tion of unlike activities and the concentratic of like functions.

The constantly changing pattern of lar use poses many problems. Near the core, lar is kept vacant or retained in antisocial slu structures in anticipation of expansion higher-rent activities. The hidden costs slums to the city in poor environment f future citizens and excessive police, fire, ar

[6] Chauncy D. Harris, "Suburbs," *American Journal of Sociology,* Vol. 49, no. 1 (July, 1943), p. 6.

ıitary protection underlie the argument : a subsidy to remove the blight. The transi-n zone is not everywhere a zone of terioration with slums, however, as witness : rise of high-class apartment development ar the urban core in the Gold Coast of ıicago or Park Avenue in New York City. On the fringe of the city, overambitious subdividing results in unused land to be crossed by urban services such as sewers and transportation. Separate political status of many suburbs results in a lack of civic responsibility for the problems and expenses of the city in which the suburbanites work.

he Form and Structure f Cities: Urban rowth Patterns

11

oward J. Nelson

modern visitor strolling down Duke of oucester Street, the main street of Colonial illiamsburg, passes mainly private dwelgs — single family detached houses large d small. But interspersed among then in an parently random pattern are various retail shops, service centers, and small manufacturers, several taverns and inns, an apothecary shop, a printing office, a milliner, a clockmaker, a bootmaker, a wigmaker, and so on. At one end of the tree shaded street, not more than three quarters of a mile long, is

printed from *Journal of Geography* (April, 1969), pp. 198-207 by permission.

the old capital of Virginia, and at the other, the college of William and Mary. Although today's buildings are merely restoration copies of the originals, many are built on the old foundations, and attention to authenticity is apparent. In all likelihood this mixture of residence, commerce, craftsmen, and public activity accurately portrays the undifferentiated structure of early American cities.

But the structure of cities in this country has changed markedly through the years. Today, even small towns with a population of 1,500 or so (the approximate colonial population of Williamsburg) show definite internal differentiation. Typically commercial establishments form a closely spaced cluster along a section of the main street, a lumber yard or other bulk handling facility is located next to a railroad or highway, and a small factory or two are attached to the same transportation lines near the edge of the town. The residential areas form their own districts, adjacent to, but apart from, the rest. And in larger cities an immensely complicated urban structure has developed, a constantly changing but delicately balanced areal organization composed of many highly specialized districts with complex linkages, the product of a variety of forces operating through several centuries.

The casual observer, in today's era of extensive travel, is perhaps first struck by the unique features of individual American cities. These characteristics that give each city a personality are often attributable to the physical qualities of the site or a distinctive historic past: the compact, hilly, water-encircled site of San Francisco, the level lake plain of Chicago, the charming Vieux Carre section of New Orleans, or the open squares in the street pattern of Savannah. But the more perceptive traveler soon begins to notice a repetitive pattern in the form and structure of our cities, and becomes almost instinctivel aware of a kind of "normal" location c specialized districts, and of associations c activities within them. For, in fact, America cities have developed a highly stylized ar rangement and characteristic, repetitive inte relations among the specialized areas tha constitute their urban anatomy.

FACTORS IN THE GROWTH PATTERNS OF AMERICAN CITIES

The form and structure of the modern Ameri can city is the result of numerous economic social, and cultural factors operating throug the many decades since the evolution of th simple forms like Williamsburg. The force contributing to the contemporary urban struc ture are many, some are obvious and strong others are more subtle, but all add an im portant dynamic quality to urban develop ment. Some of the most significant of thes factors include rapid and massive growth, heterogeneous population, the persisten desire of Americans for a single famil detached house, and the changing forms c urban transportation. The amazing affluenc of Americans in recent years has accelerate change.

American cities are the product of rapi and almost continuous growth. In 179(about 200,000 people lived in urban place (over 2,500 population), by 1890 the figur was 22,000,000, in 1960 it was 125,000,000 The population increase of our cities in th 1950-1960 decade exceeded the total popula tion of all urban places in 1900. Betwee 1930 and 1969 the urban population ha about doubled. As a result, there has been a almost frantically rapid building and rebuild ing of the structures that form our urba plant. With increasing populations cities hav only expanded areally in all directions, urban lands in general, and favored loca-s in particular, have increased enormous- in value, making rebuilding for a gher use" profitable. Rapid growth has ılted in uncommonly dynamic cities, with stant change, sorting out, and filtering vn in every section of the city.

The American urban population has not y been growing but it also has been un-ally heterogeneous, resulting in many dis-tive neighborhoods and much internal ration. In the past, much of the popula- variety was the result of a series of ves of migration. There was, for example, eavy immigration from Ireland after 1847, ther wave from Germany after 1852, and ide of immigrants from other European ntries beginning in the 1880's and con-ing until about 1920. The ethnic neigh-hood was common. Often the newer immi-nts replaced the older groups in the est sections in the inner city as the earlier vals prospered and moved further out to ver homes. During and since World War rban growth came mainly from migration m rural areas and included both whites blacks, but with the blacks in many cases ving in typical fashion into the older nes in central locations in neighborhoods ng abandoned by slightly more affluent dents. This succession of peoples results only in shifts in living quarters, but also changes in shops, schools, and churches. even where assimulation has taken place, erican cities remain heterogeneous, with regation by income replacing that of nic group or race.

The deeply ingrained desire of Americans own a single family house on a large open is a further factor influencing the struc-e of our cities. Regardless of the psy-ological origin of this drive — the difficulty European immigrants experienced in owning such a structure in Europe, the pioneer tradi-tion of a cabin in a clearing, the dispersed farmhouse familiar to the rural migrant, or notions about privacy, play space, and the nature and meaning of the family — the detached house has persisted for two hundred years in American cities. Whenever he can afford it, the American urban dweller appears to prefer ample space and a private yard over a short journey to work. Encouraged by governmental mortgage policy, perhaps as many as 15,000,000 single family homes have been built since World War II, mostly in suburban areas. These low density residential neighborhoods, covering vast amounts of space, and affecting patterns of commerce and industry, are a prominent and unique item in the American urban structure.

Urban transport not only laces the urban structure together, but it also profoundly affects the arrangement and function of ele-ments in the structure of the city. In America, the horse-drawn omnibus was important from 1830 to 1860, the suburban railroad from 1850 on in the largest cities, cable cars from 1860 to 1890, and elevated rail lines and subways from around the turn of the cen-tury. The most universal transport medium from 1890 to about 1945 in all but the largest cities was the electric "street car." But the revolutionary transportation development of the twentieth century has been the spectacu-lar rise of the automobile. There were only about 8,000 automobiles in America in 1900, less than 500,000 in 1910, about 8 million in 1920, but the number has risen spectacularly from about 25 million in 1945 to more than 70 million cars and 13 million trucks and busses today. And each automobile, on the average, is driven more miles every year. The effect on urban structure of this new private form of transportation, not confined to fixed routes, but with the vehicle requiring storage space near the driver, has been immense.

Impressed with the dynamic nature of the American city even before some of the factors just mentioned were operating at maximum strength, one geographer, Charles C. Colby, felt that two opposing forces could be identified. They were centrifugal forces that impelled functions to migrate from the central areas of the city to the periphery, and centripital forces that tend to hold certain functions in the central zone and attract others to it.[1]

Centripetal forces, Colby said, are the result of a number of attractive qualities of the central portion of the city. One of these is *site attraction*, often the quality of the natural landscape that invited the original occupance, such as a river crossing or deep water landing. *Functional convenience,* a second force, results from the possession of the central zone of maximum accessibility, not only to the metropolitan area, but often to the entire surrounding region. The concentration of one function in the central zone operates as a powerful magnet attracting other functions. This is called *functional magnetism*. Thus a large department store may attract a swarm of ladies apparel and accessory shops. *Functional prestige* stems from a developed reputation. One street may become famous for its restaurants, another for its fashionable shops, and doctors often cluster for reasons of functional prestige.

[1] Charles C. Colby, "Centrifugal and Centripetal Forces in Urban Geography," *Annals of the Association of American Geographers*, Vol. XXIII, no. 1 (March, 1933), pp. 1-21. The article has been reprinted in what may be a more accessible source: Harold M. Mayer and Clyde F. Kohn, *eds., Readings in Urban Geography* (Chicago, The University of Chicago Press, 1959), pp. 287-98.

Centrifugal forces on the other hand a not only opposite forces, but are made up a merging of influences — a desire to lea one part of the city and the urge to go another. Five forces are recognized. One the *spatial force,* when congestion in t central zone uproots and the empty spaces the other zones attract. The second is t *site force*, which involves the disadvantag of the intensively used central zone in co trast to the relatively little used natural lan scape of the periphery. Another, t *situational force,* results from the unsati factory functional spacing and alignments the central zone and the promise of mo satisfactory alignments in the periphery. Th there is the *force of social evolution* response to which high land values, hi taxes, and inhibitions growing out of the pa create a desire to move and the opposi conditions in the newly developing periphe provide an invitation to come. Finally, t *status and organization of occupance* creat a force for change, in which such things as t obsolete functional forms, the crystalliz patterns, the traffic congestion, and the u satisfactory transportation facilities of t central zone stand in opposition to t modern forms, the dynamic patterns, t freedom from traffic congestion, and t highly satisfactory transportation facilities the outer zone.

Well aware of the importance of huma choice, Colby added another factor which called the *human equation* that could wo either as a centripetal or a centrifugal forc Although today other forces may also be work, these concepts are still useful in analy ing the dynamics of cities. In addition, mo formal models of city structure have bee constructed by other students of citie sociologists, economists, and geographer ree of the most famous of these constructs low.

ASSIC MODELS OF CITY STRUCTURE

e earliest (1923) and best known of the ssic models is the concentric circle or ial hypothesis of Ernest W. Burgess.[2] The ence of this model is that as a city grows it pands radially from its center to form a ies of concentric zones. Using Chicago as example, Burgess identified five of these. the center of the city was Zone I, the *ntral business district* or CBD. The heart the CBD contained department stores, le shops, office buildings, clubs, banks, els, theaters, and civic buildings. Encircl- it was a wholesale district. Zone II, the *ne in transition,* surrounded the CBD, and nprised an area of residential deterioration the result of encroachments from the CBD. consisted of a factory district as an inner t, and an outer belt of declining neighbor- ods, rooming house districts, and generally ghted residences. In many American cities the 1920's, this area was the home of merous first generation immigrants. Zone , the zone of *independent workingman's mes* was the next broad ring, at the time gely inhabited by second generation immi- nts, and characterized in Chicago by the vo-flat" dwelling. Beyond this ring was located Zone IV, the *zone of better residences.* Here lived the great middle-class of native born Americans in single family residences or apartments. Within this area, at strategic places, were local business centers, which Burgess implied might be likened to satellite CBDs. Zone V, the *commuter's zone* lay beyond the area of better residences, and consisted of a ring of encircling small cities, towns, and hamlets. These were, in the main, dormitory suburbs, with the men commuting to jobs in the CBD.

The operating mechanism of the concentric circle model was the growth and radial expansion of the city, with each zone having a tendency to expand outward into the next. Burgess assumed a city with a single center, a heterogeneous population, a mixed commercial and industrial base, as well as economic competition for the highly-valued, severely-limited central space. He explicitly recognized "distorting factors," such as site, situation, natural and artifical barriers, the survival of the earlier use of the district, and so on. But he argued that to the extent to which the spatial structure of a city is determined by radial expansion, the concentric zones of his model will appear. Given the limited data available, the Burgess model was a remarkably astute description of the American city of the time.

A second model of the growth and spatial structure of American cities was formulated by Homer Hoyt in 1939 and is known as the wedge or sector theory.[3] Hoyt analyzed the distribution of residential neighborhoods of

First presented as a paper in 1923, Burgess restated his hypothesis at somewhat greater length five years later. Ernest W. Burgess, "Urban Areas," T. V. Smith and L. D. White, *eds., Chicago: An Experiment in Social Science Research* (Chicago, University of Chicago Press, 1929), pp. 114-23. An excellent contemporary review of the Burgess model is Leo F. Schnore, "On the Spatial Structure of Cities in the Two Americas," Philip M. Hauser and Leo F. Schnore, *eds., The Study of Urbanization* (New York, John Wiley and Sons, Inc., 1965), pp. 347-98.

[3] Homer Hoyt, *The Structure and Growth of Residential Neighborhoods in American Cities* (Washington, D.C., Federal Housing Administration, 1939). A suggestive fragment of this classic is reprinted in Mayer and Kohn, *op. cit.*, pp. 499-510.

various qualities, as defined by rent levels, and found that they were neither distributed randomly nor in the form of concentric circles. High rental areas, for example, tended to be located in one or more pie-shaped sectors, and did not form a complete circle around the city. Intermediate rental areas normally were sectors adjacent to a high rent area. Further, different types of residential areas usually grew outward along distinct radii, and new growth on the arc of a given sector tended to take on the character of the initial growth in that sector. In summary, Hoyt argued that "if one sector of the city first develops as a high, medium, or low rental residential area, it will tend to retain that character for long distances as the sector is extended outward through the process of the city's growth."

Although no geometric pattern can be superimposed upon a city to determine the position of high and low rent sectors, some generalizations can be made about their location. The area occupied by the highest income families tends to be on high ground, or on a lake, river, or ocean shore, along the fastest existing transportation lines, and close to the country clubs or parks on the periphery. The low income families tend to live in sectors situated farthest from the high rent areas, and are normally located on the least desirable land alongside railroad, industrial, or commercial areas. Rental areas, are not static. Occupants of houses in the low rent categories tend to move out in bands from the center of the city, mainly by filtering into the houses left behind by the higher income groups, or in newly constructed shacks on the fringe of the city, usually in the extension of the low rent section. It is felt by some that because Hoyt's model takes into account both distance and direction from the center of the city, it is an improvement on the earlier Bu gess effort.

A third model, the multiple nuclei, wa formulated by Chauncy Harris and Edwar Ullman in 1945 as a modification of the tw previous models.[4] They argue that the lan use pattern of a city does not grow from single center, but around several distin nuclei. In some cases these nuclei, elemen around which growth takes place, hav existed from the origin of the city, but othe may develop during the growth of the cit Their numbers vary from city to city, but th larger the city the more numerous an specialized are the nuclei.

Urban nuclei attracting growth might in clude the original retail district, a port, a rai road station, a factory area, a beach, or, i today's city, an airport. The authors identif a number of districts that have develope around individual nuclei in most large Ameri can cities. The *central business distric* usually includes, or is adjacent to the origina retail area. The *wholesale and light manu facturing* district is normally located alon railroad lines, adjacent to, but not surround ing the CBD. The *heavy industrial district* i near the present or former edge of the city where large tracts of land and rail or wate transportation are available. *Residential* dis tricts of several classes are identified wit high-class districts on desirable sites, on we drained, high land, and away from nuisances such as noise, odors, smoke, and railroa lines, the low-class districts near factories an railroad districts. Finally, *suburbs* and *satel lites,* either residential or industrial, are char

[4] Chauncy D. Harris and Edward L. Ullmar "The Nature of Cities," *Annals of the America Academy of Political and Social Science*, Vo CCXLII (November, 1945), pp. 7-17. Reprinte in Mayer and Kohn, *op. cit.*, pp. 277-86.

eristic of American cities. Suburbs are ined as lying adjacent to the city, with ellites farther away with little daily comting to the central city.

The rise of separate nuclei and differiated districts is thought to result from a nbination of four factors. 1. Certain activi- require specialized facilities, i.e., a retail trict needs intracity accessibility, a port uires a harbor. 2. Certain like activities up together because they profit from link-s. For example, retail activities may ster to facilitate comparison shopping and ancial institutions may locate in close sters to make easy face to face communiion by decision makers. 3. Certain unlike ivities are detrimental to each other. Thus ensive users of land, such as bulk storage ds, are not compatable with retail funcns, requiring dense pedestrian traffic. 4. rtain activities are unable to afford the high ts of the most desirable sites — low class using is seldom built on view lots.

Models such as these which emerge from process of analysis and generalization do t conform to the reality of any city. But yone familiar with large or medium sized nerican cities will recognize many elements each model in the vast majority of our an areas. Obviously, too, they are not tually exclusive, for the latter two models both modifications of the concentric cir- theory. Even in Hoyt's concept, resiatial areas expand outward concentrically. ere has been extensive statistical testing these models in recent years with no consive results. They remain as valuable conptual tools for analyzing the modern city, d provide a basis for cross-cultural urban mparisons.

It is obvious, too, that the American city the twenties and thirties, which provided the data upon which these models were built, is undergoing important structural changes. The traumatic effect of the automobile was not really apparent in the city studies that furnished the inspiration for these classic models. Other factors forming the basis for the urban form have been discussed previously, and some of their effects will be analyzed in the following section on the elements in the urban structure.

MAJOR ELEMENTS IN THE URBAN STRUCTURE

Central Business District

The most obvious and easy to recognize of the components in the spatial structure of American cities described by the classic models is the Central Business District. Generally located on or near the original site of the city, it became the focus of the city's mass transportation arteries, and was thus the point of most convenient access from all parts of the city. Here, in stylized juxtaposition, were found the largest department stores, women's dress shops, men's clothing stores, shoe stores, "five and ten cent" stores, jewelry stores, drug stores, and similar retail outlets. A visitor to an unfamilar city could anticipate this arrangement, and when he found it he could also be confident that he had located the heart of the CBD as well as the area of highest land values and the place of heaviest pedestrian traffic.[5]

Other groupings of activities into specialized junctional areas have also been traditional in the CBD's of large cities. These

[5] A more extensive discussion of the CBD is found in Raymond E. Murphy, *The American City, An Urban Geography* (New York, McGraw-Hill Book Company, 1966), pp. 283-316.

may include financial districts, with banks, savings and loan associations, stock and commodity exchanges, brokerage offices, trust companies, and so on. The civic center with its city and county buildings often attracts lawyers' offices and bail bond agencies. Occasionally a theater district may be present, associated with restaurants and perhaps candy shops. Hotels and office buildings are usually found several blocks from the center of the CBD. Occasionally an office building may specialize in particular services, perhaps housing doctors, dentists, and medical laboratories exclusively, or in the height of specialization, it may house the officials of a single company. The extent of an office building area, though, depends upon the headquarters quality of the city and may cover a large area or be almost nonexistent.

Although absolute areal growth of the CBD may have essentially come to an end with the invention of the elevator and the skyscraper, this area, like the others in the city, is constantly changing. Shifts in its boundaries have been recognized by Murphy, Vance, and Epstein, identifying a "zone of discard" and a "zone of assimilation" associated with this movement.[6] The area from which the CBD is migrating, the "zone of discard" often is characterized by pawn shops, family clothing stores, bars, low grade restaurants, bus stations, cheap movies, credit jewelry, clothing, and furniture stores. The CBD tends to migrate in the direction of the best residences, and in the area into which it seems to be moving, the "zone of assimilation," are found speciality shops, automobile showrooms, drive-in banks, headquarters offices, professional offices, and th newer hotels.

Not only are the boundaries of the CB changing, but due to the reduction of de pendence on mass transit and the rise of th more flexible automobile, the last severa decades have seen the movement away fro the CBD of some of its traditional function Department stores, once the exclusive posse sion of the CBD, have been built in larg numbers in outlying areas. Other retail activ ties have followed their lead. The proportio of retail sales of the city credited to the CB have gone steadily downward, and few ne retail stores have been built here. A numbe of small and medium sized cities have at tempted to reverse the declining retail im portance of their CBDs by converting th main shopping street into a pedestrian mal The success of this expedient is as yet unclea

Many students of cities feel the CBD c the future will change considerably and pe haps will consist of two centers separated b a band of parking. The financial and offic section in many large cities remains health attracted to the focus of metropolitan trans portation and the advantages of "linkages with other office functions. As the tradition retail functions of the CBD decline, perhap what remains will evolve into a separat center of dual services, comprising specialt shops serving the metropolitan area and mas selling stores supplying the needs of the inne part of the city.[7]

Outlying Commercial Centers

All cities, in addition to a CBD, have

[6] Raymond E. Murphy, J. E. Vance, Jr., and Bart J. Epstein, "Internal Structure of the CBD," *Economic Geography*, Vol. XXXI (January, 1955), pp. 21-46.

[7] James E. Vance, Jr., "Emerging Patterns c Commercial Structures in American Cities Proceedings of the IGU Symposium in Urba Geography Lund 1960, *Lund Studies in Geog raphy*, Human Geography, ser. B., no. 24 (Lun 1962), pp. 485-518.

.ety of other commercial areas, and the ;er the city the more complex the pattern the more specialized some of the ele- ıts become. The largest of the outlying pping centers, usually referred to as *re- ıal centers*, are built around one or more artment stores, with variety, apparel, and l convenience stores, and repeat in a ıned or unplanned way, the retail types nd in the very heart of the CBD. Usually e stores are smaller than the downtown nterparts and deal mainly in the staple most profitable lines of merchandise. ly centers of this class grew at intersec- ıs of public transportation lines, but most e been built up recently at points of easy ɔmobile access. Next in order of size are *ımunity centers* with a large variety or or department store, some apparel shops, ; a supermarket, drug store, bank, and ilar establishments. The *neighborhood ter*, built around a supermarket and with ıe associated stores, is an almost ubiquiti- feature. Finally, *convenience centers,* haps consisting of a "pop and mom" cery, a launderomat, and a service station, at the bottom end of the commercial hier- hy. Recently this hierarchy of commercial ters within cities has been compared to the ters of various orders in central place ɔry.[8]

olesaling and Light Manufacturing

:ated in the concentric-circle model as ng at the border of the CBD, the whole- : and light manufacturing activities have 'ted considerably in response to newer :es. As the automobile has given mobility to the worker, many firms engaged in light manufacturing have moved away from the center of the city to the suburbs where land is inexpensive and large tracts facilitate one story plants, storage areas, and parking lots. Sites near belt highways are particularly desirable. Many wholesalers, too, have been attracted to similar locations.

On the other hand, certain types of whole- salers characteristic of large cities are affected by forces that make clustering advantageous, often in or near the CBD. Clustering of wholesaling establishments normally persists when the buyers gather in person at the market, where the goods are nonstandardized, when comparison of quality or style is important, and when the establishment of the price is an important function of the market. Wholesalers of jewelry, apparel, fruit and vegetables, or cut flowers are characteristic examples.

Similarly, although most manufacturers have moved out of the central part of the city, a few characteristically remain. For example, large segments of the garment in- dustry are found at the edge of the CBD in cities where it is an important manufacturing activity. One factor involved in this location is the close linkages of the manufacturing and wholesaling aspects of apparel production. But, in addition, clustering in the garment industry results in "external economies" (economies external to the firm). For ex- ample, the presence of ancillary firms, such as textile dealers, sponging (shrinking) facilities, factors (textile bankers), trucking firms, repairmen, and suppliers of everything from pretty models to thread, provides external economies to the apparel firms.

Heavy Manufacturing

Heavy industry has almost entirely moved out of the inner ring as postulated by Burgess,

ⱥ discussion of the hierarchy of commercial enters within cities is found in Brian J. L. Berry, *Geography of Market Centers and Retail Distribution* (Englewood Cliffs, N.J., Prentice- Hall, Inc., 1967), pp. 42-58.

much of it into the specialized sites as implied by Harris and Ullman. Early located adjacent to water transport, the development of railroads and trucks permitted industries to leave the central areas. Seeking extensive sites for sprawling factories, parking lots, and storage facilities, large manufacturing complexes are now characteristically found in outlying locations. Some newer industries, such as the aircraft industry, are linked to facilities found only in non-central areas, airports in this instance.

Residential Districts

Much has been said already about residential districts in American cities. Our heterogeneous urban population has shown a tendency to sort itself out one way or another into relatively homogeneous neighborhoods. As differences in language and ethnic character have become less important, segregation of residence has been mainly by economic status, with little mixing of the homes of the rich and the poor, as in some cultures. Race remains, however, as a segregating force, seemingly more powerful than economic status.

Another striking characteristic has been the propensity of the American urban dweller for the space consuming, expensive, single family dwelling. There is some indication that isolation and privacy are qualities of increasing importance. The front porch oriented to the street from which the fami viewed the passing scene has disappear from the post World War II house. Too, t room arrangement has been reversed: nc the living room often faces the rear ya which has been designed as an addition to t living space. And in southern Californi where life-styles often seem to originate, t backyard is now enclosed with a solid wood fence or block wall, making the famil privacy complete. Continued spread of t traditional house in the middle of a lot (t now non-functional front yard remains, r quired by zoning ordinances reflecting earlier era) forecasts continued suburb sprawl.

In contrast to this long term trend towa dispersal, in recent years, there has been striking increase in the proportion of multip housing in many cities. Perhaps this is simp a temporary phenomenon reflecting t changing age composition of the populati due to the low birth rate of the 1930's; t young and old often choose apartments, tho in their thirties choose single family home To the extent, however, that this changi construction mix reflects a movement of tho who can afford to live anywhere into mo central, accessible locations, as urban di tances become greater, it may foretell weakening, at long last, of the American traditional willingness to trade commuti time for space.

he Pattern and Distribution f Manufacturing in the hicago Area

12

artin W. Reinemann

e Chicago Standard Metropolitan Area igure 1) is the second-ranking manufac-ing area in the United States. Its national minence is best indicated by the fact that 1954 it contributed more than 10½ per t of the nation's "value added by manu-ture." Some recent studies, as well as data m the *Census of Manufacturers*, reveal eral interesting relationships and trends hin this important industrial area with ard to manufacturing activities in the cen-l city as compared with the remainder of metropolitan area.

TY OF CHICAGO COMPARED WITH MAINDER OF STANDARD TROPOLITAN AREA

hough the City of Chicago comprises less n 6 per cent of the Chicago Standard tropolitan Area it contains two-thirds of population and approximately the same centage, or more, of the major economic ivities associated with the American me-polis.

Table I shows that the suburbanization vement between 1940 and 1950 resulted in a population loss of only 4½ per cent for the central city and that almost 66 per cent of the population still lived in the City of

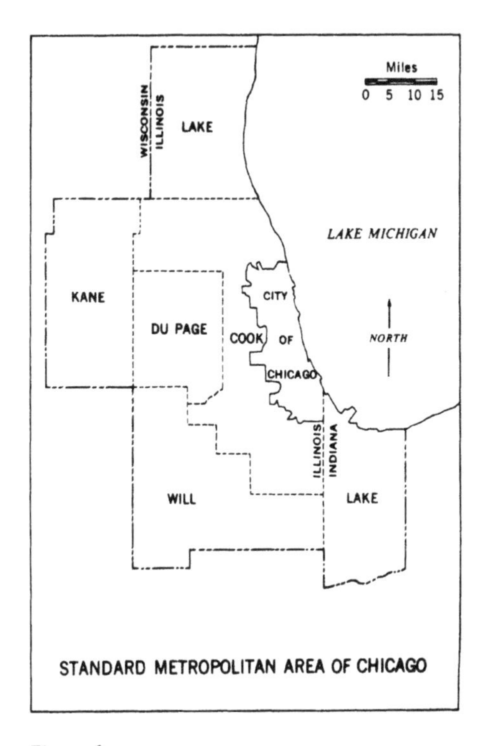

Figure 1.

printed from *Economic Geography,* Vol. 36, no. 2 (April, 1960), pp. 139-44 by permission.

TABLE I

Percentage of Selected Economic Activities in the City of Chicago Compared With Their Location in the Remainder of the Standard Metropolitan Area, 1947 and 1954

Activity	*Percentage in City of Chicago*		*Change in percentage*	
	1947	*1954*	*1939—1947-48*	*1947-48—1954*
Population	65.9 (*1950*)	—	−4.5 (*1940-1950*)	—
Manufacturing (value added)	69.2	62.7	2.2	−6.5
(establishments)	83.4	76.3	2.6	−7.1
(production workers)	70.4	65.4	−1.5	−5.0
Retail Trade (sales)	72.6	66.6	−3.3	−6.0
Wholesale Trade (sales)	94.8	90.5	−1.5	−4.3
Service Industries (receipts)	85.0	84.6	−2.1	−0.4

Chicago in 1950. While there has been additional suburbanization since 1950, it is significant that the primary labor market of the metropolitan area is in the City of Chicago and that this continues as a major factor attracting additional manufacturing establishments to the central city.

The high degree of centralization of manufacturing in the City of Chicago is also revealed in Table I.[1] Manufacturing expanded very rapidly in the Chicago Standard Metropolitan Area between 1939 and 1947 in both suburban and central areas. Industrial activity was actually slightly more concentrated inside the central city in 1947 than in 1939. This was not true for any of the other economic functions. Neither was it true fo population, which had suburbanized mo

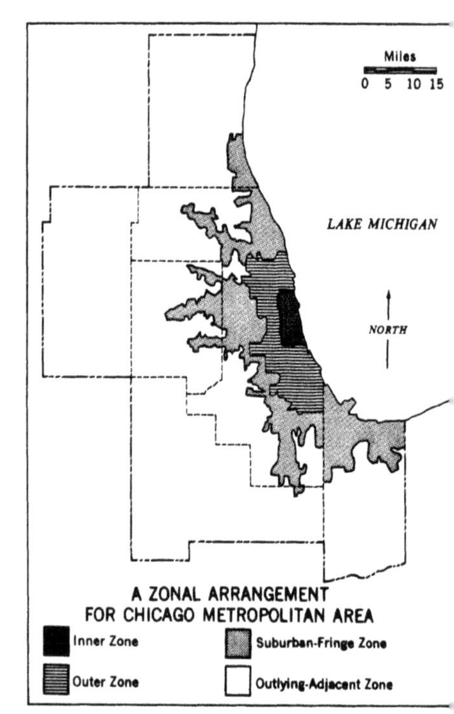

Figure 2.

[1] For a comprehensive study of suburbanization of manufacturing in all of the Metropolitan areas in the United States see Evelyn M. Kitagawa and Donald J. Bogue, "Suburbanization of Manufacturing Activity Within Standard Metropolitan Areas," *Studies in Population Distribution*, no. 9, 1955 (Scripps Foundation, Miami University, Oxford, Ohio or Population Research and Training Center, University of Chicago, Chicago, Illinois).

idly. Furthermore, the manufacturing or force, as represented by production rkers, drifted to the suburbs more slowly n the general population prior to 1948.

The strongest drift of industry to the burbs occurred after 1947, according to 54 Census data, with the greatest change ing in number of establishments (see ble I). Production workers continued to centralize less rapidly than did establishnts or production, as measured by value ded by manufacture.

A more detailed analysis of the industrial burbanization trend is afforded by outlining nes for the Chicago Standard Metropolitan ea and by using information obtained from records of the Territorial Information Detment of the Commonwealth Edison Comny of Chicago.

CENT INDUSTRIAL MOVEMENTS SED ON A ZONAL ARRANGEMENT R THE CHICAGO STANDARD ETROPOLITAN AREA

number of persons and organizations ve ascribed zones to the structure of the icago Standard Metropolitan Area for a riety of purposes.[2] For purposes of describing the pattern and distribution of manufacturing, four zones, shown in Figure 2, are suggested as follows:

Zone I, The Inner Zone

This zone has an area of about 40 square miles extending from Diversey Parkway on the north to 51st Street and Garfield Boulevard on the south, and from the lake shore on the east to Western Avenue on the west. The "Loop" of Chicago is considered a sector within this zone. In general, the zone is characterized by numerous, diverse, small, and older manufacturing concerns primarily occupying multi-storied buildings. Some very large establishments are also present.

Zone II, The Outer Zone

This area surrounds the Inner Zone and includes the remainder of the City of Chicago. Manufacturing is scattered, although many of the plants are grouped in clusters. The zone contains many very large concerns, several of them in multi-storied buildings. A few new single-storied structures have appeared in scattered locations.

Zone III, The Suburban-Fringe Zone

This zone, lying adjacent to but outside of the City of Chicago, contains a number of independent suburban municipalities that form a nearly continuous urban ring about the city. Manufacturing is characterized by both large and small concerns, either newly established or relocated, and for the most part occupying modern one-storied buildings.

Zone IV, The Outlying-Adjacent Zone

This is an area of widely dispersed, satellitic settlements that extends to the outer edge of

E. W. Burgess, "The Growth of the City," Chapter II, in *The City*, by R. E. Park, F. W. Burgess, and R. D. McKenzie, originally published as an article in *Proceedings of the Amer. Sociol. Soc.*, Vol. XVIII (1923). Another system was originated by the Chicago Census Advisory Committee and members of the Department of Sociology at the University of Chicago about 1940. See Martin W. Reinemann, *The Localization and the Relocation of Manufacturing within the Chicago Urban Area* (Unpublished Ph.D. dissertation, Department of Geography, North-western University, 1955).

Based on 1940 and 1950 *Census of Population*, 1939, 1947, and 1954 *Census of Manufactures*, and 1948 and 1954 *Census of Business*.

the Chicago Standard Metropolitan Area. Manufacturing, with few exceptions, is located in the municipalities of Joliet, Aurora, Geneva, St. Charles, Elgin, and Waukegan.

The boundary between the Inner and Outer Zones is probably less significant for differentiating local site characteristics for manufacturing than it might be for analyzing other urban characteristics. Although the Inner Zone has many advantages for manufacturers, such as nearness to market, supply houses, and good transportation, the availability of a large labor market, and the general advantage of the high density of manufacturing establishments in closely related lines, it also has many conditions that push industrial development toward the periphery. Some of the relative disadvantages of the Inner Zone are the following: scarcity of land available for industrial use; the generally unattractive and obsolete condition of buildings; and the congested, noisy, and dirty streets with attendant parking and loading problems.

The city limits as the outer boundary of the Outer Zone is important because of the differences in the costs of a variety of community services, including taxes, on the two sides of the line, and the desire to avoid the excessive frequency and undesirable nature of factory inspections, solicitations, and other related political pressures so apparent in the City of Chicago. The boundary between Zones III and IV sets apart those suburbs that are contiguous or immediately adjacent to the city from the detached satellite cities surrounded by rural countryside. This boundary, with few minor exceptions, coincides with that drawn by the Bureau of the Census to mark the outer limits of the "urbanized area." Differences in labor market, availability of public transportation, time and cost of shipping and receiving materials, supplie and products, and differences in the avai ability of land and the quality of certain con munity services distinguish the two zones.

Three publications especially pertinent t this study, all based on the same source, hav revealed the industrial expansion and subu banization in the Chicago area since 1945 From these three reports, and selected ir formation relative to the first three zone described above, one can discern the gener trend of recent industrial relocation in th Chicago Standard Metropolitan Area.

GENERAL TRENDS

Industrial expansion within the Chicag Standard Metropolitan Area demonstrate considerable areal differentiation during th period 1941-1950. There was a net loss c 120 industries in the Inner Zone during thi period (Table II), whereas the Outer Zon of the city had a net gain of 88 plants. Th Suburban-Fringe Zone, however, showe its attractive nature by a gain of 743 new an relocated industries during this interim These data emphasize a reversal in the cen tralizing trend indicated by the 1947 *Censu of Manufactures*. For the period 1941-195 the City of Chicago (the Inner and Oute

[3] Chicago Plan Commission, *Chicago Industri Study, Summary Report*, Chicago Plan Con mission (Chicago, 1952); Leo G. Reeder, *Indu trial Location in the Chicago Metropolita Area, With Special Reference to Populatio* (Unpublished Ph.D. dissertation, Department c Sociology, University of Chicago, 1952); No man E. Brown, "Location of Industries i Chicago and Northern Illinois," *Midwest Eng neer*, Vol. V, no. 2 (July, 1952), pp. 2-1 Each of these studies used information from th records of the Territorial Information Depar ment of the Commonwealth Edison Compan of Chicago.

TABLE II

ovements of Manufacturing Establishments in the Chicago Urban Area and Initial ocations of New Industries for 1941-1950*

tion A. Firms leaving Chicago and moving to Suburban-Fringe Zone

'ears	*Inner Zone*	*Outer Zone*
1-1945	18	37
6-1950	126	117
'otal	144	154

ection B. New industries moving into the three zones**

'ears	*Inner Zone*	*Outer Zone*	*Suburban-Fringe Zone*
1-1945	46	67	113
6-1950	75	78	332
'otal	121	145	445

Section C. Inter-Chicago moves

'ears	*Inner Zone to Outer Zone*	*Outer Zone to Inner Zone*
1-1950	109	12

ection D. Net gain or loss over the decade 1941-1950

ears	*Inner Zone*	*Outer Zone*	*Suburban-Fringe Zone*
	Net loss	Net gain	Gain
1-1950	120	88	743***

Includes only industries with ten or more employees.

Includes only those industries new to the metropolitan area.

Statistics for outmovement from suburbs not available for time period used but for the period 1945-1950 eleven plants moved from the suburbs to Chicago and there were six recorded moves from one suburban community to another.

Zones combined) had a net loss of 32 manufacturing plants.

The latest data available from the Territorial Information Department of the Commonwealth Edison Company of Chicago at the time of this study, show that the industrial suburbanization trend was maintained after 1950 and that it has continued unabated to the present. Between 1950 and 1953, for instance, 201 firms quit Chicago, sustaining the average of about 50 relocations per year first established during the period 1946-1949.

These data also indicate the destinations to which Chicago factories have shifted and afford a basis for showing direction and pattern of the industrial relocation. Of the 446 industries that moved out of the city during the period 1946 through May, 1954, 312 (70 per cent) moved to the Suburban-Fringe Zone, 71 (16 per cent) moved to the Outlying-Adjacent Zone, and 63 (14 per cent) moved to Illinois cities or areas outside the Chicago Standard Metropolitan Area.[4] Figure 3 shows relocations to the Suburban-Fringe Zone, where their concentration in a few suburbs is striking. Five suburbs — Franklin Park, Skokie, Cicero, Melrose Park, and Evanston — received about 53 per cent of the 312 relocated plants.[5] The general pattern of relocation has been to those suburbs adjacent to the City of Chicago, predominantly toward the north and northwest.

[4] No statistics are available concerning moves from Chicago to other states.

[5] The 1954 Census lists 127 establishments for Franklin Park, 115 for Skokie, 170 for Cicero, 75 for Melrose Park, and 130 for Evanston. Obtaining a percentage increase for the period 1947-1954 for the first two cities listed above is not possible because information on their 1947 status is not available. Percentage increases for the latter three were 27, 168, and 55, respectively.

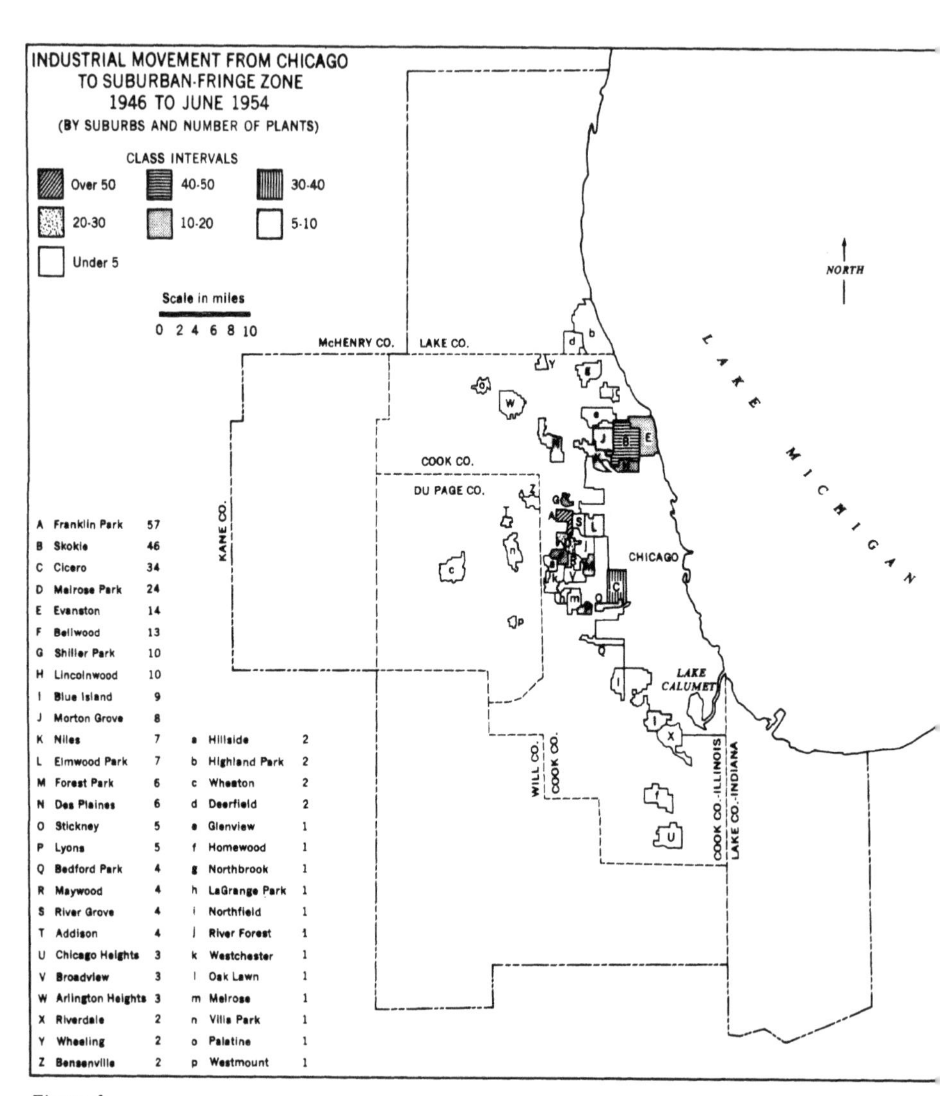

Figure 3.

CONCLUSION

Manufacturing in the Chicago Standard Metropolitan Area remains highly concentrated in the City of Chicago despite rapid suburbanization since about 1945. Although there has been considerable relocation of manufacturing plants from the Inner Zone of the city to the Suburban-Fringe Zone, the primary increase in the Suburban Zone represents new industrial establishments. A net gain of industrial establishments in the Out Zone of the city from 1941 to 1950 sugges that sufficient space and desirable facto sites were still available there during th period. The major direction of both industri relocation and new industrial developme has been toward the northern and nort western suburbs immediately adjacent to t City of Chicago.

he Distribution of the Ethnic roups in Chicago, 1960

13

ng-Cheng Kiang

e purpose of this paper is to identify and plain the distribution of the major ethnic oups in Chicago since 1960. To a large ent, it is a follow-up of two previous dies dealing with the population succession Chicago — one by Paul F. Cressey and the er by Richard G. Ford.[1] Cressey's paper vers the years 1898-1930; Ford's paper ncerns 1940. Both papers include Negroes d seven groups of foreign-born whites — lish, German (including Austrian), Italian, ssian, Irish, Swedish, and Czecho- vakian. This paper includes all these oups in addition to Mexican and Puerto can.

Following Ford's procedure, I have vided the city of Chicago into eighteen mile- de concentric circle zones from the inter- ction of State and Madison Streets. The mbers of the ethnic groups in the different nsus tracts are recorded from the U.S. nsus for 1960. Any census tract with 50 r cent or more of its area lying within a mile zone is listed as belonging to that zone. For purposes of comparison, I have grouped the tenth to eighteenth mile zones together, as Cressey and Ford did.

During 1940-60 the total population decreased from 15 per cent to 11 per cent of the total city residences for the first three mile zones, and increased from 7 per cent to 22 per cent for the tenth mile zone and beyond (Tables I and II). This indicates an outward movement of people from the Loop to the city's outskirts. The same trend is seen for all ethnic groups, except that there are no statistics showing the distribution of Mexicans and Puerto Ricans for the period of 1898-1940 in Cressey's or Ford's study. In 1940 up to 20 per cent of all the leading ethnic groups except Mexican and Puerto Rican were found in the inner zone (the first three mile zones). This reduced to 13 per cent or less in 1960. In 1940 from 2-11 per cent of all the leading ethnic groups except Mexicans, Puerto Ricans, and Negroes were found in the outer zone (the ten-mile zone and beyond). This increased to 13-47 per cent in 1960. The figures for Negroes were 1 per cent in 1940 and 9 per cent in 1960 in the outer zone. Generally speaking, the Irish moved

[1] Paul F. Cressey, "Population Succession in Chicago: 1898-1930," *American Journal of Sociology*, Vol. XLIV (July, 1938), pp. 59-69; Richard G. Ford, "Population Succession in Chicago," *American Journal of Sociology*, Vol. LVI (September, 1950), pp. 156-60.

printed from *American Journal of Sociology*, Vol. 74 (November, 1968), pp. 292-95 by permission.

TABLE I

The Distribution of the Ethnic Groups in Chicago, 1960*
(Percentage)

Mile Zone	*Population*										
	Total	*German*	*Irish*	*Swedish*	*Polish*	*Italian*	*Russian*	*Czecho-slova-kian*	*Mexican*	*Puerto Rican*	*Negr*
1	1	0	0	0	0	0	1	0	2	2	0
2	3	2	0	1	1	5	1	0	16	20	5
3	7	4	0	0	7	8	4	0	23	28	8
4	9	11	1	0	14	11	9	0	11	22	13
5	12	11	4	7	10	10	9	22	11	11	20
6	13	12	7	6	12	11	6	30	10	5	14
7	12	13	14	40	9	12	9	35	4	3	13
8	12	11	17	18	11	9	19	0	2	6	13
9	9	11	10	2	10	8	22	0	3	1	5
10	7	9	22	8	9	8	13	0	1	0	3
11	6	6	15	5	8	5	4	13	6	0	2
12	3	4	4	0	4	3	2	0	7	1	1
13	2	3	1	6	2	3	1	0	2	1	1
14	2	2	4	7	1	3	0	0	1	0	1
15	1	1	1	0	2	3	0	0	1	0	0
16	1	0	0	0	0	1	0	0	0	0	1
17	0	0	0	0	0	0	0	0	0	0	0
18	0	0	0	0	0	0	0	0	0	0	0
Total	100	100	100	100	100	100	100	100	100	100	100

* The city of Chicago is divided into eighteen mile zones from the intersection of Madison and State street The population figure in each mile zone is based on the U.S. Census. All census tracts with 50 per cent c more of their areas lying within a mile zone are listed as belonging to that zone.

farther than other foreign-born whites, who moved farther than the Negroes.

The 1940-60 trend was a continuation of the outward movement since 1898 except that the earlier movements were mainly from the inner zone to the middle zone located within 4-9 miles of the intersection of State and Madison Streets, rather than to the outer zone. As a result, the middle zone (proportionately) increased population before 1940. It declined afterward except that its Czechoslovakian population increased during 1940-60.

In 1960 the distribution of the leadin ethnic groups in Chicago was partially due t a time factor in that those of the earlies group, such as the Irish and Swedish, wer found farther from the Loop than the new comers — the Mexicans and Puerto Ricans and the Negroes. For example, in 1960 ha of the city's Puerto Ricans were found i the inner zone and only about 1 per cent i the outer zone. The corresponding figure for the Irish were almost zero or less than per cent, and 47 per cent, respectively.

Other factors related to the populatio

TABLE II

Percentage of Distribution of Ethnic Groups in Chicago*

TABLE II—*Continued*

Mile Zone	*Population*			
	1898	*1920*	*1940*	*1960*
tal:				
1-3	25	23	15	11
4-9	70	70	78	67
10 or more	5	7	7	22
Total	100	100	100	100
rman:				
1-3	44	19	12	6
4-9	51	74	80	69
10 or more	5	7	8	25
Total	100	100	100	100
sh:				
1-3	46	19	10	0
4-9	50	76	81	53
10 or more	4	5	9	47
Total	100	100	100	100
edish:				
1-3	38	10	8	1
4-9	52	78	81	73
10 or more	10	12	11	26
Total	100	100	100	100
ish:				
1-3	60	38	20	8
4-9	29	48	69	66
10 or more	11	14	11	26
Total	100	100	100	100
lian:				
1-3	85	25	15	13
4-9	12	73	83	61
10 or more	3	2	2	26
Total	100	100	100	100
ssian:				
1-3	72	30	8	6
4-9	25	66	88	74
10 or more	3	4	4	20
Total	100	100	100	100
Czechoslovakian:				
1-3	65	27	17	0
4-9	33	70	80	87
10 or more	2	3	3	13
Total	100	100	100	100
Mexican:				
1-3	—	—	—	40
4-9	—	—	—	42
10 or more	—	—	—	18
Total	—	—	—	100
Puerto Rican:				
1-3	—	—	—	50
4-9	—	—	—	49
10 or more	—	—	—	1
Total	—	—	—	100
Negro:				
1-3	66	24	15	13
4-9	33	75	84	78
10 or more	1	1	1	9
Total	100	100	100	100

* The figures for 1898, 1920, and 1940 are based on the diagrams in Richard G. Ford, "Population Succession in Chicago," *American Journal of Sociology*, Vol. LVI (September, 1950), pp. 156-60, except that there are no available data for Puerto Ricans and Mexicans during these years.

succession were the shape of the city, the reference point, and the direction of movement for each ethnic group. The city is longer from north to south than from east to west. Also, the intersection of Madison and State Streets is closer to the northern than to the southern city limits. As a result, within the city itself, the southward movement can go farther than the northern or western movement. A case to illustrate the point is that both Irish and German belong to the earliest immigrant group, but in 1960 there was a higher percentage of Germans in the inner or middle zone and more Irish in the outer zone. Actually, the Germans were dispersed as well as the Irish, but the latter moved toward the southern or the southwestern outskirts while the former were concentrated in the northern half of the city.

As a matter of fact, the whole city of Chicago contains seventy-six communities; each can be classified according to its ethnic composition in 1960 or afterward, as indicated in Table I. Two Negro districts and eight other ethnic districts can also be identified (Table II). The two Negro areas are the west (including the near west side, North Lawndale, East Garfield Park, and West Garfield Park) and the near south (including the near south side, Douglas, Oakland, Kenwood, Washington Park, Chatham, Armour Square, Fuller Park, Grand Boulevard, Englewood, Greater Grand Crossing, and Woodlawn), where the majorities have been Negroes sinc 1966. The other eight ethnic areas are th central, near north, far north, near northwes far northwest, near southwest, far southwes and far south. In 1960 the leading foreig ethnic group was German in the central an near north; Russian in the far north; Polish German in the far northwest; Polish-German Italian in the near northwest; the Polish i the near southwest (except German in Chic ago Lawn); and the Irish in the far south west. The far south was heterogeneous wher Hungarians led in Burnside; Russians, i South Shore and South Deering; Polish, i South Chicago, East Side, Hegewisch, Calu met Heights, and West Pullman; Italians, i Roseland and Pullman; and Irish in Avalo Park; here Riverdale has been a Negro com munity. While the two Negro areas may re main unchanged ethnically, the other eigh ethnic areas are subject to modifications be cause of recent developments.

Section B

Making a Living in Cities

Making A Living In Cities

Harold M. Mayer, in the opening paper of this section, overviews the characteristics of the types of economic activity common to urban places, and then deals in detail with examples of well established theories and methods of urban economic function analysis. References are made to many fundamental concepts of urban geography, primary, secondary, tertiary activities and their interrelationships; hierarchy and primate city; scale economies; trade areas; place utility. Special emphasis is given to transportation.

> . . . the location of economic activities, and hence of urban places within which these activities are performed, depends upon transportation, and transportation geography and urban geography are practically inseparable.

The bulk of the paper examines the method of functional analysis available to the geographer interested in studying the urban economic base. Again, well established theory is discussed, the basic-nonbasic approach, the input-output approach, the minimum or average requirement approach, and factor analysis. Considerable space is devoted to outlining the basis of each and the problems related to their use. All of the approaches

> . . . are concerned with the flows into and out of cities and regions and all of them develop the concept that the basic economic support of an area lies in the specialized functions which it performs for people and establishments outside of the area.

Mayer early establishes that urban socio-economic advantages

> . . . depend upon the individual making a living. The economic base, therefore, constitutes t reason for the development and growth of mc cities.

Hans Blumenfeld in *The Economic Ba of the Metropolis* argues vehemently ar somewhat convincingly against the "basi nonbasic" theory of urban economy for lar cities. He also attempts to show how, in h opinion, the theory has been abused. Blume feld maintains that

> It is thus the "secondary", "non basic" indu tries, both business and personal services, as w as "ancillary" manufacturing, that constitute t real and lasting strength of the metropolit economy. As long as they continue to functi efficiently, the metropolis will always be able substitute new "export" industries for any th may be destroyed by the vicissitudes of econ mic life.

The paper provides for the student an e ample of argument and an example of ho to substantiate argument. Blumenfeld mak his paper quite readable by using many e amples from his own experience and a liber selection of quotations from related work.

The Functional Bases of Small Towns, b Stafford, provides an easily read and unde stood application of simple statistics to ge graphic research. The economic function small urban places is particularly well cover in the study of towns as central places. Wh this paper does provide insight into makir a living in small urban places, it is the use coefficient of correlation and regression th distinguishes it. Stafford relates his work that of several other researchers, indicatir how a seemingly isolated piece of researc can be integrated with, and contribute to, t larger area of study.

e Function Of Cities

e following papers are selected as examples studies in specific economic functions of an places.

Two developments of human culture have n outstanding during the past century: The id advance of technology and the growth and ead of cities. Neither would have been pos- le without the other. Together they have made evolution in the organization and pattern of d use. Transportation is the basis of both.

With this, Mayer introduces his paper on nsportation and the city. In it he sum- rizes the development of the transportation thods, and shows how each has affected the y, morphologically and functionally. Mayer ves little doubt about his feelings on the y-making characteristics and potentialities the dominant urban modes of transport.

Voorhees, in *Transportation Planning and ban Development* concentrates on the tomobile. Congestion of cars produces traf- Traffic is discussed from the point of view factors causing congestion and the results. conclusion some generalities are made on nsportation solutions.

In *Retail Structure of Urban Economy* d *Emerging Patterns of Commercial Struc- e in American Cities,* the commercial func- n of urban places is discussed. The first per, by Kelley, focuses attention largely on 'regional shopping centre'. In so doing, wever, comparisons are made with other es of retail organization and structure. The velopment and characteristics of the 're- nal shopping centre,' the factors of location d the relations of the centre to the CBD are cussed at length in the paper. The scope of Vance paper is considerably wider. Here the author provides the student with a well organized model of subject research and presentation. After an introduction, he deals in the theory of commercial location based upon recent developments and the historical growth of the retail function in America. The last half of the paper is an application of the principles established to the San Francisco Bay Area. In this section the method of re- search is outlined and much of the data plot- ted on maps.

William Goodwin in *The Management Center in the United States* raises the question of considering functions not easily classified by writers on the economic base of cities. After establishing the importance of the man- agement function, he develops an easily understood spatial and functional analysis making excellent use of graphics in distribu- tion maps, scatter diagrams, tables, and graphs.

Reference to the central business district is made in several of the above papers. It is a spacial entity which has received considerable attention because of critical changes that have taken place and because it is the focus of major renewal activity in almost all major cities. In *Public Policy and the Central Busi- ness District* and *The Core of the City: Emerging Concepts* the functions and prob- lems of this zone are discussed and projec- tions are made for its future. Boyce raises the question of whether the CBD will or should continue to exist.

City and Region

Certainly one of the most critical areas of concern for urban geographers and planners alike has been the area into which present day cities are expanding, and the much larger sphere of influence beyond. Each city is surrounded by its hinterland, umland, suburbia, exurbia, slurbs. These, and other terms coined by researchers, relate to some aspect of the region served by and serving the city at the centre.

Spelt, in *Towns and Umlands,* defines umland — its functions, and the relationships and feelings that have developed between the city and this area. Considerable space is devoted to the problem of fixing boundaries. Many examples of devices used for boundary definition are cited. Bonn, Germany is used to introduce some thoughts on regional planning.

External Relations of Cities: City — Hinterland, by Epstein, compares the functions of city and of hinterland. The distribution of functions is dynamic. Many shifts that have taken place relate to increased individual mobility. Residence and economic functions, once tied to the city proper, have relocated outside the corporate limits. Epstein examines the centrifugal and centripetal forces opera tive today causing shifts in the pattern of lan use in city and region.

Gaffney in *Urban Expansion — Will Ever Stop?* and Clawson in *Urban Sprawl an Land Speculation* attack a universal proble of North American cities, sprawl. The con cern is for space — the amount of space con sumed and the wasteful character of its con sumption. The cause is seen to be economic Gaffney's

> . . . thesis . . . is that urban land prices ar uneconomically high — that the scarcity c urban land is an artificial one, maintained by th holdout of vastly underestimated supplies in a ticipation of vastly overestimated future de mands.

The authors describe, document and explai the problem in detail.

Some solutions are offered. Neither autho is particularily hopeful or very realistic whe one considers the record of the individuals c institutions responsible.

laking a Living in Cities: he Urban Economic Base

14

arold M. Mayer

ople live in or near cities in order to secure vantages which would not be possible der non-urban conditions. Among these vantages are the satisfaction of many sires: social, religious, recreational, and onomic, but most of them depend upon the lividuals making a living. The economic se, therefore, constitutes the reason for the velopment and growth of most cities. Even sentially non-economic functions, such as fense, recreation, and pilgrimages to his- ic and religious shrines, constitute im- rtant income-producing activities for the ies and regions within which they occur. nowledge of the economic base of cities, erefore, is an indispensable prerequisite to understanding of urban geography.

aracteristics Of Urban Functions

an Bruhnes, two generations ago, called oan land uses such as houses and streets nproductive occupation of the soil."[1] He s referring, of course, to the classical con- pt of the economist that the significant in- ts into an economic system are land, labor, and capital, and that the natural resources of the site, including its agricultural, mineral, or silvicultural productivity, are the only productive forms of land use. More recently, Chauncy Harris pointed out that cities are efficient instruments for utilizing resources productively, including labor, which must concentrate in and near cities in order to perform its increasingly specialized roles, and that, in proportion to the amount of land used, urban land is thus extremely productive, in the sense that it creates utility by the processing, transfer, and distribution of goods and services.[2]

Economists and geographers find it useful to divide income-producing activities into several categories, which they designate as primary, secondary, and tertiary.[3] Primary activities are those which produce utility, and

Jean Bruhnes, *Human Geography* (Chicago, Rand McNally & Co., 1920), pp. 74-229.

[2] Chauncy D. Harris, "The Pressure of Residential-Industrial Land Use," *Man's Role In Changing the Face of the Earth*, by William L. Thomas, Jr., *ed.* (Chicago, the University of Chicago Press, 1956), pp. 881-95.

[3] Most standard textbooks in economic geography have adopted this terminology, for example: John W. Alexander, *Economic Geography* (Englewood Cliffs, N.J., Prentice-Hall, Inc., 1963), pp. 5-6.

printed from *Journal of Geography* (February, 1969), pp. 70-87 by permission.

hence income, by extraction of the resources on the site, whether the resources are renewable, such as agricultural produce or waterpower, or are non-renewable, such as minerals and petroleum. Secondary activities are those in which goods are handled; included in that category of economic activities are manufacturing and transportation: any activity which involves the changing of the form or location of commodities. Tertiary activities involve either the performance of services or change in ownership of goods, as in wholesale and retail trade. With increasing complexity of society and its economic activities, there is an ever-increasing amount of record-keeping and "paper work"; some authors prefer to categorize these activities as quaternary, rather than tertiary.[4]

Few cities have as their most important functions the primary activities. Even though many urban places are directly dependent upon such activities, they actually function as manufacturing or service centers, utilizing the resources produced nearby or performing services for the organizations and people who, in turn, are directly involved in the primary activities. If the resource is non-replaceable, the urban center which serves the extractive operations may have a precarious economic base, and may decline or disappear when the resource is exhausted or becomes non-competitive in the market with other source or substitute resources. Such places may be come "ghost towns" unless other activities ca be introduced to replace the declining ones.

Urban land is valuable, then, not for its in herent productivity of natural resources, bu rather, because of its location with relation t the secondary, tertiary, and quaternary acti vities: the handling of goods and the per formance of services.[5] Since land is valuabl only for its present and prospective uses, an all urban functions, most of which are eco nomic in character, take place on land, it fol lows that studies of the economic base mus constitute essential parts of the field of urba geography as well as major components c the research which is an integral part of th process of city and regional planning. Be cause urban functions are highly localize and concentrated, the spatial aspects of th urban economy are the foci of the geog grapher's interest in economic base studies.

Communities which depend mainly o primary economic activities — the extractio of resources — directly or indirectly for thei support, are, of course, resource-oriente with respect to their locations: they ar located close to the resources. Those urba places which depend mainly upon secondar activities — manufacturing, assembling, an packaging of goods — may be resource oriented in location, or market-oriented, de pending upon many variables, such as th

[4] For an interesting account of the significance of these functions in metropolitan New York, see: Sidney M. Robbins and Nestor E. Terleckyi, *Money Metropolis* (Cambridge, Harvard University Press, 1960). Administrative functions may also be regarded as quaternary, as, for example, in Jean Gottmann, *Megalopolis* (New York, The Twentieth Century Fund, 1961), pp. 565-630. A discussion of business management as an urban function is: William Goodwin, "The Management Center in the United States," *Geographical Review*, Vol. LX, no. 1 (January, 1965), pp. 1-16.

[5] Location as related to urban land values emphasized in all treatises on urban land eco nomics and real estate. See, for example Richard U. Ratcliff, *Urban Land Economic* (New York, McGraw-Hill Book Co., Inc 1949); Ernest M. Fisher and Robert M. Fishe *Urban Real Estate* (New York, Henry Holt Co., 1954); and Arthur M. Weimer and Home Hoyt, *Principles of Real Estate* (New Yor The Ronald Press Company, 1954), 3rd. ed.

ations of weight, bulk, and value of the ods, and hence their ability to stand trans-tation costs as raw materials, semi-shed goods or components, or manufac-ed products, or they may be "footloose" free to locate anywhere between the rce of their resource inputs and the final rkets.[6] As transportation costs are reduced tive to other costs, there is a noteworthy dency for an increasing proportion of the nufacturing industries to become footloose narket-oriented.[7] This tends to accentuate population and economic growth of the ger urban agglomerations, and to aug-nt their attractiveness for still more activi- and population; there is thus a "multi-r effect."

Vholesale trade may be regarded as a form secondary economic activity in the sense t goods are handled and stored; orders are embled for distribution, and warehouse buildings represent forms of industrial structures. Wholesale facilities, therefore, are subject to the same effects as are manufacturing establishments. They may be located in proximity to the manufacturing establishments, but more commonly they are located close to the major markets — the larger urban centers. On the other hand, competition for centrally-located urban land by other uses generally forces wholesale establishments having warehousing requirements to the peripheries of the larger cities, while the wholesale office establishments and display rooms may remain in more central locations. In general, improvements in transportation and changes in the methods of marketing, including the rise of chain stores, standard brands, and emphasis upon reduced inventories requiring storage, has reduced the relative importance of warehousing as an element in the urban pattern. The current trend is to consider wholesaling and warehousing as an "interface" or intermediate stage in a transportation route which, in turn, is part of a total distribution system.[8] With a trend toward separation of warehouses from offices and sales rooms of wholesalers, the latter may more properly be regarded as tertiary, rather than secondary, functions. On the whole, there have been far fewer studies of the locational patterns, both theoretical and empirical, than of location of either manufacturing or of retailing establishments.

The final stage in the chain of movement of resources, whether or not they undergo change enroute, is the distribution of goods and services to the ultimate consumer. Tertiary activities thus tend to be less concentrated in location than do the primary and

'here is vast literature on location theory rela-ve to secondary activities. Some of the more nportant items include such classical studies s: Alfred Weber, *Theory of the Location of ndustries* (Chicago, University of Chicago 'ress, 1928); August Lösch, *The Economics of ocation* (New Haven, Yale University Press, 954); and Edgar M. Hoover, *The Location of conomic Activity* (New York, McGraw-Hill ook Co., Inc., 1948). A brief summary of cation theory relative to secondary functions : William Alonso, "Location Theory," *Re-ional Development and Planning* by John 'riedmann and William Alonso (Cambridge, he M.I.T. Press, 1964), pp. 78-106. For a omprehensive bibliography on the subject, see: enjamin H. Stevens and Carolyn A. Brackett, *ndustrial Location, A Review and Annotated ibliography of Theoretical, Empirical and Case tudies* (Philadelphia, Regional Science Research nstitute, 1967).

:hauncy D. Harris, "The Market as a Factor n the Location of Industry in the United tates," *Annals of the Association of American eographers*, Vol. XLIV, no. 4 (December, 954), pp. 315-48.

[8] Charles A. Taff, *Management of Traffic and Physical Distribution* (Homewood, Ill., Richard D. Irwin, Inc., 1968).

secondary activities. Some, such as food retailing, tend to locate in patterns closely resembling those of the consuming population, for everyone eats. Goods and services which are more specialized tend to be less ubiquitous, and to concentrate in larger urban centers, for they require a larger population base, or disposable income, for their support. Thus there tends to be a hierarchy of retail and service establishments, with the more specialized ones in fewer but larger urban centers, and there also tends to be a hierarchy of the urban centers which supply consumer goods and services, the larger centers, requiring a larger "threshold" of support, having larger and more specialized establishments. The "trade areas" or "hinterlands" of the larger and more specialized establishments, and consequently of the larger urban centers containing them, tend to be larger than those of the less specialized urban places. At the same time, the larger centers also have "lower order" functions within them, duplicating the functions of the smaller centers, and serving smaller trade areas than those of the more specialized functions within the same center.

The location and characteristics of the tertiary functions of urban places, and their respective service areas have been intensively studied over several decades by geographers and others, and a theoretical framework has been developed, generally known as "central place theory," following the terminology of Walter Christaller, who formalized the theoretical statement in 1933.[9] There is probably a larger volume of theoretical and empirical studies of central places than of any other topic within the entire field of urb geography.[10]

In satisfying the economic demands people everywhere, except in the few remai ing isolated regions which are self-sufficie goods and services are made available at t locations of the consumers, who constitu the market, regardless of where the resourc are originally located. In terms used by ec nomists, resources must be given "pla utility": they must be transported. Goo must be delivered and services must be pe formed either at the residential location of t consumer or at an establishment so locat as to involve a minimum of travel by the co sumer. In any event, the complex of esta lishments providing goods and services (i cluding consumption) involves clusters inter-related complementary or competiti establishments, or central places. The places, in turn, must be accessible, and th they are served by networks of transportati and communication routes. They are nod or vertices, in an inter-connected network lines or routes over which people and goo move in order to satisfy demands for goo and services. Central place theory relat tertiary functions and establishments to t transportation network. Location theor which describes and interprets the patterns

[9] Walter Christaller, *Central Places in Southern Germany*, trans. by Carlisle W. Buskin (Englewood Cliffs, N.J., Prentice-Hall, Inc., 1966).

[10] A comprehensive bibliography, listing hundre of studies of central places is Brian J. L. Ber and Allen Pred, *Central Place Studies, Bibliography of Theory and Applications, I cluding Supplement Through 1964* (Philadelph Regional Science Research Institute, 1965). recent concise statement on the subject is Bri J. L. Berry, *Geography of Market Centers a Retail Distribution* (Englewood Cliffs, N. Prentice-Hall, Inc., 1967). The subject will developed further by Bart J. Epstein in a su sequent article in this *Journal of Geograp* series.

ation of secondary functions and establishnts, likewise depends upon the existence
a transportation network[11] for assembly of
uts — raw materials, fuels, components,
d labor — and the distribution of outputs
m the establishment to successive estab-
ments for later stages in the transforma-
n of resources from their original state and
ation into forms and locations to meet
consumer demands. Thus, the location of
nomic activities, and hence of urban
ces within which these activities are permed, depends upon transportation, and
nsportation geography and urban geogra-
y are practically inseparable.[12]

Accessibility is fundamental to the per-
mance of any economic function. Since all
er-acting establishments and functions
not be located in mutual proximity, trans-
tation must be used to overcome distance.

ome geographers have applied "graph theory" o the description and interpretation of trans- ortation networks, with urban centers as the odes or vertices. See, for example: Peter Hag- ett, *Locational Analysis in Human Geography* New York, St. Martin's Press, 1966), pp. 1-86; W. L. Garrison, "Connectivity of the nterstate Highway System," *Regional Science Association, Papers and Proceedings*, Vol. VI 1960), pp. 121-37; and K. J. Kansky, *Structure of Transportation Networks: Relationships be- ween Network Geometry and Regional Char- cteristics*, Research Paper No. 84 (Chicago, University of Chicago, Department of Geog- aphy, 1963).

t is not within the scope of this series to eview transportation geography. However, see dward L. Ullman and Harold M. Mayer, Transportation Geography," *American Geog- aphy, Inventory and Prospect*, Preston E. ames and Clarence F. Jones, *eds.* (Syracuse, yracuse University Press, 1954), pp. 311-32; lso Harold M. Mayer, "Urban Geography and Jrban Transportation Planning," *Traffic Quar- erly*, Vol. XVII, no. 4 (October, 1963), pp. 10-31.

In movement of resources between origin and place of utilization, and in providing goods and services at and from central places, movement along transportation routes must take place. Movement represents a cost, and the cost may be substituted for lineal mileage in the measurement of distance in determining the effectiveness and strength of the bonds which inter-relate places with one another. The boundaries of the areas within which resources may effectively compete with alternative sources, and the boundaries of trade and service areas around central places can be ascertained, in general, by application of the concept of "distance decay," which simply means that the farther apart two places are, the less traffic is likely to be generated between them, and there is a point in every direction from a place, where some other place is equally attractive; such a place is a "traffic-divide" or a hinterland boundary. The "distance decay" tends to be a negative exponential curve; the strength of the attraction to a particular place falls off at a decreasing rate with distance. Also, the larger and more diversified a place is, the stronger the attraction it exerts. This is analogous to the physical law of gravitation, and the gravity analog or model has been very useful in predicting the traffic generation to be exerted by cities, shopping centers, and other nodes, as well as to anticipate the volume of traffic to be moved on new links in the transportation networks. Similarly, the total gravitative attraction exerted by a central place or establishment represents the total attraction of that place relative to all other places with which it has interaction. The latter represents what is called "potential." If all of the places in the United States have their relative potentials represented by an imaginary surface, such as can be shown by contours, it can readily be seen that the places and areas with highest density of population and purchasing power — namely, the cities and urbanized regions — represent peaks and ridges of maximum potential.[13] These areas are especially attractive for new market-oriented establishments and activities.[14] In other words, the larger and more accessible urban places and regions tend to attract economic activity, and hence population, to a greater extent than do smaller places and less densely populated regions. The additional concentration of people, establishments, and activities, in turn, further reinforces the market, and gives rise to still further growth, specialization, and diversity, and thereby creates still additional attractive force. With larger markets, more goods and services can be, and are, provided, and the unit costs of production, spread over more units, can be reduced, making the large urban concentrations still more attractive; this is what economists call "agglomeration ec nomies" or "scale economies." Furthermor the larger cities and metropolitan areas ca support increased specialization; a wid variety of goods and services is availab from an increasing number of specialize establishments, thereby reducing or elimina ing the necessity for many auxiliary activiti from being related to specific plants; th specialized establishment can be called upo to supply items or services which otherwis each plant would have to supply for itse This increases overall efficiency and mak establishments in larger and more diversifie urban areas better able to compete, wit lower costs, in more distant markets, thereb overcoming to a major extent the transporta tion cost disadvantages in reaching distar markets. This phenomenon is called "extern economies," and it is a major force in attrac ing many activities to the larger cities.[15] I effects constitute an essential base for centr place theory, and Mark Jefferson recog nized it in his "Law of the Primate City, when he pointed out that the largest city in region or nation tends to enhance its primac by exerting the strongest attractions for th greatest variety of activities.[16] In one sens the primate city may be regarded as a speci case: it represents the top of the hierarchy c central places. In many countries and region the numbers of urban places of various popu lation sizes seem to follow a rank-size rela tionship in which all places bear a definite siz relationship to the largest place, in regula

[13] Gerald A. P. Carruthers, "An Historical Review of the Gravity and Potential Concepts of Human Interaction," *Journal of the American Institute of Planners*, Vol. XXII, no. 2 (Spring, 1956), pp. 94-102; Walter Isard, "Gravity, Potential and Spatial Interaction Models," *Methods of Regional Analysis* (New York, Technology Press and John Wiley & Sons, 1960), chap. 11; F. Luckermann and P. W. Porter, "Gravity and Potential Models in Economic Geography," *Annals of the Association of American Geographers*, Vol. L, no. 4 (December, 1960), pp. 493-504; Gunnar Olson, *Distance and Human Interaction, A Review and Bibliography* (Philadelphia, Regional Science Research Institute, 1965), pp. 43-70.

[14] John Q. Stewart, "Empirical Mathematical Rules Concerning the Distribution and Equilibrium of Population," *Geographical Review*, Vol. XXXVII, no. 3 (July, 1947), pp. 461-62 and 471-85; William Warntz, "A New Map of the Surface of Population Potentials for the United States 1960," *Geographical Review*, Vol. LIV, no. 2 (April, 1964), pp. 170-84.

[15] Robert M. Lichtenberg, *One Tenth of a Natior National Forces in the Economic Growth c the New York Region* (Cambridge, Harvar University Press, 1960), especially pp. 56-70.

[16] Mark Jefferson, "The Law of the Primate City, *Geographical Review*, Vol. XXIX, no. 2 (Apri 1939), pp. 226-32.

scending order of rank: thus the second-ıking place would tend to be half the size the largest place, the third ranking place e-third the size, and so forth. Cities and :tropolitan areas of the United States tend have such a rank-size relationship.[17] Here, : primate city is, of course, the top-ranking e, and all others tend to be related in popu-ion size inversely to their rank; New York the primate city, Chicago and Los Angeles :tropolitan areas, respectively, are the next-ıking ones, but New York also contains re-)nal functions for the eastern third of the ıited States, corresponding to the regional ıctions of Chicago for the Midwest and Los ıgeles for the Far West. Below these re-)nal primate cities, in each instance, there a regular procession of metropolitan areas MSA's), having a rank-size relationship.

affic Flows And The Urban Economic Base

e have seen that cities constitute foci or des in networks of transportation and com-ınication, making possible the inter-actions th other cities and with the service areas or ıterlands. The urban functions depend on accessibility, which conditions speciali-:ion of land uses,[18] and give rise to inter-ange, or movement of people, goods, and :as between and among cities, and between cities and their respective hinterlands.[19] These flows, constituting what Edward Ullman calls "spatial interaction,"[20] produce traffic, the volume of which, along transportation and communication routes, is the measurable result of the gravitative "pulls" exerted by the various urban centers or subcenters: the "traffic generators."

Since cities are essentially areas of specialized land uses, facilities, and establishments, which depend upon spatial interactions with each other and with external areas, it follows that an effective method of analysis of the economic base of a city or metropolitan area would be to study the volumes and composition of the traffic flows — of people, goods, and information — between the city and each area or region oustide. These flows, of course, are two directional, and they reflect the nature of the urban functions which give rise to them.

Unfortunately, however, there are several major difficulties. One set of difficulties is the general lack of usable data on traffic flows for areas which are geographically significant.[21] There are several reasons for lack of useful data: among them are the high cost of gathering information, commercial competition in the United States where common and contract carriers are mainly private enterprises,

For a further description of rank-size distributions, see Walter Isard, *Location and Space-Economy* (Cambridge, Technology Press, and New York, John Wiley & Sons, Inc., 1956), pp. 55-64; F. W. Boal and D. B. Johnson, "The Rank-Size Curve, A Diagnostic Tool?" *The Professional Geographer*, Vol. XVII, no. 5 (September, 1965), pp. 21-23.

Walter G. Hansen, "How Accessibility Shapes Land Use," *Journal of the American Institute of Planners*, Vol. XXV, no. 2 (May, 1959), pp. 73-76.

19 Harold M. Mayer, "Urban Nodality and the Economic Base," *Journal of the American Institute of Planners*, Vol. XX, no. 3 (Summer, 1954), pp. 117-21.

20 Edward L. Ullman, "The Role of Transportation and the Bases for Interaction," *Man's Role in Changing the Face of the Earth*, William L. Thomas, Jr. (Chicago, University of Chicago Press, 1956), pp. 862-80.

21 Problems of delimitation of urban areas were discussed in the preceding article: Harold M. Mayer, "Cities and Urban Geography," *Journal of Geography*, Vol. LXVIII, no. 1 (January, 1969), pp. 6-19.

the high proportion of the traffic which is carried in private vehicles, military security, and lack of agreement on area boundaries for statistical purposes.

Geographic studies of ports are numerous, nevertheless, and it was largely as a result of the concepts of hinterlands developed in port studies that the city-region relationships have been emphasized by urban geographers.[22] Port statistics are more readily available than are statistics for other transportation movements, and especially available are international movement statistics, for people and goods involved in transit through ports enroute to or from foreign countries are subject to regulation by immigration, customs, and public health authorities, all of whom maintain statistics, and many of whom publish aggregated figures on such movements in more or less usable form.

Ports, furthermore, are of special interest to geographers, because they are generally gateways to and between complementary regions, the interactions of which, in the form of traffic flows, may be indicative of the character of development of the respective regions. Extensive statistics are available, for many nations, on the volume, direction, and composition of flows of goods and people through their respective ports, although commonly the landward points of origin an destination are difficult or impossible to deter mine without elaborate supplementar studies.

Detailed point-to-point statistics are avail able for domestic and international move ments of air passengers and cargo at airport of the United States and some other countries though here, too, the actual landward origin and destinations, beyond the respective air ports, may be difficult to determine. Never theless, the air traffic flows among urban area are indicative of the character, extent, an volume of interaction among major cities and several studies by geographers have bee useful in determining the air traffic hinter lands of cities.[23]

Railway traffic flows are generally unavail able, or are not available for geographically meaningful areas, and it is difficult, an usually impossible, to determine the volum and nature of freight movements into, out of and through, urban areas.[24] Similarly, move ments by motor truck are not generally avail able, although a few special studies have bee made to determine the motor truck hinter lands of cities; these, however, do not includ information by commodities.[25] In some part of the world, inter-city bus services carry significant proportions of the passenge

[22] Among the hundreds of port studies stressing hinterland relationships, the following represent a few of the more significant ones: Guido G. Weigand, "Some Elements in the Study of Port Geography," *The Geographical Review*, Vol. XLVIII, no. 2 (April, 1958), pp. 185-200; Donald J. Patton, "General Cargo Hinterlands of New York, Philadelphia, Baltimore, and New Orleans," *Annals of the Association of American Geographers*, Vol. XLVIII, no. 4 (December, 1958), pp. 436-55; Edwin H. Draine, *Import Traffic of Chicago and Its Hinterland*, Research Paper No. 81 (Chicago, University of Chicago, Department of Geography, 1963).

[23] For example: Edward J. Taaffe, *The Air Passenger Hinterland of Chicago*, Research Pape No. 24 (Chicago, University of Chicago, De partment of Geography, 1952).

[24] State-to-state data, by commodities, have bee available for the past two decades from on per cent waybill samples; see Edward L. Ull man, *American Commodity Flow* (Seattle, Uni versity of Washington Press, 1957).

[25] Magne Helvig, *Chicago's External Truck Move ments*, Research Paper No. 90 (Chicago, Uni versity of Chicago, Department of Geography 1964).

vement. In such instances the pattern of routes is indicative of the urban hinterds, and may be used to indicate, also, the ition of towns in the central-place hierhy.[26] In the United States, however, only nall portion, less than 2.5 percent, of the r-city personal movement is by bus, while vate automobiles are responsible for over percent of the passenger movement to, n, and between urban places. Knowledge the volumes, directions, purposes, and er characteristics of automobile trips is ispensable, therefore, not only for the nning of street and highway systems, but as an integral part of the process of city metropolitan planning generally, in which reciprocal relations between transporta- systems on the one hand and the loca- s, functional characteristics, and densities and uses on the other are studied in con- rable detail, and projections are made of sible alternative arrangements of trans- tation routes and land uses in the future. eed, such studies, integrating the func- al patterns of urban areas and the present proposed locations of streets, highways, other transportation facilities have been de mandatory by the federal government e 1965 as prerequisite for federal funding many public improvements, including ways, mass transit, urban renewal, open ce land acquisition, sewer and water supply systems, and many others. Such studies are required to extend beyond city boundaries and to include all metropolitan areas. They are generally conducted by designated metropolitan or regional planning agencies, which subsequently must indicate conformance with regional comprehensive plans if any local governmental unit is to receive financial aid from the federal government for each of a wide variety of public improvement programs. In these studies, urban and transportation geographers are playing increasingly important roles. Particularly significant is the fact that studies of origins and destinations of present and prospective trips are always major portions of such comprehensive planning operations, and economic base studies form the framework for population projections which, in turn, are highly useful in the preparation of comprehensive regional, metropolitan, and city plans.[27] The potentialities for utilization of the increasing number of such transportation-land use studies in determining the nature and extent of the economic base and hinterland connections of cities are very promising. As such studies are produced for more and more areas, intercity comparative analyses may yield significant generalizations and understandings relative to urban growth and

. Godlund, "The Function and Growth of Bus 'raffic within the Sphere of Urban Influence," *und Studies in Geography, Series B, Human eography*, Vol. 18 (1956); F. H. W. Green, Motor Bus Centers in S.W. England Considred in Relation to Population and Shopping 'acilities," *Transactions and Papers, Institute of ritish Geographers* (1948), pp. 57-68; "Bus ervices as an Index to Changing Urban Hinternds," *Town Planning Review*, Vol. XXII 1951), pp. 345-56.

[27] Examples of such comprehensive metropolitan planning studies include: *Report on the Detroit Metropolitan Area Traffic Study* (Lansing, Speaker, Hines and Thomas, Inc., 1955 and 1956), 2 vols.; *Chicago Area Transportation Study Final Report* (Chicago, 1959-1962), 3 vols.; *Pittsburgh Area Transportation Study* (Pittsburgh, 1961 and 1963), 2 vols. For a concise general description of the nature and findings of such studies, see John F. Kain, "Urban Travel Behavior," *Urban Research and Policy Planning*, Leo F. Schnore and Henry Fagin, *eds.* (Beverly Hills, Calif., Sage Publications, Inc., 1967), pp. 161-92.

development, leading to predictive models of great theoretical and practical importance. Furthermore, as subsequent studies are made, in future years, for the same areas as those subjected to earlier studies, detailed knowledge may be gained of changes and trends in the economic base, population distribution and characteristics, and travel patterns of such urban areas.

Such comprehensive metropolitan transportation studies confirm the fact that only a very small proportion of the trips originating or terminating within a metropolitan area cross the boundary of the area, as would be expected from the nature of the definition of such areas. Flows crossing the boundary, however — "external" trips — are significant to understanding the nature of city-hinterland relations. Flows of goods are less readily available statistically, but the proportion of total goods movements which are external undoubtedly far exceeds the proportion of internal trips, for urban areas consume tremendous quantities of goods and fuels, and must ship out vast amounts of products and wastes. If more detailed information relative to these movements were available, we could much better understand the nature and extent of the urban economic base.

For purposes of economic base analysis, however, the most significant classification of flows of both people and goods is the dichotomy between external movements, originating or terminating beyond the area boundaries, on the one hand, and internal movements not crossing the area boundaries, on the other hand.

Cities and regions, engaged in production of specialized goods and services, provide economic support for their populations by exchanging such production, to the extent that they do not consume it themselves, for the specialized goods and services produc by other cities and regions. Thus there a "inputs" and "outputs," the total representi the external trade of the given city or regio However, there are two additional consider tions. One is that not all of the production c be "exported," because the needs of t people and establishments within the give area must also be satisfied, and a portion the goods and services which the ar produces are for local consumption. A secor consideration is that the value of the inpu and outputs of an area may not be equal value; therefore if the value of inputs e ceeds that of outputs, the balance must b restored by outflows of money and credi and, conversely, if the area exports are higher value than the imports or input money and credit must flow into the are This, of course, has been extensively studie for nations, for which the balance of pa ments may be critical. National balance payments accounts are relatively easy determine, for records are kept and are avai able for flows of goods, services, peopl money, and credit across national boundarie For local or sub-national areas, howeve such as cities and metropolitan areas, fe usable records are available, and geographe cannot directly study the economic base such areas in the same way as the nation and international balance of payments can b investigated.[28] Furthermore, although incom and its distribution is important, geographe are also interested in the numbers and cha

[28] For further discussion of the "balance of pa ments approach," see Wilbur R. Thompson, *Preface to Urban Economics* (Baltimore, Th Johns Hopkins Press, 1965), especially pp. 6 104, or the chapter by the same author in *Th Study of Urbanization*, Philip M. Hauser an Leo F. Schnore, *eds.* (New York, John Wile & Sons, Inc., 1965), pp. 431-90.

eristics of the people supported by a city's ome, the physical flows of goods and seres, and the facilities for effectuating such ws; income flows alone tell only a part of story.

Three approaches to the study of the eco-mic base of cities have been extensively d by geographers. All are based upon the cepts of flows and the dichotomy of ex-nal and internal components of economic port. These approaches may be termed:) the basic-nonbasic approach, (2) the ut-output approach, and (3) the minimum average requirements approach. All of m are concerned with the flows into and t of cities and regions, and all of them elop the concept that the basic economic port of an area lies in the specialized func-ns which it performs for people and estab-ments outside of the area. The cliché that ple do not live by taking in each other's shing is crude, but apt.

e Basic-Nonbasic Approach

e basic-nonbasic approach has particular peal for geographic study because it pro-es a tool by which may be determined the ent of specialization within a city or region, contrasted with larger regions of which it a part, or of other cities and regions. rthermore, it can furnish an indication, ugh crude, of the number of people, as ll as their incomes, supported by each xport" activity, and it furnishes a measure the extent to which the city or region is endent upon other cities and regions. Such erminations are extremely useful in deter-ning which activities are vital, and which ve the population within the city or region ich is, in turn, dependent upon the "basic" "export" activities. In planning and devel-ment programs, it follows that stimulation of and provision for the "export" or "basic" activities would result in more leverage for growth than would the further development of the remaining functions or activities which are dependent upon the basic ones.[29] Since the population engaged in supplying goods and services for consumption outside the given region, such as a metropolitan area, must, in turn, be supplied with goods and services, the employment and income in-volved in such secondary or supporting activities represents a "multiplier effect." The population — to which must be added family dependents — and the income involved in the "export" activities are considered "basic" and the people and incomes involved in supplying the needs of the basic population is termed "nonbasic" or "secondary." This population, in turn, must also be supplied, so there is a succession of rounds of employment and in-come created, resulting in further population increases.

There are several ways in which the extent to which a given urban function is basic may be measured, and in which determination may be made of the extent to which any combination of functions contributes to the economic base of a community or region.[30] Thus, if a given activity, say steel production,

[29] A useful and concise statement of the history, assumptions, methods, and applications of the basic-nonbasic approach is John W. Alexander, "The Basic-Nonbasic Concept of Urban Economic Functions," *Economic Geography*, Vol. XXX, no. 3 (July, 1954), pp. 246-61, reprinted in *Readings in Urban Geography*, Harold M. Mayer and Clyde F. Kohn, *eds.* (Chicago, University of Chicago Press, 1959), pp. 87-104.

[30] Detailed discussion of economic base methodology is contained in a series of articles by Richard B. Andrews in *Land Economics*, 1953-1956, most of which were reprinted in *The Techniques of Urban Economic Analysis*, Ralph W. Pfouts, *ed.* (West Trenton, N.J., Chandler-Davis Publ. Co., 1960).

results in employment (or income, if the figures were available) within a metropolitan area to the extent that such employment represents twice the proportion of steel-production employment as steel-production employment percentage is of the employment in the same category of industry in a larger area, such as the nation as a whole, it is assumed that the difference, or surplus employment, represents that proportion of the workers engaged in producing steel for shipment outside the area. This is "basic" or "city forming"[31] employment, while the remaining employment in the steel plants of the metropolitan area is "nonbasic" or "city serving" employment, producing for local consumption. The ratio between basic and nonbasic employment — the "B/N" ratio — indicates the relative importance of each category of economic activity in producing employment and hence contributing to the economic support of the city or region.

There are many difficulties in applying this rather simple concept. One set of problems involves the difficulties of obtaining data. It is not always possible to ascertain the extent to which a given category of activity actually serves patrons or customers outside the region, particularly where, in large and complex areas such as metropolitan areas, there is a sufficiently large number of establishments so that individual questionnaires become prohibitively expensive and time-consuming; furthermore, because of commercial competition, a significant proportion of the business establishments would withhold such data.[32] In most such studie aggregated statistics, such as census report are used.[33]

Another problem is the selection of th appropriate measures. This is determine largely by the use to be made of the result i.e.: whether it is desired to obtain the B/N ratio with respect to employment, income present and prospective land use by cate gories of activity, or other types of finding If employment figures are to be used, allow ances must be made for part-time, seasona and other irregular employment, for the elas ticity of demand for marginal workers, suc as married women and young people wh may or may not be on the labor market de pending upon the level of economic activit and wages at a given time, commuting labo force from outside the defined area, absente population such as members of the arme forces, remittances from people and estab lishments outside which may create a marke and hence employment within the area, an many other variables.

The delimitation of the area boundarie may constitute another set of problems. Ad

[31] Gunnar Alexandersson, *The Industrial Structure of American Cities* (Lincoln, University of Nebraska Press, 1956), pp. 14-20, reprinted in Mayer & Kohn, *op. cit.*, pp. 110-15.

[32] Nevertheless, in small cities it may be possib to obtain such information, where a limite number of large establishments may be ques tioned individually, together with a carefull selected sample of smaller establishments; fc example, John W. Alexander, *Oshkosh, Wi consin, An Economic Base Study and An Ec nomic Base Study of Madison, Wisconsi* (Madison, University of Wisconsin Bureau c Business Research, 1951 and 1953).

[33] Since such sources are insufficiently detailed, th resulting economic base studies are necessari crude; for example, Homer Hoyt, *The Ec nomic Status of the New York Metropolita Region in 1944* (New York, The Regional Pla Association, Inc., 1944).

nistrative boundaries — cities, counties, d states — rarely represent functionally aningful boundaries, except for special rposes, such as public administration and cal purposes. City limits do not bound the ographically functional cities. People com- te across such boundaries, and retail and vice establishments on both sides of the e compete for their expenditures. If, how- er, a planner is concerned with the financial se, the allocation of land among potential es, or administrative policy for such a juris- tional unit, the boundary is determined. etropolitan areas (SMSA's) are generally ore appropriate functional units, but they, o, by no means have closed economies, so at people and money, as well as goods and vices, are free to cross metropolitan area undaries. Other things being equal, the ger and more complex an area, the more t it is to furnish a high proportion of its eds for goods and services; thus it would ve a higher proportion of nonbasic activi- s and employment.[34] However, urban eas generally do not produce significant antities of the foodstuffs and fuels that they quire, and they do not have adequate facili- s for disposal of wastes within their bound- es, so that an "import-export" or "basic" mponent is always present. Thus, the loca- n of an area's or region's boundaries has an portant bearing upon the B/N ratio, both total and for individual components of the onomy. The economic base of a metropolitan area is not the same as the economic base of its central city.

The basic-nonbasic approach has many limitations, and has been frequently criticized. One type of criticism is that the approach places too much emphasis upon the basic components. With increasing specialization, the growth of leisure, and growing affluence, the variety of demands for goods and services is increasing, and service functions are increasing more rapidly than is the supplying of goods. Most of the services are supplied internally, especially within the larger cities and metropolitan areas. Thus the internal or nonbasic portion of the economy is increasingly important. Hans Blumenfeld points out that much of the attractiveness of the large metropolis is not in the industrial jobs which it provides, in which the basic component is high, but rather in the attractions, including services, which the urban area itself can furnish. Such services are "non-basic" in the sense in which the term is here used.[35]

Economists and planners tend to emphasize the crude nature of the basic-nonbasic approach.[36] This, of course, is inherent in the paucity of the data which would be essential for more sophisticated methods. Other criticisms are based upon the limitations of the basic-nonbasic approach in not being able to allow for major changes in tech-

This theme is developed by Victor Roterus and Wesley Calef, "Notes on the Basic-Non-basic Employment Ratio," *Economic Geography*, Vol. XXXI, no. 1 (January, 1955), pp. 17-20, reprinted in Mayer and Kohn, *op. cit.*, pp. 101-104.

35 Hans Blumenfeld, "The Economic Base of the Metropolis," *Journal of the American Institute of Planners*, Vol. XXI, no. 4 (November, 1959), pp. 327-36.

36 For example, Charles T. Stewart, Jr., "Economic Base Dynamics," *Land Economics*, Vol. XXXV, no. 4 (November, 1959), pp. 327-36.

nology or general social, economic, and political conditions in the future. However, these criticisms are .equally applicable to more sophisticated and refined methods of study which involve projection from past trends and present conditions.[37] How, for example, could one have accurately predicted the economic base of the central-east coast of Florida — including Cape Kennedy — as recently as a decade ago, or of any region subjected to the introduction of "random variables," such as the location of major new technological developments?

The "Input-Output" Approach

A second approach to study of the urban economic base which has been developed, largely by economists and "regional scientists," is known as the "input-output" approach. Inputs are those items: physical goods, including raw materials and components as well as fuels, labor, capital investment, and credit, which make possible an economic activity within a region or an individual establishment. The sum of the inputs to all establishments, including households, the ultimate consuming unit, is the total set of inputs into a region. Similarly, outputs are the products of the region or of individual establishments. In the case of a region, such as a city or metropolitan area, inputs and outputs may, in many instances, be entirely among the establishments within the area, but some inputs and outputs will involve inter actions with external areas — other cities an regions. In this respect, the "input-output approach resembles the "basic-nonbasic approach. It evolved from the work c Wassily Leontief, an economist, who devel oped the method in studies of national ecc nomies.[38] As has been pointed out, statistic on inputs and outputs for nations are fa easier to obtain than for regions smaller tha nations. Therefore, the method has not bee widely used in studies of the economic bas of cities and metropolitan areas. However, few such studies have been produced, notabl by Walter Isard and his colleagues, who ap plied the method to determine the economi base of Puerto Rico — an island, which is a unusually readily identifiable economic uni — and to an evaluation of the probabl multiplier effects resulting from the estab lishment of a new steel plant in the easter United States.[39]

Application of the input-output approac involves the gathering, assembling, and prc cessing of great masses of statistics, an elaborately detailed matrices showing th

[37] An interesting attempt to project and predict from past trends is: Herman Kahn and Anthony J. Wiener, *The Year 2000, A Framework for Speculation on the Next Thirty-three Years* (New York, The Macmillan Company, 1967). City, metropolitan, and regional planners must consider probabilities relative to conditions fifty years ahead, which is the amortization period for many major public works projects.

[38] Wassily W. Leontief, "Input-Output Economics, *Scientific American* (October, 1951); "Th Structure of Development," *Scientific America* (September, 1963); *Input-Output Economic* (New York, Oxford University Press, 1966).

[39] For a detailed description of the input-outpu approach, see Walter Isard, *Methods of Re gional Analysis* (Cambridge, Technology Pres and New York, John Wiley & Sons, Inc., 1960) and Walter Isard and Robert Kavesh, "Ecc nomic Structural Interrelations of Metropolita Regions," *American Journal of Sociology*, Vo LX (September, 1954), pp. 152-62, reprinte in Mayer and Kohn, *op. cit.*, pp. 116-26. Alsc Walter Isard and R. Kuenne, "The Impact c Steel Upon the Greater New York-Philadelphi Urban-Industrial Region," *Review of Economic and Statistics*, Vol. XXXV (November, 1953) pp. 289-301.

ut-output interrelations of each of many
gories of economic activity both within
outside the region. The availability of
dern electronic computers has made pos-
e the processing of these masses of
istics, but the problems of availability and
ection remain to inhibit the widespread
of the method.

Briefly, the method consists of studying
effect of a given volume of inputs of
ital, labor, components, and other ele-
nts of production into each of the many
gories of activity which are present in a
ion upon the inputs and outputs of every
er category, both inside and outside the
ion. Each category is represented in a
trix, or table, by a row and a column, and
h cell — the intersection of a row and a
ımn — represents the effects of the inputs
outputs in one category of activity upon
inputs and outputs of another one. Thus,
here are five hundred kinds of economic
ivities, the number of cells will be the
are of five hundred, or 250,000. Actually,
numbers of activities of different types,
ssified in sufficient detail to be meaningful,
y be several times as great. In operating an
ut-output matrix, information is obtained
the input requirements proportional to the
puts, for each item such as labor, ma-
als, etc., in each kind of establishment
uired from each other kind of establish-
nt. Thus, a ton of finished steel, with pre-
t technology, requires a certain amount of
ut of iron ore, scrap, limestone, coke,
er, labor, and so forth. The iron ore, in
n, must be mined, concentrated, trans-
ted, and stockpiled; so must the limestone,
ile the coke production requires coal
ich, in turn, must undergo these processes.
each stage in production, transportation,
l distribution of goods, and in the perform-
ance of services, there are input requirements, which, in theory, can be quantified, and there are outputs produced, which can also be quantified. A change in input or output at any point in the change with respect to any item will be reflected in changes in inputs and outputs at many points in the chain of production. The steel workers must be fed, clothed, housed, entertained, educated, governed, and so forth, and each of these processes sets in motion other sequences of inputs and outputs, through many successive rounds, until, finally, the effects are "damped out," or subsumed in the total economy many stages removed from the original impetus. Thus, a change in labor input requirements in producing steel at Pittsburgh or Gary, because of changed market demands or improvements in technology within the steel plants will change the number of workers in the plant. This, in turn, will produce changes in the total requirements in those communities for goods and services of great variety, and each kind of changed demand will set in motion chains of input-output effects. These effects will by no means be confined to the local community. A change in demand for steel, for example, in Pittsburgh or Gary will change the amount of iron ore required from the mines in Minnesota or Quebec. This will, in turn, change the effective demand for goods and services supplying the miners in those distant areas. Furthermore, if the changes persist over a long period, it will affect the demand for transportation of ores, coal, stone, and other items constituting inputs to the steel plants, and this will, in turn, affect the number of workers required in the transportation industries and the demand for inputs into the multitudinous kinds of businesses and establishments serving them. If such changes persist, later there will be

changes in the requirements for new railroad hopper cars and ore boats, and this will change the input-output relations in railroad car manufacturing plants, shipyards, and other establishments, including the demand for steel inputs into such plants, so that some of the relationships run full cycle. The magnitude of the work required for a detailed and meaningful input-output analysis then becomes obvious, and the practicality of such a method is thereby seriously limited.

Insofar as the inputs and outputs are confined to within the area of concern, such as a city or SMSA, even though they may involve a vast and complex series of interchanges among the commercial, industrial, residential, and other establishments within the area, they are internal with respect to the area, and thus "nonbasic" or "city serving," but insofar as they involve transfers of people, goods money, or credit across the regional boundary, they are "basic" or "city forming." Thus the "input-output" approach may, in one sense, be regarded as an elaboration and refinement of the "basic-nonbasic" approach.

Minimum Or Average Requirements Approach

Another method of economic base analysis which is finding increasing popularity is the minimum or average requirements method. This is essentially comparing the city or region of concern with other cities or regions which resemble it with respect to significant attributes, such as size, age, and location. Thus, if it is desired to determine, on the average, how much land will be required for certain uses, such as industrial plants of various types, or how much employment can be anticipated, or income generated, by various types of activity, it may be useful to determine the average amounts actual utilized by comparable cities or regions. Similarly, cities and regions may be cor pared, in various categories, with respect the minimum requirements for each of t items, by finding the city or region in analo ous sets of cities and regions which has t least amount of the item, whether emplo ment, land, income, or other. All of t standard statistical measures of concentrati and dispersion, such as deciles, quartile standard deviations and others, may be a plied further to refine the analysis.

Population size appears to be the mo significant variable relative to the minimu requirements of cities and metropolitan area there appears to be a more-or-less regul progression, some requirements decreasi with size, partly, at least, the result of sca economies, while others, such as service er ployment, tend to increase with size. Popul tion size, therefore, is generally used as t determinant for classifying cities into grou within which the cities are sufficiently ali with respect to their requirements for cor parative purposes.

The minimum and average requiremen approach has been used to study a number

[40] Edward L. Ullman and Michael F. Dacey, "T Minimum Requirements Approach to t Urban Economic Base," *Proceedings of t IGU Symposium in Urban Geography Lu 1960* (Lund Studies in Geography, series Human Geography No. 24, 1962), pp. 121-4 also in *Papers and Proceedings, The Region Science Association*, Vol. VI (1960), pp. 17 94. Recent evaluations of the technique inclu Richard T. Pratt, "An Appraisal of t Minimum-Requirements Technique," *Econom Geography*, Vol. XLIV, no. 2 (April, 1968 pp. 117-24, and Edward L. Ullman, "Minimu Requirements after a Decade: A Critique a an Appraisal," *Economic Geography*, V XLIV, no. 4 (October, 1968), pp. 364-69.

es, ranging from Utica, New York, a de-
ssed area on the verge of rapid change in
composition of its industrial base,[41] to
nberra, Australia, a national capital seek-
greater diversity with respect to its eco-
mic "mix" and attempting to determine, in
preparation of a comprehensive plan,
w much employment to expect and how
ch land to reserve, for industries and other
ctions, by comparison with other Austra-
n cities of similar size.[42]

In spite of its utility, the average and mini-
m requirements method has certain seri-
s drawbacks. How can we be sure that the
es or regions selected for comparison are
y analogous? Is it not possible that there
y be independent variables which were not
en into consideration, such as, for ex-
ple, the introduction of technological
nge since the last statistics became avail-
e, or the prospects of major changes in
hnology and economic or social conditions
cting requirements in the future?

ssification Of Cities By Economic nction

order to reduce the possibilities of error in
nparing cities or regions with one another,
s desirable to take into consideration as
ge a number as possible of independent
iables, for any one or combination of them
y be significant in affecting the validity of
comparison. Fortunately, the means are
v at hand for processing vast quantities of
a, and for determining which variables are
related to each other. We can classify cities and regions by their economic functions, or by any other sets of variables for which statistics may be obtained. Modern computers have made possible the application of methods that would not have been feasible only a few years ago.

Functional classifications of cities by the distribution of employment in various categories of economic activity have been made by a number of geographers.[43] Such classifications, as well as later ones using employment distribution by kind of establishment, occupation, or activity,[44] are univariate classifications, in that they use only one variable: employment.

A new technique for comparison of cities and regions with respect to many, rather than one, attributes is available, in the form of what is known as factor analysis or principal components analysis.[45] A large number of variables, obtained from many sources, can be grouped, by a standard computer program, into a small number of groups of related variables, and then comparing the groups in accordance with the extent to which they are

ndustrial Renewal: Determining the Potential nd Accelerating the Economy of the Utica Irban Area (New York, State of New York Division of Housing and Community Renewal, 963).

G. J. R. Linge, *The Future Work Force of Canberra*, a Report for the National Capital Development Commission (Canberra, 1960).

43 For example: Chauncy D. Harris, "A Functional Classification of Cities in the United States," *Geographical Review*, Vol. XXXIII, no. 1 (January, 1943), pp. 86-99, and Howard J. Nelson, "A Service Classification of American Cities," *Economic Geography*, Vol. XXXI, no. 3 (July, 1955), pp. 189-210, both reprinted in Mayer and Kohn, *op. cit.*, pp. 129-162.

44 Otis Dudley Duncan *et al.*, *Metropolis and Region* (Baltimore, The Johns Hopkins Press, 1960), especially pp. 279 ff.; Richard L. Forstall, "Economic Classification of Places Over 10,000, 1960-1963," *The Municipal Year Book 1967* (Chicago, The International City Managers' Association, 1967), pp. 30-65.

45 H. H. Harman, *Modern Factor Analysis* (Chicago, University of Chicago Press, 1961); M. G. Kendall, *A Course in Multivariate Analysis* (London, Charles Griffin, 1957).

inter-related. It is also possible to measure cities or metropolitan areas with respect to the nature and extent of their similarities and differences. The groups of related cities then constitute classes with respect to all of the sets of variables used. Thus, many measures of economic functions can be combined, and groups or classes of cities determined with respect to each of the sets of variables, or, on the other hand, a number of different measures of economic functions alone may be used to classify cities. British towns have been classified in this manner with respect to 57 different social and economic characteristics,[46] and American cities have similarly been classified by social and demographic characteristics.[47] Functional classifications, using the techniques of factor or principal components analysis, have been published for cities in several countries, including Canada, Australia, and India, among others.[48] Generally it has been found that the characteristics tend to group in from four to six related clusters, and these can be used to determine the extent to which cities in the country or region resemble each other with respect to any or all of these characteristics. Thus it is possible to group the cities in any number of groups from one up to the total number of cities involved, and to preser cities classified into any intermediate num ber of groups based upon communalities, c resemblances. Thus, it is reasonable to trans fer the experience in solving problems in on city, if successful, to other cities whic closely resemble it, rather than to cities i which the similarities are less. Factor analysi thus becomes a tool of great potential.

Conclusion

In spite of the difficulties of analyzing the eco nomic base of cities and metropolitan area such analyses are of great importance. Plan ners, and all who are interested in the solu tion of urban problems, realize that citie exist primarily in order to enable people t make a living, and that the employment op portunities, the income and other satisfac tions resulting from employment, are th fundamental forces behind urban growth c decline. It is not possible to determine ho many, or what types of houses and othe facilities need to be provided, until one ha some idea of the present and prospective op portunities for employment. These depend, i turn, upon the nature of the economic activ ties upon which the city or region is basec Availability of resources, of course, is funda mental, and the geographer can furnish in sights different from, but no less importan than, those of the economist in assessing th prospects for economic growth and develop ment. Economic base studies clearly lead t the conclusion that no form of human occu pance can replace cities, and that, in spit of the trend toward lower densities concom mitant with improvements in transportatio and communication, modern civilization can not exist without cities. An understanding c the functions which they perform, a majc proportion of which are economic functions is therefore indispensable.

46 C. A. Moser and Wolf Scott, *British Towns, A Statistical Study of their Social and Economic Differences* (Edinburgh and London, Oliver and Boyd, 1961).

47 J. K. Hadden and E. F. Borgatta, *American Cities: Their Social Characteristics* (Chicago, Rand McNally & Co., 1965).

48 Leslie J. King, "Cross-Sectional Analysis of Canadian Urban Dimensions, 1951 and 1961," *Canadian Geographer*, Vol. X (1966), pp. 205-24; Robert H. T. Smith, "The Functions of Australian Towns," *Tijdschrift voor Economische en Sociale Geografie*, Vol. LVI, no. 3 (May-June, 1965), pp. 81-92; Qazi Ahmad, *Indian Cities: Characteristics and Correlates*, Research Paper, No. 102 (Chicago, University of Chicago, Department of Geography, 1965).

'he Economic Base f the Metropolis

15

ans Blumenfeld

THE CONCEPT OF THE ECONOMIC SE

e terms "economic base" and "basic" in-stry or employment are being increasingly ed and discussed in planning and related ds.

Sometimes the term "economic base" nply stands for "economy," considered as base of the life and growth of an area[1]; or, term "basic" is simply used as a synonym "important."[2]

Geographers frequently denote the region ich serves as market and as source of pply for a given city as its "economic base." Harold M. Mayer: "the area which appro-ately may be considered as constituting economic base of a large metropolitan y."[3] Similarly John W. Alexander defines the "Bases for the Oshkosh Economy" as "I. The Market Base. II. The Supply Bases."[4]

However, both Mayer and Alexander[5] also make use of the term "basic" in the sense in which it has become increasingly accepted, as opposed to "nonbasic." This concept claims that all economic activities of an area can and should be divided into two fundamentally different and mutually exclusive categories.

Apparently the first American planner to formulate the concept was Frederick Law Olmstead, who said in a letter of February 21, 1921: "productive occupations may be roughly divided into those which can be called primary, such as carrying on the marine shipping business of the port and manufacturing goods for general use (i.e., not confined to use within the community itself), and those occupations which may be called ancillary, such as are devoted directly

See, f.i., Economic Base Study of the Philadelphia Area, Philadelphia City Planning Commission (August, 1949).

f.i., Grace K. Ohlson in Municipal Yearbook, '. . . that furnishes the major volume of employment." Quoted by Richard B. Andrews, "The Urban Economic Base," in *Land Economics* (1953), p. 265.

Harold M. Mayer, "Urban Nodality and the Economic Base Study," *Journal of the A.I.P.* (Summer, 1954).

4 John W. Alexander, *Oshkosh, Wisconsin, An Economic Base Study* (Madison, Wisconsin, 1951).

5 John W. Alexander, "The Basic-Nonbasic Concept of Urban Economic Functions," *Economic Geography* (Worcester, Massachusetts, July, 1954).

printed from *Journal of the American Institute of Planners,* Vol. 21, no. 4 (1955) pp. 114-32 by permission.

or indirectly to the service and convenience of the people engaged in the primary occupations.[6]

Haig and McCrea, in conformance with this concept, state: "It has been urged that a distinction should be drawn between 'primary' and 'ancillary' activities: that primary activities be given precedence in the city plan."[7]

In the same year M. Aurousseau wrote: "The primary occupations are those concerned with the functions of the town. The secondary occupations are those concerned with the maintenance of the well-being of the people engaged in those of primary nature. The more primary citizens there are, the more secondary, in a relation something like compound interest."[8]

Here we find the two ideas which have determined the further application of the "basic-nonbasic" concept.

1. Planning and promotion, with preference to be given to "basic" activities.
2. Prediction, with total future population being derived from "basic" employment by application of a "multiplier."

The bias in favor of the "basic" activities, implicit in such terms as "basic," "primary," "town-building," "town-growth," versus "nonbasic," "ancillary," "service," "secondary," etc,[9] is made explicit in such statements as: "the first task of . . . Letchworth an Welwyn . . . was to secure that 'basic' ir dustries would be attracted; the inhabitan . . . could not . . . live by taking in eac other's washing."[10]

We will return to the question, if, whe and why people can or can not "live by takin in each other's washing." For the develop ment of the concept the second applicatio — for population prediction — has bee even more important. It was broadly used b Homer Hoyt in his work for the F.H.A. Th method, as developed by Hoyt, includes fiv steps.[11]

1. Calculate employment in each basi industry
2. Estimate ratio of basic to service em ployment
3. Estimate ratio of population to employ ment
4. Estimate future trend of basic employ ment
5. Derive future total employment an population from future basic employ ment.

This has become the accepted formula frequently steps 2 and 3 are omitted in favo of a rule-of-thumb formula of "population t basic employment equals seven to one." Hoyt defines his criteria for a "basic" activit in manufacturing was "basic," that all othe employment was "service," and that thei ratio was roughly one to one. For the pur poses for which Hoyt developed his formul — a quick, rough-and-ready housing marke estimate — it was serviceable. However, h soon discovered, first, the difficulties o identifying "basic" activities, and second, th

[6] Quoted in R. M. Haig and R. C. McCrea, *Regional Survey of New York and its Environs*, Vol. I, p. 43, footnote.

[7] Haig & McCrea, *op. cit.*, p. 42.

[8] M. Aurousseau, "The distribution of population," *Geographical Review*, Vol. XI (1921), pp. 567 ff. Quoted by Robert E. Dickinson, City, Region, and Regionalism (London, 1947).

[9] An exception is the use of the terms "surplus" and "domestic" in the sophisticated study by John M. Mattila and Wilbur R. Thompson, "Measurement of the Economic Base of the Metropolitan Area," *Land Economics* (August, 1955), pp. 215-28.

[10] J. H. Jones, "Industry and Planning," in E. A Gutkind, *Creative Demobilisation*, Vol. I (London, 1944).

[11] See A. M. Weimer and Homer Hoyt, *Principle of Urban Real Estate* (New York, 1939).

stence of wide local variations in the sic-nonbasic" ratio.

IDENTIFICATION OF "BASIC" WITH XPORT" ACTIVITIES

yt defines his criteria for a "basic" activity follows: "those industries and services ch produce goods for people living outside urban region being studied, and which ig in *money* (my emphasis, H.B.) to pay the food and raw materials which the does not produce itself."[12] Similarly, hard U. Ratcliff defines "primary or city-lding activities" as those "which bring into community purchasing power from out-."[13] A Swedish geographer differentiates ween "exchange (bytes)" production, ich is regarded as "primary" and "own en)" production which is considered condary,"[14] and a Swedish planner has d this distinction to develop his method of ulation prediction.[15] Perhaps the most ightforward explanation of the concept "basic" workers was given by Andrews calls them "the wage earners of the com-nity family."[16]

The concept sounds simple and convincing ugh: in order to live a community, like mily, has to earn money. The number of ilies is determined by the number of adwinners; the number of "housewives" who "service" the breadwinners, and of dependents, can be derived from the number of the former.

There are certainly cases where the concept is fully applicable. Take, for instance, a copper mining village with 1,000 miners. There will be, say, 600 people employed locally in retail trade and consumer services; if the family coefficient is 2.5, the population will be 4,000. If the company hires another 1,000 miners, it is safe to predict that they will be followed shortly by about 600 more "secondary" employed persons and that the population will increase to 8,000. Inversely, if the company lays off 500 miners, the population will in due course shrink to 2,000. It is also safe to say that no attempt to promote development of any or all branches of "secondary" activity will make a noticeable impact on the economic life or the population size of the community.

Now, let us define the specific conditions of this experimental case:

1. There is no possibility of substitution of another "basic" activity for copper mining.
2. There is no source of income from outside other than wage payments.
3. Earnings of all "basic" employed are roughly equal (or at least average earnings for any group which may be added or subtracted are equal).
4. The family coefficient of all groups in basic employment is the same.
5. None of the product of any "basic" industry can be sold locally; or, looking at the same phenomenon from the other side, all goods and services (other than those which because of their physical characteristics can be supplied only locally) are being supplied from the outside.

Iomer Hoyt Associates, *The Economic Base of he Brockton, Massachusetts Area* (January, 949), p. 4.

ichard U. Ratcliff, *Urban Land Economics* New York, McGraw-Hill, 1949), p. 42.

V. William-Olsson, *Stockholms framtida utveckng* (Stockholm's future development) (Stockolm, 1941).

red Forbat, "Prognos for Näringsliv och beolkning (forecast of industrial activity and opulation), *Plan*, No. 9 (Stockholm, 1948).

ichard B. Andrews, *op. cit.* in *Land Economics* 1953), p. 161.

It is evident that every one of these five conditions is the exact opposite of conditions characteristic of a metropolitan area. A metropolis is not simply a sum of villages, and it can not be analyzed by adding up studies of its parts.

3. LIMITATIONS OF THE "BASIC-NONBASIC" CONCEPT IN TIME AND SPACE

As has already been pointed out, the literature on the subject is pervaded by a conviction that the "basic" activities are more important than the "nonbasic" ones. Emphatic statements abound. "Basic employment is the same as . . . destiny."[17] Harold McCarthy goes so far as to call "the base . . . that group of occupations whose presence . . . is not predicated on the existence of other types of production."[18]

This is evidently untrue. No "basic" industry in a modern city could function without such services as water, transportation, and communication. Some students of the subject are aware of this. "Urban-Growth and Urban-Serving Employment . . . are both equally essential," says Victor Roterus;[19] and the U.S. Chamber of Commerce speaks of "a chicken-and-egg relationship," adding: "industrial growth stimulates the remainder of the local economy and the existen of the community makes possible industri growth."[20]

Here a new and important point is bei made: the community with its services is t basis of industry, as well as vice versa; ar it is startling to find that this point is bei made by a promotional pamphlet of t Chamber of Commerce rather than by pla ners. It is the more startling as — alongsi with the goal of "strengthening" or "broade ing" the "economic base" — the America planning profession also proclaims the go of the "self-contained community." T Greeks had a word for it: autarchy.

Evidently, the two goals are mutually e clusive. In a completely self-contained, (autarchic, community, nothing has to b bought from outside and consequently n body works to earn money for outside pa ments. There is no "basic" employment; a people live by "taking in each other's wash ing."

On the other hand, the higher the percen age of the labor force in "basic" employmen the greater the dependence of the communi on outside markets and on outside supplie the less "self-contained."

It may help to clarify our concepts to loo at extreme cases. The copper mining villag comes as close to maximizing "basic" em ployment as any community is likely to com An employed bachelor, who does not mak his own breakfast nor sew on his own button would be the perfect example of 100 perce "basic" and no "service" activity.

At the other extreme, a subsistence far — or a truly "self-contained" community lik

[17] *Working Denver, An Economic Analysis by the Denver Planning Office (1953)* (Department of Planning, City & County Bldg., Denver 2, Colo., 1953), p. 27. We will frequently exemplify our critique of the "basic–nonbasic" concept by reference to this excellent study, because it has developed the concept more completely than most others.

[18] Quoted by John W. Alexander, *The Basic-Nonbasic Concept, op. cit.*

[19] Cincinnati City Planning Commission, *The Economy of the Area*, (Cincinnati, December, 1946), p. 22.

[20] Chamber of Commerce of the U.S., Washin ton, 1954, "What new industrial jobs mean the community".

ancient Indian village — has no "basic" ployment. All occupations are "concerned h the maintenance of the well-being of the ɔple" which, according to the above-ɔted definition by Aurousseau, is the criter-. of "secondary" occupations.

Also, and perhaps more significantly, the bal community of mankind is engaged ex-sively in "secondary" or "service" activi-;. A large nation is not far from this reme; the "basic-nonbasic" ratio for the S. is probably about 1:20. The generally epted applications of the "basic-nonbasic" thod — preferential promotion of "basic" e. export) activities and prediction of ure population by applying a "multiplier" expected future employment in export ivities — would be as patently absurd for U.S. as they are sensible for a copper ning village.

We may tentatively derive from the juxta-sition of these extreme cases a first state-nt: the applicability of the "basic-nbasic" concept decreases with increasing e of the community.

Size, however, is not the only factor to be nsidered. Let us return to the case of the ɔsistence farm. By any acceptable usage ming is its "basic" or "primary" activity. the farmer or his wife engage, during the ck season, in some cottage industry, selling product for cash, such activity is to them ctly "secondary" or "ancillary." Here the ncepts appear reversed: production for n use — "taking in each other's washing" is basic, and production for sale is ancil-y. This is characteristic of a "natural" eco-my, while the reverse holds true for a oney" or "exchange" economy, which is ɔendent on division of labor. Hence our ond statement: applicability of the "basic-nbasic" concept increases with increasing specialisation and division of labor between communities.

Still another aspect may be illustrated by the ancient Indian village community, or, for that matter, by a village in medieval feudal Europe. Here a good deal of the economic activity was for "export," for the Lord of the Manor, the Church, or the King. But far from being basic in the sense of being indis-pensable for the economy of the village, this activity is the only one which contributes nothing to it. The reverse of this picture is the town which receives these payments with-out having to compensate by any "export" activity or employment. Richard U. Ratcliff quotes H. Pirenne as saying that the early medieval fortress town "produced nothing of itself, lived by revenue from the surrounding country, and had no other economic role than that of a simple consumer." Another his-torian characterizes the "economic base" of such cities as follows: "The principal, con-stituent elements of the town were those who are able by *power and wealth* (my emphasis, H.B.) to command a means of subsistence from elsewhere, a king who can tax, a land-lord to whom dues are paid, a merchant who makes profits outside the town, a student who is supported by his parents. These are "town builders. . . ."[21]

Here the "basis" for the economy of the town is not "persons employed in producing goods and services for export," as the "basic-nonbasic" method assumes, but "power and wealth." It may here be recalled that in the "tableau économique," which Quesnay, founder of the "physiocratic" school of eco-nomics, developed in the 18th century, the

[21] F. L. Nussbaum, *A History of the Economic Institutions of Modern Europe* (New York, 1933) quoted by Richard B. Andrews, *op. cit.*, (1953), p. 161.

urban middle class was called "classe stérile," as serving the ruling class rather than working for the "producing" class, the farmers. Thomas Jefferson shared this physiocratic view.

In our context it is important to keep in mind that "nonbasic" activities are supported by money gained from the outside regardless of its source, which may be "power and wealth" rather than any "basic" employment. To the extent that this is the case, the "basic-nonbasic" ratio loses its meaning.

We may therefore formulate a third statement: the greater the amount of "unearned" income (i.e. income derived from sources other than payment for work performed) flowing into or out of a community, the less applicable is the "basic-nonbasic" concept.

We will later deal with attempts to assimilate "unearned" income to the concept of "basic" activities. Leaving aside this aspect, for the time being, and concentrating our attention on the relation of "basic" and "nonbasic" employment, we may accept as valid the existence of two opposite historical trends noted by Forbat in the above-mentioned article:

1. Replacement of local crafts by large-scale industry working for a national and international market; hence greater share of "basic" employment.
2. Increase in services; hence greater share of "nonbasic" employment.

Both trends result from increasing division and specialisation of labor, the first between communities and the second within the community. It should also be noted that the increase in services refers not only to services to consumers, which are generally the result of commercialisation of functions formerly performed by the household, but also to services to business, which were previously performed as auxiliary services within other businesses, but have now become so specialised and complicated as to require special establishments.

This specialisation of business activities reaches its highest development in large and mature communities. As mentioned before the same communities also are nearest to "autarchy," because they contain the greatest number of branches of production.

We may therefore summarize:

The "basic-nonbasic" ratio is highest in small, new communities, lowest in large and mature ones.

4. MERCANTILISTIC AND PHYSIOCRATIC OVERTONES OF THE "BASIC-NONBASIC" CONCEPT

The difference between "basic" and "nonbasic" activities is the difference in their role in the balance of payments with the world outside the community. Strangely, and rather inconsistently, the "economic base" studies dominated by this concept pay practically no attention to the other side of the ledger: no attempt is being made in these studies to differentiate between those locally consumed goods and services which are produced locally and those for which payments have to be made to the outside world. Yet, rationally, the money earned by "basic" activities is merely the means to make these payments, not an end in itself.

The idea, underlying the "basic-nonbasic" method, that the acquisition of money from the outside world is the "basic" purpose of the urban or metropolitan economy has its historical precedent in the mercantilistic school of economics which regarded only gold and silver as true wealth. While in a study of the U.S. economy it is today taken for granted that increased production of

ods and services for the home market is the
ıl, in the "economic base" studies of
ıerican cities these activities are regarded
rely as supporting the "basic" ones work-
for export.

This is, of course, explainable by the role
yed by size which has been discussed
ove. If the slogan "export or die" is true for
able countries like Great Britain or Ger-
ny, it is even truer for a single city or
ion which evidently can not produce every-
ng which it consumes. In particular many
se studies stress the need of earning money
order to pay for imports of food and raw
terials. "Basic Employment . . . goods or
vices in exchange for food and raw ma-
ials . . . (is) the critical or crucial employ-
nt . . . ; without it the city ceases to exist."[22]

Here, as in many similar statements, there
the implication that the export activities
"basic" because without them the city
ıld not buy food, which is a "basic" neces-
, while New Yorkers would not "cease to
ist" without such locally supplied goods
d services as millinery or theatre perform-
ces. But they would cease to exist very
idly without water supply, which is also a
rvice" or "nonbasic" activity.

The belief that there is something partic-
rly "basic" in the production of food and
v materials also has its historical precedent;
antagonists and successors of the mer-
ıtilists, the physiocrats, believed in the
periority of farming and mining over other
ınches of production.

Incidentally, when we deal not with 19th
ıtury cities, but with the emerging, much
ger and qualitatively different form of
human settlement defined by the U.S. Census as a "Metropolitan Area," a sizable part of the food may be supplied by "nonbasic" activities; that is, supplied by residents to residents of the area. The Philadelphia Metropolitan Area, f.i., containing 2.45 percent of the U.S. population in 1950, produced 1.03 percent of all dairy products sold in the U.S., 1.17 percent of all poultry, 2.28 percent of all vegetables, and 4.68 percent of all nursery and greenhouse products. Thus, dairy and poultry production equaled almost half, and vegetable production equaled almost the entire normal consumption of the area. Altogether about 14 percent of the area's proportionate (to population) share of all agricultural products were produced locally.

If the classification of economic activities attempted by the "basic-nonbasic" concept is to acquire scientific validity and practical usefulness, it will have to discard all explicit or implicit notions that earning money or buying food is specifically "basic." It should be clearly understood that we are dealing exclusively with a difference in the market; and the appropriate terms would be "export" and "home market" activities.

There is reason to pay attention to this difference. A product or service which has to compete in the national and international market is more vulnerable than one which, like local transportation or a corner drug store, by its physical nature is protected against outside competition; it is, for the same reason, also more capable of expansion by invading outside markets. But, by and large, the share of the national product which is sold locally is just as vulnerable to competition as is the part which is sold outside.

From the piont of view of vulnerability by outside competition as well as of ability to

New York Regional Plan Association, *The Economic Status of the New York Metropolitan Region in 1944*, p. 3.

expand into outside markets, both of which we may identify as "criticality," the only meaningful distinction is between activities which, *by the nature of their product* have to and can compete with outside producers, *regardless of the location of their actual sales,* and those which do not compete; and it is just as important to measure the imported and the locally produced share of total local consumption as it is to measure the exported and the locally consumed shares of total local production.

5. DEVELOPMENT OF TECHNIQUES OF MEASUREMENT

a. Manufacturing versus services

The first studies, those made in the twenties for the New York Regional Plan and in the thirties by Homer Hoyt for the F.H.A., assigned entire activities to the one or the other category according to their predominant market; as Frederic Law Olmsted put it in the letter quoted earlier, "primary"[23] are goods for general use (i.e., *not confined to* [my emphasis, H.B.] use within the community itself). Consciously, they were satisfied with a rough approximation; subconsciously, they were guided by the criterion of competitive character rather than by that of actual markets of an industry.

Such a rough approximation by allotment of broad categories to the two classes of activities was also used — but only as a first step — by the 1944 study of the New York Regional Plan Association. As "basic — producing *in whole or in part* (my emphasi H.B.) for persons living outside of the Regio . . ." are specifically enumerated: manufac turing, wholesale trade, banking and insu ance, transportation, administrative office hotels and amusements, federal and state em ployment. As "service — producing *entirel* (my emphasis, H.B.) for persons living with in the Region" are enumerated: retail trad professions, personal services, local trans portation and utilities, construction, loca government, business and repair services, rea estate and local banking.

Parenthetically it may be noted that man of these last-named activities do not produc entirely for the local population, but als serve many persons living outside the Region

There is reason to believe that the motiv for concern with "basic" activities was thei competitive and therefore critical characte Had the authors of the New York Regiona Plan study accepted this criterion, they woul have sought further refinements along line which will be indicated later. However, the like all others using the "basic" concep interpreted it as meaning "export" and con sequently sought to refine it by measuring th portion of each particular product or servic which was sold outside the Region.

The measurement of this "exported" por tion is easy in dealing with a national eco nomy where exports and imports are counte at custom lines. In dealing with areas withi a nation, however, no comparable data ar available and other methods have to be devel oped.

b. Proportional apportionment

The method used by the N. Y. Regional Pla study and most others is to assume that th community consumes a share of the total na tional production of each category of good

[23] The concepts of "primary" and "secondary" used in this type of studies should not be confused with the concept of "primary," "secondary," and "tertiary," meaning "extractive," "processing," and "service" activities, as defined by Colin Clark and other economists.

services which is proportional to its share the national population (in some cases chasing power or other yardsticks are stituted for population). The surplus in ess of this proportional share is assumed be exported or "basic". Frequently the tion between actual and proportional re of a given category of production is ressed as a "location — or localization — tient." The "location quotient" is the centage of employment in a given local ustry of total local employment, expressed a ratio to the percentage of national ployment in the same industry of total ional employment; or $\frac{ei \,.\, Ei}{et \,.\, Et}$; e = local ployment; E = national employment; i mployment in industry; t = total employ- nt. "By means of the localization quotient the extent to which an activity is basic . . . be determined."[24]

The same method was applied by Victor terus in the Cincinnati study. "Urban- ving employment (was) calculated by as- ning that the population will consume its portionate share of the national produc- ."

Ve have used this method in defining the rvice" share of various branches of icultural production in the Philadelphia a. However, it would be quite erroneous conclude from the fact that the location tient of vegetable production for the ladelphia Area is roughly equal to one, t Philadelphians eat no vegetables grown side their area. They do, of course, and er vegetables are exported from the area. To choose another illustration: the Phil- lphia Area's share in the production of kly periodicals may about equal its share of national population and (or) purchasing power. But it does not follow that all copies of the Saturday Evening Post are consumed in the Philadelphia Area and that Phil- adelphians never buy copies of the Reader's Digest. They do (unfortunately).

The method of proportional apportion- ment is based on the completely fallacious assumption that categories of goods and ser- vices — however fine the breakdown — can ever be uniform. International trade statistics show that most countries are both importers and exporters of the same categories of goods. The same certainly holds true to an even greater extent for the exchange of goods and services between areas within the nation.

Of course, if the location quotient is very high, it stands to reason that most of the product is exported. However, in such ex- treme cases the importance of that partic- ular industry will be a matter of general knowledge. No location quotient has to be calculated in order to find out that Detroit exports automobiles or that Brockton exports shoes. On the other hand, if an area produces its "normal" share of, say, electrical ma- chinery, it would be completely erroneous to assume that this is a "nonbasic" industry working exclusively for the local market. It is entirely possible, and indeed quite prob- able, that most locally produced electrical machinery is exported, while at the same time most locally consumed electrical machinery is imported. Mattila and Thompson unwit- tingly demonstrate the fallacy of the method by presenting a completely absurd result: the "proportion of surplus ("basic," H.B.) to service ("nonbasic," H.B.) workers," cal- culated by means of the "location quotient," is given as 1:1.99 for Chicago and as 1:4.47 for Philadelphia![25] Are we to believe that

[24] Harold M. Mayer, *op. cit.*

[25] *Op. cit.*, p. 226, Table III.

one "basic" worker supports 2 "nonbasic" workers in Chicago and 4½ in Philadelphia?

This is not to say that the location quotient does not deserve careful study. By analyzing it, much can be learned about market areas and about competitive advantages and disadvantages. But as a measurement of the share of "basic" activities or employment it is completely misleading.

c. Breakdown of markets by survey

The obvious inadequacy of this method has led several researchers to embark on the difficult and time-consuming attempt to follow up the actual sales of each establishment in the area under investigation. This was apparently first done in 1943 by Fred Forbut in his study of the small Swedish town of Skörde.

In Alexander's Study of Oshkosh, f.i., establishments employing 75 percent of the labor force were asked for the percentage of their sales that was local; the same percentage of their employment was then allocated to the category of "secondary" employment. The same approach was taken by Maxine Kurtz in the Denver study.

In addition to its high cost this method obviously encounters two obstacles: first, unwillingness to disclose one's market, and second, ignorance of the location of one's customers. The first obstacle appears to have been overcome fairly successfully both in Oshkosh and in Denver. Interviewing of cash customers of retail stores and other techniques have been used to narrow the gap of ignorance. The result of the studies may be regarded as a reasonably accurate measurement of local and outside sales.

However, this is still far from finding the answer to the question: how does the community earn the money to pay for the imports it needs? Leaving aside, for the time bein the question of modifying the needs for ir ports as well as the possibility of paying f them by money derived from sources oth than export, the main shortcoming is this: v know the *gross value of the exported goo and services*. What we want to know is t *"value added" by the community*. A flour m may export 10 millions worth of flour; but it has to import 8 millions worth of grain, earns no more than 2 millions for the cor munity. Employment probably has be adopted as the only available, though e tremely rough, approximation to "val added." However, those using the methc are apparently unaware of this relation ar of its implications.

If "value added" is sought, other difficulti arise. Assume that the grain has been grow locally. It is sold in the local market, to t local mill, by definition its production is "nonbasic" activity. Yet the payment r ceived for its sale (in the form of flour) is net earning of the community, a "basic" su port of its economy. Or, to take another e ample: a community exports a million to of steel. If this steel is produced in an int grated metallurgical plant, the "value addec is the difference between the cost of ore, coa etc., and the value of the steel; and all en ployment in the plant is considered "basic But if the same steel is produced in a ste plant which buys its pig iron from an ind pendent local blast furnace, then only th value added by the steel plant is considere "basic," and the value added by the bla furnace is, by definition, "nonbasic," becau it is sold to a local customer. Thus the distin tion between "basic" and "nonbasic" is function of the inner organization of the i dustry: the higher the degree of specializatio and differentiation, the breakdown of a pr

s into parts performed by several inde-
dent establishments, the higher is the
nbasic" share. This, incidentally, is one
the reasons why "nonbasic" activities ap-
r to loom so large in metropolitan areas,
ere the process of differentiation reaches
apex.

This difficulty has given birth to the con-
t of "indirect primary" activities and emyment. "Indirect primary" are all goods or
vices sold to a local establishment which
urn exports its products. But the steel mill
s not only buy locally produced pig iron;
equally buys locally produced power,
ter, trucking services, banking services,
al police and fire protection. Moreover, it
s locally produced labor power which in
n buys locally produced bread and movie
ws. Once "indirect primary" activities are
nitted as "basic," where can the line be
wn? The economy of an area is an inte-
ted whole of mutually interdependent
ivities; the distinction between "basic"
l "nonbasic" seems to dissolve into thin air.

"Criticality" or "Balance of Payments"?

nfusion worse confounded. The more we
empt to refine the "basic-nonbasic" con-
t, the deeper we get involved in contradic-
ns. Whenever that happens, there is reason
assume that there is something wrong with
formulation of the question.

Let us return to the origin of the quest. It
y be fairly illustrated as follows. If Gen-
l Motors closes shop at Flint, no efforts to
mote the development of department
res will save the town. On the other hand,
a Flint department store closes down, but
General Motors plant continues to operate
before, it will soon be replaced by other
res. Therefore, the thing to worry about, the "base" of the Flint economy, is the automobile industry; once that works, the "services" will take care of themselves. Also, once we know how many people G.M. is going to employ, we know pretty well how many people there will be in Flint.

Unquestionably true. But why is the situation of the G.M. plant so much more critical than that of the department store? Leaving aside the difference in size which is extraneous to our problem, it is because the G.M. plant has to compete with all other automobile plants in the U.S. and in the world, while the department store has to compete, in the main, only with other stores in Flint (though its customers might purchase some goods in Detroit, or from a mail order house in Chicago).

The difference in "criticality" is determined by the extent of the area of potential competition. In actual practice this is, of course, a range of areas from the locality through ever-widening regions to the national and international markets. A development and refinement of the "criticality" approach would have to go in two directions. First, as much attention should be paid to the *actual and potential* source of locally consumed goods and services, as to the markets for local products; second, the potential area of competition should be broken down into areas of varying size.

Actually, both steps have been undertaken by many "economic base" studies. Most studies pay particular attention to industries with a location quotient smaller than one, assuming that here may be opportunities for new local industries to compete with outside suppliers. And most go into detailed analysis of their market areas; the Denver study, f.i., found that 54 percent of the sales outside the Metropolitan Area were made within the "region," defined as Colorado, Wyoming, and New Mexico.

This is contrary to the "basic-nonbasic" theory, which demands concentration of attention on the "basic" industries, those with a location quotient larger than one, and which regards all export activities as equally "basic," regardless of the size or location of the export market. Thus, the practice of the economic base studies has been generally more sensible than the theory which they claim to follow. This is, fortunately, not an uncommon occurrence in human affairs (vide the practice of American foreign policy versus the theory of "massive retaliation").

In our case the theoretical weakness lies in the confusion of the question of "criticality," that is the question concerning *potential competition* with the question of *"balance of payments,"* which is concerned with *actual sales*. Both questions are valid and important; but either can be clearly answered only if they are clearly separated.

The concept of the "balance of payments" is well understood; but the assembly of the relevant data is exceedingly difficult. Apparently the only attempt ever made was the famous "Oskaloosa versus the U.S." study undertaken by Fortune magazine in April 1938. In this study "a city of 10,000 people has been treated as if it were a little nation."

A "balance of payments" study evidently must use dollars as units of measurement, not persons. The widespread use of the categories of "basic" and "nonbasic" *employment* — a consequence of the attempt to use the concept for population prediction by means of the "multiplier" — has no place in a study of this type. What matters, is not how many persons work at supplying the outside world, but how much money they receive from it. For this reason some studies have used payrol (and net earnings of self-employed persons rather than number of persons employed. Bu payments for goods and services go onl partly into payrolls, partly into profits, ir terest, taxes, etc. The appropriate measure ment would be the one applied in interna tional trade statistics: gross value of gooc and services exported and imported. To thes would have to be added taxes and disburse ments of larger governmental units, as we as interest and dividend payments in bot directions. However, the latter are "pract cally unobtainable," according to Charles I Leven of the Federal Reserve Bank c Chicago.[26]

Nevertheless, the New York Regional Pla study of 1944 did make an estimate tha "nearly one-third of the region's basic incom was derived from dividends, rents, interest and profits." It would seem that with th amount of labor and ingenuity that went int the Denver study, f.i., it might be possible t arrive at estimates realistic enough to cor strust a model of "Denver versus the U.S.A. As Victor Roterus wisely remarks, no ecc nomic base study can achieve more than rough approximation.

The attempts to "refine" the "basic nonbasic" concept have destroyed its useful ness for the identification of "critical" indus tries, while making no more than a ver partial and dubious contribution to an identi fication of the balance of payments.

To repeat:

There is a need for two types of studies related, and using much of the same materia

[26] Charles L. Leven, "An appropriate unit fc measuring the urban economic base," *Lan Economics* (November, 1954). This is the mo concise study of the subject known to th writer.

t different in their conceptual framework:

1. A "criticality" or "variability" study, analyzing all actual and potential branches of production in the area from the point of view of the size and character of the area in which they compete and their consequent vulnerability to outside competition and potentiality to expand into outside markets.
2. A "balance of payments" study, including *all* types of payments, and giving equal weight to both sides of the ledger.

SPECIFIC PROBLEMS

Replacement of imports by local
)duction

ie of the purposes of the distinction of asic" and "nonbasic" activities "consists concentrating investigation on those indus-es and services which . . . bring in money pay for the food and raw materials which city does not produce itself," to repeat omer Hoyt's formulation. But if the city uld itself produce the goods which it now ports, the effect would be the same. Why ould not investigation be concentrated on ose industries and services, which do *not* oduce a surplus for export, but, on the con-ry, show a deficit in supplying the home rket? Evidently, if Brockton, rather than reasing its capacity to produce shoes worth million dollars annually, would build a thing plant to supply its inhabitants with a lion dollars' worth of clothing which they w have to import, the improvement of the vn's balance of payments would be the ne. It may here be noted that this might be possible in Brockton, because, the local rket may not be large enough to support efficient plant. But it would certainly be ssible in a large metropolis. The larger the community, the greater the possibility of substitution for declining industries, hence the less significant the identification of "basic" activities.

Actually even the Brockton study does examine the possibility of substituting new industries for the "critical" shoe industry. Similarly the New York Regional Plan study concentrates its attention on "industries in which New York's share of employment is far below its proportion of population and income," stating: "these industries might be explored to ascertain why they have not expanded to a greater degree in the Region,"[27] and the Cincinnati study considers specifically "local industries not meeting local demands." Charles L. Leven, in the aforementioned article, agrees that "efforts might be more profitably directed at establishment of local industries (supplying) local exporters."

The important point, in our context, is that the "basic-nonbasic" method is of no help whatsoever in identifying such industries. It rather tends to deflect attention from them and to confuse the picture. Half a loaf is certainly better than no loaf; but half a balance of payments study may well be worse than none.

The reverse substitution is no less important: the replacement of local production by imports. As Fred Forbat has pointed out, this is one of the long-term trends of industrial society, a corollary of increasing specialization and division of labor between regions. However, it still remains possible to substitute local production for imports; especially where the growth of the community creates a previously nonexistent large market, imports may be replaced by locally produced

[27] *Op. cit.*, p. 19, 20.

goods and services. In this way growth induces further growth. "He who has, to him shall be giveṇ" is a basic law of economics.

b. An extreme case of the effect of the establishment of a new "basic" industry on the balance of payments

It is generally assumed that the opening of an establishment which exports part of its products will always improve the balance of payments of an area. However, if such an establishment is a branch plant of an outside firm and works mainly with imported material, and if the part of its products which is sold locally displaces the products of a local industry working largely with local materials, then the net effect may be the opposite.

In the hypothetical case presented here, "A" represents a group of local establishments producing $1,000,000 worth of goods for the local market. "B" represents the new branch plant which produces $2,000,000; of these $1,000,000 are exported and $1,000,000 are sold in the local market, displacin the local establishments.

In case "A" the community has to pa $310,000 to the outside world in order t procure the $1,000,000 worth of goods whic it consumes. In case "B" it has to pay $410 000 ($1,410,000 minus $1,000,000 earne from export sales). Thus the establishmer of the new "basic" industry has resulted in deterioration of its balance of payments b $100,000.

Let us assume an average wage of $3,00 and an average per capita income of $1,50(Let us further assume that one-third, or $50(of this per capita income has to be spent t import goods and services from the outside Then the loss of $100,000 in means of payment to the outside has the result that th community can support 200 persons less tha before.

According to the standard formula one half of the 100 workers in the new establishment, or 50 workers, would represent "basic

Case	Item	*Spent, Total*	*Spent Locally*	*Spent Outside*	*Net Payment to Outside*
"A"	wages	220,000	220,000		
	materials	500,000	250,000	250,000	
	amortization	50,000	50,000		
	overhead	60,000	60,000		
	taxes	70,000	10,000	60,000	
	profit	100,000	100,000		
"A"	total	1,000,000	690,000	310,000	310,000
"B"	wages	300,000	300,000		
	materials	1,000,000	200,000	800,000	
	amortization	200,000	50,000	150,000	
	overhead	100,000	20,000	80,000	
	taxes	200,000	20,000	180,000	
	profit	200,000		200,000	
"B"	total	2,000,000	590,000	1,410,000	410,000

(It has been assumed that plant "B" uses more ordinary [more amortization] and fewer worker [less wages].)

›loyment. According to the rule-of-thumb
ıltiplier" of 7 persons for every one person
·"basic" employment, there should be a
ulation increase of 350 persons. But actu-
there would be a decrease of 200 persons.
'his may be an extreme and unlikely case.
as been developed to point up the prob-
atical character of the "basic employ-
ıt" method.

'arenthetically, while this case is not likely
ɔccur in the U.S.A. in 1955, it may be
ly typical of the impact of the establish-
ıt of branch plants of modern internation-
concerns in under-developed countries;
e the result is frequently aggravated by
ted effects on income distribution. The
stance of these countries to such ap-
ently beneficial improvements by foreign
estors may not be entirely due to short-
ted nationalistic prejudices.

ndirect primary activity

problem of "indirect primary" activity
attracted the attention of many students
ur subject. Fred Forbat[28] refers to a study
a new oil refinery in Aarhus, Denmark,
lertaken by the economist B. Barford.[29]
fod found that "the company's purchases
goods and services from local suppliers
e livelihood" to 70 persons for every 100
sons employed in the refinery. He classes
se as "indirect primary" and derives the
ected number of "secondary" employment
assuming that there will be 80 additional
condary" workers for every 100 new
kers in *all* "primary" employment,
ect" and "indirect" combined.

'red Forbat, "Synpunkter pa Lokaliserings-
ıultiplikatorn," *Plan* (Stockholm, 1948), No. 9.

. Barford, Local economic effects of a large-
cale industrial undertaking (in Danish); E.
lunksgaard, (Copenhagen, 1938. This writer
as not been able to locate this study.)

Andrews, in his series of articles in LAND ECONOMICS, repeatedly returns to this question. He recognizes that "linked activity . . . the chain of production . . . makes *all* activities basic" and calls this a "very serious blindspot . . . unless the anachronisms (?H.B.) of this situation can be reconciled."[30] When dealing with a concrete example, however, he says: "rigidly, we would classify the automobile-starter factory as a service activity . . . realistic(ally it) should be considered basic in that there exists only an organizational line between the starter and automobile manufacturer."[31]

Andrews seems to be unaware that "crossing an organizational line" is only a synonym for "sale" and that any method which — like the "basic-nonbasic" method — counts sales, consists in counting line crossings. He adds, however, the very pertinent remark: "the number of links involved may very well be in direct ratio to community size." But he again fails to draw the conclusion: that the applicability of the "basic-nonbasic" concept is in inverse ratio to community size.

Ullman also wrestles with the problem. He says: "a city with large basic plants might appear to produce many basic workers whereas . . . many small plants each feeding the other . . . appear to have many service workers." He then tries to compromise by stating that "some intermediate producers are classed as basic if they contribute directly to an export industry," but concludes finally that "in this light all activities appear indivisible."

Alexander is also troubled by the problem: "since these castings (made in a local foun-

[30] Andrews, *op. cit., Land Economics* (1954), p. 260 ff.

[31] Andrews, *op. cit., Land Economics,* (1953), p. 347.

dry, H.B.) are fabricated into axles which are exported, it could be said that this . . . production is for the primary market. However, this leads to complications, and the arbitrary decision has been made to classify each activity on the basis of its own direct sales."[32]

We have already shown that the complications are implicit in the ambiguity of the question. If the question is, instead, clearly directed to the balance of payments, the decision to count only direct sales is by no means arbitrary, but a matter of course. Nobody thinks of including the steel industry in the export statistics of American automobiles, because the "value added" by the steel and all other "indirect primary" industries is, by definition, included in the gross value of the automobiles. Dollars, not persons employed, are the correct yardstick.

If, on the other hand, the question is directed to the "criticality" of each establishment, i.e., to the range of its potential market and its potential competitors, the "indirect primary" activities fall into place alongside all others.

d. Inter-urban transportation

Practically all studies, while treating local transportation as "nonbasic," regard all other transportation as "basic." However, actually only those transportation activities which serve movements between two outside points earn money from the outside. A tanker, bringing oil from Venezuela to a refinery in the Philadelphia Area, exclusively serves and is paid by the Philadelphia plant. Its work might be called "indirect primary," but, as we have seen, that does not remove it from the "nonbasic" class under any consistent definition.

Normally, of the total exchange of goods between two points about half will be paid each end; thus 50 percent of it should ce tainly be classified as "nonbasic." The poi might well be made that the entire interurb transportation system performs a service fo and at the expense of, the local import ar export trade and should be classified as suc

From the point of view of competitivenes or "criticality," interurban transportation (e cept for the portion serving movemen between two outside points) is strictly no competitive; nobody but some form of tran portation can move goods and passengers in and out of the city. It is true, of course, that transportation is very poor, other branch of production may move to other areas th are better served by transportation, it th may profoundly affect the competitive pos tion of the area as a whole. It was probab this thought that caused people to classify as "basic." Yet in this respect it is not pri cipally different from supply of water, powe housing, or any other local service.

e. Public employees, students, etc.

A field in which the confusion of the "crit cality" and the "balance of payments approach has led to particularly glaring co tradictions concerns those persons whose i come, while clearly contributing to the ec nomy of the community, is derived neith from sales to the local community nor fro sales to the outside world. Andrews, Maxir Kurtz, Forbat, and others allocate emplo ment in government institutions according the population served (local or outside pop lation); and allocate the staff of universiti proportional to the number of local ar "foreign" students.

From the "balance-of-payments" point view, which these researchers are trying apply, this does not make sense. From th

[32] John W. Alexander, *Oshkosh*, *op. cit.*, p. 12.

nt of view the only thing that matters is source of the income, not who is benefited the work performed. The income of *all* eral employees is a net gain to the community, whether they deliver letters to local idents or work on projects to deliver milk or atom bombs — to the Hottentots; just all taxes paid to the Federal Government a net loss to the community. Similarly, if university professor is paid by state conoutions, by the G.I. Bill of Rights, by an side foundation, or by outside parents of students, his income is a gain to the comnity. But it is not if he is paid out of city ds, out of contributions of local alumni, or of money earned by his students in the nmunity. The home residence of his stuts has nothing to do with it.

Evidently these researchers were led astray ause in the back of their minds, but not mulated, was the "criticality" approach. a city of a certain size, post offices, local irts, elementary and high schools can ind be taken for granted; while the city has compete with other cities for a state unisity or for a regional office of the National vernment, and also has to compete with m for students at its university.

Discrepancy between "basic" employment l outside earnings

a given community with a given level of living depends for 50 percent of the goods d services which it consumes on outside irces, its size is evidently limited by the ount of money which it can pay to the tside. It is the standard assumption of the asic-nonbasic" method that in this case 50 rcent of its employment would have to be asic." The contribution to the "economic se" is assumed to be proportional to the number of persons employed in each branch of "basic employment." "Wholesale trade accounted for . . . 12 percent of its total basic employment . . . *therefore* (my emphasis, H.B.) . . . about one-eighth of the economic base.[33]

This comfortable "therefore" contains — and conceals — a number of unspoken assumptions, which should be spelled out:

1. average wages are roughly the same in all branches of production (the Denver study does touch on this question).
2. the ratio of "value added" to payrolls is roughly the same in all branches.
3. the proportion of the "non-payroll" section of "value added" going to local owners is roughly the same in all branches.
4. moneys paid or received by the community other than payments for goods and services balance out.

It is improbable that any of these assumptions correspond to the facts of life in a metropolitan area.

From the Denver study, f.i., can be seen that average annual wages in wholesale trade varied from $3,100 in "petroleum bulk stations" to $4,300 in "manufacturer's branches without stock."[34]

Sales per employee varied far more than average wages: in retail trade from $8,150 in "eating and drinking places" to $35,800 in "automotive"; and in wholesale trade from $28,000 in "auto & equipment" to $610,000 in "farm products (raw)."[35] These differences reflect largely, but hardly entirely, differences in mark-up. If we assume, f.i., that the mark-up was 20 percent in "auto & equipment" and 2 percent in "farm products," the

[33] *Working Denver, op. cit.*, p. 4.

[34] *Working Denver, op. cit.*, p. 68.

[35] *Working Denver, op. cit.*, p. 74.

mark-up, or "value added," per employee would still vary from $4,700 to $12,000.

Evidently the greater part of these $12,000 represents return on capital, and how much it will contribute to the purchasing power (the "economic base") of Denver versus the outside world will depend entirely on the share of the capital owned by Denverites.

Andrews, f.i., recognizes the importance of this factor of "absentee ownership" of capital, saying: "if a dollar-flow measurement device were employed, the loss would be clear . . . (and) the community receiving the profits would count them."

Andrews, not employing such a "device," does *not count* these losses. But he *does count* the gains in the receiving community where he classifies the income derived from investments in other communities as "capital export."

This is indeed a classical example of the confusion resulting from the attempt to achieve greater precision by refining a basically confused concept. The *export of capital* puts the community in the red; it is the *return on the capital* — and return of the capital — which produces income. Nor is this return contingent on previous "export" of capital by the community; the wealthy residents of Palm Beach derive their income from capital which was not exported by Palm Beach, but either was exported from New York and other places, or was not "exported" at all, but accumulated out of the returns of "outside" investments. Andrews' concept of unifying all sources of income of the community under the term "export of goods, services, and capital" in an unfortunate attempt to force strange bedfellows into the procrustean bed of the "basic-nonbasic" concept.

Other important factors affecting the balance of payments of a community are the ratio of payments to disbursements of state and federal taxes and the "terms of trade, i.e., the price relations between imported an exported goods. Assume, for example, tha the work of 10,000 persons employed in ex port industries is required to pay for the foo imported by a community of 150,000 per sons. If food prices were cut in half, thes 10,000 workers could pay for the food c 300,000 people.

Because of these many factors, "basic em ployment" is a very inadequate yardstick fc the measurement of the economic base of th community.

g. *The "basic-nonbasic" ratio*

As has been noted before, the attempt to dis tinguish "basic" and "nonbasic" employmen has been made primarily in order to find th ratio of the second to the first.

Jones had made the rather naive assump tion that "the majority of the town . . . ar employed in providing goods and services . . for other communities. It could not be other wise."[36] In fact, it is otherwise. In most case the ratio is considerably greater than unity but it differs widely.

Roterus and Calef, in the aforementione article, succinctly state the reason for the dif ference: "the basic-nonbasic ratio is a mea sure of the degree of economic interdepen dence."

In new communities the ratio may be ver low because they depend for most service on established neighboring communitie However, one "service" industry, construc tion, is usually over-represented in such area In Lower Bucks County, f.i., in March 195 there were for every 100 persons employe in manufacturing only 35.7 employed in serv

[36] J. H. Jones, *op. cit.*, p. 125.

industries other than construction, com-
red to a ratio of 90.4 in the Philadelphia
etropolitan Area in 1950. On the other
nd, the ratio of employment in construc-
n to manufacturing was 71.5 to 100 in
wer Bucks County as against 9.6 to 100 in
Metropolitan Area.[37]

It has been noted that the ratio is generally
gher the larger the community. The reasons
y be summarized as follows:

1. a greater completeness of all branches of production; the community is more nearly "self-contained" than a small one.
2. greater "round-about-ness" of production; the productive process is divided into a greater number of organizationally independent, though economically interdependent, units.

In addition, in most metropolitan areas
re is:

3. higher average income, commanding more consumer services.
4. a concentration of "power and wealth," drawing unearned income from the outside and spending it for local services.

The New York Regional Plan study of
44 found the abnormally high ratio of 2.2
onbasic" for every one "basic" employed.
owever, the study also states that nearly
e-third of the region's "basic" income is
rived from sources other than the export of
ods and services. If it is assumed that a
oportional number of service workers was
pported by this source of income and only
remainder is related to the "basic"
rkers, the ratio is about 1.5, practically the same as the one found in Denver, which was 1.53.

The ratio also changes over time; nor can these changes be easily explained. In Cincinnati, f.i., between 1929 and 1933, the decrease in the number of factory workers was 29 percent above the national average, but the decrease in retail sales was 6 percent below the average for the nation.

A slightly different and rather interesting approach to the question of "service" employment has been taken by Swedish planners and geographers who have attempted to find the number of service workers required to serve a given population in communities of various types. Here the distinction of "basic" and "nonbasic" is used as a tool for identifying those industries which have to be studied directly and individually, while a global average figure is used for the prognosis of all "nonbasic" employment. Forbat found that in the three towns of Kristinehamn, Skövde, and Landskrona secondary employment varied only from 20.92 percent to 22.39 percent of the total population. Even this slight variation was due entirely to variations in agriculture, construction, and domestic service; after elimination of these three categories the range was 16.37 percent to 16.49 percent.

These are three towns with a population between 15,000 and 24,000. For villages of about 2000 population, Forbat found a percentage of service workers of 15 percent, and for Stockholm of 27 percent. This correlation of percentage of service workers with size is in accord with American experience.

Sven Godlund contributed to the discussion the concept of an "index of centralization" which is defined by the percentage of the total population employed in retail trade and consumer services. He found this to vary with size from 6.5 percent in regional centers down to 3.5 percent in villages, with even

Economic Development, Lower Bucks County, Bucks County Planning Commission (February, 1954), p. 12 and table E8.

lower percentages in "special urban settlements," mainly industrial satellite towns.[38]

h. The "multiplier"

Investigation of the "basic-nonbasic" ratio is used to find the "multiplier," the ratio of total population to "basic" employment. The multiplier is determined not only by the "basic-nonbasic" ratio, but by three additional factors which are not always clearly recognized:

1. family coefficient of basic employed.
2. family coefficient of nonbasic employed.
3. percentage of nonemployed (incl. dependents) population.

Frequently a global ratio of population to employment is used. However, this ratio may vary considerably if any of these three factors change, or if their relative weights change.

In Denver, f.i., the ratio of population to "basic" employment — the "multiplier" — was 7.8 in 1940. Between 1940 and 1950, however, there were only 4.6 persons added to the population for every one person added to "basic" employment. It is evident that a population prediction based on the number of additional "basic" employed and using the "multiplier" found in 1940 would have overstated the decennial population increase by 70 percent.

Variations in the family coefficient between various "basic" industries are very significant. In the anthracite mining regions of Pennsylvania, f.i., the addition of a mining job would usually mean the addition of a family. The addition of a hosiery job generally means employment of a female former dependent of a miner's family.

[38] Sven Godlund, "Studies in Rural-Urban Interaction," *Lund Studies in Geography* (Lund, 1951).

Generally the family coefficient is low i industries with high female employment an in communities with a high rate of employ ment and with low percentages of the popula tion in the extreme age groups, i.e., childre and old people.

Forbat[39] found that the family coefficier in the Swedish countryside varied from 1.5 for textile workers to 2.49 for constructio workers; for Stockholm both figures wer considerably lower, 1.30 and 2.06 respec tively. He also found that the family coeffi cient in trade and service employment aver aged 1.7 to 1.8.

Forbat has developed a formula whic takes into account the differences in th family coefficient for different types of em ployment. The formula is:

$$P = \frac{Ep \cdot Cp}{1 - Rs \cdot Cs}$$

P = population
Ep = employed, primary
Cp = family coefficient of primary employed
Cs = family coefficient of secondary employed
Rs = secondary employment as percent of population

If Ep equals 1, the "multiplier" becomes

$$M = \frac{Cp}{1 - Rs \cdot Cs}$$

By applying this formula, Forbat found th following multipliers:

villages:	2.1 to 3.2
towns:	2.5 to 3.7
Stockholm:	2.3 to 3.3

These are large variations. They would b even larger except for the fact that high serv

[39] Fred Forbat, "Untersuchungen über den Lokal isierungsmultiplikator" (investigations on th localisation-multiplier) *Raumforschung & Rau mordnung*, no. 2 (1953), pp. 97-101.

employment and low family coefficient are erally associated, because both are corre- d with high female employment, and that r influences tend to cancel each other. s, the multiplier is low in villages, despite gh family coefficient, because the villages end for services largely on neighboring ns; it is low in Stockholm, despite high vice employment, because the family co- cient is low.

A further difficulty in deriving the "multi- r" stems from the fact that there is a size- and highly variable group in most munities which is neither in "basic" nor "nonbasic" employment nor part of the ilies of either group. These are the "inde- dent nonworkers," who may derive their ome from a great many sources: invest- ts, pensions, social insurance, relief pay- ts, etc. Forbat classifies these as "prim- " ("basic"), because their number is not endent on the number of those in other sic" groups. In the little town of Skövde r number was equal to one quarter of all er "basic" groups. In many American es it may be even higher. On the other d, in new or rapidly growing communities r number is low; in some cases practically o. Also, their number may vary widely and uptly with changes in the labor market.

Swedish statistical data make it possible to ive a separate family coefficient for this up which is, of course, lower than that for ployed persons. In Stockholm, f.i., it was 0 versus 1.63 for employed persons; in Swedish countryside 1.43 versus 2.14.[40]

American statistics lump "independents" "dependents" in the categories "unem- yed" and "not in the Labor Force." They also present no data from which family co- efficients for specific industries in specific localities could be derived. This writer, in attempting to apply the Forbat formula to American cities, has therefore substituted for both Cp and Cs (family coefficient for "basic" and "nonbasic" employed) what might be called a "community-wide family coefficient"; that is, the ratio of total population to total employed. The resultant multipliers are 5.5 for Philadelphia and 6.46 for Denver; the latter being practically identical with the 6.6 found by Maxine Kurtz.

The fact that these figures are about twice as large as those found in Swedish towns is only partly due to the different classification of the "independent nonworkers." In a letter to this writer, of September 26,1955, Forbat has recalculated the "multiplier" for five Swedish towns on the basis of the "community-wide family coefficient." The result- ing figures are between 3.5 and 4.1. The dif- ference between these multipliers and those found for Philadelphia and Denver are due first, to the fact that in Swedish cities em- ployed persons average 49 percent of total population against about 40 percent in American cities; and second, to an unusually high percentage of "basic" employment which in the five Swedish towns was 53.3 percent to 60.0 percent of all employment. This, in turn, is partly due to actual differences in economic structure, and partly to differences in classi- fication. Forbat classifies *all* "big" industry (as distinct from handicraft industry) as "basic." This appears quite permissible in small towns, but would lead to very serious distortions if applied, f.i., to the New York garment industry.

Barfod, in the above-mentioned study of the impact of a new oil refinery on the popula- tion of Aarhus, apparently ignored all these problems. He found a "multiplier" of 8.8 for

. Forbat, Untersuchungen . . . , *op. cit.*, p. 00, Table 2.

every person in "direct basic" employment and of 5.5 for every person in all (including "indirect") "basic" employment.

The enormous range in multipliers — from 2.1 to 8.8 — found by various methods shows that the multiplier is not the simple, unequivocal device for population prediction as which it appears at first sight.

7. APPLICATION OF THE "BASIC-NONBASIC" METHOD TO THE METROPOLIS

a. The "multiplier"

One of the main purposes of developing the "basic-nonbasic" method was its alleged usefulness for population prediction. Future population was to be found by multiplying future "basic" employment with a figure which could supposedly be derived from past experience.

We have seen that past experience does not and can not yield any figure applicable to future experience unless a great number of variables are known, in addition to the future number of persons in "basic" employment. The most important variables are:

1. average level of living of the community
2. percentage, in terms of value, of the goods and services constituting this level which have to be purchased from outside
3. net gain or loss to the community from money flow due to causes other than payments for goods and services
4. family coefficient of persons in "basic" employment
5. family coefficient of persons in "non-basic" employment
6. percentage of total population who are not employed, nor dependents of employed persons

These variables make it difficult to dete mine the multiplier for any community. B in metropolitan areas it is even more difficu to predict the figure which is to be multiplie the future number of persons in "basic" en ployment.

The illusion that "basic" employment better predictable than many other variabl stems from the fact that future employme of individual enterprises is, indeed, frequent known with reasonable certainty. If a ne steel work requiring 5000 workers is bei built, it is highly probable that after its con pletion there will be 5000 steel workers li ing in the community. If the community is ar will remain a company town, it is indeed po sible to find its population by adding to tl steel workers and their dependents those pe sons (and their dependents) who service tl steel workers.

But in a metropolitan area there are mar plants, big and small, which open up or sh down, expand or contract their employmen We have seen that "basic" employment, t definition, is critical, competitive emplo ment. It is the part of the economy which most vulnerable, most likely to disappear contract as a result of outside competitio and also most dynamic, most likely to sprir up or grow as a result of invasion of outsi markets. As the most vulnerable and mo dynamic part of the metropolitan econom "basic" activities are its most variable, lea predictable element.

In addition, there is the practical imposs bility of measuring "indirect primary" en ployment. Moreover, as Walter Isard an others have emphasized, a new "basic" indu try attracts not only those which supply but also those which it supplies; not on "indirect primary," but what we might ca "primary indirect" activities. Here anoth complication arises. Assume that a new ste

nt, producing a million tons of steel, at-
:ts, over the years, steel fabricating plants
ich buy half of its production. Then one
f of its workers must, by definition, be
ısferred from the "basic" to the "non-
ic" category, because they now work for
local market. In other words, the more
steel plant contributes to the community's
nomic base — in the commonsense mean-
of that term — the less "basic" does it
ome according to the standard formula.

Some, like Isard, believe that, while appli-
ion of the multiplier formula for prediction
he entire metropolitan population may not
practicable, it can be used to estimate the
ulation to be added as a result of the im-
:t of a specific known development, such
:he new steel plant at Morrisville, Pa., for
:ance.

sard's study of the impact of the Morris-
e plant clearly shows two difficulties in-
ent in this method:

1. within what area will the added population live?
2. to which figure is the "new" population
to be added?

The first difficulty is relatively minor: the
ion of impact comprises several metropo-
n areas.

The second difficulty is fundamental. Ob-
usly it makes no sense to add the "new"
ulation simply to the present one, or to
at the present one would be, as the result
natural increase, at the end of the impact
iod. The addition must be made to a figure
dicted on the basis of past trends. But
se trends reflect the dynamic nature of the
tropolitan economy, the never-ceasing
inkage of old industries and expansion of
v ones. In the Philadelphia Area, f.i., they
ect the coming and growth of the oil refin-
industry, an event closely comparable to
coming of the steel industry, and even due to the attraction of the same locational factors. Thus the figure derived from the trend already anticipated the coming of new industries, and if their impact is added separately, it will be counted twice. It is impossible thus to isolate the impact of single factors in and on the metropolitan complex.

It is worth noting that even the Denver study which had lavished so much care and ingenuity on the identification of "basic" employment, finally bases its prediction of the future growth of the economy and population of the area not on these figures, but on long-term trends and on estimates of the importance of locational factors.

We may conclude:

> As a tool for predicting the population of metropolitan areas the "basic-nonbasic" method is useless.[41]

b. Promotion of "basic" industries

The other main purpose for developing the "basic-nonbasic" method is its alleged usefulness in concentrating attention on those industries whose promotion will do the most good for the well-being of the community, which is supposed to depend primarily on improvement of its balance of payments.

We have already pointed out that the method, strictly applied, tends to divert attention from many industries which might contribute most to an improvement of the balance of payments, namely those whose products the community now imports, but might produce itself. We also noted that most

[41] Forbat, who has probably developed the "multiplier" method of population prediction more successfully than any other planner, informs this writer that he recently advised against applying it to a big city, because "in the economy of the big cities there evidently exists a different hierarchy."

authors of "economic base" studies have had the good sense to forget their theory and to give a good deal of attention to just these industries.

But suppose communities did succeed in advertising their locational advantages for all those industries which are not by their physical nature restricted to a local market and who therefore have a choice. What would be the effect on the national distribution of industry? Would it be more efficient than it is now?

There are two possibilities. If *all* communities do an equally effective job of industrial promotion, their efforts will cancel each other out and the net effect will be zero. If some communities only do a good job, industries will learn of and be attracted by their locational advantages. By the same token they will ignore and neglect equal or greater locational advantages in communities which do less or nothing for promotion. The net result can only be a less efficient national distribution than would result from the functioning of the market without benefit of local planning.

Location of industries working for the national and international market is a legitimate and most important function of national planning. Local planning organizations could make valuable, indeed indispensable, contributions to such national planning by discovering the potentialities and limitations of their areas.

Without such a national plan the value of their promotional efforts is highly dubious. Most likely they will be ineffective; if effective, they are more likely to do harm than good to the nation.

We can conclude:

> As a guide for the concentration of local promotional efforts the "basic-nonbasic" method is not a useful tool.

This is not to say that all results of th work done in the framework of this metho are to be discarded as worthless. They ca be of great value, first, for the developme of "balance of payments" studies, and secon for the exploration of the "criticality" c various industries.

c. The real economic base of the metropol

What, finally, is the relative importance c "basic" and "nonbasic" activities in a metrc politan area?

We have seen that the percentage of pe sons employed in "basic" activities decrease as a community becomes more "metropol tan" quantitatively and qualitatively: that i the larger it is and the greater the variety an differentiation of its activities.

The more metropolitan the community, th more its inhabitants do "live by taking in eac other's washing." Still, it remains depender on the outside world for many goods an services and will have to pay for the majc part of these by the products of its expo industries. It is not legitimate to worry abou these more than about the "nonbasic" ones

Certainly, from the point of view of sale there is nothing to worry about the "loca service" industries, because they cannot b replaced from the outside. For the same rea son there is everything to worry about ther from the point of view of the welfare of th consumer. If the Philadelphia subway syster goes out of business, it can not be replace by the New York subways. Inversely, fror the point of view of sales there is everythin to worry about the "competitive" industrie but from the point of view of the consumer' welfare there is nothing to worry about ther Their goods and services can be replaced b purchasing them from the outside; and th money they earn from the outside may b

·ned by other competitive industries which ·y be substituted for those which are lost. The *ability to substitute* one activity for ɔther is the crucial point. Most economic ·se studies touch on it in one form or ɔther.

"Gold mining created service jobs, but ·en it 'petered out,' catastrophe was averted local enterprise substituting for the erst- ·ile gold mine . . . ghost towns are evidence, ·wever, that the substitution did not always ·cur," says the Denver study.[42] However, ·does not inquire under which conditions ·ostitution does or does not occur.

Part of the answer is given by Alexander ·o explains that Oshkosh owed its origin to ·wmills which later disappeared, but only ·er having attracted the millwork industry. ·he reservoir of labor persisted and became ·dominant factor in the survival of the ·odworking industry."[43]

Homer Hoyt adds other factors: "What ·es Brockton have to offer as attractions to ·sting and new industries? . . . Adequate ·wer and transportation; decentralization; ·lled machine operators; a location within ·world's greatest concentration of buying ·wer; and proximity to a great pool of tech- ·al knowledge.[44]

These advantages can be summarized ·der three headings:

1. Labor force of various skills; its presence is dependent on *local consumer services:* housing, schools, stores, local transportation.
2. *Business services,* including transportation with its terminals.
3. *Markets*, local and regional.

The more developed these three factors are, the more favorable are the conditions for substitution; and it is easy to see that all three are strongest in metropolitan areas, and the stronger, the more metropolitan the area, that is, the larger and more diversified it is.

The competitive advantage of a large home market is too well known from the field of international trade to require further elaboration.

The importance of a large and diversified labor force also needs little comment. Only some special aspects may be mentioned here. Edgar M. Hoover, Jr., says: "the more abnormal the sex and age requirements or the more pronounced the fluctuations, the more (such industries must be) located near others with complementary labor demands or in a large diversified labor market,"[45] that is, in metropolitan area. Hoover also shows that the big city is the natural habitat of the small plant which is most strongly dependent on the services of other plants. The average number of wage earners per manufacturing establishment in industrial areas in 1937 was 43 in central cities, 61 in major satellite towns, and 95 in the remainder of industrial areas.

The reason for this concentration of small plants in big cities is "external economies." Hoover says: "Many of these 'external economies' are based on the availability of more and more specialized auxiliary and service enterprises, with increased concentration of the main industry . . . the availability of service enterprises makes possible a very narrow specialization of function in relatively small plants."[46]

Working Denver, *op. cit.*, p. 27.

John W. Alexander, *Oshkosh*, *op. cit.*, p. 34.

The Economic Base of Brockton, *op. cit.*, p. 6.

[45] Edgar M. Hoover, Jr., "Size of Plant, Concern, and Production Center", *National Resources Planning Board, Industrial Location and National Resources* (Washington, 1943), p. 251.

[46] Edgar M. Hoover, Jr., *op. cit.*, p. 245.

This development soon reaches a point where the "primary" industry is as dependent on the "auxiliary" ones as these are on it. "Often the number and variety of ancillary establishments clustered around some primary industry is such that the locational dependence of the primary industry on the ancillaries, though small in respect to each one, is great in respect to the total."[47]

It is this high development of "business services" and other "secondary" industries which, together with the availability of labor of all kinds, enables the metropolis to sustain, expand, and replace its "primary" industries.

It is thus the "secondary," "nonbasic" industries, both business and personal services, as well as "ancillary" manufacturing, which constitute the real and lasting strength of the metropolitan economy. As long as they continue to function efficiently, the metropolis will always be able to substitute new "export" industries for any which may be destroyed by the vicissitudes of economic life.

We have seen that such substitutions may occur even in small towns such as Oshkosh. In metropolitan areas they are the rule rather than the exception. The history of the past 40 years in Europe has given eloquent proof of the ability of metropolitan communities to survive not only physical destruction, but also the disappearance of those functions on which their existence had been based in the past. They developed new functions and survived. Vienna and Leningrad are only two particularly striking examples.

It is worth noting that this is a new phenomenon. The capitals of oriental empires soon turned into dust, once a new ruler transferred his court to a different location. Even Rome was little more than a village after the Western Empire had been destroyed. These cities were mainly centers of consumption, base on the concentration of "power and wealth. The modern metropolitan area is primarily center of production, based on a concentra tion of productive forces. It is qualitativel different from the city as it has been know throughout history. It is a genuinely new forr of human settlement, and, contrary to predic tions of its approaching transformation int "necropolis," it is showing a greater vitalit than any previous form of settlement. As fa as this writer is aware, no community tha during the last century has passed the hal million mark — a truer border line for th metropolis than the 50,000 adopted by th U.S. Census — has fallen below that popula tion level.

The basis of this amazing stability are th business and consumer services and other in dustries supplying the local market. They ar the permanent and constant element, whi the "export" industries are variable, subje to incessant change and replacement. Whi the existence of a sufficient number of suc industries is indispensable for the continue existence of the metropolis, each individu "export" industry is expendable and replace able.

In any commonsense use of the term, it i the "service" industries of the metropolis tha are "basic" and "primary," while the "ex port" industries are "secondary" and "anci lary." The economic base of the metropoli consists in the activities by which its inhabi tants supply each other.

SUMMARY

1. The concept divides all employment in community into "basic" or "primary employment, working for export, an "nonbasic" or "secondary" employmen working for local consumption.

[47] Edgar M. Hoover, Jr., *op. cit.*, p. 276.

This method purports to serve two goals:

a. Concentration of attention on the most important industries.

b. Prediction of future total employment and population, which are to be derived from future "basic" employment by means of a "basic-nonbasic ratio" and of a "multiplier."

The method seeks the answer to two different questions, which it fails to distinguish:

a. What is the balance of payments of the community?

b. What are the most "critical" industries, i.e., those most vulnerable to outside competition and most capable of expansion into outside markets?

The confusion is increased by a widespread dual bias:

a. A "mercantilistic" bias in favor of money-earning versus consumption-satisfying activities.

b. A "physiocratic" bias in favor of food and raw materials versus manufactured goods and services.

The attempt to identify "basic" activities by the widely accepted method of proportional apportionment is misleading. The attempt to do it by actual market survey is costly and ends up by revealing the inherent contradictions of the method.

The method neglects the import side of the ledger which is equally important with the export side, both from the "balance of payments" and from the "criticality" point of view.

As a result of its confusion of these two points of view, the method is unable to solve the problem of "indirect primary" activities. If a consistent "balance of payments" approach were used, the problem would cease to exist; if a consistent "criticality" approach were used, these activities would fall in line with all other activities.

8. The method fails to integrate into its conceptual scheme any payments received or made other than those for work performed.
9. Employment is not a usable unit of measurement for a "balance of payments" approach, which must use "value of product" and other value terms.
10. The proportion of "basic" activities increases with increasing division of labor between communities and decreases with increasing size of community and with increasing division of labor within the community.
11. The "basic-nonbasic ratio" is meaningful only in small and simply structured communities; the larger and more complex, that is the more "metropolitan" the community, the less applicable is the ratio and the entire method.
12. The "multiplier" varies not only with the "ratio," but also with the "family coefficient" of both the "basic" and the "nonbasic" employed, and with the percentage of the population which is not employed.
13. Because of these complexities the "multiplier" is not a useful tool for population prediction in a metropolitan area.
14. The identification of the "export" activities of each locality could be an important tool for a national agency planning industrial location. However, if local planning agencies use it as a guide to promotion, it will either be ineffective, or, if effective, result in a harmful distortion of the national locational pattern.
15. A large metropolitan area exists, survives, and grows because its business and consumer services enable it to sub-

stitute new "export" industries for any that decline as a result of the incessant vicissitudes of economic life.

These services are the constant and permanent, hence the truly "basic" and "primary" elements of the metropolitan economy; whi the ever changing export industries are th "ancillary" and "secondary" elements. Th relation assumed by the method is, in fac reversed.

16

The Functional Base of Small Town

Howard A. Stafford, J

The small town, small urban place, or small city, as designated in this study, is essentially equivalent to what Brush has included in his threefold classification of hamlets, villages, and towns.[1] The small town is of academic interest because it represents the lower end the central-place continuum. Any generaliz tions, theories, or laws developed for centr places should hold not only for large urba places but also for all the gradations to th smallest. Thomas well stated the theoretic importance of the small city when he wrot "First, logically, these small places provic

[1] John Brush, "The Hierarchy of Central Places in Southwestern Wisconsin," *Geogr. Rev.*, Vol. 43 (1953), pp. 380-402.

Reprinted from *Economic Geography*, Vol. 38 (April, 1963), pp. 165-75 by permission.

ic connection between the dispersed agri-
tural populations and the agglomerated ur-
population. For the most part, such direct
nections as do exist are through the goods
l services which are provided in these small
vns for the agricultural population sur-
nding them. Second, even if small towns
not fulfill their logical role of providing
ds and services for a dispersed farm popu-
on, the fact remains that these small places
st and that economic activities are permed in them just as they are in the larger
ces."[2] The motivation for this study is a
ire to examine the functional bases for
all southern Illinois towns and to compare
results with those of similar studies in
er areas.

Recent geographic literature contains some
studies of small urban places. Notable
ong these have been Berry and Garrison's
homish County, Washington, studies,[3]
g's comparative study of the Canterbury
vincial District, New Zealand,[4] and
omas' study of the functional bases for
all Iowa towns.[5] The present study is an
empt to duplicate Thomas' Iowa study for
ected southern Illinois towns; less rigor-
ly, it also compares the southern Illinois
dy with those for Snohomish County and
nterbury Provincial District.

Edwin N. Thomas, "Some Comments on the Functional Bases for Small Iowa Towns," *Iowa Business Digest* (1960), p. 10.

Brian J. L. Berry and William L. Garrison, "The Functional Bases of the Central Place Hierarchy," *Econ. Geog.*, Vol. 34 (1958), pp. 145-54.

Leslie J. King, "The Functional Role of Small Towns in Canterbury," forthcoming in *Procs. Third N.Z. Geogr. Conf.*, Palmerston North (1962).

Edwin N. Thomas, *op. cit.*

SELECTION OF TOWNS

Thirty-one small towns were selected as a sample of small urban places. The selection was restricted to the southern 22 counties of Illinois, random choices being made from the listing of urban places provided by the Illinois State Department of Revenue report of October, 1960, on the collection of the Retailer's Occupation Tax (Sales Tax). The spatial distribution of the 31 sample towns is shown in Figure 1. Only towns with a 1960 population of 5000 or less were eligible for selection. The 1960 population of the 31 towns ranges from 40 for Welge to 3739 for Waterloo, with an average of 552. It should be noted that the criteria used in this study differ somewhat from those used by Thomas, in that he chose only incorporated municipalities with populations of less than 2500. The 42 cities used by Thomas had 1950 populations ranging from 42 to 2333, with an average of 462 persons.

DEFINITION OF TERMS

In order that a coherent analysis can be made of the functional bases of small southern Illinois towns, the mass of data collected by field work has been summarized for each urban place in terms of establishments, functions, and functional units; the concepts are borrowed directly from Thomas' Iowa study. "An establishment is essentially the physical manifestation of an activity and is generally the unit in which an activity is performed, e.g., the building in which the office for a filling station is located or the office of a physician are examples of establishments. In contrast, the term 'function' refers to activities which are performed in the establishments. According to these definitions, it is possible for more than one function to be associated

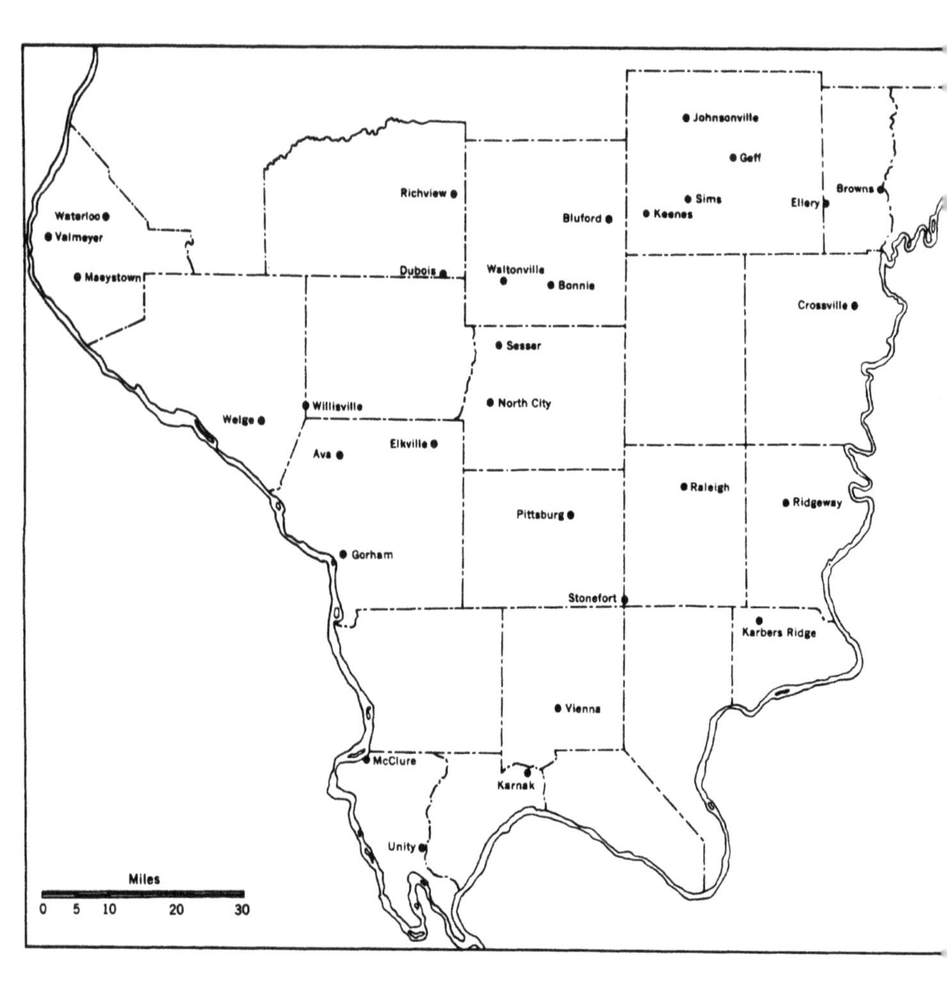

Figure 1. Distribution of sample towns in southern Illinois.

with a particular establishment. Each occurrence of a function constitutes one *functional unit*.

"Specific interest is focused on three indices by which the magnitude of the activities within an urban place may be measured. Each of these indices illuminates a somewhat different aspect of the overall distribution of activities. The three summary values which are provided for each place are (1) total number of establishments, (2) total number of functions, and (3) total number of func tional units. Differences between these value may be illustrated as follows. Let us assum that there is a place with three establish ments, A, B, C. Three functions are per formed in establishment A; it is a gasoline fill ing station, bulk oil distribution station, an used-car lot. Two functions are associate with establishment B; it is a combination foo store and filling station. Two functions ar associated with establishment C; it is a com ation food store and livestock feed store. ere are in this case, three establishments, : functions, and seven functional units."[6]

E OCCURENCE OF FUNCTIONS THIN THE SAMPLE CITIES

ta on the central-place functions permed in each of the 31 towns were obtained ough field work, rather than from second- sources. At the outset, a list was prepared all functions that might be expected to :ur at least once in the sample cities. This sic listing was altered as experience was ned. The final list includes 60 functions, ging from gasoline-filling stations and ırches, each of which occurs 96 times, to candy store, veterinarian, photographer, store, chiropractor, and taxi service, each which occurs only once in the entire group 31 towns. The list of functions used in this dy, while essentially the same, is not so ıaustive as Thomas' listing of 121 funcns. Actually, data were collected for many re than the 60 functions finally used in the alysis, but these data were not included in final tally because classification difficuls led to considerable distrust of their accu- y. The most frequently occurring funcns included in the Iowa study but not in southern Illinois study are meeting halls d insurance agencies.

At this point, a degree of similarity between frequency of occurrence of functions in all southern Illinois towns and small Iowa vns can be noted. As indicated in Table I, tally of the ten most frequently occurring ıctions in the two areas is strikingly similar. Of the ten functions on the Iowa list, eight cur on the southern Illinois list and their ıkings are not very different. That the animal feed store is not on the southern Illinois list may be a reflection of the greater importance of meat animals in the agricultural economy of Iowa. The high degree of occurrence of bulk oil distributors in Iowa may be a consequence either of the harsher winter climate or of the relatively small use of local coal or of both. That churches occur as frequently as gasoline filling stations in southern Illinois, whereas they definitely have second rank in the small towns studied in Iowa, may be an indication that church bodies in southern Illinois tend to be more fractionalized[7] and inde-

TABLE I

Rank Order of Most Frequently Occurring Functions

Iowa	*Southern Illinois*
1. Gasoline filling station	1. Gasoline filling station
2. Church	2. Church
3. Animal feed store	3. Food store
4. Auto-repair shop	4. Tavern
5. Insurance agency	5. Restaurant
6. Food store	6. Beauty shop
7. Tavern	7. Insurance agency*
8. Restaurant	8. General store
9. Bulk oil distributor	9. Auto-repair garage
10. Meeting hall	10. Meeting hall*

* Data pertaining to insurance agencies and meeting halls were not included in the subsequent analysis due to the difficulty of accurate classification in some of the towns; however, the summations appear to be reasonably accurate.

[6] *Ibid.*, p. 11.

[7] Small independent churches are often thought of as being characteristic of denominations, such as the Baptist, which have large rural memberships. For an indication of the relative strength of the Baptist Church in Southern Illinois, see Wilbur Zelinsky, "An Approach to the Religious Geography of the United States: Patterns of Church Membership in 1952," *Annals Assoc. Amer. Geogrs.*, Vol. 51 (1961), pp. 139-93.

pendent, thus giving rise to a larger number of small congregations. Another interesting difference between the two listings is the relative importance of the general store in southern Illinois. It is possible that this is related to the diversified rural economy in southern Illinois and also to the less prosperous nature of the economy, the latter resulting in lower buying power and therefore relatively less opportunity for retailers to specialize.

The interesting deviations notwithstanding, the overall impression gained from a comparison of the two lists is that, assuming that the functions which occur most frequently are the activities which provide the economic bases for most of these towns, small Iowa towns and small southern Illinois towns exist for essentially the same reasons. As expected, considering population size, the goods and services provided by the small towns are frequently used and relatively standardized (convenience goods and services). In the present era, it appears that the economic bases for most of these small towns center on two general demands. The first demand is created by the insatiable appetite of the ubiquitous American automobile, giving rise to the frequent occurrence of gasoline filling stations and auto-repair garages. The more specialized demands created by the automobile, such as new-car dealerships, are not often found in the small town. As Thomas points out, it is ironic that one of the reasons for the decline of the small town, namely improved transportation, provides it, at the same time with a substantial portion of its present *raison d'etre*. The second general function performed by the small town appears to be the provision of facilities for religious, social, or purely recreational gatherings. Since people seem to prefer to have contact with groups consisting c close friends and acquaintances the sma town is in reality performing a convenienc service in providing for meetings of neigh bors. In the lists of the ten most frequentl occurring functions, both churches and meet ing halls are certainly catering to the tendenc to congregate. It also appears that one of th major functions of the restaurant and th tavern in the small town is to help to satisf this same demand.

RELATIONSHIPS BETWEEN THE THREE INDICES AND POPULATION

It is obvious, even to the most casual ob server, that large urban centers perform mor functions than do small urban places and tha they also have more establishments and func tional units. That a positive relationship exist between functions performed and populatio size is generally assumed. However, thre questions might be raised. First, are change in the indices directly proportional to change in the population size from town to town, o are there changing rates of increase over th population range? Second, to what exten might there be disruption of the normal re lationships in areas of general populatio decline, owing to the survival of certain func tions as a result of inertia? Third, is it possibl that the rate of population increase relativ to the increase in functions from town t town is greater in some areas than in others This section is concerned with examining th relationships in southern Illinois betwee population size and (1) number of establish ments, (2) number of functions, and (3 number of functional units; the results ar compared with those for similar analyses i Iowa.

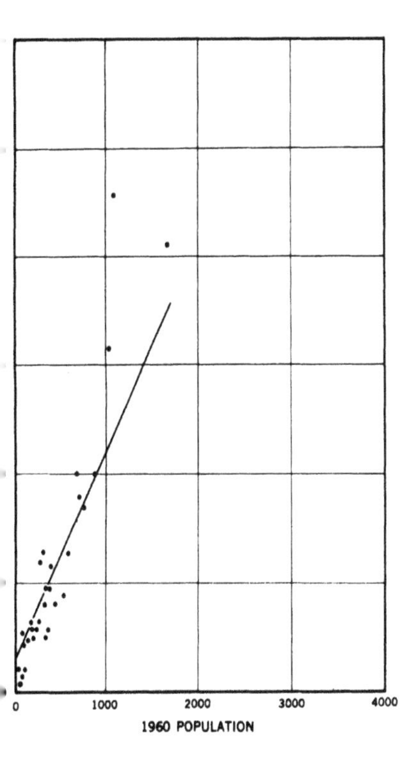

re 2. Relationship between town populations number of establishments.

ationship between number of blishments and population

relationship between the number of blishments and population size in the ll towns of southern Illinois is very close. high degree of direct variation is indid on the scatter diagram (Figure 2), the efficient of simple correlation being .929. Thus, the wide variations in popula- size are matched by a wide range of blishments per town (from 1 for Welge 29 for Waterloo).

he close positive correlation indicates , even in areas where towns are generally declining, the number of establishments is quite responsive to population change. That inertia will seriously disrupt the expected population-establishment relationship appears at this point to be an unsupported hypothesis, *unless* the degree of inertia is quite similar for *all* the towns. That this latter possibility is of consequence seems unlikely. Rather, it appears that population changes over time are quickly reflected by changes in the number of establishments.

Both in southern Illinois and in Iowa, the small towns exhibit a high degree of correlation between population use and number of establishments. Is it possible that, whereas the degree of association is quite similar, the nature of the relationship is significantly different in the two areas? A comparison of regression equations is interesting in this regard:

Southern Illinois: $y' = 5.49 + 3.8x$
Iowa: $y' = 9.60 + 6.6x$

In both cases, the relationship is linear. However, the Iowa line has not only a greater y-intercept, but also a steeper slope, indicating an establishment increment from town to town of 6.6 for each increase of 100 persons; this compares with an increment of 3.8 establishments for southern Illinois. Since *twice* as many classes of functions were tallied for the Iowa study as for the southern Illinois study (121 to 60), both constants in the southern Illinois regression equation should be doubled for purposes of comparison. When this is done, the resultant equation is very similar to that for Iowa. The conclusions, then must be that (1) the number of establishments is directly proportional to the number of people per town, and (2) that there is no significant difference in the relationship between population size and number of establishments for small towns in southern Illinois and Iowa.

Relationship between number of functions and population

The number of functions per town in southern Illinois ranges from 1 for Welge to 51 for Waterloo. In 27 of the 31 towns, the number of functions is less than the number of establishments. This is due to the tendency for a number of establishments to perform the same function in a given town, e.g., three or four filling stations. On the other hand, in the very small towns (less than 200 or 300 persons) the number of establishments is usually nearly equal to the number of functions; this is because of the small town's inability to support more than one establishment of a given type, and the fact that some duplication of establishments is offset by multi-functional establishments.

The relationship between population size and the number of functions in southern Illinois is shown in Figure 3. As expected, the degree of association is positive and quite high ($r = +0.892$). An examination of the best fit regression equation ($y' = 24.52 \log x - 46.43$) reveals a curvilinear relationship, which is positive over the entire range. The curve increases at a decreasing rate. Thus, variations in town size in the very small centers (in this case, below approximately 500 persons) are associated with disproportionately large variations in the number of functions performed. Conversely, in the larger towns comparatively fewer functions are added. One suggested explanation for the curvilinear relationship between functions and population is that "there may be a definite limit to the functional complexity of urban places. As cities become larger, greater numbers of establishments and functions are formed within them. Once a certain level reached, however, establishments are adde much more rapidly than functions. This su gests that to a considerable extent greate numbers of people found in larger places d not desire different kinds of functions, b merely convenient access to the same ones.'

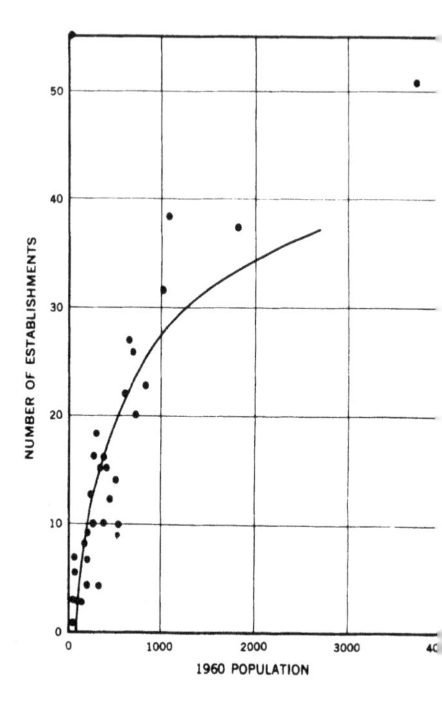

Figure 3. Relationship between town populatio and number of functions.

The regression equation between numbe of functions and population for Iowa is $y' = 39.91 \log x - 66.31$. The nature of the associa tion is very similar to that for souther Illinois. Again, the differences in consta values are probably explained by the diffe ence in the number of functions eligible f inclusion. There appears to be little evidenc in this regard at least, that small souther

[8] Thomas, *op. cit.*, p. 15.

ıois towns are significantly different from ıll Iowa towns.

'ationship Between Numbers of Functional its and Population

e number of functional units in each of southern Illinois towns surveyed ranged m 1 for Welge to 141 for Waterloo. This npares with a range of from 4 to 201 for Iowa towns. Generally, there are, per 'n, more functional units than functions or ıblishments. By means of the same reason- used in dealing with establishments and ulation, it can be hypothesized that there . close, positive relationship between num- s of functional units and population per '. Figure 4 and a correlation coefficient of $+0.934$ substantiate the hypothesis. The ression equation ($y' = 6.18 + 4.2x$) in- ates that, on the average, a change in town of 100 persons will call for a change of htly more than four functional units.

The association for small Iowa towns is ɔ positive and very close, and the regres- n equation ($y' = 15.03 + 8.0x$) exhibits same linearity as does the southern Illinois ıation. Again, the differences in *y*-intercept slope values are tentatively explained by iations in the data collection procedures ıer than by any fundamental difference in functional bases of small towns in the two as.

lationship Between the Functional Unit- 'ablishment Ratio and the Number of 'ablishments

omas computed for each of the Iowa towns 'unctional unit-establishment ratio which dicates the number of functional units that associated with each establishment ated in the city and provides an approxi-

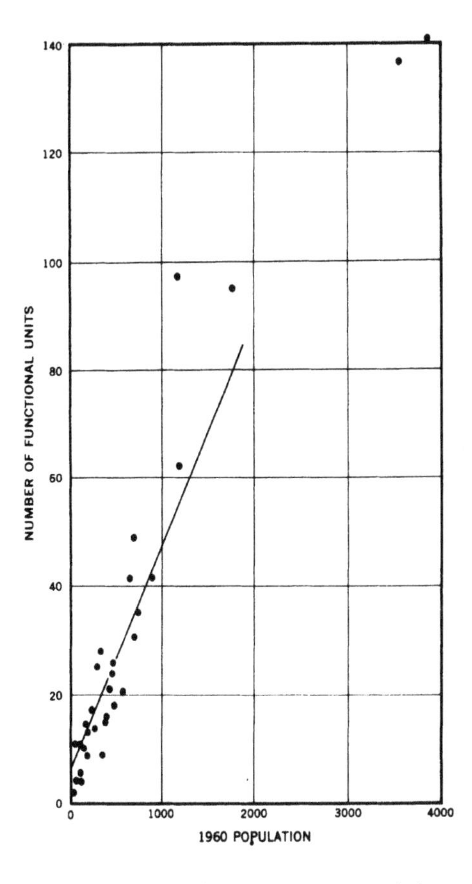

Figure 4. Relationship between town populations and number of functional units.

mate index of the degree of specialization of those establishments."[9] He found the association between this ratio and population size (Figure 5) to be significant, negative, and curvilinear ($y' = 1.75 - 0.266 \log x$), so providing evidence "that the establishments which are located in cities with fewer establishments are less specialized than the establishments which are located in cities with greater numbers of establishments."[10]

[9] *Ibid.*

[10] *Ibid.*

The nature of the relationship between the functional unit-establishment ratio and the number of establishments for southern Illinois (Figure 5) is very similar to that for Iowa. The degree of association, $r = -0.414$, is not high; but it is significant. The best-fit regression line ($y' = 1.39 - 0.19 \log x$) is curvilinear and negative over the entire range. This relationship suggests that in southern Illinois, as in Iowa, there is a tendency in the smaller towns as compared to the somewhat larger towns, for the establishments to be less specialized and their functions less segregated.

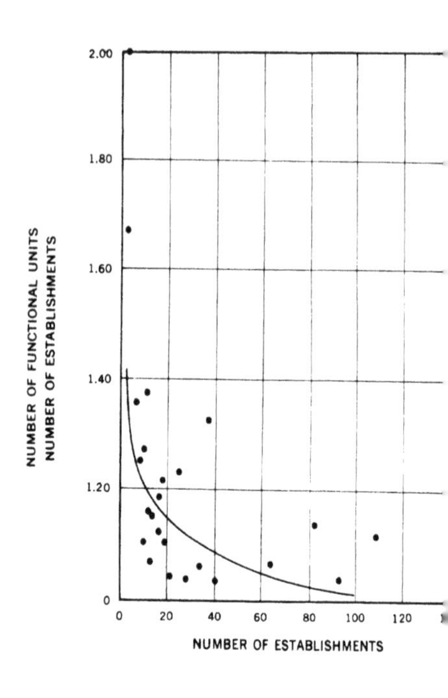

Figure 5. Relationship between number of establishments and the ratio of number of functional units to number of establishments.

COMPARISONS WITH DATA FROM SNOHOMISH COUNTY, WASHINGTON, AND CANTERBURY PROVINCIAL DISTRICT, NEW ZEALAND

An examination of the data provided by Berry and Garrison for Snohomish County, Washington, reveals that the type and frequency of functions found in the urban places surveyed are very similar to the functions found in places of similar size in southern Illinois and Iowa. The coefficient of correlation between population and number of functions for Snohomish County is high and positive; the correlation is similar for the Canterbury Provincial District of New Zealand.[11] These compare to high positive coefficients between population and number of functions for southern Illinois and Iowa. Furthermore, the nature of the relationship is very similar (curvilinear) in the four areas.

King, in his New Zealand study, states that "the high correlation obtained between population size and number of functiona units confirms the belief that the majority o the small towns in Canterbury act as servic centers."[12] The coefficient of $r = +0.930$ i almost identical to $r = +0.934$ for souther Illinois. These are both similar to the hig positive association found by Thomas in Iow and an $r = +0.789$ computed from the Snohomish County data.[13]

Table III presents the ten functions tha appear most frequently in the towns studie

[11] The coefficient of correlation for Snohomish County is $r = +0.751$, compared to $r = +0.892$ for southern Illinois and $r = +0.823$ for the Canterbury Provincial District.

[12] King, *op. cit.*

[13] The correlation coefficients between populatio size and number of functions are $r = +0.9$ and $r = +0.77$ for the Canterbury District an Snohomish County respectively. However, thes used only data on "variates," whereas thos correlations computed for this study use dat pertaining to "attributes" as well.

TABLE II

The Functional Bases of Selected Central Places in Southern Illinois

Town	*1960 population**	*Number of establishments*	*Number of functions*	*Number of functional Units*	*Functional unit establishment ratio*
aterloo	3739	129	51	141	1.09
sser	1764	83	38	94	1.14
enna	1094	91	39	97	1.03
dgway	1055	64	32	67	1.05
ossville	874	40	23	41	1.03
ville	743	33	20	35	1.06
lmeyer	709	36	26	48	1.33
rnak	667	40	22	41	1.03
a	665	26	27	32	1.23
llisville	532	17	14	20	1.18
tsburg	485	16	10	18	1.12
Clure	400	23	15	24	1.04
ltonville	394	19	15	23	1.21
ford	388	19	16	21	1.10
aham	378	16	12	16	1.00
ns	376	12	10	15	1.25
ello	362	10	5	10	1.00
nefort	349	26	18	27	1.04
ff	330	24	16	24	1.00
hview	255	12	10	14	1.16
owns	251	13	13	18	1.38
Bois	229	12	5	13	1.08
leigh	225	10	7	10	1.00
nnie	215	13	9	15	1.15
acystown	158	9	8	11	1.23
enes	114	8	6	11	1.37
ity	110	3	3	5	1.67
nsonville	96	10	7	11	1.10
ery	80	2	3	4	2.00
rbers Ridge	50	4	3	4	1.00
lge	40	1	1	1	1.00

opulation figures are from the U.S. Census of Population, 1960, except in the case of unincorporated owns for which estimates were made.

the Snohomish County and Canterbury strict areas. A comparison of its lists with ose in Table I reveals a number of similari- s as well as a number of differences.[14] The ost striking similarities are in the food store egory and in that for the services catering the demands created by the automobile.

[14] A large measure of the dissimilarities between the Canterbury rank order and the others is due to the fact that King, in this section, relied on secondary data whereas the Iowa, Snohomish and southern Illinois data were collected in the field. King points out that his source (*The Canterbury Business and Trade Directory*, Auckland, N.Z.) does not consistently list all functions, notably halls and churches, and that, therefore, many functions were omitted.

TABLE III

Rank Order of Most Frequently Occurring Functions

Snohomish County	*Canterbury Provincial District*
1. Filling stations	1. Insurance agency
2. Food stores	2. Motor service station
3. Churches	3. Grocery
4. Restaurant	4. Garage and motor engineer
5. Taverns	5. Library-lending
6. Elementary schools	6. Builder
7. Physicians	7. Hotel
8. Real estate agencies	8. Carrier and cartage
9. Appliance stores	9. Restaurant-milkbar
10. Barber shops	10. Engineer-general

There are two types of deviation from the general trends which should be mentioned. The first is that there are deviant towns *within* each of the study areas. Thomas offers two simultaneously-operating factors, nearness to a much larger town and an unusual amount of activity, to explain his outstanding deviate case, the town of Buffalo.[15] Berry and Garrison cite the towns of Beverly Park, Lowell, Lake Stevens, and Edmonds as exceptions, and point out that they are becoming dormitories for the Seattle area.[16] King indicates Kaiapoi as one of his deviant cases and concludes that "this undoubtedly reflects the importance of a large manufacturing component in the town's economic base."[17] He also indicates that the towns of Lincoln and Pareora are exceptions which "would seem to support Berry's contention that 'well defined population: function ratios characterize a system c central places only where the major economi base of these centres consists of central plac functions.' "[18] In southern Illinois, the tow of Vienna has, for example, more establish ments, functions and functional units tha expected on the basis of population. Thi might be explained by the relative isolatio of this county seat town from towns of com parable size.

The second kind of deviation is exempli fied by the differences among the four areas i the type and importance of various function found in the small towns. These difference are tentatively explained by (1) variations i the data collection techniques, and (2) b regional economic variations between th four areas — areal variations in the basic eco nomic activities which these towns serve. Tha this second hypothesis is plausible is sug gested by the fact that the areas which ar most similar in terms of the functional base of their central places, Iowa and souther Illinois, are the two which are closest to gether.

However, these deviations and difference do not negate the major conclusions of thi study: (1) that the economic bases of smal towns are highly predictable; and (2) that th economic structures of small towns i southern Illinois, Iowa, Snohomish County Washington, and Canterbury Provincial Dis trict, New Zealand, are very similar.

SUMMARY

Data collected from 31 small, souther Illinois towns indicate that these towns ar almost exclusively service centers. Practicall all the employment opportunities in thes

[15] Thomas, *op. cit.*, p. 12.

[16] Berry and Garrison, *op. cit.*, p. 154.

[17] King, *op. cit.*

[18] *Ibid.*

ll urban places are in the service occupa-s such as those represented by store ks, barbers, teachers, and insurance its. The only type of manufacturing found any degree of regularity is typified by grain mill, an activity which has many of attributes of a service function. Further-e, the towns provide only standardized frequently-used goods and services. As ry and Garrison have indicated, this is to expected, since the "threshold popula-s" of these towns and their trade areas not sufficient to support the more speciali- types of functions.[19]

comparison of the types of goods and ices offered in small towns in southern iois with those offered in the other areas rred to reveals that the types of function each list are almost identical.[20] However, frequency rankings of the functions differ n area to area; these differences appear to eflections of differences in the economic es of the areas. This observation leads to the hypothesis that small towns in "open country" areas such as Iowa and southern Illinois, are local service centers, the frequency and magnitude of the functions performed depending on the basic economic structure of the relevant regions.[21]

The correlations between population size and the three indices of functional size and complexity are, as was to be expected, high and positive. There appear to be no significant differences in the strength or the nature of the associations when the southern Illinois data are compared with the Iowa data. For both areas, changes in the number of establishments and the number of functional units are proportional to changes in population size from town to town. However, the number of functions added increases at a decreasing rate as towns become larger. An examination of the data indicates that essentially the same relationships exist in Snohomish and Canterbury.

Brush has stated that "an interesting field of comparative geographical research is open in the study of central places."[22] The present study, in conjunction with those that have preceded it, lends empirical support to Brush's statement that "small towns and villages in agricultural areas of Anglo-America exist mainly because of their function as central places for the exchange of goods and ser-

rian J. L. Berry and William L. Garrison, A Note on Central Place Theory and the ange of a Good," *Econ. Geog.*, Vol. 34 1958), p. 50.

rian J. L. Berry and Harold M. Mayer have ecently (February, 1962) completed a report, Comparative Study of Central Place Systems," or the Geography Branch, U.S. Office of Naval esearch. In each of five study areas, south-estern Iowa, northeastern South Dakota, outhwestern South Dakota, a portion of sub-rban Chicago, and a portion of central Chicago, ttention has been focussed on the simple func-ons, and spatial patterns of central places. The ections of these studies concerned with the unctioning of central places reveal results hich compare quite favorably with the results f the present study. This is especially true when he data of the present study are compared ith the data from the southwestern Iowa and he two South Dakota areas; Berry and Mayer ppropriately point out that central place regu-arities are most pronounced in rural areas.

21 An alternate, or additional, explanation of the variations from area to area in the frequency and magnitude of functions performed in the small towns is suggested by Berry and Mayer *op. cit.* It is probable that in areas of very low population density the small towns do not have a sufficient clientele in their effective trading areas to support certain functions which might be expected in such towns. In these cases, it is suggested that the affected functions will move to larger urban centers.

22 Brush, *op. cit.*, p. 402.

vices, each for its local farm trade area."[23] By building one similar study upon another in different areas, progress is made towar valid generalizations concerning the ec nomic functioning of central places, so mak ing possible increasingly precise prediction.

[23] *Ibid.*, p. 380.

17

Cities, Transportation and Technolog

Harold M. Maye

Two developments of human culture have been outstanding during the past century: the rapid advance of technology and the growth and spread of cities. Neither would have been possible without the other. Together they have made a revolution in the organization and pattern of land use. Transportation is a basis of both.

About 4 percent of the United States population was urban in 1790. Now about 7 percent of the population lives in metropol tan areas and other urban places, and the pro portion is increasing rapidly. There also ha been a substantial growth of rural nonfar population, which depends on urban areas fo employment or on the passing highway traff from the cities.

The growth of cities has accounted fo

Reprinted from *Land, The Yearbook of Agriculture* (1958) pp. 493-502 by permission of the author and the United States Department of Agriculture.

st of the increase in the population during 20th century, while the farm population declined.

Agricultural areas, because of improve-nts in farming, have furnished a substantial tion of the immigrant population, which, led to the high net reproduction rate in es, has been responsible for much of the rease in urban population.

The largest metropolitan areas have been wing, in general, at a faster rate than aller cities.

The location of the areas of most rapid oulation growth has shifted significantly ce about 1920 with respect to their situa-s within the metropolitan areas. Until out four decades ago the cities were being jected to increasing populations at ever her densities, but since the end of the First orld War the maximum rates of growth erally have been outside of the central es of metropolitan areas. Suburban com-nities and unincorporated areas have wn much faster than the central munici-ities in recent years. Metropolitan areas in ny instances have had substantial increases population but actually have shown de-es in the populations of their principal or tral cities.

Part of the reason is that municipal bound-es rarely coincide with the limits of the lt-up urban area.

Cities once could rather easily annex near-areas that became urbanized. Extensive as could be annexed in advance of the ead of urban development. The formation many small incorporated municipalities t to the central cities more recently has de annexation difficult or impossible. ese small cities, towns, and villages have veloped local governments and vested in-ests in their perpetuation. Many people ove to the suburbs with the expectation of being able to have a more personal and inti-mate relation to their local affairs, and subse-quent merger with the big city is therefore almost invariably resisted.

The average population density of most cities has dropped sharply in recent years. This reflects a demand for land that is increas-ing at a much faster rate than even the spectacular rate of increase of urban and metropolitan population. Ranchhouses or ramblers have become popular and 60- and 80-foot lots are replacing 30- and 40-foot lots. Single-story industrial plants, with exten-sive areas for parking and with substantial setbacks from the highways and access streets, are replacing the multiple-story in-dustrial and loft buildings of the congested central parts of cities. The modern planned outlying shopping center includes at least three or four times as much area for auto-mobile parking as it does floor area of selling space.

The increasing demand for land for urban uses has been met by an accelerating expan-sion of cities into the rural areas.

It has recently been estimated that urban areas in the United States occupy slightly more than 18 million acres — a little less than 5 percent as much as the total land area occupied by railroads and highways. The urban land area is about 1 percent of the total land area of the United States.

The outlook is for a faster rate of conver-sion of agricultural land into nonfarming use, particularly for urban expansion. Many of the metropolitan areas may be expected to double the amount of area they will occupy within the next two or three decades.

Much of the land they will occupy is crop-land that is used for intensive production of specialty crops and has a higher value per acre than the average value of all agricultural land. These croplands, being devoted to intensive cultivation of specialized crops, are characterized by small farms. Thus they have a population ,much denser than the average for all agricultural areas.

Certain important areas of specialty crops, such as the truck-farming areas of New Jersey, the citrus areas of southern California, and perhaps the fruit belt of southwestern Michigan, may be expected to be invaded by the urbanization from nearby large cities.

While the total loss of cropland may not represent an actual net loss of agricultural production nationally, the loss of specialty crops, involving conversion to nonagricultural use of some of the best land for such crops, could well become significant.

Most cities exist primarily to satisfy economic needs. Growth of population occurs in response to economic opportunities. People live where they can earn a living. Since economic opportunities are greatest in number and variety in the larger metropolitan areas, it is those areas which have been experiencing the fastest growth. Great concentrations of economic opportunity depend on concentrations of labor force, which in turn produces additional incentive for population growth. Thus the metropolis expands.

Urban and metropolitan concentrations could not exist were it not for transportation facilities. Many of the outstanding technological advances have been in transportation, which affects the size, functions, structure, and growth of cities and metropolitan areas.

Functional specialization of areas — the differentiation of one land use from another — is made possible by the availability of facilities for the movement of goods and people between those areas.

Cities produce manufactured goods and perform certain services which are "exported" to other areas, in return for the gooc and payments that are brought into the urba areas from other urban areas and from th countrysides.

The interconnections between cities an between individual cities and their respectiv service areas or hinterlands — as well as th interconnections among the various func tionally specialized parts of city and metrc politan area — depend on efficient system of transportation.

Streets alone in most cities account for 2 to 35 percent of the total built-up urban are The building and maintenance of facilities fc internal circulation, including streets, cor stitute a sizable part of the budgets of a cities.

Most urban street patterns have been ir herited from the past and are inadequate fc the needs of modern traffic. The obsolete pa terns have been extended to newly develope areas on the outskirts of cities.

While street traffic faces delays because c insufficient numbers and capacities of arteri routes, an excessive proportion of the area of most cities paradoxically is devoted t local access streets.

The largest parcels of land devoted to single urban use and under single control i most cities are the airports, which may cove several square miles and influence land use far beyond their own boundaries.

The construction, maintenance, and opera tion of transportation facilities and equip ment directly contribute substantially to th employment base of urban areas. More tha 20 million Americans are employed in publi and private transportation. Automobiles an trucks account for a large proportion of thi employment — nearly one-third of the tota employment in the Nation.

A study by the Port of New York Authorit indicated that about one in every four job

Greater New York is attributable directly ndirectly to the port function, which repre-ts only a part of the multiplicity of basic nsportation functions performed by the w York metropolitan area.

The effects of the transportation industries felt through the entire economy, because nsportation uses vast amounts of ma-ials and equipment, the manufacturing d supplying of which create other millions jobs.

The automobile industry used 22 percent d the railroads 11 percent of the steel pro-ced in the United States in 1957, a typical ır. Large tonnages of steel also were used building highways and ships. Most of the and rubber and a major share of the al used in this country are used in trans-rtation.

The rapid changes in transportation are lected in the changes in the growth and ucture of cities: each major innovation in ercity and local transportation has been lowed by significant changes in urban as.

The uses of urban land are related closely their "circulatory" systems. The relation-p is reciprocal. Land uses — other than ch uses as agriculture, forestry, and mining, ich depend on primary production on the e — are where they are largely because of ferences in the availability of transporta-n from place to place. On the other hand, d uses (individually and in combination) nerate varying amounts of movement of ods and people that in turn make it neces-y to provide varying amounts and kinds transportation.

Each type of nonagricultural use of land s a different set of requirements as to loca-n. For some — as, for example, the bulk-ceiving industries that use raw materials in ipload amounts and therefore need loca-tions along navigable waterways — the choice of location is narrow and rather inflexible. A much greater variety of locations is suitable for other uses, such as one-family homes.

Transportation in one sense is a substitute for nearness. Other things being equal, the best locations for interrelated activities are close together in order to reduce the amount, and hence the cost, of the transportation of goods and people.

Transportation costs — whether measured in money, or distance — are incurred because it is physically impossible and sometimes un-desirable to place the activities and uses of land in the best locations for each because other activities and uses that require similar sites bring competitive pressures.

The increasing size and complexity of cities widens even more the separation of the urban functions and increases the amount of trans-portation that is needed. Separation of places of work from places of residence gives rise to the daily journeys to work, which are responsible for half of the total number of trips made in metropolitan areas.

The competition among all urban functions and land uses that could operate most effec-tively near each other engenders a high de-mand for centrally located sites. The demand drives up the values of such sites. Not all urban land uses or functions can afford cen-tral sites. Indeed, some functions can better be carried on at some distance from the urban centers if adequate transportation is available. Thus, in the normal operation of the real-estate market, urban land uses are sorted out in accordance with their relative ability to pay high costs for sites that are most desirable because of proximity to other uses or because of the accessibility provided by the convergence of local transportation in the central parts of cities.

For any given type of use, a balance exists

between the costs of competitive sites and the costs of overcoming the friction of distance. The uses that depend on maximum accessibility can afford the high costs of central locations. The other select locations at varying distances from the points of maximum accessibility in accordance with their ability to pay site costs. Transportation in most instances is the factor that makes possible the concentrations of land values, because it converges and produces maximum accessibility at the urban core.

The forces that affect the patterns of land uses in urban areas may be described as centrifugal, or outward, and centripetal, or inward. The resultant of these forces is reflected in the degree of decentralization or deconcentration of any individual land use or groups of land uses in urban areas.

When the centripetal forces are stronger, the city develops with heavier concentrations at higher densities.

When centrifugal forces are stronger, the average densities are lower. The relative importance of the forces varies for each type of urban land use and for each establishment, whether industrial, commercial, residential, or institutional.

The development of improved transportation generally has strengthened the centrifugal forces by making greater and more extensive areas around cities accessible for urban expansion. On the other hand, however, transportation has increased the numbers and the strength of "linkages" among establishments and so has created a demand for increasing concentrations of business activity in the larger cities, especially in the central parts, where face-to-face contacts are maximized.

The concentration of people in cities, the rapidly increasing number and complexity of urban functions, and the resulting competition for space have brought about an ever increasing separation between places of employment and places of residence.

In the medieval city, manufacturing, commerce, and residence were usually on th same parcel of land or in the same structure The craftsman produced and sold his good and lived with his family in one building With the Industrial Revolution, these func tions had to separate because of the develop ment of the factories, which formed nodes o nuclei in the urban pattern, and of markets which later became the central business dis tricts.

Thus the modern city has many nuclei the central business district, which generall has the heaviest concentration of employ ment, industrial areas, and outlying com mercial developments, which have othe concentrations of basic economic activities.

The increasing complexity of the land us and functional patterns of cities has attracte the attention of many economists and sociol ogists, who have tried to make generalize descriptions that would fit most cities.

Ernest W. Burgess developed the concen tric zonal hypothesis, based upon the wor of J. H. von Thünen, a German economist in the early 19th century.

Burgess described the city as consistin roughly of concentric zones. The central busi ness district is the nucleus. The land uses i each successive zone outward from the cor are sorted out in order of their relative abilit to benefit from (and pay the costs of) proxi mity to the center. As a city grows, land use and people successively "invade" each zon outward from the center. This creates a suc cession of land uses in each zone, and eac succeeding group of uses is developed a higher density as a result of increasing com petition for centrally located land.

Homer Hoyt, then of the Federal Housin

ministration, later propounded the wedge, ector, theory.

t describes the process of urban growth l expansion in terms of differentiation of d uses and functions along wedges radiat- out from the central core. The general racter of the uses along each radial or lge is similar in nature from the core to iphery.

Chauncy D. Harris, of the University of cago, and Edward L. Ullman, of the Uni- sity of Washington, described the city as a es of nuclei — generally concentrations of ployment. The various urban land uses located with relation to relative proximity each of the multiple nuclei.

None of these generalized descriptions fits cities. All are based upon the concept of balance of proximity to the core and other an nuclei and the availability of trans- tation to overcome the lack of proximity ulting from the impossibility of locating land uses with maximum mutual prox- ty.

Whatever the specific patterns of urban d uses and internal functional organization cities may be, the specialization of areas their separation from one another are de possible by the availability of trans- tation.

Each successive form of urban transporta- has had significant effects in accelerating h the expansion of cities, on the one hand, concentrating industrial and commercial vities in the nodal portions of cities, on the er.

Before urban transportation was me- nized, the extent of a city was limited by se-drawn transportation, at an average ed of 3 to 4 miles an hour. Cities had to small and compact so that all parts could reached in a reasonable time. Factories e relatively small, and little need existed for wide separations of places of work and of residence.

The horse-drawn street railway car was the dominant form of urban transportation from the period immediately before the Civil War until nearly the end of the 19th century. Although placing the vehicles on rails reduced friction in comparison with the free-wheeled vehicle, speeds were limited, and cities, though expanding, remained crowded and compact.

The steam railway, with commuter schedules, offered opportunities for urban expansion during the latter part of the 19th century in the vicinities of some of the larger cities. Beyond the main urban mass, with its radius of 3 or 4 miles from the commercial core, the steam railroad, with its higher speeds, made possible the development of nodes of suburban growth.

The result was a moderately densely developed series of outlying settlements, clustered about each suburban railroad station, the railroad forming an axis. The pattern that developed resembled beads on a string, with nonurban land lying along the railroads between the stations.

Since the railroads radiated from the urban core, the resulting pattern consisted of radial strings of suburbs, each radial separated from the next by open country, and each suburb along a rail line separated from its neighbors by open country between the railroad stations. Beyond each railroad station, urbanization was limited by the range of horse-drawn transportation. Since the practicable commuting time in each direction to and from the core of the city was about 1 hour, the distance from each outlying station at which urbanization took place was limited by the combined time of rail trip and connecting trip by horse-drawn vehicle, or, in a few instances, by local electric car.

The development of electrified railway transportation in the early years of the 20th century expanded the areas available for urban development. The electric street railway lines were extended beyond the limits that were possible for the horse-drawn streetcar, because of the higher speed. The speed was still limited by urban congestion, however. Along the street railway lines, land values (and hence density of development) were concentrated.

The street railway made it possible for the main urban mass to expand along the routes that were in operation. The resulting pattern of the expanded urban development was roughly in the shape of a star, whose points developed along streetcar lines. Within the urban mass, the densest development was also along streetcar lines.

The main lines in most cities were radial, focusing on the central business district, where most of the employment was located. In some of the larger cities, circumferential or crosstown routes were in operation to provide service to factories and offices not directly associated with the commercial core. At the intersections of the radial and circumferential routes, major outlying shopping centers tended to develop at the transfer corners. Some of them became almost small-scale reproductions of the central business districts and created problems of traffic congestion and competition of commercial land uses to get nearest to the major intersection.

The application of electric power to suburban transportation beyond the main urban mass took two forms. One was extension of the street railway into suburban areas. The first two decades of the present century marked the heyday of the interurban electric railway. Because a car or train could stop at any place along the line, suburban development was freed from dependence c proximity to outlying railroad stations.

The electric suburban or interurban rai way represented a considerable advance i opening up new areas for urban expansio Rapid acceleration and deceleration permi ted more frequent stops. The operatio of several cars in a train related its power an speed to the fluctuations of traffic from day day and hour by hour more readily than coul the steam railway train. The result was th on the fringes of many cities electric railway were built parallel to the earlier steam rai ways in order to secure initial traffic fro preexisting suburbs. These lines permitted filling in of the areas between the steam rai road stations. The radial tentacles of sub urban development filled in and became mo or less continuous. Farmland was subdivide and converted into suburban residential lan more rapidly than in the previous period.

The second form of application of electri power to suburban and urban passenge transportation was by the electrification c steam railroads near some of the larger citie The advantages of the railroad as a heav mass carrier of passengers on high-densit routes was combined with the advantages c multiple-unit operation.

The electrified steam railroad, howeve did not approach the flexibility of the inte urban electric railway, which usually repre sented less investment and could be extende more easily into newly developing suburba areas.

Five large cities — New York, Chicago Philadelphia, Boston, and Cleveland — developed rapid transit elevated and subwa railways for internal transportation when th concentrations of traffic exceeded the capaci ties of the streets.

The rapid transit line, unlike other form

ırban transportation, is separated from all er traffic. It is on a reserved right-of-way l has no conflicts with street traffic. Most id transit lines are operated with multiple-t trains at relatively high speeds and with :ances of one-third mile to several miles ween stops. A busline operating on a right-way or lane reserved for its own use would ɔ be a rapid transit line. The capacity of a id transit line exceeds that of any other m of local transportation in terms of the nber of passengers that could be moved in :rtain period.

Electric surface transportation has almost e full cycle. The electric interurban rail-y has nearly disappeared in the United tes being largely replaced by the auto-bile. The local street railway survives in y a few places, having been replaced the motor bus. Only in the rapid transit : and the electrified suburban steam rail-y does the application of the electric power vive in rail passenger transport of daily ne-to-work movements.

The development of the automobile and tortruck has produced the most rapid and -reaching changes of any technological in-ᴧation in transportation in the rate, direc-ı, and scale of urban expansion. No longer :d urban development be tied to the limited nber of routes feasible for rail transporta-ı. The flexibility of the individual privately ned vehicle opens up vast areas beyond former limits of cities and suburbs for an development.

Our cities have been building up around automobile. Many newly developed areas ɔend entirely on automobile transporta-ı, for they are beyond the range of public riers. The areas between the older radial ngs of suburban growth are filling in, because the automobile can go anywhere where passable roads exist.

The areas of countryside available for urbanization are several times as extensive as the areas that could be developed when people had to depend on public carriers for the journey to work. Many industries no longer need to locate near the convergence of public transportation in order to assemble workers, who increasingly come by automobile.

The truck makes possible the assembly of raw materials and semifinished products from many sources and the delivery of manufactured goods—for which sometimes railroads now are not used at all. Many industries do not need railroad sidings, for they can truck their shipments to the nearest rail freight station. Piggyback — the transportation of motortruck trailers on railroad flatcars — combines flexibility of motortruck transportation and the economy of the railroad as a large-scale hauler.

Factories more and more are tending therefore to locate away from congested industrial districts, which were built when railroads provided the only intercity freight transportation.

Highways and motor vehicles also are opening up opportunities for lower urban densities. Thus residential areas can develop free from some of the disadvantages and limitations imposed by the need to be near mass transportation.

The effect of the new flexibility is generally to reduce the emphasis on relatively few focal or nodal areas and to spread the demand for land over larger areas.

Lower densities — if there is proper planning — provide opportunities for more open space and for many amenities that are lacking in the older sections of many cities. Parks, playgrounds, ample backyards, and larger sites for schools can be provided.

The amount of service provided by the mass carriers is less in most urban areas than ever before. Local transit systems face prospects of further cutbacks in service as their costs rise and patronage declines.

Some form of public transportation is essential in most cities, however. Central business districts still are the major foci of employment and shopping in nearly all cities, where parking has become the biggest problem of all. The larger the city, the more dependent is it on mass transportation, even though the relative dependence is declining.

The result is that mass transportation, instead of being the basic general intracity and suburban form of transportation as in the past, is increasingly specialized in function. It is best adapted to the transportation of heavy volumes of passenger traffic along high-density routes and during the peak hours of the day. Since the highest densities exist in the older sections of cities and the peak volumes are to and from the central business districts, mass transportation is most used for the journey to work in the central business districts of residents in the older and more densely developed sections of cities.

In outlying areas of sparser population, combining the flexibility of the automobile and the economy of the mass carrier sometimes is feasible by providing outlying parking facilities along the transit lines and at suburban railroad stations.

The expressway is a new element of increasing significance in the evolution of future urban land use.

An expressway — or freeway or thru-way — a specialized traffic artery for the high-speed movement of vehicles, is free of the delays and hazards of conflicting cross traffic. It is separated from other traffic routes. The ordinary arterial street combines through movement, local movement, parking, and loading and unloading of vehicles. The expressway has only one function — to speed up through movement.

Several hundred miles of expressways have been completed in cities and metropolitan areas. A number of cross-country expressways, some of them turnpikes, connect major cities. The program of Federal interregional highways, authorized by the Congress in 1956, provides for 41 thousand miles of modern highways. A substantial part of their mileage will be in metropolitan areas.

The effects of the expressways will be tremendous. The new routes will be basic elements in the entire structure of urban and metropolitan uses of land. Because the rights-of-way are 250 to 300 feet wide, each expressway in an urban area removes from other uses of a strip at least a city block wide for the entire length of the route. At the interchanges between expressways and between expressways and other arteries, vast areas of land must be taken and hence made unavailable for other development.

In the areas that must be taken for the rights-of-way are thousands of business establishments and hundreds of thousands of residences, which must be relocated. The selection of the relocation sites will strongly influence the future patterns of the cities. Relating the major transportation routes to comprehensive city and regional plans becomes more important than ever.

The expressways are being located primarily with reference to their ability to move vehicular traffic. That is their function. But too little thought is given to the relationship of the routes to the present and future patterns of commercial, industrial, and residential areas they serve.

Several vital questions need answers.

Will the new traffic facilities cause addi-ıal concentration and congestion in al-dy congested areas?

What additional parking facilities will be ·ded to accommodate the vehicles after y arrive in the congested areas?

What effects will their routes have on the idential communities through which they l pass?

Will they increase neighborhood and com-nity cohesion by forming barriers at the ındaries of the neighborhood and com-nity areas — or will they disrupt existing nmunities by causing the removal of sub-ntial populations and by creating barriers ween the residences and such community i as the schools, churches, parks, and shop-g centers?

Time will provide answers of sorts to some the questions. Right now we need to study ectively and thoroughly the existing phy-al and social patterns of cities that the ex-ssways will affect.

We can foresee some of the effects of the ›ressways. By providing high-speed trans-rtation for both the private automobile and motortruck, they will increase further the ficulty of providing mass transportation ilities for peak loads to and from the cen-l business districts. At the same time they mselves will not provide complete solu-ns to the problems of transportation to and m such districts.

Integration of planning of expressways and ss transportation — in other words, think-; about the movement of people and goods rather than just the movement of vehicles, and thinking about cities rather than about transportation alone — is essential.

Thus the new forms of urban growth, like the new technological inventions, produce new problems and accentuate the urgency of solving old problems.

They also produce new challenges. Among the most urgent challenges is the one represented by the lag of our social and political institutions behind the increasingly urgent problems which they are being called upon to solve. Our metropolitan areas, for example, are fragmented into dozens or hundreds of small political units — cities, towns, villages, school districts, park districts — each concerned with its own functions or its own limited area of jurisdiction.

Cities expand — but without equal expansion of the horizons of social and political organization. Some groups of the population, attracted to cities by the greater employment opportunities, meet resistance in some cities. Schools in most newly developed suburban areas are not planned and built sufficiently in advance of the population growth, and their problems are complicated by the small size and financial inability of many of the political jurisdictions.

Few metropolitan areas have adequate machinery to plan for the new conditions systematically and comprehensively.

Technological advances therefore must be paralleled by social, political, and economic advances if their full potentialities for the benefit of man are to be realized.

18

Transportation Planning anc Urban Developmen

Alan M. Voorhee

There are, of course, many ways that we can look at traffic. Certainly we should consider it from the individual's point of view. We should get right down to his basic needs for travel. If we do this, we find that the average person makes about two trips a day. However, this will vary depending upon his status in life — whether or not he is a breadwinner, or a student, or just how old or young he may be. His economic position will also have an impact upon this average. Of course, there are other factors involved in trip frequency which are not solely related to the individual; for example, the geographical position, climate, adequacy of the transportation system, and the patterns of urban development.

TRANSPORTATION REQUIREMENTS

Trip Production

Although this approach is the best way to look at transportation needs, we have found it easier to analyze them in other terms. For example, how many trips are made by an auto in a given day? As you might expect, we have found that this is highly correlate with auto ownership. In fact, a great deal o research has been undertaken recently to ex plore the other elements that are involved i "auto trip production." Although these ex plorations have not been as revealing as w had hoped, they certainly indicate that man other factors are involved — family size, th number of retired persons in the family, th number of school age children, etc. Howeve it is very clear that car ownership is an excel lent indicator of the trips that are made by family and the mode of transportation that i used. Generally we know that about fou daily trips per car are made in the large cities — 200,000 or more with more trip being made in smaller cities. We also know that once the level of one car per family i reached, transit usage becomes very low Therefore, it is very important to understan what influences car ownership.

Figure 1 shows what we feel are the ke factors influencing car ownership; namely type of residential area and income. It show that with a rise in income there is a rise in ca

Reprinted from *Plan Canada,* Vol. 4, no. 3 (1963), pp. 100-110 by permission of the Town Planning Institute of Canada.

ership. This seems to "ceiling out" at ıt $8,000 to $10,000 income per family. ownership is particularly sensitive as ilies rise in the $3,000 to $5,000 income :ket. At this range there may be a 10 per- : change in income and a 30-40 percent ıge in car ownership. Many European ıtries are moving through this range at present time. We all went through this in 30's and 40's. Right now, car ownership person is going up quite slowly. By 1980 ıould stop climbing. In fact, if we could ɔ the birth rate down, we might have the ·le problem licked.

'his curve also indicates that if we could ɔ everyone living in higher densities, we ıld be a lot better off. As shown by Figure s the density increases the car ownership ɔs. Of course, this reflects many factors :h influence transportation requirements. t of all, as the density increases there are :rally errands that can be accomplished on , thereby eliminating the need of an auto- ɔile. But unfortunately, many of our new ɾtment house areas do not provide for this : of development — too often they are sterilized from commercial use and walking trips to the grocery store or laundry are almost impossible. So I am not too sure if these patterns will hold true for the new apartment areas. But certainly with higher densities we can gain a higher level of transit service, since there is a greater market.

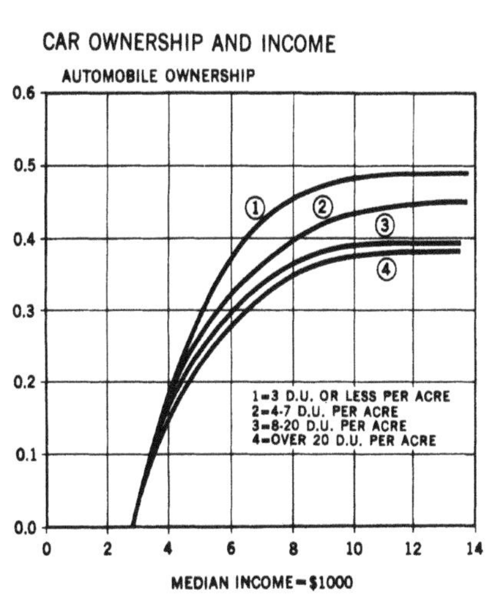

ıre 1.

Just how important the level of transit service is in this whole picture is still not clear. However, stepped-up transit service by itself cannot be assured usage. If parking is readily available people undoubtedly are going to have a car, and once they have a car they will tend to use it. The only time they will not use it is if the parking rates where they are going are exceedingly high, or if the transit service is much better than auto. Perhaps the zoning ordinances that we have which require sufficient off-street parking in apartment house areas have a greater impact than we realize upon the mode of travel. We have never adequately evaluated our zoning requirements, particularly in light of the impact they have upon the mode of travel that people use.

Trip Length

Another important element in transportation — probably more important even than the question of trip production is distribution and trip length. Surely we know that work trips must go to places where there are jobs and that shopping trips must go to shopping centers — but do they always go to the closest place of employment and the closest shopping center?

In reality we all know that they do not. People require and demand variety — not only in terms of where they shop, but where they work and where they spend their leisure. This desire is really the culprit behind our traffic problem. This can be made clear by a simple example. If trip length could be cut in half, the traffic loads on all our streets could be reduced by half. If the trip length in Los Angeles were similar to that in Chicago, there would be one-third less traffic in Los Angeles. So trip length characteristics are of vital importance.

The next graph, Figure 2, shows the range of trip length for different purposes that have been found in various studies throughout the country. As might be expected, the work trip is the longest. In considering the various cities and localities in the country, it is quite evident that the communities that have had the greatest fluctuation in the job market and have the highest level of transportation service have the longest trip length. Los Angeles is a good example of this. However, areas like the New London Area in Connecticut have similar trip length mainly because of fluctuations in the economy. People are willing to travel great distances if necessary to get a satisfactory job. This has been the case in the New London region, since some of the basic industries have moved to the South. In other areas where social patterns and local traditions had made these areas more self-

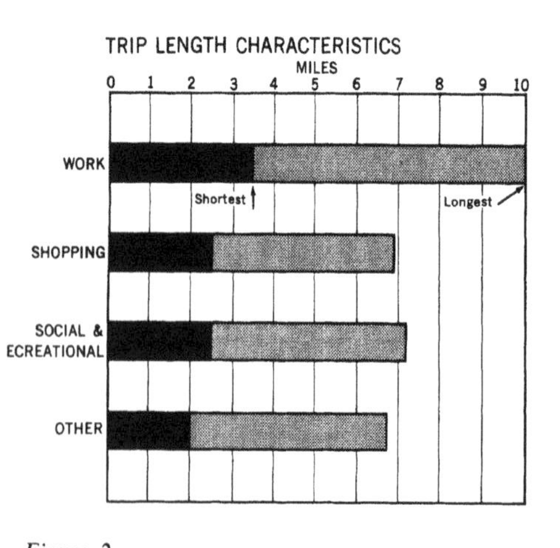

Figure 2.

contained, the work trip may be very sho by comparison. This is true in some cities Iowa where over the years people have liv and sought employment close to home.

Unfortunately, we have not been able quantify all these things, but it does appe that an area that is subject to employme changes is likely to be subject to greater tran portation problems than a stable communit therefore, a community that is expecti fluctuations in its economy should probab try to provide for greater flexibility in i transportation system than a more stab community.

Work Trips

The only solution I can see to shortening th work trip is by carefully planning and pr gramming the development of our commun ties. As new industries or activities move in the area, their development must be tied with housing programs. Adequate housir for their employees must be nearby. Th must be available at the time industrial deve opment occurs, and it must be at a price th worker can afford. Some of our faster grov ing metropolitan areas are plagued with th problem. For example, while industries we expanding in the Los Angeles Basin, a gre deal of the housing was being provided in th San Fernando Valley twenty miles away. Th naturally generated a long work trip. I fe that with proper coordination of public an private interests we can do much better i this direction than we have in the past. Ce tainly we should give more consideration t the work trip in our urban renewal progran and try to reduce the work trip whereve possible.

The other types of trips, such as thos related to shopping, are going to be ir fluenced by dispersion of our retailing activ and the transportation services to them. hough in many ways we should attempt to uce them as much as possible, they do not 'e as great an impact on traffic since the jority of these trips do not occur at the .k hour. From a planning point of view, it probably much more important to have ong commercial centers that have a depth variety in merchandise than to have them se to everyone, which really doesn't efit anyone.

As for the social and recreational trip, I that with our increased standard of liv-, we can expect nothing but additional in-ase. If our transportation facilities of the ure permit it, we will undoubtedly see ger and longer trips — they may even eed the work trips in length. This un-btedly will put greater strain on our trans-tation facilities, particularly those that are ving the key recreational areas. I wish re were ways that we could reduce these lengths, but on the other hand a good n for an urban area certainly would call many types of social and recreational ersion and they should be accessible to all. is is undoubtedly what the people will nt, and it is something that we will have to n for.

We undoubtedly should be doing more nsportation planning for our recreational ds than we have done in the past. Along line, several recent transportation studies developing a special analysis of recrea-nal travel and plan to take recreational ds into consideration in designing for all ure highways.

e City

enever we talk about transportation ds, I think we cannot gloss over the city lf. After all, transportation needs are related to the city, and the exact form that the city takes in the future will have a terrific impact upon transportation requirements. Thus, we are justified in emphasizing a few things that are occurring in urban growth, which provide a full perspective on our trans-portation problems.

The first one that I should like to stress is the flexibility that has been brought about by the automobile. As we all know, the auto-mobile has provided a fantastic flexibility not only to the businessman but to the individual. The flexibility has permitted the individual to live almost any place that he chooses in the metropolitan area, and it has permitted the industrialist and businessman to move about freely and locate wherever they desire.

By providing this flexibility we have given people the opportunity to consider many other factors in making the decision as to where to live or work or locate their business. The individual is given the opportunity to consider the quality of the schools, residential amenities where the area has prestige, hous-ing costs and many other factors in choosing his home.

This means that in making a decision as to where to live transportation is not having the influence that it once had when our cities were built around a rapid transit system. The individual under those conditions had little choice, but now with our extensive highway systems in urban areas and with the auto-mobile he is given a great deal of freedom.

Furthermore, as he rises on the income scale, he seems to put more emphasis on larger lots and the amenities associated with such lots — as illustrated in Figure 3. This, of course, is somewhat related to the fact that people's values would change with a rise in income, and it also shows up in various other aspects about urban living. A curious fact that we have observed in connection with some of the attitude surveys we have been conducting is that as income goes up the proximity of commercial development become less and less important. The reasons involved are not established, but it appears that as income rises people would prefer to have "pure" residential areas, and hence travel several miles farther to a store rather than to have commercial areas mixed in with their residential neighborhoods.

This may be due to not having any example of how commercial areas have been effectively woven into residential development. But it is probably related to the fact that once people accept the fact that they have to use a car for an errand, they are not too concerned whether the trip is four miles or two miles. This is illustrated by Figure 4 which shows the satisfaction of people with various public services located at different distances from their home.

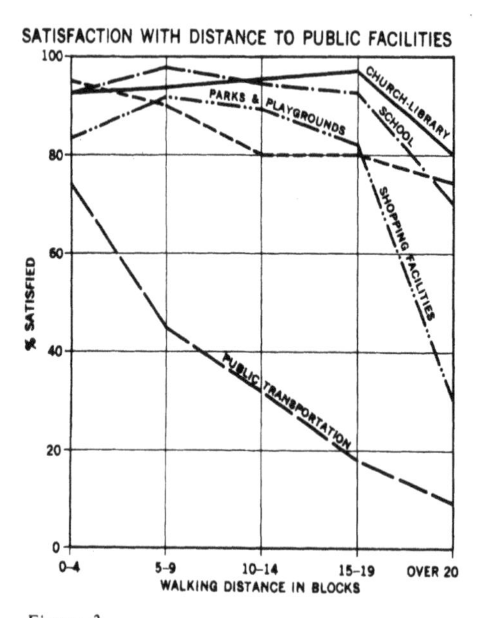

Figure 3.

This was obtained simply by asking peopl if they were satisfied with the distance to thei schools or shopping centers, and then deter mining the actual distance and rating the pro portion of people who were satisfied at particular distance from these various facili ties. With the exception of mass transit, mos people are not too concerned about havin some of these facilities close at hand. It wa found that they really would like to hav mass transit close by in order that they coul walk to this facility. For the other activitie they more or less expect that they will hav to go by auto and are not too concerned abou additional distance. They are more con cerned, in my opinion, at what they get at th end of their trip — is it the right doctor; is the right store; is it their favorite hairdresser So, in effect, their desire for variety in choic has magnified our transportation problem.

I don't think it is correct to say that "th spread city" is caused by the automobile, al though it has made this possible. After al some of the fast-growing downtown areas lik Toronto or Montreal, and Houston, Texa

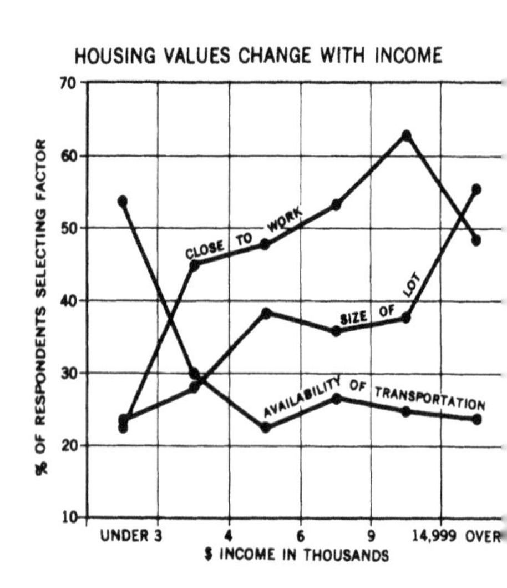

Figure 4.

auto-oriented cities. The fact that Los geles has a relatively small downtown area orimarily related to the nature of the Los geles area — that it developed as an instrial complex of many downtown areas — sadena, Hollywood, Long Beach, etc. The g trip that is familiar to most Los Angeles idents is due to the way the city grew — development of a large number of small, lependent cities that have grown into one ge metropolitan area. I cite these cities ly to point out my basic thesis that there much more involved in urban growth than nsportation facilities, and that as planners have too often over-emphasized the role transportation in urban growth, while der-playing some of the key issues, such as cial and technological change. Until we derstand these latter factors better, we will t be able to truly plan for the future.

The technological change that has been curing in urban areas is very remarkable. hink that one of the best indicators of this ange is that with all the growth that we are ting in the United States we have had no rease in manufacturing jobs since 1955. industrial city like Baltimore has less anufacturing jobs today than it did in 1948. ring this period, the population has grown 45,000. All this can mean is that service os have been growing at a fantastic rate to -set the lack of industrial growth in emyment. This undoubtedly will continue — aning that more job opportunities will be ated in our commercial centers in contrast our industrial parks.

In attempting to look at the social changes at have been occurring, we have been study- leisure activities. Figure 5 represents a mmary of these findings. This shows that th change in income we get change in sure activities. Television becomes less imrtant. Even in the higher income groups it consumes 40 percent of their time, and in the lowest income groups it consumes about 70 percent of their time. I think that the startling thing from a planning point of view is that 80-90 percent of the time is spent around the house. The gay life just doesn't exist in the typical family. This, I think, influences the values that people have about their community. It certainly places a high degree of emphasis on the home and the areas immediately around it. It certainly indicates that as planners the way to make the greatest impact is to concentrate on the immediate neighborhood surroundings. We need to know more about these and other social changes. This will call not only for the kinds of studies that have been made but for more studies in depth than ever have been done before. We certainly have to watch changing technological development so that we can more adequately anticipate the needs of the future.

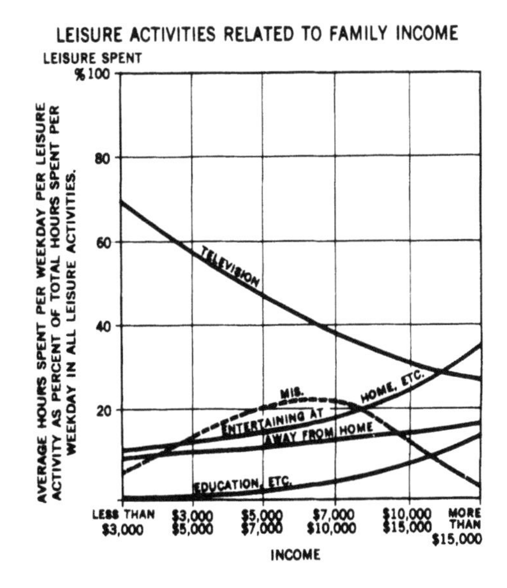

Figure 5.

TRANSPORTATION SOLUTIONS

These are some of the conclusions that can be derived from studies that have been made on urban transportation and urban growth. These points are seldom brought out in the report, but I think they are the key factors behind our planning problems. Unfortunately, we are not really using them as effectively as we could in our planning for the future.

Metrotowns

Here are a few areas where some of this knowledge is being applied. One of the most commendable is the work being done in the Baltimore Region where they have developed a concept of "metrotowns" which take into consideration the way people live in metropolitan areas today. This concept is aimed at reducing transportation requirements and obtaining greater amenity and personal identity. It attempts to eliminate the suburban sprawl by intelligently arranging the land so that it takes into consideration social as well as economic goals.

At the heart of these metrotowns is a commercial core which will house between 10,000 and 20,000 jobs. This core will be fed by freeways, and, it is hoped, by mass transit. The advertising values provided by freeways is almost essential if you are to provide strong commercial development in any particular section. The development of transit stations to this area will further strengthen this pattern.

A large portion of the high density development in these towns will be located nearest these centers. The density will decrease to the edge of the metrotowns where open space will be provided. This open space will help define these metrotowns and force commercial development to the center, for one of the best ways to prevent commercial spread is t create open spaces which automatically de fine market areas.

The type of metrotown center required i being planned for a community outside c Washington. It contains a subway station tie in with an improved arterial street systen that tends to strengthen the whole core Direct highway access is provided throug the retail area; transit service is highl coordinated with the high density employ ment areas. It is felt that by this genera pattern of development transit usage will b maximized and that the center will attain it greatest strength.

We hope that these metrotowns can b built through a cooperative relationshi which will tend to assure that the right type c residential development is built for the job that are created in the towns. This will mak it easier to find work within the town in whic you live. However, it is well recognized tha people are going to find jobs beyond th metrotown in which they live. It is believe by the intelligent planning of these towns an the programming of their development tha perhaps 40 percent of the people within town might work in it. This is in contrast t the 10 percent to 20 percent that is found i existing suburban complexes of similar size

The one thing different about the concep of these "new towns" is that they are reall part of the metropolitan area. The wide rang of choice found in a large metropolitan are is still available. Our attitude studies in citie of less than a million indicate that people ar generally quite unhappy with job opportuni ties in such areas.

The size of these metrotowns should b between 100,000 and 200,000, which is ade quate to serve most of their daily and weekl needs. Current studies indicate that as a uni like a metrotown grows in size in a metropoli area it becomes more and more depend- upon itself until it reaches 100,000 or :e. This, of course, reflects the fact that we d 100,000 people to support the variety ıctivities that we desire — enough hair- ssers, dentists and doctors to give your : a choice.

'et Patterns

those of you who do not live or work in ırge metropolitan area you may be inter- :d in knowing some of the thoughts we e been developing on street patterns for ıller cities. In such cities we have, for ıe time, recognized that streets should per- n various functions and that the design racteristics should reflect these functions. have talked about expressways, arterials, ector streets and local streets. In our ıller cities the expressway, or freeway, /es a unique function. It may bypass the ımunity or it may provide important ac- ; to the downtown area or other key indus- l areas, but the bulk of the traffic will have be carried by the arterial streets. The cing of these arterial streets will vary de- ding upon the density of development and location of the streets in relation to other ior traffic generators. In lower density ıs — those with 3-5 families per acre — :reet spacing of about one mile apart has n found satisfactory, but if the density gets ıer these streets must be closer together. en we find that it is desirable to have these ets spaced at one-half or one-quarter mile :rvals in the more congested areas.

With regard to collector streets, we are inning to recognize that streets which ve over 100 homes (100 homes generate 00 auto trips a day) have to be wider than 30 feet generally recommended for resi- tial areas. In such cases a 36 to 40 foot street of higher design standards is required.

If you accept that collector streets in residential areas should not have commercial development along them, every attempt should be made to keep the traffic below 10,000 trips a day. It should be less than 5,000 per day, because as soon as you get volumes of this magnitude commercial developers will think about developing along them — then your only hope is zoning.

The Neighborhood

As we all know, the classical neighborhood building around the elementary school does not tie in with the social patterns of the teenager or the adult today. In fact, the only thing that you can say about the neighborhood is that it is the sphere of interest of the elementary school child. For example, our studies have shown that 90 percent of the socializing people do beyond their home is done more than one mile from their home and only about 2 percent of the socializing is done on foot. Therefore, it would appear that the kind of neighborhood that made sense is one at a walking distance scale, which is probably less than 1,000 feet. And, instead of 1,000 or 2,000 families the only type of neighborhood unit that makes sense today is one of 100 to 200 homes which are tied together by unique topography, an original design concept, a common recreational area such as a swimming pool and tennis courts, or just a wooded area left untouched.

In considering this neighborhood concept — of 100 to 200 homes — we should attempt to keep through-traffic from piercing it wherever possible.

This certainly is a lot easier to do than laying out a classical neighborhood. It may be necessary to have a collector street in them, but often this can be avoided by keeping collectors on the boundaries of such neighborhoods. However, sometimes it has been found that collector streets in such neighborhoods can help knit the area together, particularly if the volumes do not exceed a couple of thousand trips a day.

If this type of neighborhood concept makes sense, then all that one needs to do is set up standards as to what type of activities should be located on various streets. For example, "collectors" might only serve

Elementary Schools
Minor Recreational Facilities

while "arterials" should serve

Commercial Areas
Industrial Areas
Major Public Facilities.

There are many other concepts that w could discuss which would show how to inte relate transportation and city planning, but hope that these examples have given yo some insight into how this can be done. W must analyze our traffic patterns mo thoroughly and evaluate people's value Until we understand the nature of our prob lems more clearly it is going to be difficu to develop the type of planning principles th we need to plan adequately for the future.

etail Structure of
rban Economy

19

igene J. Kelley

e spatial structure of American retailing eing changed by three forces familiar to se concerned with traffic and highway tters. These forces are the suburban popu- on movement, the increasing dependence the consumer on the private automobile shopping, and the growing number of h speed roads enabling consumers to travel es from home even for convenience goods chases.

The response to these forces has produced stantial changes in consumer behavior and rchandising practices. It has been esti- ted that 50 percent of the automobile driv- public will travel thirty minutes to reach hopping center when assured of satisfac- y merchandise assortments and parking ditions. But so slowly does our thinking pt to change and so slowly do adjust- nts develop, that the full impact of these ces on the metropolitan economy may not be fully appreciated.

The average American city is still twenty twenty-five years behind adjustment to the omobile. So when it is remembered that the three forces mentioned have gained momentum only since the end of World War II, it is understandable that some have not grasped fully the extent of the marketing revolution currently underway. As recently as 1950, only a handful of business men and economists visualized the change in spending-patterns that would be brought about by the suburban population movement and other forces. A statistical overview of some retailing results of this "painless" revolution was given by McMillan. These numbers in parentheses (1) refer to notes at end of article.

This article presents a conceptual scheme for analyzing the retail structure of the metropolitan economy and offers some guides to the placement of regional shopping centers in the structure. Regional centers are one of the most spectacular recent evidences of the dynamism of the retail structure and of the American economy. The use of space as a business resource is commented upon here prior to discussion of the retail structure and placement of controlled shopping centers in the structure.

printed from *Traffic Quarterly*, Vol. 9 (1955), pp. 411-30 by permission of the Eno Foundation for nsportation Inc.

SPATIAL POSITIONING IN MARKETING

Sellers of goods have generally been preoccupied with the task of creating demand for their products. Certainly, the greatest amount of marketing managerial time, energy, and imagination has been focused on the product and its promotion.[1] The spatial and temporal conditions influencing the sales of the product typically have received less study than creating demand. Yet sellers are concerned with the creation of time and place as well as possession utilities. Creation of space utility is an area in which traffic engineers, architects, and planners have much to contribute to business. All of these specialists are interested in the intelligent use of space in the metropolitan economy.

A seller has four decisions to make about the ideal spatial relations or positioning of his product in the market.[2]

1. He must first select the area or areas in which he will offer his goods. These areas are his markets.

2. He must make a choice among the types of distributive agencies selling space in the market. Will space be preferred in drug or hardware stores, mail order catalogs or department stores? This is selecting the channel of distribution.

Institutions to supply the chosen retail outlets must be selected. Will a service wholesaler offering full decentralized spatial services be used or will a limited function middleman such as a manufacturer's agent represen the better channel?

3. He must select within competing reta and wholesale institutions of the same type Will a policy of exclusive, selective, or intensive distribution be followed? Will Chain A or Chain B or both be used? Are urban suburban, or rural positions preferred?

4. Finally, there are questions of the desired internal positioning of the goods withi the outlets. This involves questions of layou and display.

This article is concerned with an aspect o the third level of spatial decision. Specifically the nature of controlled shopping centers i examined and some impacts of these center on the retail structure are suggested.

IMPACTS OF LOCATIONS ON RETAILING

Business men are generally quite aware of the importance of the right location to marke oriented plants and stores. Some study ha been given to the effect of different location on the volume of goods sold. But many of the other relationships between locations and the creation of possession utilities have not bee explored.

Location is important not only as it affect the volume of goods sold, but as it influence other variables of marketing transactions. Fo instance, what are the effects of different locations on: the quality and type of good offered and sold, the degree of sales servic required, and the amount of promotion an information needed to complete marketing transactions? What effects do different loca tions have on the time people buy, the fre quency with which they purchase, the price

[1] The art of marketing is the manipulation of temporal, spatial, and possessory forces to achieve an objective in management. As a discipline, marketing is the study of the temporal, spatial, and possessory forces influencing economic transactions, and of the interacting efforts and responses of traders (buyers and sellers) in the market.

[2] The writer is indebted to Dr. Lincoln Clark for this concept and for other guidance given during the preparation of a Ph.D. dissertation at the Graduate School of Business Administration of New York University.

y pay, and the cost of sales? What are the pacts of a new location on business done other locations?

In terms of regional centers, how will the ablishment of regional centers affect mar- ing transactions in other elements of the tropolitan retail structure? These ques- ns concerning the effect of position in ice on consumer behavior seem relevant ether one is concerned with increasing the ofits of a particular enterprise or advanc- science in business.

Merchants have recognized some differ- ces between customers shopping downtown d in regional center stores. The two groups customers are from different sections of the tropolitan area, from different income oups, have different tastes and attitudes, d even may be of different sizes. Speci- ally, shopping center customers buy more orts clothing, casual wear and children's thing than do patrons of downtown stores.

It is probable that retailers experimenting h suburban locations will continue to find at merchandising problems vary between ations, even though basically the same es may be carried. But what will the nature d extent of the difference be? How will the me-owning child-raising, casual living, do- yourself families of suburbia differ from eir central city cousins shopping exclu- ely in downtown stores?

These questions are important since sub- banites represent the most important single rket in the country. Forty million people mprise the suburban market today, but this ure alone does not tell the whole story. The burban market contains more than its share middle income consumers in the 25–45 ar-old age bracket. *Fortune* reported that e average family unit income of the sub- ban population in 1953 was $6,500 or 70 rcent higher than the rest of the nation.(2)

As a starting point for analyzing the above questions the elements comprising the retail structure are identified along with certain characteristics which may aid in formulating hypotheses about the questions raised. The shift of retail sales from the cores of the larger cities to other elements in the structure makes it more important than ever to consider as a market unit the entire metropolitan area rather than just the central city or any political sub-division.

In main outline the retail structure of the 168 standard metropolitan areas follows the pattern suggested in this article. In 1950 the areas had a population of 84,500,680 — more than half the people enumerated in the continental United States.

There is a wide variety in the distribution of the approximately 1,748,000 retail outlets in this country. Yet classification into groups for locational analysis is possible. Duncan and Phillips (3) maintain that in their main outlines the retail structure of cities and their surrounding areas is generally similar. These authors identify a central shopping district, secondary or outlying shopping centers, neighborhood business streets, and scattered individual stores or small clusters of stores.

ELEMENTS OF THE RETAIL STRUCTURE

Brown and Davidson (4) offer a five-fold classification of store locations found in most metropolitan areas; central shopping district, secondary shopping districts, string street locations, neighborhood clusters, and isolated locations. Weimer and Hoyt classify the retail structure into business districts, outlying business centers, and isolated outlets and clusters. (5) Other analyses in the literature of marketing and real estate follow a similar pattern.

A new classification of elements is suggested here, integrating controlled shopping centers into the retail structure of metropolitan areas. Some of the key relationships between elements are summarized in Table I.

THE RETAIL STRUCTURE

1. Central business district
    - A. Inner core
    - B. Inner belt
    - C. Outer belt
2. Main business thoroughfares (string streets)
3. Secondary commercial sub-districts (unplanned)
    - A. Neighborhood
    - B. Community or district
    - C. Suburban or outer

3a. Controlled secondary commercial sub-centers
    - A′. Neighborhood
    - B′. Community or district
    - C′. Suburban or outer

4. Neighborhood business streets
5. Small store clusters and scattered individual stores
6. Controlled regional shopping centers.

1. Central business district

A Commerce Department study in 1935 used terminology that can be helpful in considering the structure of central business districts. The terms are "inner core," "inner belt," and "outer belt." Figure 2.

The inner core of the central business district is typically the point at which all intra-city traffic converges, the center of shopping and specialty goods activity and the home of the large department stores. (6) In the inner belt are found communication agencies, banks, law offices, the administrative offic of political, recreational, religious, and oth services. The inner core and belt comprise th heart of the retail structure and also of the other activities as well. Through these offic the "manifold activities of the community a directed and integrated. The special functio of the principal center is that of dominance c control . . ." (7)

The first two elements of the central bus ness district are typically the home of th largest stores, both in floor space and volum Some convenience-goods retailers are locate in the central business district, but the shop ping and specialty goods stores are the mag nets which draw customers from the entir metropolitan area to shop downtown. Th inner core of the central business district ha the highest concentration of pedestrian traff in its relatively small area. Because of thes factors land values are highest here so th only high volume retailers can ordinari compete for premium locations in this are

In the inner belt immediately surroundin the core, land values are lower and pedestria traffic much less concentrated than in th inner core. The separate but related functior of government, finance, professional service cultural, entertainment, and wholesale activ ties are found here.

The third element of the central district i the outer belt. This generally includes less de sirable commercial structures and dwelling and some residential areas that have ru down and are on the verge of becoming slum

2. Main business thoroughfares

Leading out of the central business distric are streets lined with all kinds of retail ou lets and services. These thoroughfares ar described as "string streets." Such streets ar heavily traveled by automotive and pedestria fic. Retailers on these streets do not de-d on the residents of their immediate area patronage but are favored mostly by ple using the street as a thoroughfare.

ome of these streets developed when etcar routes from the central business dis-t were laid out on fixed rails, and various es of commercial enterprises lined up ng both sides of the streetcar system. omobile dealers, furniture stores, and rly every other type of consumer goods chandiser can be found along the main iness thoroughfares of most American es.

Secondary commercial sub-districts

nmercial sub-districts develop as the ulation of the central city increases. It becomes more convenient for people in hborhoods away from the downtown area hop closer to home more often, instead of neying downtown to the central business rict. Merchandise sold in secondary com-cial sub-districts is similar to that sold ntown. However, the breadth and depth lines carried is more limited, the stores ller, and customers are drawn from a ller area. A larger proportion of conveni-e-goods stores is located in these areas n in the downtown districts.

ypically secondary commercial sub-ricts are located on heavily trafficked tes between residential areas and the cen-city. On the basis of parking facilities, two es of secondary shopping areas can be dis-uished. The first is situated on or off the n business thoroughfares. In these sub-ricts only curb parking is available for automotive customer. Newer and modern-l secondary shopping areas attempt to pro-off-street parking for customers. All perly controlled neighborhood, commun-ity, or district centers offer this service. The great majority of commercial sub-centers are uncontrolled.

3a. Controlled secondary sub-centers

Structurally each type of controlled center is placed in relation to the trading area it is de-signed to serve. Controlled neighborhood shopping centers are built near the areas occupied by neighborhood business districts. Community or district centers of the con-trolled variety are located in appropriate secondary shopping areas. Controlled sub-urban shopping centers are situated farther out near suburban cities. Controlled regional

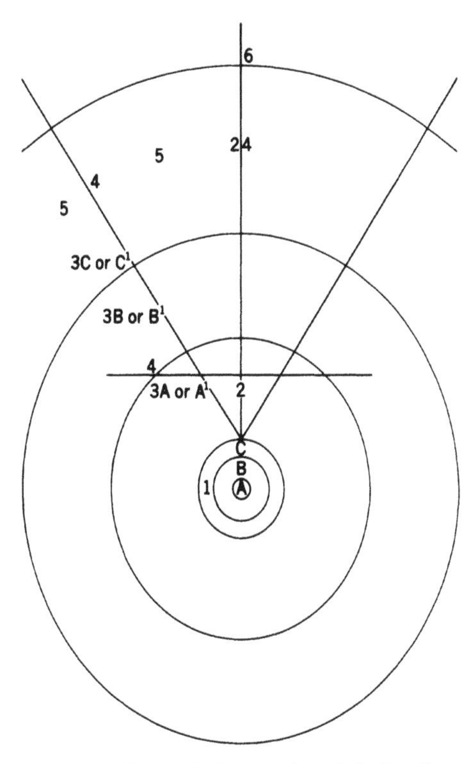

Figure 1. Schematic Presentation of the Retail Structure of Metropolitan Areas.

shopping centers are planned either in a suburban location or at a point beyond the typically suburban. In this matrix the next significant retail element is the neighborhood business street.

4. Neighborhood business streets and areas

Neighborhood business streets contain convenience-goods stores with primarily a neighborhood appeal. These streets typically include a small cluster of several kinds of retail establishments on heavily traveled streets or at an intersection of two or more main thoroughfares. The principal trade comes from neighborhood residents.

5. Small clusters and scattered individual stores

Clusters and scattered individual stores also deal basically in convenience goods. The clusters are made up of two or more complementary, rather than competitive convenience-goods stores. Individual stores and small clusters are scattered throughout the residential areas of cities and towns surrounding the central city wherever population density invites more convenient shopping facilities than provided by neighborhood centers.

6. Controlled regional centers

Controlled regional shopping centers have nearly all been built in suburban areas. But the centers are intended not only to attract patrons from immediately surrounding areas, but from the entire region or quadrant of the metropolitan area in which they are located. A region may include all customers within a given driving-time distance, usually thirty or forty minutes from the site. Regional centers are situated far enough out in suburbia for the land to be relatively inexpensive. Trac can be used without the expense of demolish ing too many existing structures.

A basic reason for preferring the suburba location is the large amount of non-sellin space needed for a regional center. The cen ter's layout is designed to provide amp parking facilities even for peak periods. Park ing space may occupy from three to nin times as much area as the floor space devote to the sale of merchandise. (8)

CHARACTERISTICS OF CONTROLLED CENTERS

Controlled shopping centers are considere to have the following characteristics:

1. The land on which the center is situate is owned by a single agency. Typically th buildings are also owned by the develope but concessions in this respect might be mad to department store tenants. The factor c single ownership makes possible an unusu measure of control of architectural, parking service, and other features of the center b the developers.

2. An assortment of different types of re tail outlets offering a balanced representatio of goods and services is featured. The store are on one integrated site designed for one stop shopping at the level of trade area bein serviced, i.e., neighborhood, community, sub urban area, or region.

3. Planning is done in advance of con struction. The completed shopping center i designed as an integrated, harmonious uni as an efficient merchandising machine. Th controlled center is developed according t specifications by architects, market analyst and other types of consulting specialists.

On the basis of the trading area serve four types of controlled centers have bee distinguished. These are the neighborhood nmunity or district, suburban or outlying tral city, and regional centers.

ı. A neighborhood shopping center is one nprising seven to fifteen retail outlets sell- primarily convenience goods. A super- rket is the core of this type of center. ighborhood centers serve a minimum of) families or about 3,000 people. Larger ghborhood centers may serve as many as 000 people. Store groups of six or less are inarily better described as small store clus- ; than as shopping centers.

). Community or district centers draw ater numbers of customers from wider as than do neighborhood centers. The ding area usually includes several neigh- rhoods within one to three miles of the :. A population of between 15,000 and 000 is typical of the community center's ding area. The community center includes, addition to the service stores found in ghborhood centers, a complete range of ivenience-goods outlets, shopping and spe- ty goods stores emphasizing apparel and ne furnishings in the middle price ranges, fessional offices, and usually a branch ık.

It is characterized by a greater depth of rchandise than the neighborhood center. tomotive traffic is more important than in neighborhood center and off-street park- facilities generally have to be provided. ɔm sixteen to thirty-five stores are found he usual community center.

:. Suburban or outlying central city cen- s normally serve a population of 30,000 100,000 people and are commonly built ɔund a department store branch or two and eral large supermarkets. Generally, except unusual specialty items, an assortment of rchandise adequate to serve all needs of trading area is offered. The centers serve large suburban area consisting of many communities and neighborhoods. From twenty-five to fifty or more different retail outlets typically comprise the suburban center.

When the uncontrolled suburban or outlying central city center serves over 100,000 people, it begins to assume in miniature the retail structure of larger cities. It tends to develop its own "downtown" district and the resultant parking and traffic problems. When this occurs, the suburban center loses some of its attraction as a site for additional expansion, particularly by department stores.

d. The regional center services a trading area of from 100,000 to 1,000,000 or more people. In addition to convenience and specialty goods stores, it contains at least one department store branch. Usually the branch has between 100,000 and 300,000 square feet of selling space. A full line of shopping and specialty goods is featured in the center. Regional centers are in effect decentralized substitutes for the downtown center featuring forty to one hundred or more stores on a site of at least fifty acres.

CENTERS ARE POSTWAR OUTGROWTH

The controlled center is largely a post World War II development, though isolated controlled neighborhood and community centers were in operation as early as the 1920's. Various estimates place the number of controlled centers at probably between 2,000 and 3,000 controlled shopping centers of all four types, completed or in advanced planning or construction stages. Only six regional centers were in operation as of January, 1955.

The reasons for the emergence of large controlled centers have been discussed by Duncan and Phillips (3), Brown and Davidson (4), and others. There is agreement that the centers evolved to meet the needs gen-

TABLE I

The Retail Structure of the Metropolitan Economy

Retail Element	General Character	Source of Customers	Store Types	Parking	Traffic	Goods Sold
1. Central Business District A. Inner core B. Inner belt C. Outer belt	Inner core and belt solidly commercial. The business and recreational heart of metropolitan economy. Residents fill in back streets. Typically, residential areas are blighted.	Come from all parts of city and tributary area. Sites are most accessible to most consumers. Intra-city transportation converges in this element.	Largest in floor space and volume. Multi-story department store is symbolic. Home of specialty shops. Outer belt activity less intense. These stores do smaller volume per unit.	Totally inadequate in inner core and belt. Trend to provide public lots and commercial parking lots to supplement limited curb parking in inner belt and outer belt.	Extremely heavy. Congested during peak periods.	Shopping and specialty goods emphasis. Area is center of apparel, home furnishings, other department store lines. Service and other commercial activities found in belts.
2. Main Business Thoroughfares (String streets)	Mixed zone of retail and light industrial enterprises and working class homes. Featured by long series of miscellaneous stores.	Basically trade is transient, consisting of commuters, suburbanites, and inter-city automotive traffic. Some patronage from neighborhood residents.	Concentration of larger food stores, automobile dealers, and supply houses, service and convenience-goods stores.	Usually dependent on curb parking. Inadequate during most periods.	Streets are main traffic arteries. Usually heavy, but particularly so during commuting peaks.	Essentially business streets. Stores are widely spaced over length of artery.
3. Secondary Commercial Sub-districts (unplanned) A. Neighborhood B. Community or district C. Suburban or outer	More residential than first two elements. Owner-occupied residences increase with distance from central business district. The sub-districts tend to appear. island-like along string streets	Come basically from A, B, or C trade areas. The districts developed as city grew at focal points of intra-city transportation. Dependent on traffic brought by public carriers.	Unplanned competition featuring convenience and shopping goods. B and C tend to be miniatures of central business districts.	Mostly curb, plus some off-street parking provided by individual merchants.	Since stores are typically clustered at key intersections and transfer points of public carriers, the traffic is heavy.	Convenience-goods featured in A. Increasing shopping goods emphasis in B and C.

Secondary Commercial Sub-centers A′. Neighborhood B′. Community or district C′. Suburban or outer	marginal stores at a minimum. Found near more prosperous residential areas. Unified architecturally. Most built after World War II. New, fresh appearance compared to 3.	on automotive traffic. Parking provided so customers are drawn from greater distances than in case of unplanned centers. Generally found in suburban district.	of supplementary stores possessing aesthetic appeal. Centers stress convenience and service, not price appeals.	co-operative basis within the center. Parking and other facilities related in size to surrounding trade area.	automobiles key consideration. Even so, automotive traffic heavy.	present an integrated retail organism to customers coming from A′, B′, or C′ distances. A′ stresses convenience goods. B′ and C′ feature shopping and specialty merchandise.
4. Neighborhood Business Streets	Residential with commercial use distinctly secondary.	Neighborhood is primary source. Most customers come from within walking or 5 minutes driving distance.	Usually rows of convenience-goods outlets found in center of neighborhood community.	Mostly curb. Due to convenience-goods nature of most items sold, parking turnover is rapid.	Heavy during peak hours. Otherwise not a handicap to trade.	Emphasis on food and drugs. Grocery store-drugs combination frequent. Service stores common.
5. Small Clusters and Scattered Individual Stores	More thinly populated residential areas. Neighborhoods tend to be middle class.	Come from homes not within easy reach of larger elements in structure. Many walk to stores.	Smallest outlets in structure. Many are marginal. This classification dominated by food and general.	Curb and small lot parking usually adequate.	Usually not a problem. The lack of traffic congestion, plus the availability of parking, represents an appeal of this element to customers beyond neighborhood range.	Usually supplementary and not directly competitive.
6. Controlled Regional Shopping Centers	Overall unity obvious at a glance. Landscaped frequently. Off-street parking. Harmonious effect is objective. May be equipped to serve as area's civic and cultural center.	Drawn from families within 30-minute driving range. Customers typically come from a number of suburban communities. Pull varies with effectiveness of central business district retailers and competing centers.	Attempt made to duplicate shopping facilities of central business district with minimum of overlapping. "One-Stop Shopping in the Suburbs."	Usually best facilities in metropolitan area. Adequate for all but occasional peak periods.	Problem usually under control as a result of coordinated planning.	One or two department store branches and satellite stores offer widest range of merchandise and services outside central business district.

erated by changing environmental factors in the economy. These factors include increasing urban population decentralization, increased use of the automobile, increased congestion in central business districts, the lack of economical and convenient parking facilities in downtown areas, and changing consumer buying habits. The important point is that these forces and others favoring retail recentralization seem to be increasing rather than diminishing.

A population of 221,000,000 by 1975 has been predicted. The increase in population alone will be enough to support dozens of large new shopping facilities. In addition, higher living standards brought about by automation and rising productivity, increased leisure time, and a rising educational level, all seem to favor further population decentralization and retail recentralization in large shopping centers.

Other forces, such as an increasing national income, are also likely to result in a response by entrepreneurs of more and better decentralized shopping facilities. It may well be the controlled shopping center movement is in its infancy. Of course, a saturation point for shopping centers can be reached. And merchants in shopping centers can fail. For location is still only one ingredient of a successful retailing operation. Many other questions of planning, financing, and operation must be answered before a center is assured of success.

LOCATING REGIONAL CENTERS

The ideal regional center site is the one from which the largest number of automotive customers in a trade area can be served at the minimum of transfer costs. Consumers' transfer costs include not only money costs but the expenditure of time and physical and nervous energy that must be made t purchase goods. The reason controlled re gional shopping center sites are preferre close to the center of the suburban popula tion areas is that consumers' transfer cost are usually minimized at such sites.[3] In locat ing market-oriented facilities, the ideal is t choose a site as close as possible to the scarc est factor and the one having limited mobilit — the customer.

It is believed the choice of a site at whic transfer costs will be minimized for the larg est number of automotive customers will b facilitated by using the following factors a criteria of site selection.[4] The factors ar classified as either regional or site factors Regional factors are those of population, pur chasing power, growth, and competition. Th site factors are access, traffic, size, expansion parking, cost, terrain, and utilities.

REGIONAL FACTORS

Population

Regional centers are best located in concen trated residential populations in outlying sec tions of large metropolitan areas. Ordinaril a minimum of 500,000 people should resid within thirty minutes' driving time distance o the site.

[3] In economic theory the reason why sites at th center of urban activity are considered mos desirable is found in the labor savings involve in a central location. This valuation is reflecte in the form of higher rents.

[4] The criteria were prepared during a study c the processes used in locating the followin regional centers: Shoppers' World, Framinghan Massachusetts; Cross County, Yonkers, Ne York; Roosevelt Field, Hempstead, New York Garden State Plaza, Paramus, New Jersey Bergen Mall, Parmus, New Jersey; and North land, Detroit, Michigan. The writer is gratefu for the cooperation of the developers and man agers of these centers.

chasing power

ional centers should be located in an area after an analysis of the purchasing power stability of income and expenditures of dents indicates the trade area is sufficient upport a regional center of the size con-plated.

wth

section of most rapid population growth probable future expansion within the tropolitan area is normally the most mising sectional choice for a suitable re-al center site.

npetition

location of competition as it affects ential sites should be investigated both ntitatively and qualitatively. A regional ter should be located in an area only n proof exists that operating and planned il facilities are inadequate.

E FACTORS

ess

egional center site should be easily acces-e to automotive traffic. The site should be a prominent location and be served by a tem of primary and secondary roads, offer-convenient, safe, and free flowing means ccess and egress.

ffic

ficient road capacity should be available to dle existing traffic around the site, traffic ly to be produced by future expansion in area, and traffic created by the additional icular activity the center will generate.

Size

The site should be large enough to provide the desired amount of store and service facilities, and parking at a parking-space to floor-space ratio of at least three to one and preferably, four to one. Sufficient land should be acquired to serve as a buffer and possible expansion area. With regional centers these specifications usually require a minimum site of fifty acres.

Expansion

Provision should be made in the earliest planning stages for expansion after the center is established. The developers should attempt to build with expansion provisions for five and ten year periods ahead. Excess space can be used for landscaping and recreational purposes until needed for commercial use.

Provision for expansion may be necessary to hold a planned position in the event of an increase in population and trade after the center opens. If the center can not expand as needed, other shopping facilities will develop in the area pioneered by the first center.

Parking

The tract should be of a size and shape to provide parking in at least a three-to-one ratio of parking to store space. Shoppers should not have to walk more than four hundred feet from their automobiles to the nearest store. The ideal ratio of parking to store space increases with the size of the center.

Site cost

The cost of acquiring the site, preparing it for construction, and any extraordinary maintenance costs must be carefully measured and considered. In general, land costs are not to be economized upon at the expense of losing a premium site.

Terrain

The terrain should be thoroughly examined by architects and engineers in advance of purchase to ascertain conditions which might affect the locational decision. In general, level ground and solid earth represent the preferred terrain conditions.

Utilities

Utilities should be available to the site at the time of acquisition or at completion of the center. Regional centers will ordinarily maintain some of their own utility services, but power, water, and sewage facilities should be available to the property line.

These factors have been incorporated in the following site-ranking chart. The ranking chart assumes the availability of the rated site, together with its possibility of rezoning. The ideal site might be unavailable. In such a case, the next most promising site on the market should be rated. Similarly, rated sites should be zoned or rezonable for shopping center development. It is self-evident that a developer should not go too far in his planning about a particular site unless he has reason to believe the site is available commercially and legally.

The site-ranking chart is designed to serve only as a general guide in deciding on the site selection. It is recognized that an entrepreneur choosing sites might find it difficult, if not impossible, to focus on a particular site all of the information that might be relevant. Yet, the use of a site-ranking chart should produce more valid site ratings. Table II.

CONTROLLED CENTERS AND CENTRAL BUSINESS DISTRICTS

It appears probable that population ar transportational forces responsible for the i crease in regional centers after World War will continue unabated for at least the ne few years. But the central business distri will continue to dominate retail trade in mo metropolitan areas, although the proportic of business done by suburban centers wi increase. Suburban volume will increase b cause of suburban residents' dissatisfactic with existing retail facilities in the major ce tral business districts.

The chief complaint is transportational. is a wearisome chore for suburban shoppe to overcome the friction of space and reac most central business districts through co gested traffic and crowded streets. On reac ing the central business district, finding a co venient parking space is typically a problen Public transportation is available, but it ca do only part of the job in a culture wedded the private automobile.

Some evidence exists that central busine district retailers are becoming concerne enough about suburban competition to tak action to increase downtown trade. In som instances merchants are cooperating to obtai improved transportational and parking fac lities for central business districts. Their caus is hopeful. There is no reason why centr business districts served by efficient highway bringing suburbanites to the city and to th downtown areas, perhaps on expressway and to adequate parking facilities, should n be able to compete successfully with region centers.

But even with the adoption of every typ of palliative advanced, it appears the facto of population increase and migration alone

TABLE II

Controlled Regional Shopping Center Site Ranking Chart

scription of Site Being Rated ....

....

tor being rated	*Ranking of site* (in order of relative preference)				
REGIONAL FACTORS	1	2	3	4	5
oulation					
vithin 15 minutes					
6-30 minutes					
chasing power					
mount and stability					
listribution					
owth of population					
mount					
legree					
mpetition					
mount					
uality					
SITE FACTORS					
e of tract					
ninimum size					
ndivided					
uffer area					
cess and egress					
rimary roads					
econdary roads					
ffic					
resent pattern					
uture pattern					
king					
mount					
earness to stores					
st					
cquisition					
naintenance					
rain conditions					
grading					
ubsoil conditions					
lities					
roximity					
pansion-Environment					
xpansion					
nvironment					

enough to favor the creation of more recentralized retail facilities in the hinterland.

In both theory and fact central business districts have attractions that will insure their continued importance as centers for commercial and recreational life of metropolitan areas. But they will encounter more competition from other elements in the retail structure. However, the central district has a natural locational advantage of being the point which the greatest number of people in the metropolitan area can reach most economically. This advantage ordinarily should suffice to assure its dominant position.

Notes

(1) Samuel C. McMillan, "Decentralization of Retail Trade," *Traffic Quarterly* (April 1954).

(2) "The Lush New Suburban Market," *Fortune* (November, 1953), p. 128.

(3) Delbert J. Duncan and Charles F. Phillip *Retailing Principles and Methods*, 4th ed. (Chicag Richard D. Irwin, Inc., 1955), pp. 79, 82.

(4) P. L. Brown and W. P. Davidson, *Retailir Principles and Practices* (New York, Ronald Pres 1953), p. 75.

(5) Arthur M. Weimer and Homer Hoyt, *Pri ciples of Urban Real Estate* (New York, Rona Press, 1948), p. 138.

(6) U.S. Department of Commerce, *Intra-Ci Business Census Statistics of Philadelphia, Pennsy vania* (Washington, Bureau of the Census, 1937 p. 25.

(7) Amos H. Hawley, *Human Ecology* (Ne York, Ronald Press, 1950), p. 270. The influenc of metropolitan centers was also examined b Donald J. Bogue, *The Structure of the Metr politan Community: A Study of Dominance an Subdominance* (Ann Arbor, University of Mich gan Press, 1949).

(8) Ernest M. Fisher and Robert M. Fishe *Urban Real Estate* (New York, Henry Holt an Company, 1954), p. 317.

merging Patterns of ommercial Structure in merican Cities

20

mes E. Vance, Jr.

e traditional view of the retail trade pat- n of cities needs reexamination in light of r growing awareness of the dynamic nature urban structure. Particularly in the United tes, the geography of urban commercial ivity has ceased to be the study of the cen- l business district with only minor attention en to small outlying business centers. In the st-war period virtually all growth in the nmercial structure has taken place outside city core. Concentration on that area alone uld, at best, give us a picture of the realities the pre-war world. Development outside central business district is frequently cited evidence of the decay and disintegration of core region. Rather more accurate, it uld seem, would be to view the newer wth as coming from a change in urban namic features, which calls for a reassign- nt of function between central and out- ng districts. In truth the core is changing her than decaying but until we understand t metamorphosis it is easily mistaken for line.

The purpose of this paper is to attempt to ablish the location factors in commercial land use. First the aspects of urbanism which lead to change are considered. The main cur- rents in commercial structure over the years are briefly described and analyzed in order to discover trends in location. All these trends when carefully assessed provide us with an insight into the dynamics of urban commer- cial structure and allow us to formulate a body of simple location factors at work within that structure. To test commercial location factors a brief sample study of the San Francisco Bay Area is presented based on field investigation carried out during the last eighteen months. In conclusion, the changes that may be antici- pated in urban retail structure in the near future resulting from current dynamics are suggested.

Dynamic Factors Affecting Commercial Structure

Without question the most important dynamic factor affecting commercial structure has been the changing means of personal transporta- tion. Cities came into existence at a time of mass transport and the office and commercial

printed from *Lund Studies in Geography* (1958) by permission of the author and the Department of ography, University of Lund, Sweden.

functions at the core are the creation of these fixed lines of movement. So long as individual transportation was not available there was little locational choice in the establishment of shops selling to a mass market. The focus of set routes necessary for the creation of office districts was equally as necessary for the formation of a major shopping district. But the functional linkages among diverse offices are stronger and more ramified than the linkages among diverse shops. When the automobile came into wide ownership it had far less effect on the convergence in the office quarter than it had on the focussing on the retail quarter. The stronger linkages in the former were not easily broken. In retail districts the tie between stores was largely one of similar location requirements rather than functional ties one with another, so that changed conditions of access could bring quite different location requirements.

A second dynamic factor is found in the matter of changing purchasing power and tastes. In the United States personal consumption expenditures, on a per capita basis, have gone up from 880 "constant dollars" in 1929 to 1,295 "constant dollars" in 1955. This 47 per cent increase has meant that the average individual's purchases have risen from $1,012 in 1929 to $1,527 in 1955, with the figures adjusted to 1955 prices. Any discussion of commercial structure must take into consideration this 50 per cent increase in per capita expenditures during the past thirty years. Also during this period there has been great change in consumer preferences among goods. This is not the place to investigate the matter of tastes other than to remark that a more mobile population with a greater influence of mass media of news and entertainment has shown an increasingly standardized taste in goods bought.

New patterns of housing have influenced commercial structure. Although these pa terns of housing are dependent upon the wid ownership of automobiles, which has bee proposed as the most important dynamic fac tor, it is best to emphasize that a virtua revolution has taken place in the housing c recent additions to American urban popula tion. Perhaps this should be called a "counte revolution" as it has done much to turn awa from the course of urban physical develop ment introduced with the modern industria city. A romantic desire to abjure the city ha seized a large part of American urban dwell ers. The desire for single family housing an disassociation from the heart of the city ha led to a vast urban sprawling. Particularl since the end of the Second World War addi tions to the total of urban housing have bee (1) characteristically related to our auto mobile "civilization" in being mass produced homogeneous over large areas, and character ized by price-class orientation; (2) typicall found in areas where the initial land cost will permit medium density housing; and (3 usually given to a checkerboard or salien pattern of expansion which results in a greate extension of distance from the core of the cit than is required by the absolute need fo building land or the craving for the bucoli life. The considerably increased distance o this housing from the city core when tied i with the homogeneity of the areas has create a fertile ground in which to plant the nev mass selling integrated shopping center. An these centers rather than supplementing th central business district, as did the older out lying shopping districts, have tended to as sume part of the previous commercial func tion of the core.

A factor of great importance in the creatio of new patterns of commercial land use ha been the imposition of land use plannin through zoning. Today virtually all area ch may be considered urban are covered zone ordinances so we may assume that er commercial locations reflect planning ls as well as the operation of location nomics. These planning goals have sought lo away with the isolated shop and have forced the economic trend toward inte- ion of shopping facilities.

final dynamic factor affecting commer- structure is found in changes in mer- ndizing. The deification of size among erican retailers and their craze for high ume of sales has, within a restricted local rket, led to a fuzzing of the lines of demar- on among types of shops. The American g store is the classic and incredible mple. Here the expansion of sales volume a local market with a relatively inelastic nand for goods has necessitated the com- ation of many "lines" of merchandise and creation of the modern equivalent of the neral store". We have had to borrow the tish term apothecary for the rare phar- cy which sells neither toys nor hiking ipment. Along with this commodity- nbining which has greatly recast the form commercial districts we should note the id rate of reconstruction of commercial ldings. On the average American commer- buildings are demolished and replaced at end of two generations. This combined h the fluid boundaries among establish- nt types has led to a continual transforming commercial districts.

e Elements in the Commercial Geography American Cities

e geography of commerce in American tropolitan areas seems at first view appal- ly complex. Without denying the absence obvious organization in this land use, it is sible to establish a typology of commercial districts through a study of the history and function of these areas. Malcolm Proudfoot (1937, p. 425) distinguished five types of "retail structure": (1) the central business district; (2) the outlying business center; (3) the principal business thoroughfare; (4) the neighborhood business street; and (5) the isolated store cluster. His classification was concerned primarily with the morphology of these districts. To understand how the physical form is changing requires that we look into its origins.

The mercantilist basis of several of the British North American colonies brought trade and commerce at an early date. By the time of the American Revolution there was sufficient demand in the few major port cities to support a considerable artisan class which provided for a number of limited-appeal wants. The Boston silversmith Paul Revere is best known but he was certainly not alone in this realm. The mass-appeal goods were supplied in the leading towns of the new republic by merchants' establishments selling notably imported foodstuffs, cloth, and crockery. Naturally, so long as the rural areas of the country were tied to subsistence agriculture, little trade existed outside the port cities. With the rapid growth of industry and the equally rapid expansion of commercial farming in the early nineteenth century trading was carried out of the town to the agricultural village and the mill town. The vast scale and freehold nature of farming in America when commercialization was introduced brought about the creation of the first and most widespread of the commodity-combining store types. These "general stores" were for more than a century the center of rural commerce and held sway until the early part of this century when mail order merchandising and the building of farm-to-market roads introduced larger scale competition. Even as late as 1929 (Table I) the general store was basic to retailing, though by 1954 the competition of more diversified commercial districts which could be reached along the improved rural roads had destroyed the general store as a type.

By the mid-point of the nineteenth century the older port cities such as Boston, New York, and Philadelphia had grown greatly in size and become dominantly proletarian. Diversity of goods was desired by shoppers of much greater number than the small well-to-do group of merchants who had supported the artisan's shop. To meet this demand a second type of commodity-combining shop, the department store, was developed. By locating these stores in the heart of the city where the largest possible market for any specialized good would have its center and by combining all the specialties in a single establishment in order to maximize the total profit rather than the profit from each line of merchandise it was possible to offer a more diverse group of goods at a lower unit profit. Demand was greatly increased thereby, and the urban mass market came into existence for goods beyond the most staple sort. These department stores ultimately assumed such dominance in volume of sales, and attraction of customers, that they became the anchor of the central business district and, until the last fifteen years, were unique to that area.

Even the department store could not care for the most proletarian of demands and in 1879 F. W. Woolworth opened his first store in Lancaster, Pa., to sell variety merchandise at the lowest possible price by developing the mass market. The fact that the mass market was international is amply demonstrated by the rapid spread of the department store and "5 and 10" variety store to western Europe and Japan, in fact, anywhere that industrial society had introduced a money economy.

The American Civil War added still a other mass-appeal shop to cities where joined with the "5 and 10" and the depar ment store to form the first and domina nexus within the central business district. Th demand for uniforms within the Union Arm brought the first mass production of read made clothing. In the years after 1865 th producers of uniforms for the first mode conscription army sought to continue produ tion by shifting their operations to mass-sa clothing for men. Only much later, after th turn of the century, was the women's read made clothing industry well developed, but the case of both men's and women's clothir the introduction came in the mass mark before the specialty market, so that the ce tral business district location was essential. remained for the mail order firms to demo strate in the 1890's that clothing along wi other mass produced and demanded iten could be sold outside the core of the city. Th American drug store came into existence soo after the "5 and 10" to carry standardizatio of product and combination of merchandis lines into those realms untapped by its pr decessors. Although the four mass-sellin establishment types, department stores, va iety stores (5 and 10), ready-made clothin and drug stores, have been in the van development outside the core, they still ac count for 60 per cent of the sales in the centr business district. As they brought the dow town commercial area into being it is n surprising that they also initiated the outlyin shopping center.

Proudfoot writes that in the outlying bus ness centers which grew up around inte change points on mass transit lines with th introduction of the street railway in th 1890's "are found shopping goods outle such as men's and women's clothing store niture stores, shoe stores, jewelry stores, : or more large (branch or junior) depart-nt stores, and an admixture of convenience-ods stores". This catalogue reads like an entory of the core area except for the mat- of specialty shops. With the passage of e, and increasing competition from out-g centers, the central business district be-ne distinguished for price and commodity iety. The mass-appeal goods which had ught the downtown shopping area into stence could, like the earlier projection of ding into rural areas through the general re, be carried out closer to the consuming lic. What could not be decentralized was ding in specialties, which continued to d the uniqueness of location to tap the ire metropolitan market that the core vided.

With the introduction of automobile travel street railway interchange point lost much ts locational advantage because cars were t of greater use in the journey-to-shop than he urban journey-to-work. As women had omobiles for shopping the principal busi-s thoroughfare grew up in the form of ps of commercial land use along major erial roads. There the universally needed d store combined with the mass-appealing res selling clothing, furniture, and other oles. The blaze of neon signs along these ps of commerce attests to the fact that queness of offering is far less common n strong competition for the mass market. e neighborhood business street and isolated re cluster were a logical elaboration of the 's commercial structure stemming from reasing city size and prosperity. The mini-m tributary area for a store selling goods universal appeal is very small. The recent lacement of many of these business streets a single supermarket tells us mainly that in American retailing there is constant pressure to expand the unit size of shops, which can only be done within a minimum tributary area by reducing what specialization may have existed in the past.

"In the decade following World War II the universal use of the automobile in the United States and Canada has superimposed a new retail structure upon the retail pattern developed in the previous decades [under mass transit] . . ." (Hoyt, 1958). The first large automobile-oriented stores were those built by Sears Roebuck and Co., when that mail order house entered the retail store field in the early 1930's. At first these stores were thought of by the owners as places for men to shop so that central location seemed unnecessary and the range of choice implied by the designation "style goods" was avoided. Emmet and Jeuck (1950, p. 490) in their authoritative study conclude "that Sears does, at best, a mediocre job in style merchandise but an outstanding job in staples". In addition it seems that Sears also understood, well before most retailers, that the outlying market can only be developed as a mass market as there unique location is impossible and there must be a sharing of the limited-appeal market among a number of competing sites. Although the Sears' stores were almost the only outlying department stores before the Second World War, since the close of the war the major part of store construction and virtually all the increase in sales has taken place in outlying areas. In the 48 larger Standard Metropolitan Areas of the United States, between 1948 and 1954 retail sales increased 32.3 per cent but the central business district rise was only 1.6 percent or a loss in relative terms.

The core of all the larger integrated outlying shopping centers has been the branch department store usually having at least 100,000 square feet of sales area. Like the Sears' stores built in the 1930's these branches have dealt almost exclusively with staples. Hoyt (1958) remarks that, "The largest regional shopping centers do offer a complete selection of both fashion and convenience goods, but in these new centers, built at today's high costs, the most profitable lines are selected, leaving for the older (downtown) stores, built at lower costs, the types of merchandise that have a less rent-paying capacity." By 1957 there were 36 centers with 100,000 square foot department store branches and since then the number has probably doubled. In the San Francisco Bay Area, where the type is well represented, the increase in three years has been from six to eleven. The largest of these regional shopping centers may have one million square feet and probably require a minimum tributary area of 200,000 people.

In addition to the regional centers there are smaller types. The community center usually has at its core a large variety (5 and 10) or "junior department store", that is a store dominated by the selling of readymade clothing and household linens, that ranges in size from 25,000 to 90,000 square feet. The center would normally total 100,000 to 400,000 square feet. The community center is largely competitive with the older outlying business districts at interchange points on the street railways or suburban railroads. The tributary area normally would have at least 100,000 people.

Finally, there is the integrated equivalent of the older business street or isolated store cluster, usually called the neighborhood center. At the core is a supermarket with up to 50,000 square feet and the tributary area may be as small as 10,000 people due to the dependence upon the universal need for foo shopping.

The planned shopping center differs fror the older outlying shopping center in severa important aspects. The most significant c these is the continuing ownership of the lanc and often of the buildings, by the develope of the center, a fact which usually leads to considerable control over the types of business conducted. This situation creates a carefully determined assorting of store types an makes possible a rigid application of the tes of maximized profit in the selection of occupants as the land owner normally receives share of the profits of the renting firm. I Proudfoot's classification the outlying shopping center was an area of individual ownership and chance assortment of shops, ofte with quite varying profits.

The planned centers also normally have control on size because of the fact that th original site is usually restricted and cannc be enlarged because of zoning provisions These centers, which depend on automobil access, are tied to highway junctions rathe than to the older town centers and for thi reason, along with the maximization of rent seldom have any other appreciable use element than commerce. Being planned th centers are normally located in a section o the suburbs which is without competin centers, though up to the present the necessary tributary area population and distanc over which customers could be induced t travel to shop there have been open to differences of thought. Because of the fact tha these centers must depend upon the full exploitation of the immediately adjacent marke there has been little interest on the part o developers for Hoyt's "low rent-paying merchandise lines". Thus, if these centers have single dominant characteristic it is strikin

formity. The mass market in its very nature
ıniform and the integrated shopping center
ly reflects this. Obviously, uniformity leads
complete competition among centers and
restriction of their tributary areas largely
the basis of distance.

cation Factors in the Geography Commerce

ıas been traditional in discussing the loca-
n of commercial activity to distinguish be-
een "convenience goods" and "shopping
ods" on the assumption that certain goods
purchased so frequently that their sale
ıst take place on the local level while other
ods are infrequent purchases for which the
stomers engage in planned "shopping". On
ser consideration, however, it must be
ed that locationally the simple distinction
meaningless. What appears to have been
erlooked is the fact that any aggregation of
ople requires convenience goods sales, even
hin the central business district, and that
class usually called shopping goods is so
ad as to obscure the difference between
ods of mass appeal and those of limited
peal.

In terms of location a much more analyti-
distinction is provided by visualizing two
ıtinua of specialization. It is possible to
k classes of establishments along lines of
ational specialization and *commodity spe-
lization* and thereby establish the geogra-
cally important relationship between trade
ter and tributary area as well as an hierhy of trade centers on a local level. Loca-
nal specialization is a simple concept seek-
to quantify the distinction among shops as
the economically minimal tributary area. It
lows that a shop with a small minimum
port area, measured more in terms of population than geographical area, has little locational specialization. A grocery store may prosper on the patronage of the inhabitants of a few city blocks but a rare book dealer may require the support of an entire nation. The point might be made that the rare book dealer would be classed as a shopping goods store in any event and we need not think about locational specialization. But it is obvious that locational specialization is a much more refined measure. In such a convenience good as food there are certain luxury or ethnically associated items which may be secured only in the largest metropolitan center and there only in a single locale.

The general conditions of locational specialization may be inferred from the data in Table I which shows the national totals of establishments for the 41 types of retail shops in the Censuses of Business carried out in 1929 and 1954. At the time of writing the data for the 1958 were unavailable. Between 1929 and 1954 the total number of shops decreased by one-sixth while the total of sales, in "constant dollars", increased three times as much. Of the 41 categories more than half, 23, declined in their percentage contribution to the total. These types displaying a *concentration of sales* into a smaller relative number of outlets were all types of food stores, all but one of the clothing types (women's ready-to-wear), all but one of the general merchandise types (variety stores), and approximately half the remaining types. Integration of several lines of merchandise in a single type, such as the supermarket, has gone on at the same time that the size of individuals has increased, as also in supermarkets. This results in a reduction in the number of specialty food shops and of food shops in general.

In the group of shops exhibiting an expansion of relative numbers the growth, with

TABLE I.

Rank of Retail Trade Establishments, by Total Numbers, for the United States 1929—1954

1929		*1954*	
Type of Establishment	*Number*	*Type of Establishment*	*Number*
Grocery, without meat	191,876	Food stores	279,440 —
Gasoline stations	121,513	Gasoline stations	181,747 —
Grocery, with meat	115,549	Motor vehicle dealers	61,666 —
General stores	104,089	Drug stores	56,009 —
Candy, nut, confections	63,265	Hardware stores	34,858 —
Drug stores	58,258	Dry goods, gen. merchand.	34,113 —
Motor vehicle dealers	45,301	Liquor stores	31,240 —
Meat markets	43,788	Lumber, bldg. material	30,177 —
Dry goods stores	38,305	Women's clothing	26,893 —
Cigar stores and stands	33,248	Furniture stores	25,475 — 1
Men's and boys' clothing	28,197	Jewelry stores	24,266 — 1
Lumber, bldg. material	26,377	Shoe stores	23,847 — 1
Furniture stores	25,854	Meat markets	22,896 — 1
Hardware stores	25,330	Household appliance	21,974 — 1
Shoe stores	24,259	Variety stores	20,917 — 1
Fruit and vegetable stores	22,904	Candy, nut, confections	20,507 — 1
Tire, battery, auto acc.	22,313	Men's and boys' clothing	19,247 — 1
Jewelry stores	19,998	Tire, battery, auto acc.	18,845 — 1
Women's ready-to-wear cl.	18,253	Farm equipment	18,689 — 1
Radio stores	16,037	Florist shops	16,279 — 2
Second-hand stores	15,065	Second-hand stores	14,364 — 2
Millinery shops	12,433	Fruit and vegetable stores	13,136 — 2
Farm equipment	12,242	Gift and souvenir shops	12,149 — 2
Variety stores	12,110	Family clothing stores	11,056 — 2
Delicatessen shops	11,166	Paint, glass, wallpaper	9,249 — 2
Family clothing stores	10,551	Sporting goods stores	8,396 — 2
Newspaper dealers	10,285	Delicatessen shops	8,132 — 2
Household appliance	9,329	Farm-garden supply stores	7,262 — 2
Florist shops	9,328	Newspaper dealers	7,178 — 2
Paint, glass, wallpaper	8,870	Children's clothing stores	7,024 — 3
Fish markets	6,077	Cigar stores and stands	6,859 — 3
Farm-garden supply stores	5,740	Music stores	5,810 — 3
Gift and souvenir shops	5,186	Radio and TV stores	5,800 — 3
Department stores	4,221	Stationery stores	5,473 — 3
Stationery stores	4,047	Millinery shops	5,473 — 3
Office-store machine equip.	3,498	Fish markets	4,458 — 3
Book stores	2,809	Floor covering stores	4,335 — 3
Music stores	2,232	Camera and photography shops	2,896 — 3
Sporting goods stores	1,930	Department stores	2,761 — 3
Floor covering stores	1,503	Book stores	2,642 — 4
Children's clothing stores	1,309	Office-store mach.-equip.	2,216 — 4
Camera and photography shops	710		— 4
Totals	1,342,072		1,115,014

Sources: U.S. Bureau of the Census 1929—1954.

or exceptions, has occurred in newly intro-ed lines of merchandise or in the selling of ıry items where mass selling and low unit fit are not characteristic. This contrast veen staple and luxury goods leads us to conclusion that an increased scale of blishment size, to enhance the total rather ı the unit profit, has led to an enlargement he economically minimal tributary area.

ommodity specialization is distinct in n and operation and concerns the mini-m range of merchandise types necessary to port a shop. Thus, we may progress from single commodity shop, such as the ıben Glass Shop in New York, to the local olworth store which also sells glass but of ery different sort and then only as part of a t array of goods. Admittedly, the Wool-th company has stores in several thousand ns in North America and Europe and ıben a single shop in New York, so that tional specialization also enters the pic-. But it is important to note that outside of v York Steuben glass can be sold only as ıdjunct to the selling of a diversity of other ıs.

he relationship between locational special-ion and commodity specialization is essen-y direct. *As commodity specialization eases so must locational specialization.* e, however, we must introduce the con-t of mass-appeal and limited-appeal goods. ational specialization is a simple con-um from the least to the most restricted commodity specialization has two series, for mass-appeal goods and another for ited-appeal goods. It is this fact which kes it necessary for us to consider both es of specialization even though they oper-in somewhat the same manner. Without distinction between mass- and limited-eal goods we could not logically explain different location factors shaping the eral types of commercial districts.

Prior to the Second World War American suburbs tended to fall into two classes rather distinct from each other. There were strictly residential areas of generally high income housing and there were industrial satellites of the central city with worker's housing. It may be seen that the two classes of suburban housing worked against the creation of a mass market with its need for standardization. The ties of each segment were with the central city rather than with each other so that the journey-to-shop was either on a very local scale, with some small specialization in limited-appeal goods based on a restricted but socially and economically homogeneous market focussed on an outlying shopping center, or a metropolitan scale which brought all journeys together in the central business district. Only there was mass selling possible. After 1945 the conditions in suburban areas were radically changed. Vast areas of post-war housing grew up at the urban fringe and even though economic and social homogeneity held within individual development tracts these new residential areas were joined together in being independent of pre-war patterns of shopping trips. The distance to the older outlying shopping centers was sufficient to impede the tying of the newer housing tracts to these places. As will be shown subsequently, local journey-to-shop movements are highly limited in distance. For this reason new residential areas meant new shopping districts and the considerable mixing of tracts of contrasting socio-economic nature ruled out the creation of new shopping facilities on any but a mass base. The individual mobility provided by shopping by car and the increased size of tributary areas stemming from the joining of all classes of housing made the major regional center practicable when it had been impracticable under mass transport and an economic segregation by area. A test of the contrast between the mass-appeal and limited-appeal shopping is provided by the fact that in the San Francisco Bay Area only three limited-appeal types of shops have grown greater growth in sales in the central business district than in the suburbs (San Francisco CBD Bulletin, 1954).

In general terms, we may say that the lower the position of a shopping district in the hierarchy of commercial centers the smaller will be its tributary area and the greater its domination by mass-appeal goods. Even though the central business district is the largest seller of standardized merchandise, it is less dominated by uniformity of goods than are the outlying centers. And it seems certain, though impossible to quantify at this time, that the central business district is changing from mass-selling to specialization to the extent that its importance today for the metropolis as a whole is that of a specialty shopping area. Just as the mass selling of food was removed from the core in the 1930's the mass selling of clothing is departing today, at least in terms of relative sales totals. Because of the broad nature of census classifications it is impossible from published statistics to test the trends in central business district sales between mass and specialty lines without an independent study. This should be undertaken but until it is, it may be conjectured that the bare maintenance of absolute sales volumes in the downtown area is the result of increasing specialty sales which counteract the decrease in mass sales. As specialty sales can never equal mass sales, the central business district has been unable to maintain its relative sales position.

William J. Reilly (1929) writing about the boundaries of trading areas concluded that retain centers attract customers in direct proportion to their relative population and in inverse proportion to the distance to each center. His "retail gravitation" is a characteristic of any commercial district so we mu ask whether these conclusions are valid fc outlying centers as well as the central busine districts he considered? The substantive pa of this paper corroborates his findings wit respect to the effect of distance but leads u to doubt that Reilly's formulation concernin population is applicable. Regional shoppin centers in particular do require a minimu population but also appear to have a max mum supporting population due to the fa that an overly large trading area is quite su ceptible to sub-division through the constru tion of a second, competing center. No doul the density of population within the tributa area will affect the profitability of a center b any attempt to tie an unduly large area to single center would probably be unsuccessf in the long run. Within the metropolitan are with dispersed selling of mass-appeal gooc the concept of gravitation has been replace by one of dismembering the market amon equal and competing centers. Still u answered is the degree of dissection possibl in other words, how small may the tributa area be?

If we accept that Reilly's concept of reta gravitation has been modified by the growt of the suburban integrated center, we are le to ask if the concept of equilibrium of loca tion put forward by Lösch is also limitec Anticipating the answer, it should be note that the generality of Lösch's ideas mak them fit the situation much more closely. H first condition of locational equilibrium not that "The location for an individual must b as advantageous as possible" (Lösch, 195 pp. 92-100). In retail location this has take the form of dividing retail establishments in the two groups noted, the specialty shop i the core area necessarily dependent upon th entire metropolitan market, and the ma goods store, often combining several stan ized merchandise lines, in outlying centers ere proximity to the customers assures the ximum patronage from any given residen- area. Locational advantage is quite a dif- ent thing for mass and specialty selling so t locations differ. The second condition of uilibrium that "The locations must be so nerous that the entire space is occupied" borne out by the very proliferation of re- nal and other shopping centers. In the e of the highly specialized store at the core limitation of profitability in condition 1 tricts location to a single central site. sch's third and fourth conditions are inti- tely interrelated in commercial location. saying that "abnormal profits must dis- pear" (Condition 3) we imply in the case ributary areas dominated by a single shop- g center, usually with single establish- nts of a retail type, that "the areas . . . of es must be as small as possible (Condition ". Otherwise, we would be hard put to plain why central department stores build nches as a single super-department-store the core could yield a higher profit than a ng of branches, where the fixed costs must a greater part of the cost of selling. The uation is different in the case of specialty ops as a single central location does not oduce an "abnormal profit." Any other ation would most probably produce no ofit. Lösch's fifth and final condition for ational equilibrium that "At the boundary economic (tributary) areas it must be a tter of indifference to which of two neigh- ring locations they belong" requires that butary areas be clearly defined. The study San Francisco which follows suggests that broad measure this condition is met in the se of outlying mass shopping centers by a gmentation of the suburban areas and in the se of specialty shopping in the central busi- ss district by the uniqueness of this area within the metropolitan region. Thus, it appears that the present situation in a region highly developed with integrated shopping centers in outlying areas, such as the San Francisco Bay Area, is one of locational equi- librium under the current conditions of (1) residential distribution, (2) transporta- tion, and (3) mass demand for goods.

The Commercial Structure of the San Francisco Bay Area

It is unfortunately impossible in the limited time available to dicuss the origin of the commercial structure of the Bay Area in any detail. This story is one affording interesting insights into the transformation of a primi- tive to an urban landscape in the course of one century. During this metamorphosis there was much rivalry among germinal cities for the commercial domination of the area but San Francisco was never seriously threatened in its leadership. One other city from the Mexican period, San Jose, and a latter day rival, Oakland, did manage a sufficient devel- opment by the time of the First World War to have their own central business districts. At that time these three foci of street railways were distinguished one from another mainly in terms of their degree of specialization, with San Francisco serving as the site of most of the limited-appeal goods shops and Oakland and San Jose as the site of mass selling alone. Each was tied to the surrounding area of resi- dential towns by a well articulated street rail- way system and no adjacent commercial area could hope to compete in tapping the mass market. The five-mile-wide Bay, crossed at that time only by ferries, provided some pro- tection to downtown Oakland's shopping area in relation to San Francisco, and San Jose was separated from San Francisco and Oakland by 45 miles of scattered residential and farm land. The Bay Area at the time of the First World War was an economic unit but not a morphological one (Figure 1).

A number of local shopping districts came into existence as residential land use spread onto the Berkeley Hills from the initial core in Oakland. At the place where the local street car lines in Berkeley came together at Shattuck Square shops were opened where those returning from work in Oakland and San Francisco could purchase perishable food daily. Here also the more commonly available dry goods and stock clothing items were on sale in a place which could easily be reached from all parts of the residential community. This pattern was reproduced in a number of places in the Bay Area where local street car routes came together and were connected to express routes to San Francisco. These major suburban business cen ters formed without any predetermined pla in Berkeley, on Alameda Island, and in Eas Oakland in the East Bay and at the mor important stations on the suburban railroa south of San Francisco on The Peninsula. N one of them could assume a dominant pos tion as the street car lines fed in only fro the immediate vicinity. Each suburban cente reflected the character of that vicinity so tha there were book stores and young people' clothing stores adjacent to the University c California in Berkeley and Stanford Uni versity in Palo Alto, shopping areas for work ing class people in South San Francisco an Redwood City, and more expensive but nc very specialized districts in Burlingame an San Mateo. This class orientation of outlyin commercial districts reinforced the isolatio imposed by a transit system joining all limb together only at the city core. Under thes conditions mass selling was possible only i the core.

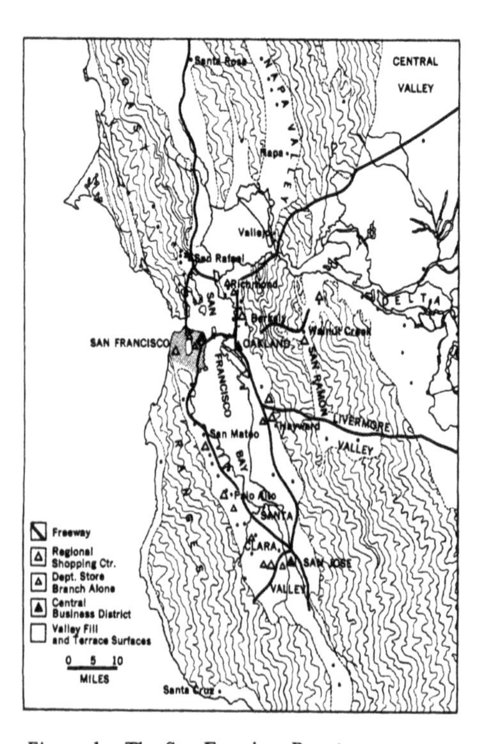

Figure 1. The San Francisco Bay Area.

There were many types of stores in th outlying business districts before the Firs World War because the unit size was smal But the diversity was actually within a nar row range controlled by a test of restricte local support. The supermarket of the presen combines the goods of at least six indepen dent shops at the turn of the century. Th "specialization" found within the 1900 busi ness district was primarily a function o minute division of a few staple lines amon a number of family operated shops, not of then smaller minimum for tributary areas The assembling of formerly separate foo and staple clothing lines into larger stores ha been mainly to maximize profits rather tha to carry the selling into smaller markets. To day the specialty food shop or narrowl stocked clothing store can exist only i central business districts or the largest shop g centers where a volume of sales sufficient profitable operation within a single type sales can be secured.

Outside the major outlying business cens isolated store clusters and individual res were widespread. These morphological ms were given over mainly to the selling food as the absence of mechanical regeration made daily food shopping more cessary and the absence of individual nsportation made easy access more imrtant. The imposition of land use zoning dinances since 1916 along with the induction of refrigeration and automobiles s spelled the doom of many of these small nmercial districts and has prevented their production in new housing areas. Today re are fewer but larger business districts.

Throughout the years after the First World ar the number of automobiles increased pidly, particularly so in California which ly became and still remains the site of the rld's highest per capita automobile ownerp. Today there is one car for more than ery second Californian. For some time after omobile ownership became common in 1920's there seems to have been little ange in the commercial pattern. The lag ween the gaining of individual transportan and its impact on shopping resulted from inherited residential patterns. As long as ople lived where the street car had put m, the pattern of commerce inherited m that era continued. But with the rapid wth of residential areas in what had merly been open land, which took place a small degree before the Second World ar and to a much larger degree after the r, new commercial areas had to be develed. Only then could the revised location tors operate. To understand the revision must appreciate that (1) population nsities in the center of the Bay Area, and of cities in general, have tended to decrease, (2) the newer housing here as elsewhere was built in areas without preexisting commercial districts as the typical pattern of post-war housing developments was one of massive tracts of residence placed in formerly farming land rather than concentric additions to older towns, (3) the distance from the newer residences to the central business district was very much greater than for the housing they replaced, and (4) particularly in California there was no attempt to locate housing tracts so that mass transit could be used, which enforced an automobile journey-to-shop. Under these conditions there was a loosening of the ties of the growth element of the population with the central business district. The people left in the heart of the city were mostly the young childless or the elderly. This fact in itself helps to explain the exchange of functions between central and outlying business districts as the mass market tends to fall between these two age extremes. In California, and in other sections of the United States where suburban development has been on a large scale in recent years, the developers of residential tracts commonly included in their plans what we have called neighborhood business centers even during the initial phase of suburbanization but in the beginning there was little realization that more was needed. The neighborhood center was essentially an updated version of the staple goods business centers which had been in pre-war towns. What was needed in addition was the projection of the mass selling component of central business district activities to this new mass market now far removed from the core. For the five years between 1945 and 1950 a situation of locational "disequilibrium" existed. A very rough measure of this is the fact that between 1945 and 1950 all retail sales increased by 96 per cent whereas for department stores alone, which were then largely in downtown areas, the increase was only 36 per cent. Between 1951 and 1955 total retail sales and department stores sales showed a closer correspondence in their increase, 17 and 11 per cent respectively. During this period the disequilibrium of concentrating department stores in the downtown area was considerably overcome by the building of branches in the regional shopping centers.

In this context we may view the building of regional shopping centers as a restoration of equilibrium, which between 1939 and 1950 was lacking because of changed settlement and transportation patterns but unchanged commercial patterns. For an answer to what sort of commercial pattern has emerged in the decade of the 'fifties we may turn to the San Francisco Bay Area where evidence is strong that adjustment has been restored.

Due to the absence of separate data on retail trade within the central business district before 1948 it is impossible to trace earlier changes in the relative position of the district and outlying shopping centers. In the period 1948 - 1954, while the San Francisco-Oakland Standard Metropolitan Area was experiencing the most rapid growth of any large city (over 500,000 population) metropolitan region, an increase of 53.3 per cent, the central business district of San Francisco remained completely stable in its retail sales (1948=$408 million and 1954 =$410 million). The relative loss is apparent, resulting in a decrease for the central business district from 18 to 14 per cent of the metropolitan total of sales. Oakland's central business district differs only in having suffered an absolute loss during this six year period from $187 million to $182 million. Only San Jose, in the burgeoning Santa Clara Valley, experienced a rise in central sales, from $96 million in 1948 to $115 million i 1954. Thus, out of a total of $2,579 millio sales in the two SMA's (San Francisco-Oakland and San Jose) in 1948 the three centra business districts accounted for $691 millio or 27 per cent. In 1954 the total sales of th two SMA's were $3,418 million with $70 milion or 21 per cent in the three centra business districts.

It is apparent from the diminution of sales both absolutely, in terms of constant dollars and relatively, in terms of percentage contribution, that the post-war commercia development in the Bay Area has been peripheral. This has been particularly true wit respect to department stores. San Jose's central business district has no true departmen store and those in Oakland cannot be assessed because of the Census "disclosure rule but in downtown San Francisco the department stores dropped in sales, measured i constant dollars, from $98 million to $9 million in the period 1948 - 1954.

Prior to the end of the Second World Wa there were no integrated regional shoppin centers in the Bay Area. The only department stores outside the central business districts were a single "A type" Sears Roebuc store in each of the two core cities and much larger Montgomery Ward store in thei branch mail order plant in Oakland. Eac of these was a product of the initial stages o the mail order companies' expansion int retail operation after 1927. The three wer outside the central business district but sufficiently close to it so their position was no truly outlying. Only the Montgomery War store was above the minimum of 100,00 square feet established in the late 'forties a the size for outlying stores. At the time o construction the fact that these stores wer built by mail order houses caused most observers to conclude that their constructio resented no strong trend away from the tral business district in mass selling rations.

Vhen the pent up demand for housing endered by the rapid growth of the Bay a metropolis during the Second World r could be met after 1945 there was a five r lag in commercial construction, thus, figures for retail trade in 1948 may be ught fairly typical of the summit of tral business district trading. Between 8 and 1954 the San Francisco and Oak-d central shopping areas had a ten per t drop in sales measured in constant dol-. At the same time the retail trade of the ole metropolis increased 15 per cent in stant dollars.

'he increase in the mass selling of goods er than food has obviously tended to con-trate in the regional shopping centers and larger community centers. Of the regional pping centers the first to be completed was Stonestown center in the southwestern ion of the city of San Francisco. This tion within the central city is unusual and lted largely from the construction in the a of two large "high-rise" apartment elopments, both oriented toward middle upper income groups. In addition, this a, though served by a long street car tun-connecting to the San Francisco central iness district, was rather isolated by the n range of hills forming the spine of the Francisco Peninsula all the way to its nination at the Golden Gate. At the pre-: time this center receives a surprising part ts patronage from pedestrians and street riders, quite in contrast to the conditions other outlying shopping centers. In many 's Stonestown Center represents a rather ative initial experiment in the decentrali-on of mass selling, to a point outside the e but not outside the central city. This center has been set up to care for the needs of a prosperous area in the central city and its patronage is largely restricted to city dwellers as is shown by the map of tributary areas. Stonestown has succeeded in the eight years it has been open in developing a total of sales approximately one-tenth as large as the San Francisco central business district.

Before continuing, the origin of the data used to evaluate the tributary areas of out-lying centers should be discussed. In under-taking the analysis of the relative functions of regional and central business districts it is of critical importance to discover the form and size of areas tributary to each type and to individual regional centers. The transitional nature of a center such as Stonestown makes it necessary for us to determine whether it is functionally in the group with integrated re-gional centers or more precisely an adjunct of the city's central business district. In the case of the central business district numerous studies of San Francisco and other large cities have shown that patronage is metropolitan-wide in origin, though demonstrating a con-centration of staple goods customers within the closer residential areas. Thus, we may generalize to the extent of saying that the metropolitan core has the entire metropolis as its tributary area. In the case of the re-gional center data have been lacking for any comparison among centers.

To overcome the absence of reliable data with which to compare regional centers a sample of the registration number plates of cars parked in the customer parking spaces of the eleven regional centers of the Bay Area was carried out between December, 1958, and August, 1959. In addition, similar samples were secured from two community shopping centers, two independent branch department stores, and six of the older out-lying shopping districts which were established before the period of integrated development. In preliminary tests a sample was taken five or six times during the last three days of the shopping week to determine any differences in residence pattern among shoppers visiting centers at different times. But careful analysis suggested that there is no appreciable divergence among the patterns on the basis of time. Thus, for nine of the regional centers and the majority of the other classes of commercial districts the data stem from a single sample time. The size of the sample varied between 100 and 500 registrations depending upon the size of the commercial district. The total sample contained approximately 5,000 registrations for which the owner's address was secured on IBM punch cards from the Registry of Motor Vehicles of the State of California. This was necessitated by the fact that there is no areal assignment of registrations in California and the total number of automobile registrations currently exceeds seven million. When the punch cards were sorted by the commercial district where the registered automobile was found parked and by political sub-division of the owner's residence the location of the residence was plotted by street address on large scale maps, one for each center. The tributary areas for the places derived from this plotting of residence of shoppers are presented in somewhat generalized form on the figures 2 - 4. Before discussing the findings it should be noted that although care was taken to avoid "sampling" cars parked by employees or salesmen visiting the center their exclusion was not completely successful. The inability to exclude also implies the inability to measure the inclusion. With some knowledge based on the multiple-time samples, however, it seems probable that the area of labor-shed is coterminous with the normal tributary area and the number of em-

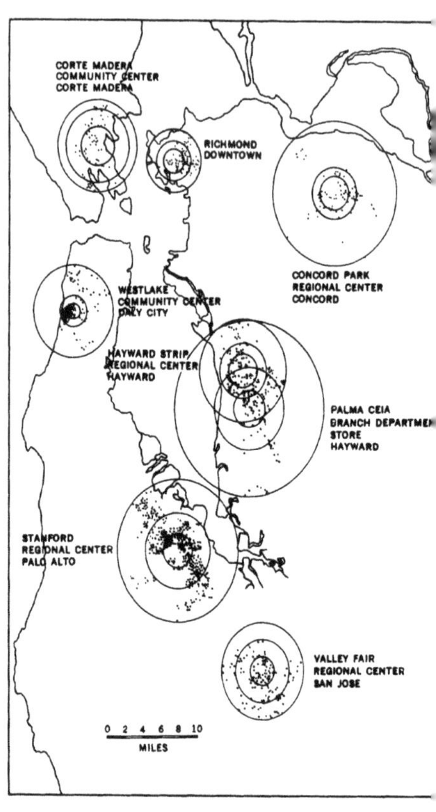

Figure 2. The trading area for shopping centers in the San Francisco Bay Area. The spacing of concentric circles is established so that each successive circle encloses one quarter of the total shoppers using the center. Data derived from parking surveys, 1958-1959.

ployee cars should not exceed three per cent of the total. It also seems that the degree of refinement afforded by larger samples is not worth the labor and considerable expense. A sample of 300 cars for a regional shopping center appeared adequate to apportion by residence location the shoppers at the center.

If the assumption is made, as here, that the dynamics of urban commercial structure is comprised of differing degrees of locational and commodity specialization, then it follows that to test this hypothesis we must conside

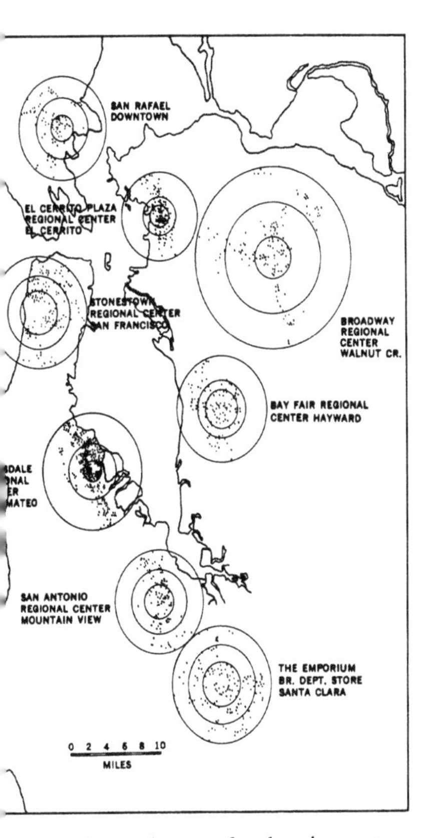

ure 3. The trading area for shopping centers he San Francisco Bay Area. The spacing of centric circles is established so that each cessive circle encloses one quarter of the total shoppers using the center. Data derived from king surveys, 1958-1959.

) the nature, size, and hierarchical relans of areas tributary to the several types commercial center, and (2) the number 1 diversity of commercial establishments hin these same types of commercial dis-:t.

A tool for establishing the characteristics tributary areas has been developed through registration sample method described. ble II summarizes the information secured m this sample. The most notable result of s investigation is quantification of the lial extent of the tributary area. Within the 21 places investigated, regional shopping centers, community shopping centers, "free-standing" stores, and older business districts, there are 16 department store branches (all but two with 100,000 square feet or more) whose aggregate sales are approximately $130 million, a total which is essentially equivalent to the total sales in department stores in the central business districts of San Francisco and Oakland. Thus, in the fifteen years since 1945 there has been a doubling of the sales of department stores but the entire increase has come in outlying branches. Sharply contrasting with the great increase in sales at the periphery is the narrowing of the tributary base there. Most certainly San Francisco, and to a lesser degree Oakland, in the pre-war years tapped the whole metropolis. In the regional centers and free-standing department stores two-thirds of the customers come from within a tributary area which extends no farther than 4.5 to 5.5 miles from the commercial place. As might be anticipated, an even greater concentration of

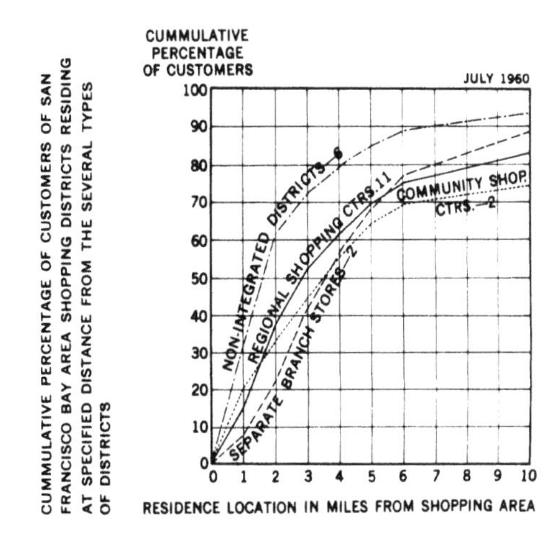

Figure 4. Cumulative percentage of customers of San Francisco Bay Area shopping districts residing at specified distance from the several types of districts.

TABLE II

Distribution of Customer's Residences by Distance Zones from Various Commercial Areas

Type of Commercial Area	*Estimated Annual Sales*	*Per Cent of Total Number of Customers Residing Within Band (in miles)*							
Regional Centres (11)		*0.00—0.99*	*1.00—1.99*	*2.00—2.99*	*3.00—3.99*	*4.00—4.99*	*5.00—5.99*	*6.00—9.99*	*10.00—*
"Stonestown" (1952) San Francisco	$42,000,000	15.64	16.94	14.98	9.78	10.42	7.82	3.57	20.85
"Hillsdale" (1955) San Mateo	60,000,000	21.60	19.40	12.10	8.60	8.40	6.20	13.70	10.00
"Broadway" (1955) Walnut Creek	25,000,000	7.98	15.54	10.08	10.08	2.52	5.46	13.86	34.45
"Stanford" (1956) Palo Alto	40,000,000	14.00	27.60	18.40	8.60	11.20	7.00	9.90	2.90
"Valley Fair" (1956) San Jose	15,000,000	14.22	17.99	18.83	10.04	7.95	5.44	9.63	15.90
"Bay Fair" (1956) San Leandro	20,000,000	11.56	21.33	16.45	11.12	8.00	6.22	7.56	17.78
"San Antonio" (1957) Mountain View	17,000,000	15.60	31.90	10.60	20.50	6.30	6.30	6.30	2.10
"Hayward Strip" (1957) Hayward	30,000,000	14.54	23.79	16.74	7.44	7.04	7.05	4.40	19.38
"Concord Park" (1957) Concord	22,000,000	6.92	34.62	13.08	3.85	5.39	3.85	6.93	25.38
"El Cerrito Plaza" (1959) El Cerrito	25,000,000	26.06	23.51	11.54	5.56	7.69	7.26	5.32	13.25
"Palma Ceia" (1958) Hayward	unknown	10.82	17.75	14.28	9.52	6.49	4.76	11.26	25.11

"Emporium" (1958) Santa Clara	6,500,000	8.29	12.90	16.59	17.51	9.67	9.67	9.21	16.13
"Sears Roebuck" San Jose	14,000,000	7.90	15.80	23.30	12.00	14.20	9.30	13.50	5.60
Community Centers (2)									
"Westlake" (Daly City)	—	27.51	15.34	10.59	11.11	6.87	4.24	8.99	15.34
"Corte Madera" (Corte Madera)	—	13.23	10.29	12.50	12.50	8.83	6.62	0.00	36.03
Non-Intgr. Older Ctrs. (6)									
Burlingame		45.70	35.50	8.40	5.00	1.70	3.40	1.70	0.00
San Mateo		38.00	22.00	13.00	3.00	4.00	8.00	8.00	4.00
Redwood City		32.70	36.00	8.20	6.50	3.20	4.80	6.40	1.60
Palo Alto		32.30	41.50	7.70	3.00	12.30	3.00	0.00	0.00
Richmond		21.97	28.03	16.67	6.07	6.82	2.27	3.79	14.39
San Rafael		19.72	18.30	11.27	17.60	4.93	2.11	4.93	21.13
Average Percentage Values by Class of Commercial Area									
Regional Centers		15.21	22.76	14.28	9.55	7.40	6.12	8.40	17.01
Separate Branches		8.10	14.35	19.94	13.76	11.94	9.49	11.36	10.87
Community Centers		20.37	12.82	11.55	11.81	7.85	5.43	4.49	25.69
Non-Intgr. Older Ctrs.		31.73	30.22	10.87	6.86	5.49	3.93	4.14	6.85
All Types of Areas		18.85	20.04	14.16	10.50	8.17	6.24	7.10	15.11

customers is found for the older nonintegrated outlying business areas. The restriction of suburban tributary areas is highly significant in relation to Lösch's conditions of equilibrium of location, particularly the condition that sales areas will be reduced to the minimum (constantly sub-divided by new sales places) and the condition of indifference at the boundaries. The mean spacing of centers one from another, within the Bay Area, is 8 miles and the median spacing is 9 miles. The correspondence between the radial extent of tributary areas and one-half the mean and median spacing is worthy of emphasis. Also this relationship supports the belief that the boundaries do represent remarkable "indifference lines" as positted by Lösch in a condition of locational equilibrium. The fact that half the customers, on the average, come from within three miles of the center supports the contention that the socio-economic character of this immediately adjacent area must be directly reflected in the composition of the establishments in the center, and even more significantly, that within such a limited trading area all possible patronage must be secured. Mass selling alone would make this possible. With so small a base a partial market would be uneconomic. This is the most significant contrast between the central business district and the outlying center. The core draws widely but selectively whereas the shopping center draws narrowly but necessarily as completely as possible. The outlying center does not, in any meaningful way, replace the core *as a unit* despite the fact that it may nearly completely replace it *with respect to individual establishments or individual lines of selling.*

It should be noted that the areal extent of tributary areas differs with the spacing of the centers. In the case of centers more removed than average from competition (Walnut Creek) the percentage contribution from the inner ring of customer residence is less. This condition supports the belief that the parcelling of the metropolitan market among centers is effectively one of mutual exclusion. If we consider the number of discrete locations where department stores or their branches exist within the Bay Area, 20 places, in relation to the total population of the whole region, we discover that an equal sharing of the population would result in an exclusive tributary area for each department store of 180,000. The average population within five miles of the 11 regional shopping centers is 237,000 but the range among individuals is great. Stonestown within San Francisco itself has 640,000 people within five miles while the shopping center at Concord at the very edge of the metropolis has no more than 82,000 people within five miles. Despite this great difference in population the two centers have 69 and 64 per cent of their patronage from within five miles. In general terms the volume of sales at a regional center increases as the density of population in its vicinity increases but its attraction at a distance does not appear to be greatly enhanced. All regional centers are primarily of local importance.

Turning to the establishment content of regional centers, which provides the obverse view to the size of tributary area in evaluating commercial location theory, it is possible to analyze a recent inventory of outlying shopping centers issued in 1959 by the *San Francisco Examiner* (1959). Table III contains the totals of establishments, by type, found in the 150 integrated centers of all sorts within the Bay Area. Of these, 11 are regional centers, 15 community centers, and 124 neighborhood centers. Recasting this simple inventory to correspond with the classification of commercial establishments used by

TABLE III

mber of Integrated Centers in the San Francisco Bay Area with Specified Types of Establishments

e of Establishment	*No. of Centers*	*Type of Establishment*	*No. of Centers*
ermarket	147	Furniture store	18
aning and laundry shop	120	Confectionery store	17
ber or beauty shop	119	Music store	17
g store	113	Post office	15
iety store	96	Sporting goods and hobby shop	15
taurant or bar	95	Stationery store	14
rdware or paint store	77	Weight reducing shop	13
ery or doughnut shop	66	Auto accessory store	12
uor store	66	Department store branch	12
men's apparel store	64	Junior department store	12
vice station	54	Pet shop	10
urance or real estate off.	54	Family clothing store	8
e store	52	Bowling or amusement	7
elry store	41	Sewing machine sales store	7
k branch	41	Book store	5
ildren's clothing store	40	Public Library branch	4
rdage, yarn, rugs	39	Stockbroker's office	4
e repair shop	36	Movie theatre	2
n's apparel store	33	Employment agency office	2
pliance or TV store	33	Automobile sales establishment	2
t shop	31	Automobile washing establishment	2
dical or dental office	29	Trailer rental agency	1
ces not otherwise counted	25	Office equipment store	1
shop	24	Newspaper publication office	1
nera and photography shop	23	Lawn mower repair shop	1
n or finance office	20	Church	1
rsery or florist shop	20		
ician office	19	*The total number of centers of all*	
d specialty shop	19	*types was*	*150*

rphy and Vance (1955. Appendix B) in study of American central business dists we may compare the overall content of lying and core shopping districts. Important contrasts show up in the percentage he total space used for retail and service les in the two types of commercial district l in the nature of the retail and service nponents. In the nine central business dists only 40 per cent of the total floor space s given over to retail and service space ereas in the 145 outlying centers over 95 per cent of the space was in retail and service uses. The office, transient residence, public and organizational, and manufacturing and industrial components of the central districts are little represented, with transient residence unrepresented in outlying centers and the other non-retail and service elements virtually absent. In the retail group food stores take up nearly three times as much space in the outlying centers as in the core even though restaurants fall within this class in both places. Household goods sales space is nearly twice as large in the downtown districts as in the outlying business areas mainly because of the large number and large size of downtown furniture stores. This class of establishment is unusual in integrated centers because of its low "rent-paying" character. The fact that automotive districts are part of the core area and usually separate in the suburbs accounts for the contrast between downtown and exterior with six times as much space used at the center. Miscellaneous sales, the group that includes most of the non-clothing specialty shops, is three times as large proportionally in the central business district as in shopping centers. Variety store space is 60 per cent larger in proportion to total retail and service space in the outlying areas mainly because of the great dominance of department store branches in the larg integrated centers.

Within the eleven regional centers th department store branch is notably dominan accounting for 45 per cent of all space. Ap parel and shoe stores rank next with 15 pe cent of regional center space. Other types commodity-combining stores, mainly variet and drug stores, account for 13 per cent the regional center space so that nearly thre quarters of all the floor space in these larg planned facilities is taken up with mas selling stores merchandising fashion but n "high-style" clothing, non-durable househo goods, and small durable household item Food stores are much less important in re gional as opposed to all outlying centers con prising only 11 per cent of the space in th

TABLE IV

Average Size of Establishments, by Type, for 150 Centers in the San Francisco Bay Area

Establishment Types	*Aver. Size*	*Establishment Types*	*Aver. Size*
Department store	155,133	Liquor store	1,787
Junior department store	29,774	Stationery store	1,684
Bowling or amusement pl.	23,913	Yardage, rug, or yarn	1,621
Supermarket	13,905	Music store	1,554
Furniture store	11,542	Gift shop	1,438
Automobile washing estab.	9,368	Camera or photography shop	1,269
Variety store	8,664	Appliance or TV store	1,265
Bank branch	4,900	Food specialty store	1,184
Drug store	4,476	Bakery or doughnut shop	1,137
Women's apparel store	3,652	Loan or finance offices	1,134
Men's apparel store	2,910	Insurance or real estate	1,129
Shoe Store	2,861	Confectionery store	1,122
Hardware or paint store	2,819	Jewellery store	1,115
Auto accessory store	2,600	Cleaning or laundry store	1,098
Post Office	2,454	Book store	1,045
Toy shop	2,268	Family clothing store	969
Sporting goods or hobby	2,180	Pet shop	820
Restaurant or bar	2,162	Barber or beauty shop	738
Children's clothing store	2,044	Shoe repair shop	596

Average size is given in square feet of rented building space
Source: *San Francisco Examiner* (1959).

ger centers. In number of establishments her than size apparel stores are dominant with a quarter of all shops. Food stores make up 15 per cent, shoe stores 12 per cent, and

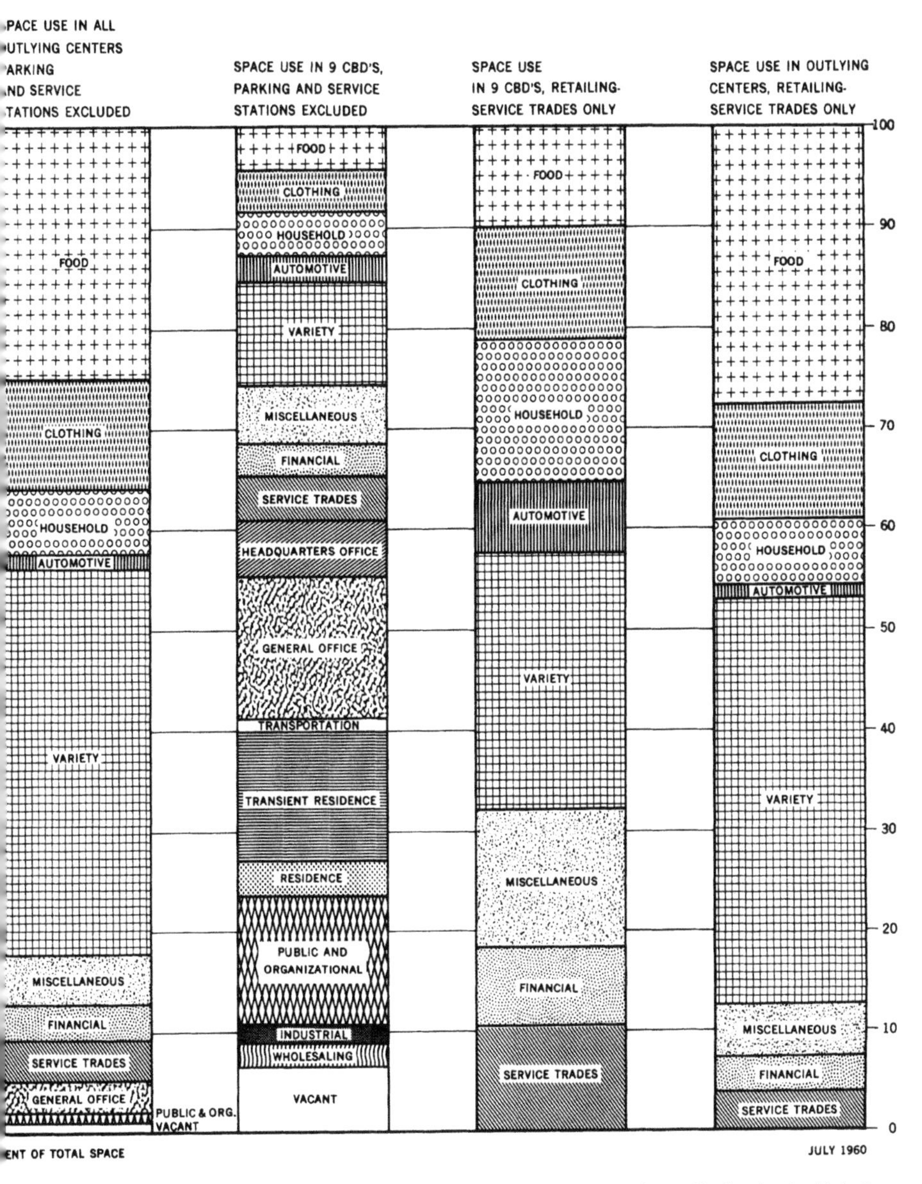

ure 5. Apportionment of space by major land use types in nine central business districts in the United es and eleven outlying integrated shoppng centers in the San Francisco Bay Area.

service trades shops such as barber and beauty shops, shoe repair, and laundry-dry cleaning stores are numerous. Only medical and dental professional offices are at all numerous and there are very few general offices. Each center has but a single department store, much as it usually accounts for half the floor space in the center.

The outlying center, even in the case of regional centers, is a place for selling highly standardized and rapidly disposed of goods to a local market. Within the service trades only most commonly sought personal and clothing services are found and there is no appreciable amount of business or office service. Financial services are restricted to personal loan companies, small branch banks, and a few stock broker's branch offices, along with a great number of real estate and insurance offices aimed at the youthful and home-buying inhabitants.

Conclusions

The purpose of this paper has been threefold: (1) to consider briefly what factors have in the past influenced the location and nature of American commercial districts, (2) to undertake to formulate the general location factors affecting commercial development today, and (3) to test this formulation against actual conditions in the highly diversified commercial structure of the Bay Area.

Throughout the history of commercial development in the United States there has been a constant effort to develop economically viable units of selling as close to the residence of customers as possible. If we relate this effort to Lösch's five conditions of equilibrium, we may say that there has been a continuing reduction in the size of trading areas to the point that we seem to have reached the minimum size economic area in the case of mass selling with the regional shopping cente It seems probable under the conditions c automotive transportation that much reduc tion in the size of tributory areas from tha presently encountered in a region such a the Bay Area would result in unprofitabl operations.

The consequence of this progressive cor traction of trading areas, or dispersion of sell ing, has been that only mass demands fc goods can be met within the basic (suburban trading areas. This is fully demonstrated b the absence of specialty selling on this loca level. The emerging pattern of commercia geography in the United States is the separa tion of mass selling from specialty selling wit the removal of the former from the centra business district save for that part whic serves the local population within the centra city. The result is that the expansion of sale and the consequent morphological growt which we would anticipate from expandin population, has occurred at the peripher where the condition of minimum tributar area for each sales place would dictate th siting of increments to the commercia structure.

The central business district will continu to exist but shorn of much of its former pu pose. A growing body of urbanists in th United States views the core as a place c office rather than of large stores. The focus c metropolitan arteries along with the func tional convenience of geographical integratio for offices tend to maintain the financial sub district at the core. This focus is in som places being reinforced by the introduction c rapid transit which meets the needs of th office district. But in the case of the retail sub district mass transit appears to hold much les promise. Shoppers at most visit the downtow infrequently and prefer the flexibility of driv ing to shop. The result is that the "best loca ı" for the specialty shops which remain in core is away from the focus of the journey-work in offices with its congestion and pre-pted parking. Although this is not the ce, much study should be given to the con-st between the two downtown functions, a trast which appears to introduce two her divergent trends within the core to the ent that today the two parts may be "pull- apart" in location to be separated by a nd of parking facilities severing the two. If trend continues, the single focus which in the past characterized the core will be t and we may find ourselves a generation m now restricting the term central business trict to the office area and coining a new ne, perhaps metropolitan specialty district, what we still by tradition consider the wntown business district.

In sum it seems that the central business trict has become the mass seller to the er part of the metropolis, the specialty ler to the geographical city, and the office a for the region. In turn, the regional inte-ted center has become the mass seller to individual suburb alone, with no other portant function, And through this change it seems that there has been an adjustment to the urban dynamics of transportation and settlement which has restored locational equilibrium in commercial structure after two decades of instability and doubt.

References

[1] Emmet and Jeuck, *Catalogues and Counters* (Chicago, 1950).

[2] Hoyt, H., "Classification and Significant Characteristics of Shopping Centers," *Appraisal Journal* (April, 1958).

[3] Lösch, A., *The Economics of Location*, trans. by W. H. Woglom (New Haven, 1954).

[4] Murphy, R. E. and Vance, J. E. Jr., *Central Business District Studies* (Worcester, 1955).

[5] Proudfoot, M., "City Retail Structure," *Economic Geography*, Vol. XIII (1937).

[6] Reilly, W. J., "Methods for the Study of Retail Relationships," University of Texas, *Bureau of Business Research*, Research Monograph No. 4 (1929).

[7] San Francisco CBD Bulletin, *Census of Business* (1954).

[8] *San Francisco Examiner*. Unified Shopping Centers, San Francisco Bay Market Area, 1959 (San Francisco, 1959).

[9] U.S. Bureau of the Census, *Retail Trade Bulletins* (1929, 1954).

21

The Management Cente
in the United State

William Goodwi

In the literature on the classification of cities little or no attention has been given to management per se or to management centers. A recent bibliography of central-place studies,[1] which lists the significant studies of tertiary activities, does not contain among its more than five hundred entries a single reference explicitly to management centers; and neither this work nor the extensive bibliographies in Isard's "Methods of Regional Analysis"[2] make any mention of management functions.

Among the better-known functional classifications of cities, none appears to regard management as a separate function. Harris[3] states that his classification "is based on the activity of greatest importance in each city. Functional importance, in the Harris classi fication, is measured mainly by the numbe of people employed in each industry. Harri recognizes nine classes of cities but does nc include management among their activities.

More recently, Nelson has also presented functional classification of American cities, based, like Harris's, on United States censu categories. Although Nelson includes a wide range of activities than Harris, he makes n mention of management as a separate cate gory. Alexandersson,[5] in his comprehensiv analysis of the functional role of America cities, which utilizes the concepts of "cit forming" and "city serving," evidently doe not regard management as either and there fore does not include it in the classificatior Hart's study of the cities of the America

[1] Brian J. L. Berry and Allan Pred, "Central Place Studies: A Bibliography of Theory and Applications," *Bibliography Ser. No. 1*, (Philadelphia, Regional Science Research Institute, 1961).

[2] Walter Isard, *Methods of Regional Analysis* (Cambridge, Mass., New York and London, 1960).

[3] Chauncy D. Harris, "A Functional Classification of Cities in the United States," *Geogr. Rev.*, Vol. XXXIII (1943), pp. 86-99; reference on p. 86.

[4] Howard J. Nelson, "A Service Classification c American Cities," *Econ. Geogr.*, Vol. XXX (1955), pp. 189-210.

[5] Gunnar Alexandersson, *The Industrial Structur of American Cities* (Lincoln, Nebr., and Stock holm, 1956).

Adapted from *Geographical Review*, Vol. LV, no. 1 (1965), pp. 1-16.

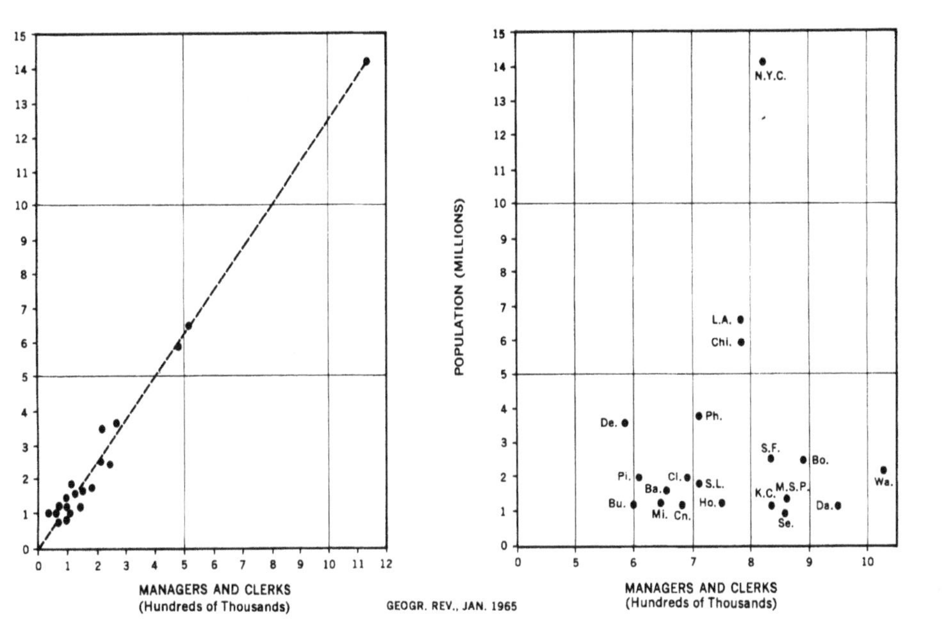

ıre 1. Managers (except farm managers) clerks in relation to total population for the ıty largest urbanized areas in the United States. rce: 1960 Census of Population and 1958 sus of Manufactures.

Figure 2. Managers (except farm managers) and clerks per 100,000 workers in relation to population for the twenty largest urbanized areas in the United States. Key: N.Y.C., New York City; L.A., Los Angeles; Chi., Chicago; De., Detroit; Ph., Philadelphia; Pi., Pittsburgh; Bu., Buffalo; Mi., Milwaukee; Ba., Baltimore; Cn., Cincinnati; Cl., Cleveland; S.L., St. Louis; Ho., Houston; K.C., Kansas City; Se., Seattle; M.-St.P., Minneapolis-St. Paul; S.F., San Francisco; Bo., Boston; Da., Dallas; W., Washington, D.C. The horizontal scale has been expanded to permit clarity in reading the graph. Source: 1960 Census of Population; 1958 Census of Manufactures.

th[6] follows the general pattern set by rris; it likewise makes no mention of the ortance of management and fails to ıtify any city as a management center.

n "Metropolis and Region" Duncan and associates classify standard metropolitan as with 300,000 inhabitants or more in 0 into seven categories according to etropolitan functions and regional rela-ships."[7] Although these investigators did not consider management as such, their grouping of the SMA's appears to correspond closely with the grouping resulting from the work presented in this paper.

The omission of management in the classifications reviewed above is readily explained

ohn Fraser Hart, "Functions and Occupational tructures of Cities of the American South," *nnals Assn. of Amer. Geogrs.*, Vol. XLV 1955), pp. 269-86.

)tis Dudley Duncan and others, *Metropolis nd Region* (Baltimore, 1960), pp. 259-75. 'he seven categories are as follows: National Ietropolis (N); Regional Metropolis (R); Re-ional Capital, Submetropolitan (C); Manu-acturing, three classes (D, D-, and M); and Special Cases (S). These are based on a scattergram that plots per capita value added by manufacture against per capita wholesale sales; the third dimension of population is indicated by the size of the circle (pp. 264 and 271).

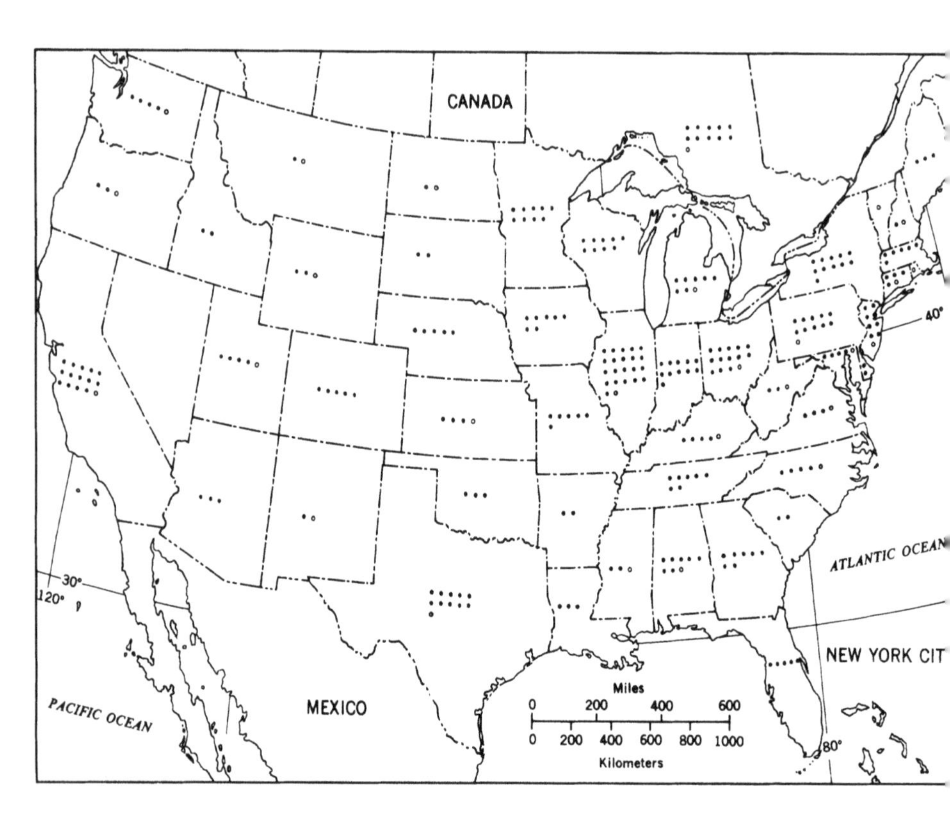

Figure 3. The number of companies, by states, whose headquarters offices are located in New York. A solid dot represents two plants; a circle, one plant. Source: *Moody's Industrials,* 1962; *Thomas' Register of American Manufacturers,* 1961.

by the fact that their basis was the most important "function" as measured by employment. By this criterion the number of people engaged in management is not large enough to be of significance. Moreover, although it is possible to extract "managers" from the census data, the category is too inclusive to be satisfactory. In any case, it is questionable whether or not employment is a proper measure of management activities.

It is hoped that the present paper may contribute toward filling the gap in the literature by identifying the cities in which management is important. Admittedly, the methods used are not wholly satisfactory, but it is believed that they are a step in the righ direction.

Management and Management Centers

Management is an idea-handling, not materials-handling, function, and as such is somewhat intangible. Vernon[8] points o that "whereas manufacturing, transportatio retail trade, and wholesale trade are eco nomic activities whose existence is easi

[8] Raymond Vernon, "The Changing Econom Function of the Central City," *Supplementa Paper No. 1,* Area Development Committee CED; Committee for Economic Developme (New York, 1959), p. 55.

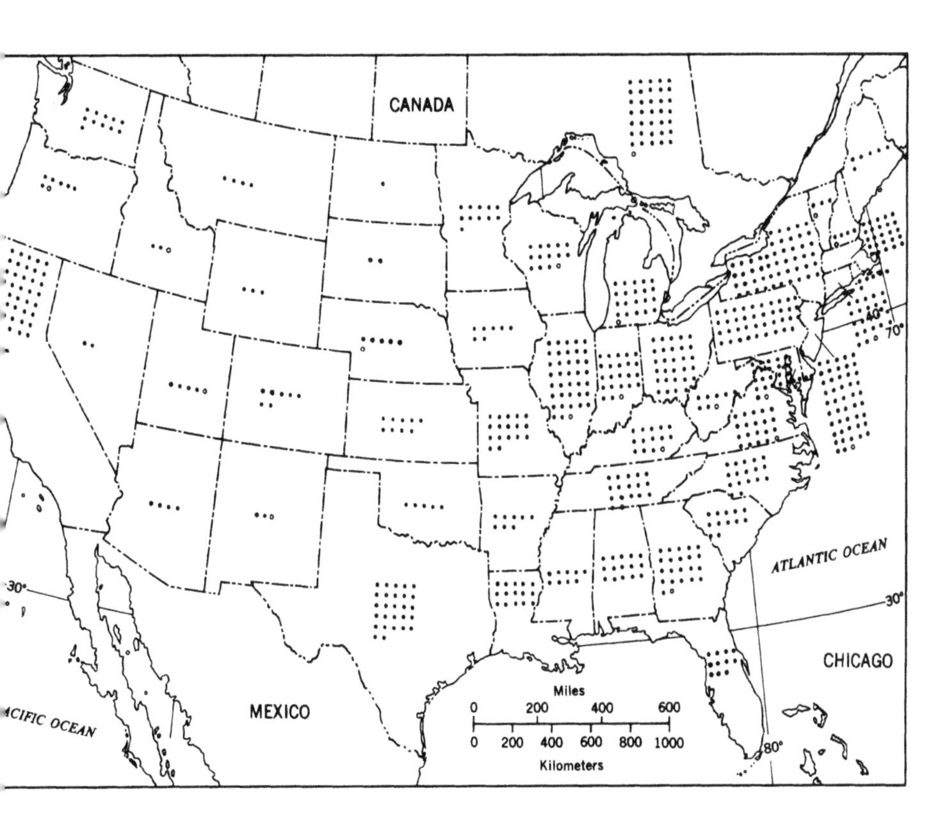

ure 4. The number of companies, by states, whose headquarters offices are located in Chicago. olid dot represents two plants; a circle, one plant. Source: *Moody's Industrials,* 1962; *Thomas' Register* *American Manufacturers,* 1961.

ognized and catalogued, many aspects of ce activity are more difficult to classify." nagerial operations are office activities if hing else. As part of the national and local ban scene, managers, the offices they cupy, and the distribution of these offices ite investigation.

Vernon also observes:[9] "To the extent that office function grows, therefore, the wth may well occur to a disproportionate ent in the office districts of the *larger tral cities*, at the expense of the regional centers. The possibility [is] that only the largest cities may be the principal beneficiaries of continued office growth — indeed, . . . they may be the only beneficiaries." If, as it appears, office functions and, particularly, the headquarters offices of nationally important companies are to continue to gravitate to the already existing office centers, it is pertinent to establish which are the presently important cities.

In the past twenty years the electronic computer has grown from a curiosity to a much-used tool of management. Many routine decisions are programmed for electronic computers; many data are processed by punch

Vernon, *op. cit.*, p. 60. Italics are the present writer's.

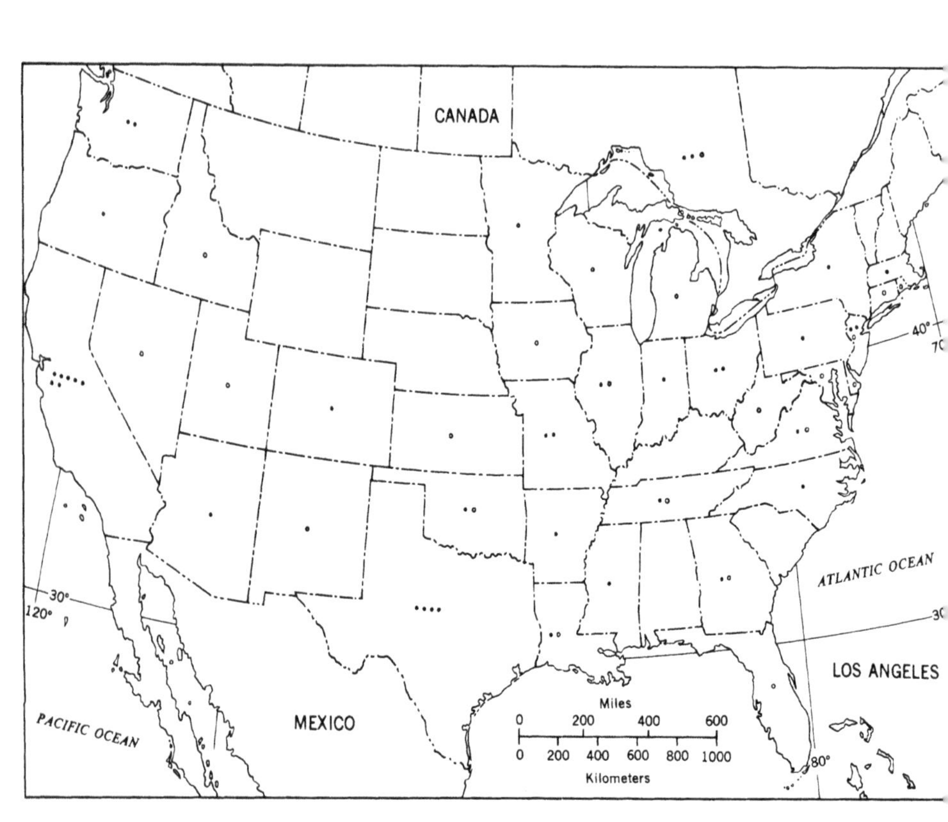

Figure 5. The number of companies, by states, whose headquarters offices are located in Los Angeles. A solid dot represents two plants; a circle, one plant. Source: *Moody's Industrials*, 1962; *Thomas' Register of American Manufacturers*, 1961.

cards rather than by pencil, paper, and desk calculators. This "revolution" has reduced the number of clerks needed to prepare the raw materials for decision making. The change now taking place in the mechanics of decision making may have either of two opposite results with respect to the location of management centers. On the one hand, the reduction in the number of employees needed to staff a headquarters office may hasten the concentration of decision making in a few locations; on the other hand, it may well mean dispersion of headquarters because of the flexibility of data flows through a computer. At the moment it is not clear just what effect the rapid introduction of electronic data processing will have on the concentration o[f] office functions, but one must be aware of it[s] great potential for changing the pattern c[of] "office" cities.

Although the making of decisions is th[e] function of only a very small part of th[e] American labor force, the influence of th[e] decision makers on the social and economi[c] welfare of the nation is enormous. The day[-]to-day decisions of the executives of the larg[-]est businesses, together with the decision[s] made in Washington, D.C., determine th[e] course of economic events in the country — and, to no small degree, in the rest of th[e] world as well.

A management center may therefore b[e]

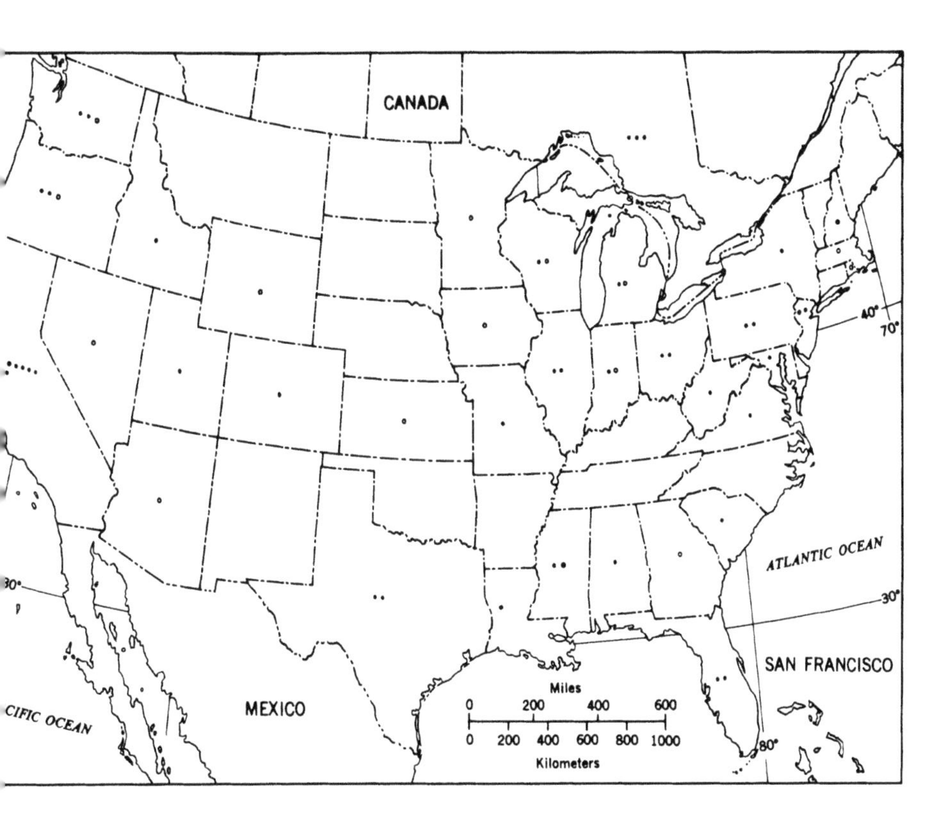

ure 6. The number of companies, by states, whose headquarters offices are located in San Francisco. olid dot represents two plants; a circle, one plant. Source: *Moody's Industrials,* 1962; *Thomas' Register American Manufacturers,* 1961.

ined as a city in which there is a *concentration of headquarters offices of nationally portant companies.* It is a place apart from production centers. The people identified "managers" are those who sell their manerial talents irrespective of the nature of company that employs them. A management center is a reservoir of managerial talent ailable for hire.

anagement and Population

w York City would unquestionably be acrded the position of prime management nter of the United States, but which other ies exhibit the same general characteristics New York?

Intuitively, and probably by common assumption, the relative importance of a city as a business-management center is considered to have a more or less positive linear relation to the size of its population. The most readily available measure of management would be the count of managers and clerks made by the census. Figure 1 plots the total number of managers and clerks for the twenty largest (according to population) urbanized areas of the United States against the populations. Clearly a straight-line relationship exists.

However, when the relation of managers and clerks to all other workers is calculated and the result plotted against the population (Figure 2), no such straight-line relationship

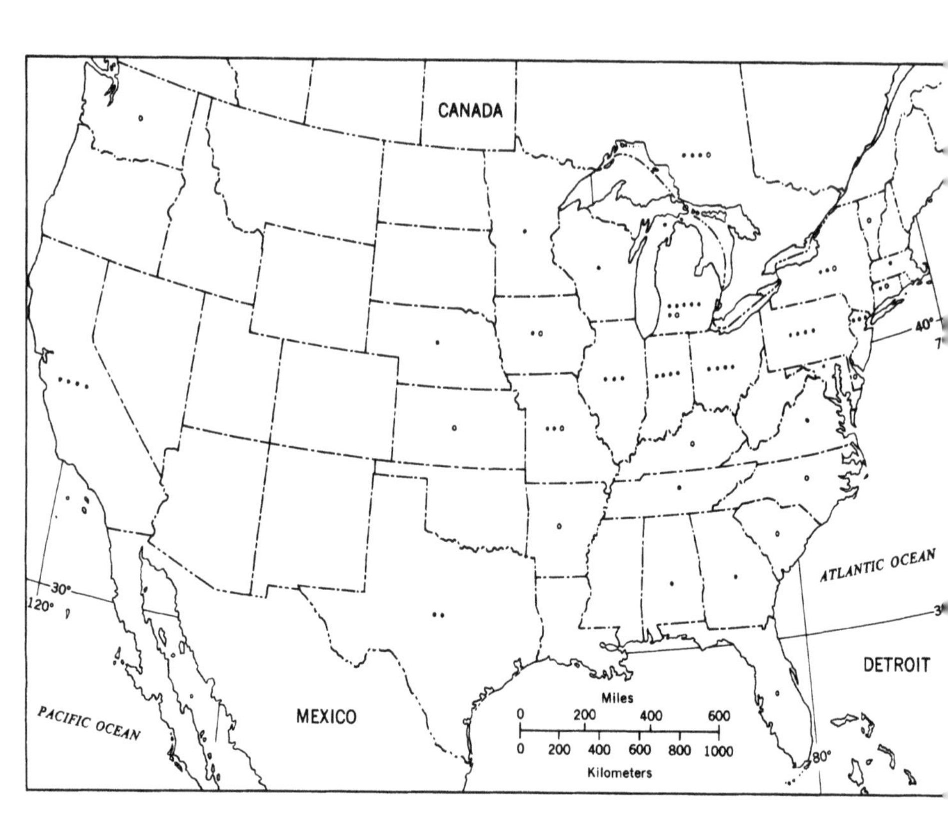

Figure 7. The number of companies, by states, whose headquarters offices are located in Detroit. A solid dot represents two plants; a circle, one plant. Source: *Moody's Industrials,* 1962; *Thomas' Register of American Manufacturers,* 1961.

emerges. Rather, a distribution of cities can be observed from left to right across the graph, from the industrial centers of Detroit, Pittsburgh, and Buffalo to the purely administrative center of Washington, D.C. Thus it would seem that the intuitive view is both confirmed and questioned, and one is led to further consideration of the question.

Sources of Data

It can be assumed that large companies require a large managerial personnel, and therefore the concentration of large companies in a city is indicative of managerial concentration also. Relevant information about individual companies was assembled and summarized, city by city. Publicly owned companies are required to publish annual reports of their financial status, and these reports carry a wealth of additional information. The business publications, such as *Moody's,*[10] and the *Standard and Poor* investors reference manuals[11] publish annually the information available about companies. *Fortune* has made a

[10] Published by Moody's Investors Service, New York.

[11] Published by Standard and Poor's Corporation, New York.

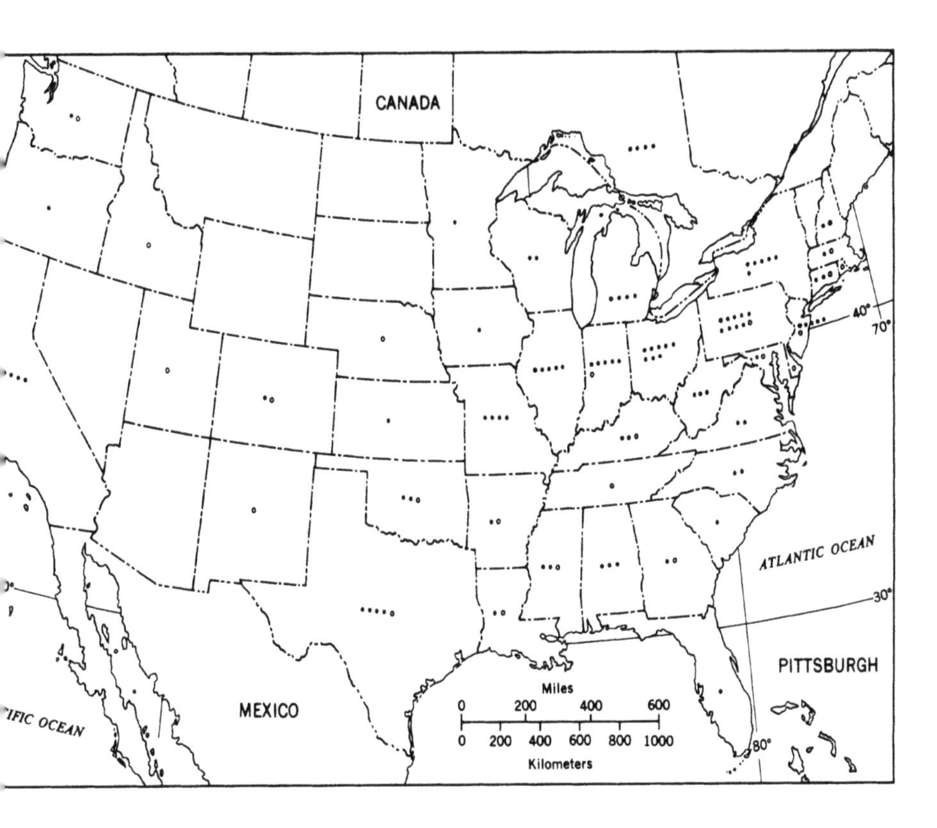

ure 8. The number of companies, by states, whose headquarters offices are located in Pittsburgh. olid dot represents two plants; a circle, one plant. Source: *Moody's Industrials,* 1962; *Thomas' Register merican Manufacturers,* 1961.

vey of America's largest companies, both ustrial and nonindustrial, and for the past e years has published annually (July and gust) a list of America's five hundred larg- industrial companies, and for the last five rs a list of the fifty largest commercial ks, insurance, transportation, utility, and rchandising companies respectively.

ince it is presumed that only companies national importance can collectively create anagement center of national importance, 750 companies listed in the *Fortune* sur- served as a base. Each company was cked against one or both of the standard estors manuals and also against *Thomas' Register of American Manufacturers,*[12] 1961 edition, for accuracy and to determine the location of the production facilities of the company. The data used in this study refer to the 1962 status of the companies. This list, published in 1963, was first compared with the lists published in earlier years[13] to deter- mine whether there had been significant shifts

[12] Published annually by the Thomas Publishing Company, New York.

[13] A careful company-by-company comparison was made with the annual surveys for 1955, 1957, 1959, 1961, and 1963. It was disappointing that no significant trends in concentration were dis- cernible.

in the composition. Only the companies in the lower ranks appear to have changed completely, though naturally there has been some shifting of relative importance over the years. It was felt, however, that changes over the last ten years had not been sufficient to warrant taking earlier years into account, because the appearance or disappearance of companies from the list did not modify substantially the relative importance of a city as a management center. Also, comparison of the two years 1954 and 1963 did not reveal any substantial geographical shift in management centers.

Results of the Analysis

The results of this investigation are most easily presented in a series of tables. Because of the limited scope of the data, these results are certainly neither final nor conclusive, and undoubtedly other means of attacking the problem need exploration.

Table I presents data concerning the 500 leading industrial companies of America aggregated according to the locations of their headquarters cities. Three reasonable, if arbitrarily chosen, criteria served to determine the cities that seemed to meet the test of national importance as management centers: (1) at least ten headquarters offices, (2) at least two billion dollars in assets or sales in 1962 (about 1 percent of the total assets and sales of the 500 companies), and (3) at least 100,000 employees (about 1 percent of the total employment of the 500 companies). Centers included in the original compilation but eliminated from the table are Akron, Minneapolis-St. Paul, Wilmington, Bethlehem-Allentown, Seattle, Cincinnati, Dallas, Toledo, Milwaukee, and Bartlesville (Oklahoma). Although these cities had one or both of the other qualifications, each failed to have more than ten offices.

It is not enough that a city has a large co
centration of offices; these offices should re
resent widespread control, both spatially ar
industrially, and separation from productic
facilities. To determine the extent of the "en
pire" controlled from each city, a tabulatic
was made of the states in which each con
pany with headquarters in that city ha
located one or more plants, and the tota
were summed. For example, 83 compani
with headquarters in New York City had o
or more plants in California. However, f
determining the extent of geographical co
trol of a city the total number of plants w
not considered necessary, merely the numb
of companies with branch plants in the var
ous states. A city whose control did not e
tend to at least twenty-four states and Canad
was not regarded as a management cente
Only New York City and Chicago compani
are represented in all forty-eight states of th
conterminous United States and in Canad
the most poorly represented is Clevelan
with twenty-nine states and Canada. Table
summarizes this information.

A more satisfactory view of the extent c
the control of a city can be gained from a ma
on which have been plotted the basic dat
from which Table II was constructed. Si
such maps are presented, for New York, Ch
cago, Los Angeles, San Francisco, Detroi
and Pittsburgh (Figures 3-8). The impo
tance of New York and Chicago as head
quarters cities emerges clearly from thes
maps and contrasts strongly with that of th
West Coast centers of Los Angeles and Sa
Francisco. Detroit and Pittsburgh, notab
manufacturing centers, appear to have the
control principally in the industrial section
of the country.

Although a city need not necessarily hav
a diversity of industries to qualify as a man
agement center, undoubtedly a larger pool c
broader managerial talents would be availab

TABLE I

Concentration of Industrial Headquarters Offices

;ity	Number of Offices	Sales (In thousands of dollars)	Assets (In thousands of dollars)	Number of Employees
v York	163	84,355,806	124,477,751	3,650,089
:roit	13	28,801,564	19,319,678	1,594,487
cago	51	19,798,326	14,300,775	717,541
sburgh	21	10,760,815	12,031,841	439,313
Angeles	16	8,361,378	5,698,259	307,056
Francisco	14	5,690,201	6,973,327	206,227
ladelphia	16	4,423,190	4,372,831	171,613
Louis	12	4,229,250	3,177,712	185,734
veland	15	4,172,215	3,512,877	188,902
ston	12	2,140,787	1,503,882	118,358

npiled from *Fortune*, July, 1963 and *Moody's Industrials*, 1962 edition.

TABLE II

Extent of "Empire" by City of Control, Conterminous United States and Canada*

Headquarters City	No. of States with Branch Plants	Total No. of Plants	Headquarters City	No. of States with Branch Plants	Total No. of Plants
w York City	48	1,455	San Francisco	34	97
icago	48	528	Wilmington[a]	34	51
nneapolis-St. Paul[a]	43	128	Detroit	30	112
:sburgh	40	212	Philadelphia	30	82
Louis	38	125	Cleveland	29	99
Angeles	38	98	Boston[b]	21	48

rce: *Moody's Industrial*, 1962 edition; *Thomas' Register of American Manufacturers*, 1961.

\ll cities listed have a plant or plants in Canada.

Iot included in Table I because of failure to meet requirements, but included here for completeness nd the fact that it is industrially significant.

'he small extent of its industrial empire eliminates Boston as an industrial management center, in ccordance with the criterion.

a city with many different industries than a one-industry city. Other things being ıal, a city with a wide range of nationally portant companies is better qualified to im a position of rank among managerial iters. An attempt was made to classify all npanies studied according to the United States Standard Industrial Classification (Table III), but only at the two-digit level.[14]

[14] In the Standard Industrial Classification (SIC) all industries are divided into 9 major groups at the one-digit level, into 99 groups at the two-digit level, and into 999 groups at the three-digit level.

TABLE III

Number of Companies According to United States Standard Industrial Classification

Management Center	Sic Number																				
	20	21	22	23	24	25	26	27	28	29	30	31	32	33	34	35	36	37	38	39	U
New York	19	5	9	2	1	1	10	5	30	6	4	0	2	13	12	11	12	7	5	0	9
Chicago	14	0	0	1	0	0	2	0	4	1	1	0	1	2	5	4	9	3	1	0	3
Pittsburgh	1	0	0	0	0	0	0	0	1	2	0	0	1	6	2	3	2	0	0	0	3
Los Angeles	2	0	0	0	0	0	0	1	2	4	0	0	0	0	0	1	0	4	0	0	2
Philadelphia	0	0	1	0	0	0	2	0	2	2	0	0	1	0	0	1	2	2	0	0	3
Cleveland	0	0	0	0	0	0	0	0	4	1	0	0	0	4	1	4	0	1	0	0	0
San Francisco	3	0	0	0	2	0	1	0	1	0	0	0	0	1	1	1	2	0	0	0	2
Detroit	1	0	0	0	0	0	0	0	1	0	0	0	0	2	1	2	1	5	0	0	0
St. Louis	4	0	0	0	0	0	0	0	3	0	0	2	0	0	0	0	2	1	0	0	0
Boston	1	0	1	0	0	0	0	0	2	0	1	0	1	0	1	1	1	0	2	0	1
Minneapolis-St. Paul	3	0	0	0	0	0	1	0	0	0	0	0	0	0	1	1	1	0	0	1	0

[a]Unknown.
Code: 20, Food; 21, Tobacco; 22, Textiles; 23, Apparel; 24, Lumber and wood products; 25, Furniture and fixtures; 26, Paper and allied products; 27, Printing and publishing; 28, Chemicals; 29, Petroleum and coal; 30, Rubber and plastics; 31, Leather; 32, Stone, clay and glas 33, Primary metals; 34, Fabricated metal 35, Machinery (except electrical); 36, Electric machinery; 37, Transportation equipment; 3: Instruments; 39, Miscellaneous manufacturing.

The internal diversity of many companies makes it impossible to use a more detailed category, and for a few it was not possible to use even the two-digit categories; these companies are listed as unknown. Where possible, the company was assigned to a class according to its principal product. A summary appears in Table III. Clearly, certain cities that rank high in Tables I and II have little diversity in production, and consequently little diversity in management talent.

Ranking of Industrial Management Centers

If it can be assumed that the four criteria of Table I have equal value as measures of management concentration and that the extent of areal control and diversity are also of about the same value, the cities can be tentatively ranked (Table IV) by a simple averaging of the total rankings of each city. Akron, Wilmington, and Bethlehem-Allentown, with sma "empires" and little diversity but importar for particular products, are not general management centers, even industrially.

Nonindustrial Position

The importance of a city as a managemen center has so far been confined to its impor tance as a center of industrial management but this is inadequate, for it does not take int account other business activities. For example, the large commercial banks and th insurance companies exercise an all-pervasiv influence as major sources of funds that recognize no visible bounds. Legal restriction and banking and insurance laws place limit on the tangible evidence of control, but th size of assets alone is indicative of relativ importance. Few transportation companie are national in scope, but the larger airline

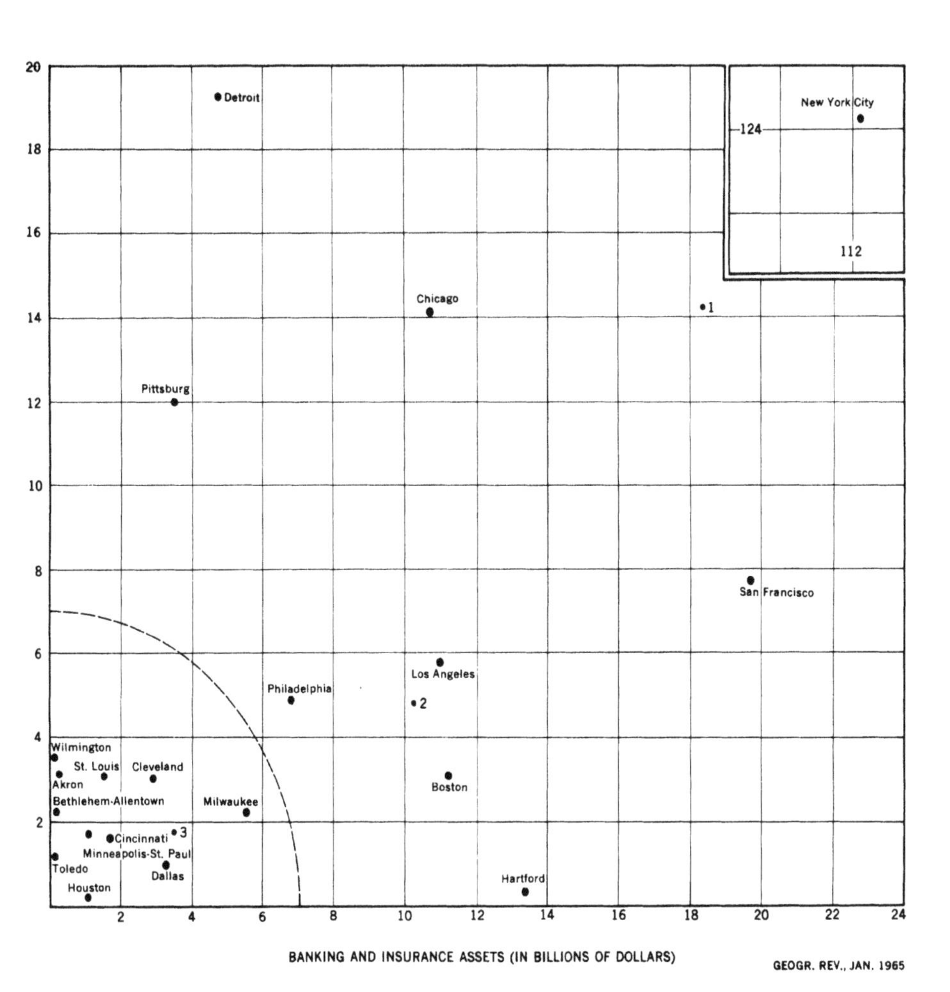

ure 9. Industrial assets plotted against banking and insurance assets. The numerals 1, 2, and 3 represent positions that Chicago, Philadelphia, and Minneapolis-St. Paul respectively would have had if the ets of transportation companies had been included in the nonindustrial assets. The arc in the lower left ner roughly delimits the nationally important centers from the centers whose influence is more restricted.

d railroads serve large sections of the coun- and may therefore be considered of na- nal importance. Utility companies, with e exception of American Telephone and legraph and Western Union, are commonly cal, both in service area and in outlook; erefore only the national service utilities re included in the computations. Inasmuch the sole common denominator among the various kinds of business activity is assets, this criterion was used, though "policies in force" or "policyholders" might have provided a better measure for insurance companies. Table V presents a compilation of the nonindustrial assets of the cities in the same fashion that the industrial data were summarized.

Many of the important industrial cities are

also important nonindustrial centers, but Hartford, Milwaukee, Houston, and Dallas, as measured by assets, are obviously more important as financial centers than as industrial centers. Thus a second criterion of ran[k] is suggested, namely the total of nonindustria[l] assets. In addition to repeated proof of th[e] importance of New York City and Chicago a

TABLE IV

Ranking of Cities as Industrial Management Centers

City	Rank						
	No. of Offices	*Sales*	*Assets*	*No. of Employees*	*No. of Plants*	*Diversity*	*Averag[e] Rank*
New York	1	1	1	1	1	1	1
Chicago	2	3	3	3	2	2	2
Pittsburgh	3	4	4	4	3	4	3
Detroit	8	2	2	2	5	6	4
San Francisco	7	6	5	6	8	4	5
Los Angeles	4½	5	6	5	7	9	6
Philadelphia	4½	7	7	9	9	4	7
Cleveland	6	9	8	7	6	7	8
St. Louis	9	8	9	8	4	9	9

TABLE V

Nonindustrial Assets, Fifteen Largest Centers
(In billions of dollars)

City	Assets				
	Bank	*Insurance*	*Transport*	*Utility*	*Total*
New York	49.0	63.6	4.9	36.0	153.5
San Francisco	19.9	—	2.8	—	22.7
Chicago	9.9	0.9	6.0	0.7	17.5
Hartford	—	13.5	—	—	13.5
Boston	2.0	9.2	—	—	11.2
Los Angeles	8.8	1.6	0.1	—	10.5
Philadelphia	3.6	3.3	3.2	—	10.1
Cleveland	3.5	—	2.2	—	5.7
Milwaukee	0.9	4.5	—	—	5.4
Houston	0.9	—	—	3.9	4.8
Detroit	4.6	—	—	—	4.6
Pittsburgh	3.6	—	—	—	3.6
Dallas	2.5	0.9	0.1	—	3.5
Minneapolis-St. Paul	0.7	0.4	2.4	—	3.5
St. Louis	1.5	—	1.9	—	3.4

Source: *Fortune*, August, 1963; *Moody's Financials*, 1962 edition.

trol centers, the great banking wealth of Francisco is pointed up, together with insurance centers noted above.

Vith the two rankings of importance be- us, it is now necessary to attempt to ıbine them into a single measure. The one of data common to all economic activities to all cities is assets. Assets, whether of a iness firm or of a bank, are commonly ›gnized as a measure of importance. Table combines all assets, industrial and non- ıstrial, for each city with assets of more than one billion dollars and ranks the cities accordingly.

The relation between the industrial and nonindustrial assets can be analyzed both arithmetically and graphically. The simple ratio of nonindustrial to industrial assets was computed (Table VII), and the relative importance of the cities was graphed (Figure 9). The graph serves as a basis for qualifying the cities as balanced in management control or more important industrially or financially. If the graph is read clockwise from the vertical

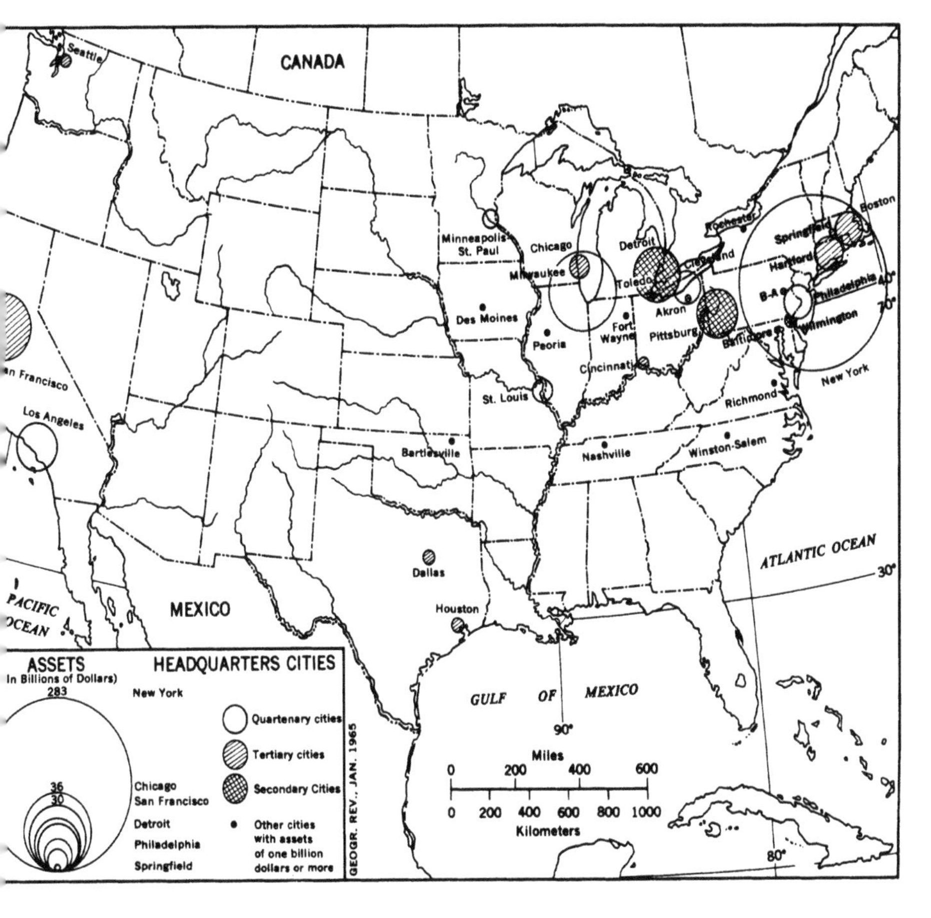

ıre 10. Headquarters cities. This map is based on the total assets held by major companies in the cities shown. The cities have been differentiated into three major categories according to the source of r power.

TABLE VI

Total Assets Controlled by Companies in Management Centers*
(In billions of dollars)

Rank	*City*	*Assets*
1	New York	283.3
2	Chicago	36.5
3	San Francisco	30.2
4	Detroit	24.7
5	Los Angeles	16.4
6	Pittsburgh	15.7
7	Philadelphia	14.4
8	Hartford	13.5
9	Boston	13.0
10	Cleveland	9.2
11	St. Louis	7.0
12	Houston	6.7
13	Milwaukee	6.6
14	Minneapolis-St. Paul	5.4
15	Dallas	4.5
16	Seattle	3.8
17	Wilmington	3.5
18	Cincinnati	3.3
19	Akron	3.1
20	Springfield (Mass.)	2.7
21	Bethlehem-Allentown	2.4
22	Winston-Salem	2.0
23	Des Moines	1.9
24	Bartlesville	1.7
25	Fort Wayne	1.6
26	Richmond	1.5
27	Toledo	1.2
28	Rochester	1.2
29	Peoria	1.14
30	Baltimore	1.12
31	Nashville	1.0

Compiled from *Fortune*, *Moody's*, *Standard and Poor*.

* Includes assets held by national merchandising companies but not used in previous calculations. Their inclusion here is to indicate full economic power and thus the magnitude of management control centred in each city.

TABLE VII

Ratio of Nonindustrial to Industrial Assets

City	*Ratio*
Akron	0.16[a]
Toledo	0.16[a]
Bethlehem-Allentown	0.21[a]
Detroit	0.26
Pittsburgh	0.31
St. Louis	1.06
Cleveland	1.48
Cincinnati	1.54
Chicago	1.58
New York	1.82
Los Angeles	1.89
Minneapolis-St. Paul	1.98
Philadelphia	2.35
Houston	3.00
San Francisco	3.34
Milwaukee	3.86
Dallas	4.00
Boston	7.35
Hartford	13.50

[a]Estimate based on assignment of nonindustria assets as determined by summing the assets c banks and the like within these as recorded i *Moody's*. Nonindustrial assets include assets c nationally important merchandising companie

axis, the cities are seen to be arrayed in orde from those dominated by industry to thos which are primarily financial centers. In addi tion, the graph shows the relative importanc of each city, at least with regard to assets hel by its major companies, and thus, if assets c large companies are a reliable measure o management, their importance as manage ment centers. A similarity is apparent be tween Figures 2 and 9. It would seem that th relation of managers and clerks to all worker is indicative of a management center — a assumption confirmed by the analysis in thi paper.

pes of Management Centers

ree distinct groups of cities emerge: (1) eral management centers of two sizes, ional and national, called by Gottmann[15] *aternary centers;* (2) financial cities of ional and national importance, here label- *tertiary centers;* and (3) industrial cities, which the bulk of the wealth and control n industry, here called *secondary centers.* Figure 10 identifies all cities with assets of billion dollars or more held by nationally portant companies (Table VI). The nty-one largest are identified according to ir relative importance as quaternary, terti- , or secondary. It is clear that few cities be called national management centers. e dominance of New York is overwhelm- ; its total assets are nearly eight times se of its nearest rival, Chicago. No other city can approach the diversity of operations performed, and it must be remembered that using only the five hundred largest industrial companies and the fifty largest of each kind of major nonindustrial company has not revealed the full importance of the city. In the quaternary group, New York, Chicago, Los Angeles, and Philadelphia can be considered national management centers (though the case for Los Angeles is weakened by its heavy dependence on the aircraft industry for the bulk of its industrial assets), and Cleveland, St. Louis, Minneapolis-St. Paul, and Cincinnati regional management centers. In the tertiary group, San Francisco, Hartford, and Boston must be classed as nationally important; Milwaukee, Dallas, and Springfield (Massachusetts), all with important insurance companies dominating the pattern, are of lower rank, as is Seattle. The secondary centers are Detroit, Pittsburgh, Akron, Bethlehem-Allentown, and Wilmington.

ean Gottmann, *Megalopolis* (New York, 1961), . 576.

22

Public Policy and the Central Business District

Ronald R. Boyce

The nature, future place, and importance of the *central business district*, or CBD as it is commonly called, is a subject of extreme controversy. Some say it is no longer necessary — indeed, is an anachronism. For example, the geographer Edward Ullman said, "If we were to apply private enterprise depreciation principles to the inner portions of cities we would write them off, just as machinery is scrapped, and throw them away. But where would we throw them?"

Others claim that the CBD is so vital an organism in the metropolitan anatomy that any city without a healthy CBD is dead or in danger of dying. They argue that the CBD should be restored to its former and rightful place as the heart of the metropolis — indeed, should surpass anything it was in the past. Charles Abrams, a planner, recently claimed that "without the CBD the suburbs cannot exist" because they are not viable without it. In this light the CBD is looked upon as an opportunity to build a truly representative symbol of our urban civilization.

These two different conclusions result largely from two very diverse perceptions as to what the city is. The social scientist, on th one hand, views the city as a laboratory fo analysis, as a phenomenon which primaril serves and reflects man's needs and tech nology. He sees the city as population clus tered tightly together in order to serve bette the assembly, production, service, and distri bution needs of its inhabitants; he sees it as tightly knit web of spatial, economic, an social interconnections. Melvin Webbe former president of the American Institute o Planners, recently stated that "the history c city growth, in essence, is the story of man' eager search for ease of human interaction. Viewed in this light cities, and the CBD, ar expected to change and to adjust to man' changing technology and needs.

This perspective of the city is vastly dif ferent from that of those who view the city a an artifact, or as an ideal expression of ou civilization. Thus, the architect-designer i ever proposing utopian, or ideal, urba designs. In this context the city must hav order, beauty, harmony, and symbolic mean ing as an entity. Each structure should com plement all others in a vast symphony of con

Reprinted from *Journal of Geography* (May, 1969), pp. 227-32 by permission.

:te and pattern. The city is viewed as a gle expression with finite boundaries and cernible internal sub-units. Urban sprawl therefore treated as a disease. This phil- ɔphy was perhaps best expressed by the e architect-city planner Eliel Saarinen, who d, "Just as any living organism can be althy only when that organism is a product nature's art in accordance with the basic nciples of nature's architecture, exactly for : same reason town or city can be healthy physically, spiritually, and culturally — ly when it is developed into a product of ın's art in accordance with the basic prin- ɔles of man's architecture."

Almost all concerned are in complete reement that the CBD is unsuited to pre- ıt needs and, furthermore, has been rapidly sing its monopolistic and dominant position the metropolis. This is demonstrated by the pid and continuing decline in CBD retail les, by the dilapidated and dis-functional ndition of many downtowns, and by the ntinual erosion and decentralization of tivities to outlying locations.

This article attempts to do three things: to ɔk objectively at the major assets and de- its of the CBD in terms of the two stated rspectives, to attempt to pose some reasons r the many problems and trends which are ecting the CBD, and to present conclusions to the public policy which I think should adopted for the American central business strict.

ssets Of The CBD

ıe central business district is an outstanding set for at least two reasons: first, it con- ins a concentration of activities and em- oyment, and it exerts control over the urban bric. Second, it has symbolic, cultural, and storical value. The first asset is one perhaps ost appreciated by the geographer; the second one is most appreciated by the architect-designer.

The concentration of activities and employment is demonstrated by the fact that well over one-half of all employment in most central cities occurs, in, or very near, the central business district. About 80 percent of all department store sales, and about 90 percent of all banking occurs in the CBD of even the largest metropolises. The small sub-CBD nodes of Wall Street, LaSalle Street, and Market Street undoubtedly control the finances of much of the nation. The daytime population in downtown Chicago, for example, amounts to almost 300,000 persons — more than the total population of Greater Des Moines, Iowa. Such concentration is reflected in the value of land and buildings in the CBD, which often amounts to upwards of 15 to 25 percent of the physical value of the entire city. Over one million dollars an acre was recently paid merely for air rights in downtown Chicago; $10,000 a front foot is not an unusually high price for CBD land in our largest cities.

The cultural and historical value of downtown is equally impressive. It contains the great hotels, restaurants, night clubs, movie houses, and theaters. In addition, it represents the initial beginnings of the city and contains the historical buildings and places. It is the area which most people associate with any given city. In short, it is the distinctive attribute of the metropolis. The subdivisions and industrial parks look much the same from city to city, but the downtown is different. It is most representative of the character and nature of any given city. Because of this and other reasons, a business location in the CBD carries with it great prestige. By the same token, the vitality of the downtown is often taken as the bellwether of a city's growth by the casual visitor.

Deficits And Problems

If the assets of downtowns are impressive, the deficits are even more so and are surely the reason for paying so much attention to the CBD. The major difficulties or problems of the CBD are primarily related to its inability to adjust to new needs. This inability is most clearly reflected in the problems of obsolescence. The buildings of most downtowns date back a half century — before the motor car — as do their streets and general physical layout. As a consequence such structures and blocks are not suitable for many of today's space needs.

One has only to observe the space now being used by outlying business and industry to note the disparity. Many industries occupy the equivalent of 10 or 20 downtown blocks. The new Prudential office building in Houston alone occupies some 28 acres. Yet many of our largest CBD's contain only about 50 acres. It is not unusual for a new regional shopping center to cover 100 acres.

The general appearance of downtowns is also a severe handicap. The dilapidated and unesthetic appearances are partly the result of the age or the structures. In addition, there is a great lack of landscaping and general architectural style as well as a void in the physical coordination of structures. Most buildings have been placed with little regard as to how they would fit into the general scheme of things. Finally, many downtowns are characterized by a great deal of broken frontage where buildings have been torn down and used for parking lots, thus making great gaps in the business pattern. There has been little thought as to the best arrangement of functions inside of the central business district, and consequently many institutions such as banks, insurance companies, and the like are so located that a shopper must walk farther to get between stores than would be necessary if conscious thought had been given to the order and arrangement of functions, as in modern shopping centers.

The most talked about problem in the central business district is, of course, traffic and parking. Congestion has reached huge proportions in many downtown areas. There is a great lack of parking space, and many people visiting the downtowns have to walk considerable distances or pay very high parking fees. Although high-rise parking ramps are being built, they are still inadequate to serve the needs in most cities. Mass transit has been continually deteriorating in both service and quality, while the price has been increasing and it is no longer nearly as convenient as formerly. Moreover, transit does not truly serve many of the outlying residential territories adequately.

Many of the problems in the central business districts are the result of the extreme governmental fragmentation in our metropolises. Central city municipalities, which often contain only about half the total metropolitan population, are greatly concerned about their central business districts and do, in fact, undertake various renewal and redevelopment schemes which would not ordinarily be undertaken if the metropolis were under one municipal government. This, of course, creates a tax burden on the population within the city limits. It also causes the central city government to become gravely concerned about the decentralization and new placement of functions and activities which would otherwise be welcomed. The decentralization of retailing to outlying locations is in many regards a real asset and benefit to the consumer. The problem of the central city government is that such relocation generally occurs outside of its particular municipal boundaries.

\ lack of progressiveness is also evident in st central business districts. This is re-ted in decor and general appearances, as lined previously, as well as the parking blems, which have been referred to. Al-ugh many downtowns have now devel-d various "save downtown" associations, of these are of significant value. Most concerned with promotional and superfi- schemes rather than with obtaining a d base on which to make decisions.

asons For CBD Change

t what are the major reasons for such tral business district problems? Let's ex-ine some of the changes which have been urring in the central city — the municipal- which contains the central business dis-t. A metropolitan area includes the central , the county in which it is located, and er surrounding counties which are signi-ntly associated with the central city. tropolitan areas increased in population percent between 1950 and 1960, whereas central city has barely held its own, popu-ons averaging an increase of only one and -half percent during this time. Many cen- cities of large metropolises have actually t population in the last fifteen years. Such oulation decline has had a major impact the downtown area.

Moreover, such population decline has not n offset by increases in nonresidential acti-es, as was formerly the case. In fact, the oulation remaining has become far less uent than that which preceded it. The zone mediately surrounding many downtowns is en characterized by slum conditions.

The factors which have caused central siness district decline are reflections of the v mobility of the population as represented the automobile, the increased leisure time, and the general technological advancements made in construction since World War II. Such changes are reflected most clearly in what is commonly termed suburbanization. Subdivisions, planned industrial parks, and planned shopping centers have augmented the population decentralization. The great increase in the importance of and territory occupied by municipal airports during the past decade have, in turn, sparked outlying residential, industrial, and commercial development. Development of freeways is exerting tremendous decentralization pressures by providing outer circumferential highways and encouraging people to live even further from their places of work.

As monopolistic effects have been broken, the end result is that the central business district has continued to become more off-center. Until the past decade, most cities occupied very small territories. Today, however, with large subdivisions and the generally more generous use of land, the location of the central business district has become problematical. Although it is still the focus of the major transit routes and even the interstate freeway system, distance has become a major factor in determining whether people will patronize or work in this center. Generally, as a city grows the central business district tends to become more off-center inside of the metropolitan complex. As cities continue to expand into new rural territories, the location problem with regard to CBD's will surely become of even greater significance.

Developing A Public Policy For The CBD

Given these few facts, what should one conclude about the CBD? All signs point to a continuation of the rapid decline in the CBD and a continuation of rapid growth in most other parts of the urban complex. If current trends continue, the CBD will become but one of the many nodes of commercial activity in the metropolis; and perhaps not even the dominant node. It is also clear that the architectural thesis that the city is dead without a healthy CBD is unjustified. In fact, Los Angeles, the city in search of a CBD, is one of the most rapidly growing in the nation and now is the second largest metropolis in the United States and the sixth largest in the world.

This kind of argument perhaps obscures the real policy questions, however. The first question is not really whether the metropolis *can* effectively operate without a CBD, but whether it should or must. It clearly can. The second question is whether deliberate intervention is necessary in order for the city to operate effectively. The CBD is truly tied to other urban components, and, if not operating effectively, can have a deleterious effect on the entire urban system.

My conclusion with regard to the first question is that, given existing conditions and investments in the CBD, almost every metropolis should probably continue to have one, but not to the present extent for any given sized city, and surely not an augmented and symbol-laden CBD as envisaged by many architect-designers. Although presently there are many functions exclusively limited to the CBD, I can think of no single function, or activity, which must of necessity be located here in the future. While the CBD might be the best location, given the location of complementary activities for many functions, especially in smaller cities, there is no compelling reason why such functions should be encouraged and promoted here. The variety, pedestrian contacts, and other generally desirable urban features can be created in outlying locations in perhaps better form than is possible in remodeling our central busine districts.

My conclusion with regard to the secon question is that the CBD has indeed become drag on the urban system, inasmuch as it overbuilt, and requires a catalyst for chang which will diminish its prominence. It seem abundantly clear that without major govern mental intervention the CBD will never agai regain its former high position of value, pre tige, and general importance in the metr polis.

But if one accepts these premises, wh specifically should the CBD be like in th future? What specific catalytic actions an public policies are necessary in order t achieve this?

Lest the reader think that I am going untie the Gordian Knot, let me hastily assu you that I really do not know the answer. can partly describe the role of the future CB by describing what I think should *not* be th nature of things in the future. First, I do nc think that the city should be looked upon an artifact of mankind for the simple reaso that I think the city is far too important to b used as a monument or a museum. Neither d I think the future CBD should be the captiv promotional device of special interest terr tories or groups such as the central city mun cipality or various "save downtown" group It is necessary that goods and services be dis tributed throughout the metropolis in a wa which best serves the total metropolitan citi zenry, not just a selected few. Many dowr town functions might best be decentralized a soon as possible so as to be within close range of the consumers. Finally, I do nc think plans for downtown or the central cit should be made independent of the entir urban complex, and perhaps even the com posite urban interests of the nation. This, i ıple terms, means that most city planning ›artments which serve merely central city /ernments are obsolete.

Γhis, in turn, leads to a major public policy tement: planning should be done on a ›er-metropolitan basis. This necessitates tropolitan government or some such alter- :ive. Despite the many pitfalls and bitter ›eriences connected with attempts at tropolitan government, this policy must be ıtinued with renewed energy. Indeed, one ght argue that the metropolitan unit is al- ıdy far too restrictive a concept for today's 1 especially tomorrow's urban residential :terns.

Second, the U.S. Government should ablish a more comprehensive department urban affairs than that currently envisaged, that major and comparative research on : city complex can be undertaken. It is a tional disgrace that there is probably as ıch information available on the swamplands of Florida as is available on the cities of the United States. For example, it is not even known within thousands of acres how much land is now occupied by urban residents in the United States. Almost nothing specifically is known on a comparative basis about the location and extent of many major components of the metropolis. Until more information is available, and more comprehensive studies are made, we must continue to operate partly in a vacuum.

References

1 Gruen, Victor, *The Heart of Our Cities: The Urban Crisis, Diagnosis and Cure* (New York, Simon and Schuster, 1964).

2 Horwood, Edgar M. and Boyce, R. R., *Studies of the Central Business District and Urban Freeway Development* (Seattle, University of Washington Press, 1959).

3 Vernon, Raymond, The Changing Economic Function of the Central City (New York, Committee for the Economic Development, 1959).

23

The Core of the City: Emerging Concepts

Edgar M. Horwood

Malcolm D. MacNair

For the past five years there has been increasing interest among urban researchers and planners concerning the core of the city. A large body of literature on the subject has been developed[1] and this topic has been on the agenda of nearly every recent major city planning conference.

Amid this sea of literature it becomes very difficult to grasp central issues both in regard to research constructs and the development of planning policy. Most articles and books deal with only certain segments of CBD (Central Business District) study or planning. In fact, we are at the point where only a summary article in the *Reader's Digest* can save us from being devoured by our own words. This article will attempt to present such a summary and at the same time touch upon a few ideas which the authors believe will absorb attention in the near future.

One could classify literature on the CBD into the following broad groups:

1. *Studies based on defining and describing the CBD.* This group tends to define what we mean by the CBD, how to measure it, and in general how to define it in certain universal terms that will be applicable from city to city.[2] These studies are done mostly outside of city planning agencies. They are usually made by university scholars from one of a number of disciplines relating to urban analysis and financed by some agency or foundation with broad research objectives. The importance of studies in this group is that they paint a very broad picture of the nature of the CBD to serve as a framework for the conceptualization of local agency studies. Their disadvantage, from a local agency point of view, is that they do not answer specific questions for CBD planning in specific cities.

[1] See bibliography in Shirley F. Weiss, *The Central Business District in Transition, Methodological Approaches to CBD Analysis and Forecasting Future Space Needs* (City and Regional Planning, University of North Carolina, 1957).

[2] Perhaps the most noteworthy of these are the studies by Raymond E. Murphy and J. E. Vance, Jr. in *Economic Geography*, Vol. XXX (July, 1954), and Vol. XXXI (January, 1955).

Reprinted from *Plan Canada*, Vol. 8, no. 3 (1961), pp. 108-14 by permission of the Town Planning Institute of Canada.

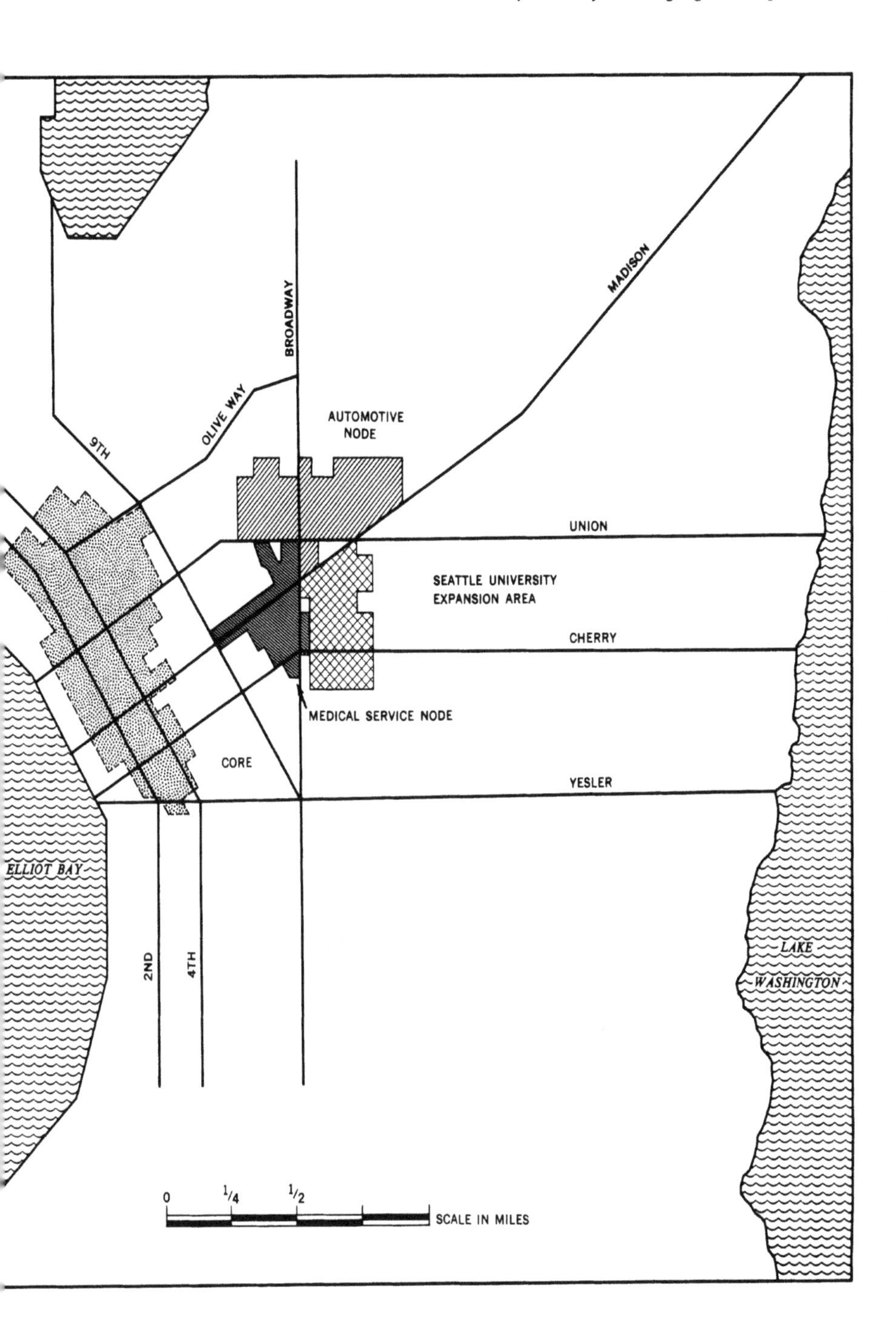

attle Central Business District, Functional Centres, 1959.

2. *Planning agency studies dealing with the CBD of specific cities.* These studies are typically the counterparts of those just mentioned. They tend to present specific inventories of what the CBD contains in different space uses, they project trends into the future, and they generally throw in a little graphic art depicting an anticipated future condition such as a pedestrian mall, open plaza, or other manifestation of esthetic improvement.[3] The strength of these studies is that they are oriented to a specific problem and geographic locale, and their weakness tends to lie in their lack of being based on a sound conceptual framework. Also, there is rarely enough across the board information presented upon which local forecasts can be tempered, such as the general parameters of certain types of uses in cities of varying sizes.

3. *Market analyses relating to central land use.* This type of study deals with central land use marketability and other factors relating the economic base to the demand for central land. They are usually carried on by public agencies or consultants in regard to specific urban renewal programs, and arise out of the demand for reuse appraisals of land coming under the clearance axe.[4] The strength of these studies is that they are generally realistically founded in terms of a local real estate market. Their weakness, on the other hand, arises out of the fact that they are typically oriented to coming up with predetermined answers to support a renewal program, or they bring facts together only regarding the marketability of land apart from the planning for an improved urban form.

There are, of course, many specialized studies not mentioned in the above categories such as analyses of daytime population, transportation and goods movement in the core of the city, and studies of the core of a city in terms of its regional market, to mention only a few. In the latter instance, many specific questions can be answered by studying the entire region, in regard to CBD planning which cannot be answered from studying the internal characteristics of a particular CBD itself. For example, the central place importance of a city has much to do with the capacity of its CBD to cater to particular specialized functions.

Work by the authors in CBD analysis has been mainly in regard to the first of the three groups mentioned, namely, those dealing with problems of definition, measurement and description which are common to all cities. At this time the authors desire to present some emerging ideas on the organization of the core of the city which have not generally been dealt with in the city planning literature. It will be necessary to recapitulate for only a moment on some general universals recognized by all students of the CBD.

Almost all observers of CBD phenomena have recognized that the central region of the city is not one of uniform intensity, but contains an exceptionally limited area of very high land value upon which most of the downtown retail and office uses exist. This area is generally referred to as the core area of

[3] This group includes the large number of reports published by local planning agencies on the subject of plans and studies for specific CBDs. The best reference on these studies is the annual index pages of the *ASPO Newsletter*, American Society of Planning Officials, 1313 East 60th Street, Chicago, Illinois.

[4] Larry Smith, Real Estate Consultant of Seattle, and Homer Hoyt, Land Economic Consultant of Chicago, have contributed materially to Studies of this classification over several decades. Most of these studies, however, are private consulting reports for specific client agencies and not generally available for public circulation. See a compilation of some of Larry Smith's findings in "Space for the CBD's Functions," *Journal of the American Institute of Planners*, Vol. XXVII, no. 1 (February, 1961), pp. 35-42.

city, and often as the CBD itself. Further, as been recognized that this core area is directly a function of city size, but is ted more to the human scale of convenient king distance between establishments. For mple, cities of quite disparate size, one aaps ten or twenty times the other, have ntially the same size of pedestrian core sured in ground area.

t is this intensively used core area which been the focus of most of the recent rest in the resurgence of the city center dynamic force in the urban scene. Good, or indifferent, the fact remains that busi- s interests in this portion of the city in most as of western culture have adopted a deter- istic attitude as to the value of a strong D, and are supporting their attitudes by h private actions and co-operative group rt through their downtown associations. a matter of fact, it is no longer a per- king question as to whether or not the tral core will survive in this age of ap- ent decentralization. Recent events in st of the regional capitols of North erica give testimony to the fact that the e of the city is not withering away and ng. Although retail sales in the core are inishing as a percentage of citywide sales, hey have been doing for the past 50 years, tral office space is growing sufficiently to e than counterbalance the outward flow etail activity.[5]

)ne might conceive of the structure of the ter of the city as being composed of a e laterally restricted core given over to the il sales and office functions (people, erwork, and parcels), with a relatively large frame surrounding this core given over to a variety of business and commercial functions as well as institutional and high density residential uses, all of which are a little bit less apparent in their organization than the spectacular core area itself with its tall buildings. This observation has come to be accepted as the Core-frame theory of the CBD region, and it is with the frame that the authors have now become quite concerned in terms of analysis and planning.[6]

The main proposition to be presented here is that the CBD is substantially more than a highly intensive core where face-to-face contacts make possible the social, economic and political leadership of the urban region. We have come to equate the CBD to this very apparent core. The core is vocal. It is urbane. It is what we have come to recognize as the character of the city, more than any other part of the city. And yet in relation to the entire central area the core is something like the seventh part of the iceberg, which is the only apparent part as far as the surface observer is concerned.

The CBD frame, on the other hand, constitutes landwise the six-sevenths of the central area, including many vital organs of the city, and generally either neglected or casually dealt with in CBD studies. This area has hitherto been identified in the sociological literature as the area of transition of the city center. At one time it was thought that core functions would reach out and occupy the surrounding space. History has proved otherwise. The core area grows vertically rather than horizontally. Although there are many transitional uses in the frame, certain strongly identifiable functions have been growing in these areas in most western cities for the last half-century at least. The frame is the neg-

dgar M. Horwood and Ronald R. Boyce, *tudies of the Central Business District and 'rban Freeway Development* (Seattle, Univer- ty of Washington Press, 1959), Chapters 3 nd 4.

[6] *Ibid.*, Chapter 3.

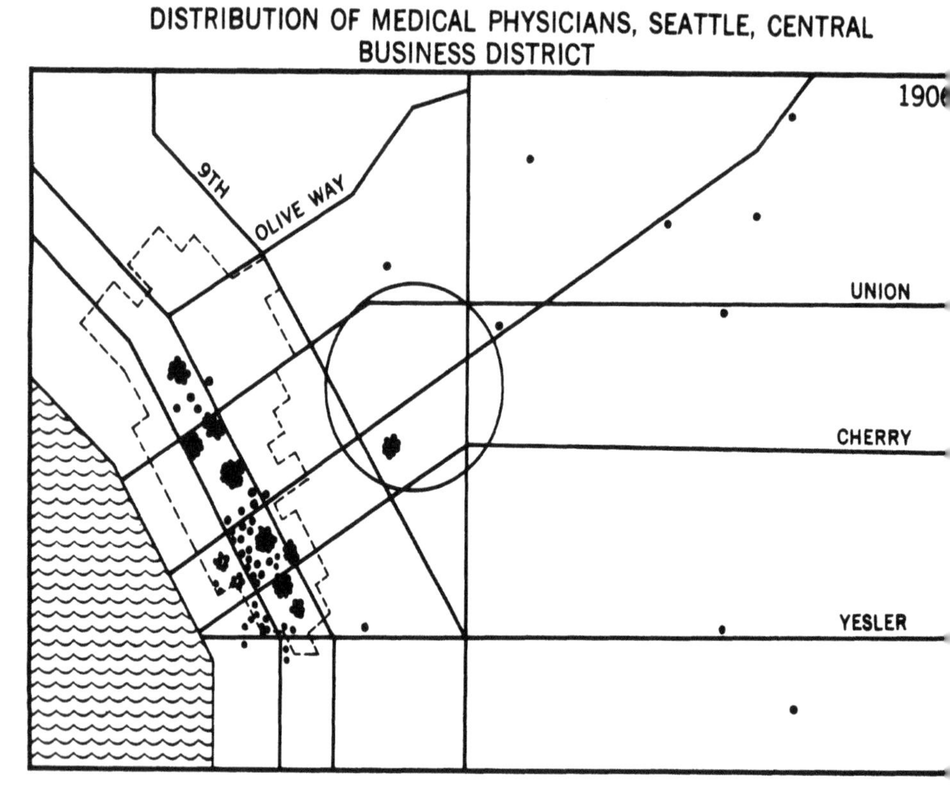

One dot represents one physician.

lected area of urban analysis and planning. It is more difficult to classify than other areas of the city. Many parts of it, such as the warehousing and wholesaling portions, lack esthetic interest. It invariably contains slums. It does not fit into the planners' neighbourhood configuration. Its telephone poles and wires are not buried, as in the core, and its pedestrian amenity is lacking. We tend to sweep this piece of urban real estate under the carpet in city planning effort. It has taken a decided second place to the more spectacular core as well as the more apparent neighborhood units on its outbound side.

From preliminary or windshield observation the frame is difficult to classify. In fact, its true identity may never become known conventional land use classifications are use It must be studied first in a functional sens through an establishment survey. From th beginning its functional nodes may be ident fied and subsequent work may be done t clarify the different component parts of th frame, to measure them, and to test the dynamic changes over a time span.

There are invariably about a half-doze distinct functional sub-regions within th CBD frame areas of our North America metropolises, and in fact including Europea cities as well. One typically finds within th region such activities as wholesaling wit stocks, transportation terminals, medic

DISTRIBUTION OF MEDICAL PHYSICIANS, SEATTLE, CENTRAL BUSINESS DISTRICT

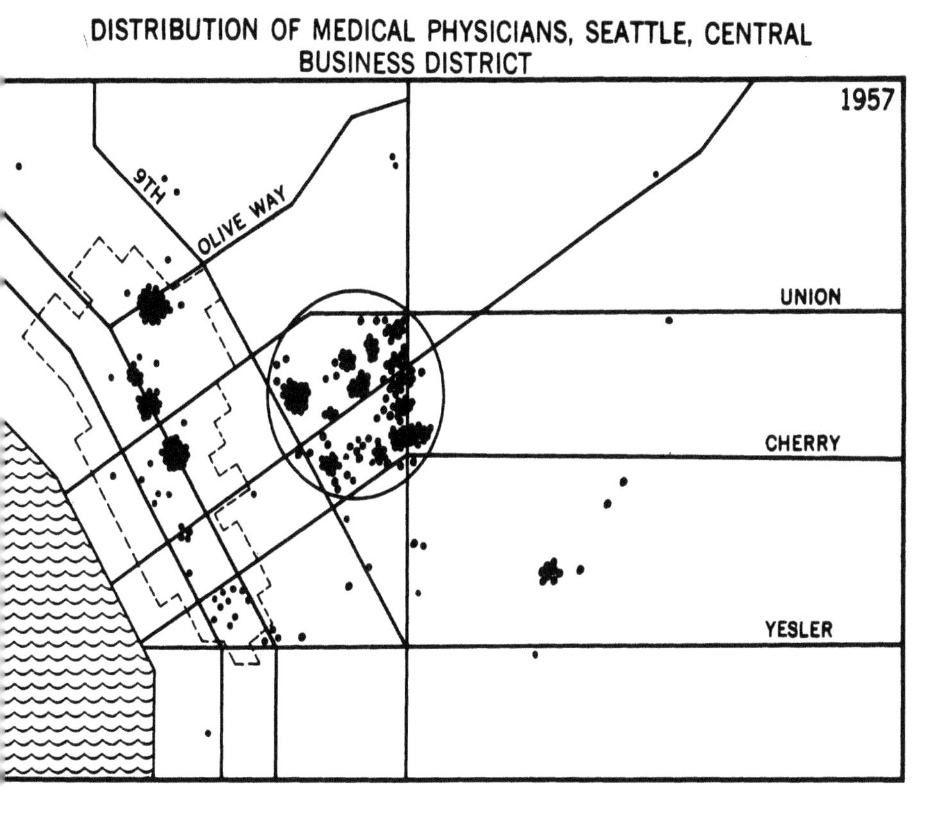

hin a 50-year period the medical service cluster had virtually deserted the core area of the CBD and ved to the frame. This change in both scale and location reflects not only changes in the technology nedical practice, but improved regional transportation as well. The completion of the emerging urban ional freeway system will further accelerate the trends in the medical service area by vastly expanding the graphical limits of the service. That is to say, the *commuter shed* for medical services will be doubled ize by the new transportation facility. Although many more interesting facts were ascertained about this ticular part of the Seattle frame, one of the most revealing findings was the complete lack of realization he cluster and its needs in public policy.

vice centers, automobile services and often ne manufacturing or assembly of goods. ese uses are all interspersed with both old l new housing and many non residential s which are related to the functional nodes a symbiotic way. The latter tend to obscure basic function of the establishment cluster ler study, much as trees obscure some of geologic features of the earth's surface.

n spite of certain observable trends toward entralization, there are a great many activities which, by virtue of the need for transport economies must have locations within the frame area.[7] For example, one rarely finds truck terminals outside the CBD frame because of the need for these concerns to minimize short haul pickup and delivery costs which are extremely high in relationship to the relatively low cost of moving goods via consolidated shipments in intercity transport.

[7] *Ibid.*, Chapters 5, 6 and 7.

A verification of the observations regarding the CBD frame as discussed above obtains from a series of studies under way at the University of Washington concerning the frame of the three largest cities in the state, Seattle, Tacoma and Spokane. An effort is being made in this program to analyze systematically the major functional nodes of the CBD frames of these cities with a view toward sharpening the theory and advancing techniques for its analysis. Space here permits only a limited description of these studies, and rather than becoming overextended in the presentation, an example will be presented instead. The example deals with the medical service node of the Seattle CBD frame, as one of several functional areas examined within that frame.[8]

The methodology of this study was a detailed examination of all establishments of a medical service nature within a general area recognized to be accommodating growth of these activities. Linkages were then studied between the establishments to determine the scale of exchange of goods, persons and communications. These linkages gave an interaction concept of the concentration. Another important examination was changes over time. By means of old medical directories and phone books the development of the node was reconstructed for a time span of 50 years. Finally, ancillary and competing uses were examined, as well as the constraints imposed by city zoning policy. In the process of these studies indices were developed to delimit the node, which should be applicable to other cities. The entire procedure required many interviews, as well as the examination of both internal records of the establishments them selves and documents of the type customaril not used in land use studies. An importan distinction was made between arriving at con clusions from what could be seen by visua observation and what could be compre hended from the examination of the establish ment activity itself.

Among the important findings of this stud was the amazing growth of the concentratio of specialized medical services. Well over hal of the specialized medical practitioners i Seattle have offices in this area. Within a 50 year period the medical service cluster ha virtually deserted the core area of the CBI and moved to the frame. This change in bot scale and location reflects not only change in the technology of medical practice, but im proved regional transportation as well. Th completion of the emerging urban regiona freeway system will further accelerate th trends in the medical service area by vastl expanding the geographical limits of the ser vice. That is to say, the *commuter shed* fo medical services will be doubled in size by th new transportation facility.

Although many more interesting facts wer ascertained about this particular part of th Seattle frame, one of the most revealing find ings was the complete lack of realization c the cluster and its needs in public policy. O Seattle's comprehensive plan the area show up as a high density residential area, with ribbon of commerce along one of the princi pal streets cutting through it. No studies hav ever been made by the planning agency t ascertain growth trends of medical servic space. In fact, twelve-storey apartment builc ings are permitted in the node, wherea clinics are restricted to two storeys. Furthe wholesaling establishments specializing i medical supplies are prohibited by zonin code from locating either in this area or its en

[8] Malcolm D. MacNair, "The Medical Service Area, Seattle: A Functional, Spatial and Linkage Analysis,", Unpublished Master's Thesis, University of Washington (1959).

ons. In the face of these restrictions the ed for medical service and hospital space the Seattle frame is expected to double by 80, and is doubling every two decades.

Similar techniques of study were also aped to other activity clusters in the Seattle me and findings of similar scale made. The y of Tacoma studies, conducted by the city nning agency staff using the general format the Seattle studies mentioned above, o corroborated the same findings.[9] The Spokane studies are not yet complete, but will be materially enhanced by the use of both electronic data processing techniques as well as the electronic mapping of land use data. It is hoped that by the completion of this group of studies there will arise new theories of central urban organization or the corroboration of old. Findings to date lead the authors to believe that the Core-frame construct is a valid one, and worthy of attention in all local agency planning studies.

Tacoma City Planning Commission. *The Central Core*, Central Business District Studies (June, 1959).

24

Towns anc Umland

Jacob Spel

If modern town planning wants to measure up to its task, it will have to plan for both city and umland as one unit. Town planning can no longer permit itself to be limited in scope by municipal boundaries, as was the case in the past, when the field was mainly concerned with architectural forms and artistic points of view. Town and umland are, so emphasizes Hans Carol, only two different aspects of one and the same coherent whole.

THE TOTALITY OF TOWN AND UMLAND

Much of the discussion in this category has been familiar to geographers for a long time. On the other hand, it is gratifying to see it appear in a planning journal and even more so to see it used as a basis for planning.

At one time town and country each had its distinctive forms of production and people lived where they worked. Now a daily and in part a weekly migration between place of work and place of residence has become typical. The 1950 census of West Germany revealed that one out of every seven gainfully employed persons leaves his municipality every day to work elsewhere. In total, mor than 3,000,000 commuters were counted o the enumeration day. Sixty-seven per cen originated in places with fewer than 5,00 inhabitants; in other words, they came fror the umlands of the larger cities and town which now have become the greatest concen trations of employment opportunities.

The central city depends on its umland no only for labor supply but also for a stead influx of people to maintain a steady popula tion growth. In his article "Das Stadt-Um land-Verhältnis in seiner planerischen Pro blematik" Dittrich points out that migrant came from the umland, but that now a larg city draws many of its people from a muc wider area, sometimes covering the entire na tion. On the other hand, the newcomers n longer settle exclusively in the central city but also in the suburbs and towns in the um land. As such, umland and city have a com bined drawing power and the absorptiv capacity of the umland is of vital significanc to the city. A city, therefore, may have na tional or world-wide connections, but thi does not mean that its roots in the umlan could be severed.

Reprinted from *Economic Geography,* Vol. 34, no. 4 (1958), pp. 362-69 by permission.

)ne of the most important functions of umland is to provide and maintain open :e for the benefit of the population in the tral city. There the city population can opportunity for recreation and relief n crowded and congested conditions.

)n the other hand, the city performs many er services for the umland besides employ-nt opportunities. At one time, weekly n markets played a vital part in the provi- of goods and supplies for the population. w, stores in varying degrees of specializa- offer a great diversity of goods and have laced the old markets. Dittrich points to importance of the fact that these stores located in the city, and that modern ins of transportation make it possible to ve the population of both umland and city a day to day basis. In addition, the city is focal point for the umland in matters of nce, education, health, administration, ertainment, and others. In many ways the is the exclusive provider of such services.

'he relationships between city and umland so intimate that the one puts its imprint the other. Dittrich reaffirms that the cities towns of Europe are not standardized uniform. Each of them is unique, an ividuality forged not by the city alone but by its umland. The natural conditions of umland, the character of its people, and ir ways of life contribute greatly to the ioning of the city.

Jnfortunately, however, city and umland e generally not been aware of the signifi-ce of this interdependence. On the con-y, often tensions have developed and usations have been flung back and forth, netimes justified, sometimes not.

The city became a powerful economic con-tration and in part acquired this position collaboration with its umland which pro-ed it with workers, maintained its popula-tion growth, and provided open space for suburban expansion. At the same time its population patronized the city stores.

The central city accumulated considerable wealth and big financial differences arose between it and the predominantly residential municipalities in the umland. The umland considered itself exploited and impoverished. Many of the services and functions of its villages and towns were usurped by competitors in the central city. In Germany, as in other European countries, a system of tax equalization has helped the umland municipalities whose citizens are employed in the central city to pay for their educational, social, and other services. By means of this equalization the communities in the umland share in the industrial and commercial tax yields of the central city.

The city is inclined to cite the expenditures it makes to maintain facilities which are also used by the inhabitants of the umland, but which are not supported by them financially. In Germany this applies in particular to hospitals, certain forms of education, museums, and theaters. Klüber in his article "Kommunalwirtschaftliche Verflechtungen zwischen Stadt- und Landkreis" calculates that the Landkreis Offenbach[2] is saving over 320,000 DM annually on its educational expenditures, because pupils in the Kreis attend schools in Offenbach a.M. The same is true for hospital costs. The city of Offenbach (104,800 population) has three private and one municipal hospital with 794 beds. Only the latter is supported financially by the city. The Landkreis (154,200 population) maintains only two small hospitals with a total of 225 beds. Over 25 per cent of the patients in the Offenbach municipal hospital come from the Landkreis, constituting an extra annual expenditure of about 380,000 DM for the city. The insurance authorities object to higher fees for out-of-town patients and the Landkreis refuses to pay subsidies towards the care of its population in the city hospital. On the other hand, hydroelectricity, gas, and water for the Kreis population are supplied by companies in the city which operate on a profit basis. For them the inhabitants of the Kreis are important customers. The Landkreis took 5.3 per cent of the water, 28 per cent of the gas, and 55 per cent of the hydroelectricity sold by the companies in Offenbach.

However, only in certain cases is it possible to calculate a numerical value for the degree of dependence of umland on city and vice versa. There are many other services and expenditures for which values cannot be calculated. The inhabitants of the umland come to the city for a great variety of reasons and invariably spend money, thereby stimulating the economy of the town. On the other hand, city people frequently spend money in the umland, while seeking recreation there.

Klüber emphasizes the vital significance of the umland as a source of labor for the central city. But then he also points out that the umland municipalities are able to compete with the central city for new industries. They can offer lower wage rates and lower taxes. In towns close to the central city these advantages are not offset by poorer transportation facilities, and consequently the central city may experience serious competition. It is especially painful for the urban tax payer, when the children of those competitors come to attend city schools and their employees are admitted to city hospitals.

Both Dittrich and Klüber consider it unlikely that it will be possible to draw up a balance sheet of the relationships between towns and umland, but this does not reduce the significance of the fact that umland and town constitute economically, culturally, and socially a closely integrated entity. Existen of one without the other is virtually impo sible. Furthermore, action taken in one ma have far-reaching repercussions in the other

Recognition of the umland-town entity b planners has resulted in a new approach planning problems in Germany. In certain in stances this has crystallized in the form planning satellite towns in order to organi the umland in a functional relationship wi the city. Roads may be built into the umlar to open up new areas for recreation or bring fruit and vegetable growers closer to t city market. The first steps toward mode land use planning in Germany were measur to preserve open space in the Rhine-Wes phalian industrial district. Specific attemp may be made to counteract undue "bleedin of the umland by the city. Tax equalizatic is a step in this direction, but Dittrich sugges the possibility of moving certain establis ments from the city to the umland. Caref analysis of the city-umland relationships ma reveal that a city does not utilize the poter tialities of its umland to the fullest extent ar plans can be formulated accordingly (Klöp per). In brief, planning in both city ar umland must be founded on the premise the fundamental unity of town and umland

DRAWING THE BOUNDARIES

The matter of boundaries cannot be ne lected. This involves the boundary of th city complex and also the umland boundar

The City Complex

So far it has not been possible to develc in Germany a concept similar to the Urba ized Areas in the United States, the Conurb tions in Britain, or the Tätorter in Swede Horst Fehre in his paper "Zur Abgrenzur der Stadtregion" emphasizes that a clear d tion is urgently needed in view of the 1960 sus. Attempts before 1945 display a great l of unanimity in the selection of criteria h as settlement forms, population data nsity, rate of growth), transportation, nomic relationships, and the movement of nmuters. Geographers especially em- asized the trend and mode of settlement hout attempting, however, to apply an ective system of measurement. In general agreement was reached on the relative nificance of the criteria or which of these uld be combined. However, the more the blem was studied, the clearer it became t no sharp boundary could be drawn be- en city and countryside. The outcome of se earlier attempts was a twofold ap- ach. One group, mainly geographers, rched for a solution in the form of transi- nal belts or zones between city core and ntryside. The other, the statisticians, pre- red one single line, which could only be wn after much generalization. Some ad- cated simply the drawing of a circle with ertain radius. In both cases it was a dis- vantage that the method evolved lacked a oretical structure, and could be applied y to the city studied.

More recently Voigt and Boustedt have d only economic data as criteria for undary definition. Voigt took the propor- n of the total population engaged in farm- (20 per cent) and Boustedt the portion of the total gainfully occupied pulation engaged in farming (35 per cent) a criterion to draw the boundary of the city a. The boundaries, as arrived at by Voigt d Boustedt, of course do not coincide, but he umland is divided into an inner and an ter zone, then the 35 per cent line of Voigt incides with the 50 per cent line of ustedt.

Voigt and Boustedt considered also the number of commuters working in the central city, although again in a slightly different manner. Voigt took the proportion of the gainfully occupied population of the umland community which commutes to the central city (40 per cent) and Boustedt took the proportion of commuters in the gainfully oc- cupied population in non-agricultural em- ployment, which corresponds to 20 per cent in the Voigt system. Boustedt supplements the two aforementioned criteria by those of population density and house types. A municipality has to have at least 500 people per square kilometer and not more than 33 per cent farm houses before it can be con- sidered as an extension of the central city area.

Sirp tried to apply the Voigt-Boustedt method to the Cologne area, but found it unworkable, partly due to short-comings in the method and partly because of special conditions in the Cologne district. To the southwest of Cologne lies the Erft lignite district which complicates considerably the pattern of commuter flow. There are some very extensive municipalities for which the census does not supply data on the basis of the various residential quarters within them. This was the main reason for the failure of the method in this area. Large numbers of commuters go to the central city but also to other neighboring centers. The problems in the Cologne district are, first, to draw a boundary between the central city complex and a district with a non-agrarian economy and, second, to decide whether this district should be added to the central city. Sirp found that the number of dwelling units per residential structure could be used to dis- tinguish between truly urbanized municipali- ties and those with only a non-agricultural economy. Districts with 1.8 or more dwelling units per structure may be considered as truly urban. Sirp has come forward with a fresh idea, but according to Fehre he did not succeed in finding a real solution.

Other workers such as L. Fisher, K. Hook, and Eicher resumed the earlier attempts of others to define the boundary by drawing concentric circles around the center of the city. The difficulty of this method lies in determination of the length of the radius. There are three possibilities: a) one radius for all cities; b) radii in proportion to the size category of the town; and c) a separate radius for each city. Fehre suggests that the third approach probably would offer the most satisfactory solution. There are, however, some serious shortcomings in the circle method. Most cities are star-shaped and thus entirely contrasting areas may be included in one particular ring. Then there is the problem of overlapping circles between neighboring towns. Fisher suggests that the common chord could be taken as the boundary. However, a boundary arrived at in this manner would probably not depict reality and might easily separate what really belongs together. According to Fehre the method remains at best a temporary solution.

According to Rudolf Klöpper, the problem of whether a city has an umland and how far it extends can only be determined after a survey in the field.

THE HIERARCHY OF CENTRAL PLACES

In Germany the study of the city as a central place was initiated by Bobek with his examination of Innsbruck, Austria, in 1928. A whole series of similar studies, entitled "The City X and its Lebensraum," followed. A few years later, this type of research was given new stimulus by Christaller's theory on the location of central places (1933).

Of special interest in this field is the contemporary contribution of Hans Car (Sozialräumliche Gliederung und planerisch Gestaltung des Groszstadtbereiches — Da gestellt am Beispiel Zürich). Carol mad studies in Switzerland and South Africa du ing the late forties and early fifties and fro these he recognized central places of uppe middle, and lower order, distinguished on th basis of services provided. The central place of lower order are characterized by 12 se vices such as certain professions, bank shops, cinemas, newspapers, farmers' ma kets, railway connections, and postal del veries. Cartographically, the quantity of eac of these indicator services is shown by mear of radials in clockwise arrangement aroun the symbol of the central place. In order t avoid an impractical length of the radials, th latter are composed of three types of dc symbols, each the equivalent of a certai number of units. Depending on the numbe of indicator services present, the centra places of lower order are further classified a fully central, semi-central, and sub-centra Carol intends to develop a more valuabl quantitative classification by making use c the average number of persons required t support a certain central service. This shoul provide a unit to summarize such varie criteria as physicians, clothing stores, an cinemas.

Central places of middle order have a main indicator secondary schools, supplemented with eleven other services, such a book shops, automobile dealers, bank wholesalers, technical schools, and dail newspapers. Depending on the number c services available there is again a classification into fully central, semi-central, and sub central places.

The main criterion for the central place of upper order is a university. To the remaining eleven criteria belong such services a eign legations and consulates, inter-canton ;anizations, large daily papers, medical iters, etc. There is a similar breakdown in three categories. Switzerland has five fully itral and several semi-central and sub- itral places of upper order.

Carol considers it the task of planning to prove the existing distribution pattern of vice centers. He is an advocate of decen- lization and a controlled growth of Zürich, ich he visualizes as gaining a stronger tropolitan function as the umland experi- ces greater economic development. In ticular he calls for better conveniences for rural population, thus slowing down the gration to the cities. Carol proposes to ect only a few well established market iters at one time for development into well lipped towns (5,000-20,000 pop.) with itral services of lower order. In areas ere such small centers do not exist, new es should be created. Carol also suggests t the desired goal may be reached by im- ving the transportation facilities between places to be developed and the metro- itan city, in this case Zürich. Other steps gested are advice for industries on matters location, the creation of conditions which uld appeal to new and expanding indus- s, and greater financial help for the centers be developed.

ANNING FOR TOWN AND UMLAND

nning in different cities gives an excellent istration of the foundations upon which dern town planning in Germany is inded. The new federal capital of Bonn y be taken as an example.

Bonn, situated in the apex of the Gulf of logne, has always tended to expand up- eam along the Rhine River instead of king the open spaces of the widening low- land in the opposite direction. The expansion of the city complex followed generally the transportation routes and extends at present over a distance of about 15 miles along the Rhine River. After the selection of Bonn as the federal capital, the building density in- creased considerably and the population of the city proper grew from 101,000 before the war to 142,000 at present. Suburban Bad Godesberg almost doubled its population and has now over 58,000 inhabitants. Not much land remains for further building. The great building density already there does not leave enough open space for recreation, a problem which will become more pressing with the gradual shortening of the work week. These and other problems can only be solved by drawing the umland within the framework of local planning.

The umland still offers many possibilities for recreation. However, roads built to open up areas for this purpose are often considered by municipalities as means to promote in- dustrial, commercial, and residential develop- ments. This attitude of the municipalities in the umland counteracts the aims of the plan- ners in the city.

In addition to space for recreation, Bonn needs land for residential expansion, not only on account of Bonn's new administrative function, but also because of its importance as a university center and cultural center (Beethoven), its tourist industry, the settle- ment of refugees and last but not least its residential function (Bonn is a pleasant place in which to live, attracting persons as far away as Cologne).

The local planners do not look favorably upon satellite towns as a solution. This would reduce the freedom of the individual to change his place of work if he so desires, since such a move would probably involve a change of residence. If the aim is to promote pride in home ownership and neighborhood, and if one wishes to avoid a home becoming a commodity lightly to be disposed of, then the largest number of employment opportunities must be easily accessible. The planners, therefore, intend to guide the new expansions to sites attractive as places of residence and with easy access to the main transportation arteries. They hope to avoid as much as possible restrictions on the freedom of movement for the individual.

Such considerations led to the decision to locate the new extensions on the northern boundary of the city. This decision was also influenced by the results of climatological studies in the Bonn area. The new suburbs will house 20,000 to 15,000 people, and will consist of two parts separated by a greenbelt. The greenbelt extends right into the center of the city proper and constitutes an attempt to improve the climate of the city. The belt or wedge is oriented in the direction of the prevailing winds and should serve as a huge breezeway. With these expansions, the rebuilding projects, and the filling in of still available open spaces, Bonn will be able to accommodate 200,000 inhabitants.

Of vital importance is a wise planning of new transportation routes. Not only will the success of the new suburbs depend on a judicious layout of a new network of roads, street car lines, and bus lines, but there are also other implications which reach far beyond the municipal boundaries of Bonn. Th new routes can be used to stimulate the ecc nomic prosperity of both city and umland.

A new east-west connection is needed t relieve traffic congestion in the central part c Bonn. The best location of the new Rhin bridge would be to the north of Bonn. Thi would make the large airport which serve both Bonn and Cologne much more accessible to the capital. At the same time industrial towns across the river would fall withi easy commuting distance of the new suburb in north Bonn, thus enlarging the employment opportunities for the residential popula tion in these suburbs. A measure of stabilit would be achieved, especially significant if th capital were to return to Berlin. The ne bridge would also provide greatly improve market opportunities for the fruit-growers t the north of Bonn, on the other side of th Rhine River. Finally a bridge at this poir would open up the Sieg Valley as one of th new recreation areas for Bonn's populatior Water sport is rather difficult on the Rhin River with its heavy traffic and pollution. Th Sieg River opens entirely new vistas.

Thus there are many possibilities for th creation of a harmonious development of th area, but this can only be realized if city an umland are taken as an integrated unit. It i only on this basis that Bonn and its umlan could create the most desirable milieu fo their inhabitants.

ternal Relations of Cities: ty—Hinterland

25

rt J. Epstein

Harold Mayer has demonstrated in the article of this series,[1] consideration of an problems and relationships necessitates efinition of major terms. This is no less essary in considering the external relations ities.

ince the "real city"[2] extends to the limits n urban density of settlement, a focus on t is beyond in relation to what is within es little other than agricultural produc-, extractive industry, and specialized nts of manufacturing or recreation. These important functions and are often vital to existence of the city, but are hardly as l and interesting as a more introspective v.

he urban population of the U.S. and of ny parts of the world[3] is increasing at a spectacular rate. Since most of our cities in the U.S. (the major focus of the article) are past the days of great annexation possibilities,[4] growth has continued outside the corporate city limits, giving rise to the phenomenon we describe as urban sprawl.[5] Sprawl is hardly a new phenomenon; it is only recently that its problem value has been recognized and attempts made to control it. To some degree external relationships of cities are expressed in the process of sprawl and its physical and socio-economic characteristics. Our focus in this discussion will be on the corporate rather than the real city.

HINTERLANDS

A city's hinterland[6] is a vast physical area

arold M. Mayer, "Cities and Urban Geography," *Journal of Geography*, Vol. LXVIII, . 1 (January, 1969), pp. 6-19.

or a brief discussion of the "Real City" or eographic City," see Raymond E. Murphy, *he American City* (New York, McGraw-Hill ook Company, 1966), pp. 13-14.

tatistics are available in United Nations Publitions. Note also Gerald Breese, *Urbanization Newly Developing Countries* (Englewood liffs, N.J., Prentice-Hall, Inc., 1966), pp. 1-32.

4 Major exception is Jacksonville, Florida, which on October 1, 1968, spread its corporate boundaries to the Duval County Line, making it the largest areal city in the United States.

5 Planning literature on sprawl is abundant, for a geographer's viewpoint see Jean Gottman and Robert A. Harper, *eds.*, *Metropolis on the Move: Geographers Look at Urban Sprawl* (New York, John Wiley & Sons, Inc., 1967).

6 For a brief discussion of hinterland concepts see Raymond E. Murphy, *The American City* (New York, McGraw-Hill Book Company, 1966), pp. 52-71.

inted from *Journal of Geography* (March, 1969), pp. 134-67 by permission.

compared to the city proper, and it performs a variety of functions.[6] Its major development is governed by the economics of space.

The size and complexity of the city will govern the size and complexity of its hinterland and the relationships between them.

Perhaps the simplest example of external city relationships to be found today involves settlements in agricultural areas. In the string of communities that dot the Rio Grande Valley in Texas, communities such as McAllen and Weslaco perform urban functions for extensive areas devoted to farming. These areas abut the urban settlements themselves.

The fact that there may be dominant cities which may extend their influence to lesser developed cities,[7] in no way detracts from the individual hinterland relationship.

At the other extreme is *the* megalopolitan city, New York, whose hinterland may well include not only Gottman's *Megalopolis*,[8] but possibly the U.S., North America, and, perhaps, the world.

HINTERLANDS AND URBAN FUNCTIONS

Somewhere between the two extremes used for illustration lies "the city" that will permit consideration of function and process at a useful and manageable scale. Omitted fro consideration will be agriculture and extra tive activities. Therefore, major concern wi be with

(1) Residence
(2) Manufacturing
(3) Distribution
  (a) Wholesale
  (b) Retail
(4) Recreation
(5) Personal and professional service (including government)
(6) Communication and transportation.

Throughout each discussion, either stated implied, will be the effects of the maj catalysts for growth and functional different ation — communication and transportation.

RESIDENCE

Commencing at the corporate boundary lir of our "city," residential development de scribes the most obvious land use. This res dential function was the earliest expression c extension of the city into its hinterland. Pre sure on the land to accept greater numbers urbanites is present to day and shows no sig of diminishing. On the simplest of terms – the hinterland is the catch-basin for the grow ing city as well as for its suburbs (both ol and new).

This residential push, a centrifugal force, is the result of many pressures. Our citie today are growing from both natural increas and immigration from rural areas and, to minor extent, from abroad. Where the natura increase and immigration have increase Negro or Latin population, there has been a additional racial force which may ultimatel

[7] A useful source for central place studies is Brian J. L. Berry and Allen Pred, *Central Place Studies, A Bibliography of Theory and Applications, Including Supplement Through 1964* (Philadelphia, Regional Science Research Institute, 1965). For a review of central place systems, see Brian J. L. Berry, *Geography of Market Centers and Retail Distribution* (Englewood Cliffs, N.J., Prentice-Hall, Inc., 1967), pp. 1-25. Note also that this latter publication points to a lack of theory development within the metropolitan complex.

[8] Jean Gottman, *Megalopolis: The Urbanized Northeastern Seaboard of the United States* (New York, The Twentieth Century Fund, 1961).

[9] Charles C. Colby, "Centrifugal and Centripet Forces in Urban Geography," *Annals of th Association of American Geographers*, Vo XXIII (1933), pp .1-20.

ect substantial changes in city-hinterland ationships.

In addition to biological and racial pres- es, there are also physical, socio-economic, d governmental pressures assuring a con- ued bedroom function for the city's hinter- d.

Atmospheric conditions such as smog in Los Angeles metropolitan area, the fouled of New York City, the acrid odors of the mical plants near Niagara Falls provide a ysical incentive to escape the city and the ected parts of the metropolitan areas. Lack enforcement of zoning and building codes y also produce urban building deteriora- n.[10] Temporarily, the slum may house a ater density than ever before, but ulti- tely there must be a physical spillover, ich is combined with the desire to escape the hinterland. An additional social force isiders association with the periphery of slum undesirable, and thereby adds to the ward movement.

Another centrifugal residential force is io-psychological — the search for the od life." In middle class American terms s is the struggle to have one's own single nily home on a green plot of ground, away m the "city" (nevertheless, seeking the ature comforts and services of the city). is outward movement, in which man at- pts to upgrade housing and improve his ironment, is also evident in the movement m suburb to exurb.

Regardless of motivation or of physical or ial impetus, residence in the hinterland is sely tied to the city; very often the tie is ral, even physical; i.e., via utilities. Population pressure develops in the city and the close-in hinterland, and the outer hinterland supplies the escape area. No doubt there are residents of the hinterland with only loose ties to the central city, but a trip on the Bayshore Freeway (San Francisco) or the West Side Drive (New York City) or the Southeast Expressway (Boston) any weekday morning will vividly demonstrate the tie to the central city. A lesser flow of traffic in early evening hours is also obvious, indicating yet another functional association. (See section on Recreation.)

The dynamics of residence in a metropolitan area is not a one directional phenomenon. It is not only rural immigrants who settle in the central city, but also the affluent and the aged. Housing for the aged is commonly placed within the corporate city limits, very often near the central business district. Akron, Ohio, has just completed one group of apartments for the elderly within walking distance of 100 percent point;[11] St. Joseph, Michigan, has also completed an "old age" high rise apartment at the very edge of the central business district. Further, such housing developments as the West End project in Boston and Society Hill in Philadelphia have displaced low and moderate income families in substandard housing with higher income families in luxury apartments and town houses. A river setting and proximity to both the central business district and financial districts, and to all other central city amenities have made these developments a counteractant to the usual outward flow of population, albeit they constitute a relatively tiny force.

Nevertheless, the net effect of all forces is outward, thereby making residence the single

n addition to urban renewal, planners are eeking to rehabilitate the city by enforcing building codes. Proper code enforcement might have initially acted as a great deterrent to slum development.

[11] The 100 per cent point is usually regarded as the point of highest land value in the city. It is most often at the key traffic intersection of the central business district.

most outstanding, most important urban function performed by the hinterland for the city. This may be an oversimplification of a complex relationship, but until we can exercise complete control over our environment and over urban development, we may look forward to more manifestations of the same phenomenon.[12]

MANUFACTURING

Historically, manufacturing has been a centralizing force (centripetal). From the early days of handicrafts and the Guilds through the Industrial Revolution, there was a banding together for purposes of production of goods. This phenomenon was common to agricultural production as well. Strong forces operating in the distant past through recent times kept a large share of manufacturing a captive of the inner city.

Development of ports and the spread of railroads permitted and encouraged a degree of decentralization, but did not develop the freedom of modern times. The automobile, the truck, and improved highways have indeed permitted and encouraged the diffusion of the manufacturing function beyond the corporate city and beyond the dominant influence of navigable water and railroad lines. There has been a liberation from restrictive locational factors.

It should not be inferred that there is only one set of forces at work. Surely, the development of Bethlehem Steel at Sparrows Point, in Baltimore's hinterland, was but slightly influenced by existing overland transportation facilities. The deep water port and the economy of water transportation in addition to proximity to substantial markets were prim location factors. Nonetheless, Baltimore' proximity was the ultimate key, since thi was the home base of the labor force, (despit the construction of a company town). Th recent development of the aluminum indus try, oriented to power, provides a somewha similar situation.

These hinterland phenomena are bein overshadowed today by the shifts taking plac within the urban metropolitan areas. Ther are a variety of forces at work in decentraliz ing industry. Some of these follow.

Modern Design

There is a stress in most manufacturing prc duction lines on single story operation (breweries are a major exception). Thi creates a need for extensive acreage for builc ings alone. In Everett, Washington, Boein has recently built a plant to produce the 74 jet; it is the largest building in the world, cor taining 160 million cubic feet.[13] Added to th land used by the structure is a substantia additional area for employee parking, truck ing, and storage. This type of land need coul not be satisfied in Seattle; consequently, th move was made to the hinterland (even with in Everett the move was westward, periphera to the built-up part of the city).

Modern Technology

At least two striking phenomena have bee evident in recent years. Industrial engineer have been successfully diminishing air an water pollution and have also abated nois This has made previously noxious industrie into acceptable neighbors.

[12] It may be fair to ask if much of present day zoning is not in effect producing controlled sprawl!

[13] Seattle-First National Bank, Annual Reviev *Summary of Pacific Northwest Industries* (Ma 1967).

There has also developed a new type indus-
' including cybernetics and electronics
ich can be housed in aesthetically pleasing
uctures which produce no smoke and no
ise — only traffic and payrolls. There are
o the research parks which resemble
npuses rather than industrial complexes.
These developments have made it easier
decentralize production jobs, and have
de production facilities more welcome in
hinterland.

vernment Policy and Land Economics

ning is an instrument of public policy. By
ans of restrictive codes, manufacturing has
en forced to areas within the city that are
st desirable for any other use.[14] Combining
s restriction with relatively small land hold-
s in the city and the relatively high prices
land acquisition, there results a great
entive to escape the city and to seek
endlier, more available opportunities out-
e — both outside the city and outside of
hinterland.
Taxation is inescapable, but its impact is
nageable. Manufacturing is so desirable
m the point of view of the community tax
plicate, that there will often be incentives
move it to a particular place. At least part
the motivation for the desertion of New
gland by the textile industry was a favor-
le taxation climate in the South. In any
se, the hinterland often escapes the probns and the "costs of doing business" of the
y, thereby permitting a friendly tax environment, enriching the hinterland and possibly adding to the financial problems of the city.

Labor

There is one other major force at work between the locus of manufacturing and a city hinterland spatial arrangement — labor. The simple act of increasing population in the hinterland, as described in a previous section, provides a varied labor force for wandering industry. For example, what better attraction might there be for needle trades to leave the city than a large potential female labor force in the suburbs? Skilled workers today earn enough to break the city bond. It is no longer necessary to locate manufacturing facilities in the midst of densely built-up urban areas. The labor force is dispersed and perhaps most important is mobile.

In many cities there is a reverse commuter rush with the hinterland supplying the jobs for the city dweller. Metropolitan areas in the extreme east and west provide vivid examples: Route 128 in the Boston Standard Metropolitan Statistical Area is not only a Boston circumferential by-pass, but is a new industrial belt complete with industrial parks and more independent types of facilities; e.g., Polaroid. Most of the major arteries originally aimed at Boston can now be considered as spokes in the 128 wheel and carry the heavy flow of outward bound traffic that eventually feeds into the circumferential. Equally striking has been the development of automotive industry in Fremont, California, in the southern part of the Bay Area. The Bayshore together with the Nimitz and the more local San Jose roads feed large numbers of commuters to these plants. These workers are in addition to the local inhabitants who have recently moved to be close to work and

Manufacturing has rated very low on the scale of land use quality. In old zoning codes it was possible to build higher rated uses on lower rated land. It is more common today to find exclusive manufacturing zones which may be used for nothing else. Nevertheless, manufacturing is still delegated to lands undesirable for residence or commerce.

to the general developmental excitement in this new area.

DISTRIBUTION

The distribution of goods is a function shared by the city and its hinterland. In the recent past, the city had a virtual stranglehold on wholesaling and an even stronger hold on sales of general merchandise to the consumer (especially department store sales). The post-World War II era has witnessed noticeable changes in both re-distribution (wholesaling) and direct distribution to the consumer (retailing).

Wholesaling[15]

In many respects wholesaling's history resembles that of manufacturing; perhaps it was responsive to physical and transportation centrality to an even greater extent. So long as urban relationships remained simple, wholesaling was keyed to primary transportation networks. Goods arriving by sea or rail were assembled at or near the central place and then organized for re-distribution. (Witness the old wholesaling areas associated with a port, railroad yards, or the central business district itself.)

Additional pressures due to increased population that was more dispersed than ever, a cost price squeeze experienced by all business, an accent on truck transportation within metropolitan areas, congestion within the city, taxation problems (real property and inventory) — all of these forces were at work and decentralization has taken place.

Just as there is reverse commuter traffic, there is a reverse wholesale traffic. Modernization and automation of goods handlin facilities have created the need for "assembl line" type efficiency. Indeed, the moder wholesaling or distribution center today re sembles the modern manufacturing plan Sprawling single story buildings with sub stantial service area acreage is the rule toda and land (and other) pressures have brough this redistributive function to the hinterland The city has not abdicated its redistributiv role completely. Defensive action such as th South Philadelphia Distribution Center an the new Boston Market attempt to reverse th new tide. There are, indeed, advantages to central city position, but the "lock" is broke and the hinterland now shares this vital func tion.

Retailing

Perhaps no other phenomenon of the post World War II era is as striking as the reta revolution. Its basic element has been de scribed in the previous section on residence and since most people with their greater dis posable incomes to spend have been caugh up in its wake, they have experienced th changes and played an important part in it.[1

The central business district is, and ha been, the epitome of commercial develop ment in our cities.[17] Its dominance in an are can be appreciated by shopping there or b

[15] For a definition of wholesaling, see Bureau of the Budget, *Standard Industrial Classification Manual* (Washington, D.C., United States Government Printing Office, 1957), Code 51.

[16] For a discussion of the relationship betwee retailing, population and sprawl, see Bart . Epstein, "The Trading Function," Jean Got man and Robert A. Harper *eds., Metropolis o the Move: Geographers Look at Urban Spraw* (New York, John Wiley & Sons, 1967), p 93-101.

[17] A landmark study of the central business distric by American geographers is Raymond I Murphy, James E. Vance, Jr., and Bart . Epstein, *Central Business District Studies* (Wor cester, Massachusetts, Economic Geograph 1955).

king a simple map analysis. In almost all es the central business district can be ated by inspection of the transportation work. Certainly, most main roads lead re and often a railroad confluence is also rby. This commercial hub was most easily essible to the city's population and to its terland as well — as long as the hinterland relatively close-in. Here we find the super es — department stores — with the wid- range of goods and services offered in the l market place. These stores are almost ays the highest sales producers in a metro- itan area. They are the "unreproduce- es," at least on a store-to-store basis. Here, , is found the widest range of speciality es, the greatest concentration of furniture es (at the periphery), and the greatest centration of services. The city and its terland have provided the market upon ch these developments depend. Most of goods purveyed here are commonly called opper's goods."[18]

Together with the spread of population rawl) has come substantial retail develop- nt. Early retail development outside the tral business district (described by Proud- t)[19] has been accompanied by a phenome- of the forties and more recent days — planned shopping center. Historically, iness has developed in conjunction with markets. (Witness the proliferation of es along old street car lines such as Lake in Minneapolis and the present day exten- along major roads.) Most of the development centered on "convenience goods"[20] with an occasional shopping goods concentration at points where major transportation lines met and crossed (Coolidge Corner, Boston). This built-in hinterland potential has been met in a variety of ways, among them the planned neighborhood shopping center and the planned community shopping center. These developments are little more than a response to meet a mobile, mechanized market potential as close as possible to where it resides. The city dweller rarely crosses into the hinterland for this type of shopping trip, with one possible exception — the discount department store with its discount supermarket associate.

The discount store is a "child of the city." It originated in the deserted mill buildings of the central cities of New England, attracting its business in much the same fashion as a small central business district.[21] The rapid acceptance of the "new" type of merchandising (surely influenced by the super-market) encouraged discount store entrepreneurs to leave for the hinterlands, where they could get cheap land and escape restrictive building codes (and perhaps the union as well), and started a helter-skelter pattern of store facilities in the hinterland. It did not take long for competition to weed out the poorly financed and the poorly located.[22] After greater and

hopper's goods satisfy long term needs and re bought relatively infrequently — as shoes nd apparel, furniture and household appliances.

Malcom J. Proudfoot, "The Outlying Business Centers of Chicago," *Journal of Land and Public Utilities Economics*, Vol. XIII (1937), p. 57-70.

20 Convenience goods are necessities which are bought frequently as food, kitchen supplies, liquor, and notions.

21 The discount store has evolved into the self-service department store, and has been accounting for an increasing share of the general merchandise sales dollar.

22 *Discount Store News*, a weekly newspaper, chronicles the events of this new retailing industry. In recent years there has been a great deal of news on corporate mergers. Previously, equal space was concentrated on business failures (for a variety of reasons).

greater acceptance, the discount store became respectable and sought better locations, even within the city itself (e.g., Alexander's and E. J. Korvette in downtown New York). These latter moves are restoring an equilibrium, keeping city business potential within the city (away from the hinterland) and even attempting to capture some hinterland sales potential when the suburban dweller "comes to town."

Bureau of Census data[23] and Urban Land Institute[24] studies have described the most revolutionary retail force of the sixties and its effect on the city — the planned regional shopping center. Although the central business district is still the single largest, highest sales volume shopping centrality in the standard metropolitan statistical area, it has not kept its share of market and is being overshadowed by the regional centers. Some of the loss in position is no doubt due to lack of central city growth, especially compared to the hinterland. (But rising incomes have at least balanced the loss, and the total effect is still a loss of position.) Whereas the neighborhood and most community centers are self supporting within the hinterland, the regional center taps both city and hinterland mark potential.

Most regional shopping centers are sit ated at key access points and provide fr (and adequate) parking for all customer For many shopping trips this ease of acce draws trade from a wide variety of areas.

Further, the location of department stor within a short walking distance, combine with attractive walkways where frontage occupied by specialty stores makes shoppi more efficient than in the central busine district where distance, traffic, vacancy, ar the elements provide deterrents to shoppi ease.

Regional centers which started by featu ing one department store (e.g., Shoppe World — Framingham, Massachusett originally Jordan Marsh) have now work their way up to as many as four (natural passing through the 2- and 3-store stage department stores (e.g., Yorktown Mal Lombard, Illinois: Montgomery Ward, J. C Penney, Wieboldt's, and Carson, Pirie, Scott occupying centers with over one millic square feet of space.[25] Further enhancing th regional center is its most recent innovatio the enclosed, air-conditioned mall. This sing development may provide the most serio impact on the central business district. Citi such as Rochester, New York, and Gree Bay, Wisconsin are already geared to me this attack, but most central cities will not b as fortunate.[26] The enclosed, air-condition

[23] Bureau of the Census, *Major Retail Centers, 1963 Census of Business* (Washington, D.C., United States Government Printing Office, 1965). This is a series of reports on SMSA's, CBD's, and MRC's for selected areas.

[24] The Urban Land Institute has published a series of reports on shopping centers, some of which are:

1) Technical Bulletin No. 33, Homer Hoyt, *A Re-examination of the Shopping Center Market.*
2) *The Dollars and Cents of Shopping Centers* — Parts I and II.
3) *The Dollars and Cents of Shopping Centers* — 1963 and again 1966.

Note also that Saul B. Cohen prepared an annotated bibliography on shopping centers for the Kroger Company which was published for limited distribution in the mid-1950's. These may still be available on a limited basis.

[25] Yorktown Mall will not have a great impact Chicago proper, for it will attract only th occasional shopper who may be drawn to th shopping center by curiosity.

[26] Rochester, New York has already built a dow town, enclosed mall on a small scale. Gre Bay, Wisconsin's downtown urban renewal pl calls for creation of an extensive downtow enclosed mall with a special circumferenti traffic pattern including tunnels for north-sou through traffic (under the mall).

ll provides a controlled environment both ter and summer, as well as a wide range goods and services. With strategic location l vigorous sales promotion, these centers l keep more business out of the central siness district and will also attract present D business into the hinterland. This does suggest the death of the central business trict (it is still virtually impossible to dupli-e), but may yet force a new function upon In any case, the regional shopping center s, in its way, reversed the traditional flow retail dollars.

CREATION

consideration of recreation within the ur-n complex once again illustrates centrifugal l centripetal forces. As in retailing, the ces are powerful, but perhaps not as com-itive (except for the competition for the al recreational expenditure in the market).

ntripetal

major portion of nighttime activity in the ntral business district is associated with reation — entertainment. New York's roadway" may be too over-powering an mple to cite, but surely it is no less typical n San Francisco (with its North Beach nplex as an additional lure) or Toronto ith its recent addition of the O'Keefe nter to an already highly developed night-complex including Bay-Bloor and the tral business district), or "any city" with concentration of movie theaters, night bs, restaurants, and hotels. (Las Vegas is haps the gaudiest example of this phenom-on.) As in shopping and in personal serv-s, the central business district is also the ght light mecca for the standard metro-litan statistical area. Indeed, the trading a for entertainment centers may even be ional in extent — Broadway and Bourbon Street certainly serve broader areas than the New York and New Orleans standard metro-politan statistical areas.

Centrifugal

Just as the city provides a concentration of recreation facilities unmatched elsewhere in the Standard Metropolitan Statistical Area, so does the hinterland have its peculiar char-acteristics. Whereas there is rarely a concen-tration of facilities, there is a great variety dispersed through the hinterlands, parts of which cannot be matched by the city.

Recreational opportunities and facilities will often occur in a considerably larger hin-terland than has been used to this point. Home recreational facilities such as those that lawn and garden may provide are important, but highly localized, and not comparable to the commercial aspect of recreation found in the city.

Perhaps the *cash* factor provides an initial point from which to progress. Open land, ex-cept for occasional green strips within the city, is the most important and striking feature offered by the hinterland. Most often it is here that the city dweller can find camping, boat-ing, swimming, and other outdoor activity. It is here that scenic views can be found. Here "nature" is displayed for the citizen accus-tomed to a manmade environment. Some of this may be free, but there is a commercial-ized segment of this phenomenon that attracts sales as well as participants.[27] This pull out-ward generates sales as well as traffic, and also supplies a service virtually impossible to obtain in the city.

The city's hinterland should not be infer-red to be barren of "city functions." The

[27] Usually, sight seeing traffic is carried by "old type" roads which have had limited control of roadside development. The souvenir shop, gas station, restaurant, and produce stands all cap-ture city dollars.

differentiation is one of concentration. We can all find examples of the fine restaurant nestled in a rural setting or even associated with a new shopping center. We are familiar with the summer theater and the outdoor performing centers such as the Blossom Center between Cleveland and Akron, Ohio. Although dispersed spatially, they nevertheless draw "trade" from the city, where these facilities are not available (despite the seasonal nature of some of these recreational activities).

Today there is a rebirth of the motion picture industry and the movie theater. The TV doldrums which resulted in the abandonment of many downtown movie houses have finally been passed. Where Hollywood and others are supplying film matter impossible (at least temporarily) to present on the TV screen, the cinema entrepreneurs are providing new, "slick" theaters for the viewing public, but mostly outside the central city. We may even find twin theaters, very often within shopping centers which once discouraged their association.[28] This, of course, is a variation of an older phenomenon which has been the property of the hinterland — the drive-in theater. These developments have had their greatest effect where the central business district did not have a well-developed or modern entertainment center.

There is two-way traffic in recreation. The stream of cars coming into the city at night is partially balanced by the Sunday exodu each area performing the services that it ca do best, but there is an overlap despi specialization.

[28] Theaters were once on the undesirable tenant list in shopping centers, especially where parking was a problem. Long term parking during peak shopping periods was the chief problem created by the theater. Today, shopping centers are attempting to perform more community functions and encourage establishment of theaters in order to foster additional trips to the center. The General Cinema Corporation of Boston has been a leader in this recent theater surge, in many cases, establishing twin theaters within the planned shopping complex.

SERVICES (PERSONAL, PROFESSIONAL, AND GOVERNMENTAL)

The dominance of the central city has bee challenged on many fronts. Medical specia ists, lawyers, and craftsmen are leaving thei traditional locations in order to be closer t (or more convenient to) their growing mar kets. It is no longer unusual to see large offic buildings under construction in even the mos distant parts of the suburbs.

Again, it is the residential sprawl from th city which creates concentrations of sufficien size to warrant establishment of hospitals an their associated medical facility concentra tions outside the central city area. It is th increased efficiency in communication tha permits attorneys to leave the central busines district and also permits insurance companie to erect large complexes in the hinterlan (American Mutual Insurance Co. — Wake field, Massachusetts). Despite the outwar move, the central business district and th central city still have the greatest concentra tion of office space and the greatest concen tration of all types of service specialists. Th ambulance with an emergency case mos often is inbound, as is the man with busines problems. Even those specialists who leav the central area often do not cut their tie completely, for they operate dual office facili ties. Regardless of proliferation of facilitie outside the city, there is still a basic centri petal power which is exercised on the whol metropolitan complex.

Just as the concentration of hospitals in th inner city draws the medical profession, s does the concentration of government draw the attorney, the accountant, the realtor, th e company, and the architects, engineers, l other business service professions. Even hout metropolitan government, the seat of wer is most often in the city. Where else one find the legislative, judicial, and exe- ive functionaries of government together h the economic giants such as banking and ance? Whether the smaller communities in etropolitan complex admit it or not, they dependent on the city. The hinterland ellers may believe they escape physical itact with city problems, but it is an illu- n. This has been recognized by the federal vernment and the metropolitan areas in the ablishment of planning areas for transpor- ion studies underway in almost every tropolitan complex. Toronto has estab- ed a form of metropolitan government, as ; Dade County, Florida (Miami). In many er areas metropolitan commissions are icerned with a variety of problems such transportation, law enforcement, and sup- ing public utilities.

City planners have been most sensitive to needs for supra-government in our metro- itan areas. Suburbanites guard their gov- ing rights most zealously, but they are still ected by the actions of the city and one y must yield to its dominance. Perhaps a t step is the income tax (variant of a head) now levied on "outsiders." Lest the resi- nts of the hinterland think there is escape, y may well read a short article in *Atlanta gazine,* an organ of the Atlanta Chamber Commerce, by Opie Shelton entitled "The y That Atlanta Disappeared."[29]

The historians later called it the greatest unex- plained phenomenon in the history of the world. No one disputed their claim.

Imagine one day there sat Atlanta, dynamic and bustling, one of the nation's greatest cities. The next morning it was gone, having disap- peared during the night without leaving the slightest smidgen of a clue behind.

Gone were the towering office buildings, the banks, the stores, the public utilities, the fac- tories, the hospitals, the schools, the colleges and universities, the centers of culture, the stadium, the homes, the people, the churches, everything.

The suburbs were left untouched. But when early risers attempted to turn on the water, nothing happened. Neither would the electric lights turn on. Their battery-powered radios offered a hint of things to come. Only a hand- ful of the radio stations were on the air, but there was nothing on their news tickers about what had happened during the night.

It was only when these suburbanites headed to their downtown jobs that they actually became a part of the nightmare. As they reached the city limits they suddenly were jolted by the scrub trees and weeds which stood where only yesterday had been a majestic city.

Panic followed.

By word of mouth, the news spread. ATLANTA HAS DISAPPEARED!

Some of the less enchanted chorused, "Good Riddance!" A few said, "Well, we finally got rid of all the niggers."

But the masses didn't join in the chorus.

Those who went to their jobs in the suburbs found the doors of most establishments locked. Lockheed had posted a notice, saying that all of their skilled engineers and scientists had resigned en masse, leaving to accept their choice of dozens of other jobs in other areas. The same was true of the automobile plants and of most of the other job producers of the suburbs.

It took a few days for the original shock to wear off. Then came the mass exodus.

The young and talented pulled up roots and headed for other cities to seek their fortune. The displaced college students began knocking on other college doors, seeking admittance. Doc- tors and dentists packed up their tools and left. Anyone who could joined the exodus. Soon all that was left were the very old, the feeble, the untrained and the unskilled.

Within one year's time from The Day That Atlanta Disappeared the once prosperous sub- urbs lay prostrate. Homes were boarded up and abandoned. The automobile had become the only means of transportation available. No longer was there the sound of jetliners or trains. The few stores that were left did little business. No one was moving in, only out.

In short, these once flourishing suburbs had taken on the same desolate look which marks so many of the decaying and dying towns over the nation.

There has always been a dispersion of service activity concomitant with a dispersion of population (the market). Despite the growth (relative and actual) of medical, legal, and personal services in the hinterland, the basic thrust and orientation is inward. The central business district and the central city have strong attractive forces, and overshadow the equivalent activities in the hinterland.

COMMUNICATION AND TRANSPORTATION

Communication

Communication is the unifying influence on the metropolitan or even megalopolitan area. Its role in permitting functional differentiation both within and without the city has been demonstrated amply. Nevertheless, there are special points to be made regarding this extremely complex phenomenon.

Communications are possibly the most centralized urban function. Recognizing that NBC in the Los Angeles area operates from "beautiful downtown Burbank,"[30] it is more common to find radio and television studio centralized in the city.

The centralizing power of airborne com munications should be recognized by all of u who live outside the corporate city limits, o for that matter, outside the urbanized area.[31] New York's radio and television signals in vade Connecticut and New Jersey; Los An geles stations can be heard in Redlands; cabl television brings in signals from even greate distances. City news and city phenomena ar distributed outward. In many areas the bi city stations, both radio and television, eve present the farm news! It would be difficul to measure the effect of these media in estab lishing a metropolitan identity, but surel they must create at least a metropolita awareness and character.

Newspapers function the same way in thei preparation and publication. Although ther may be a proliferation of daily and weekl newspapers throughout a metropolitan area the major media are central city citizens Within the past 10 years this writer has wit nessed the movement of two large news papers, the Boston *Globe* and the Bosto *Times Herald,* from the central business dis trict to other close-in central city locations Employment in this medium is drawn fron all parts of the metropolitan area. The prod uct originates centrally and is dispersec through wide areas.[32]

Newspapers, radio, and television are al

> There were acrimonious afterthoughts.
>
> "Perhaps we had more stake in the well being of the core city of Atlanta than we were willing to admit," some said.
>
> But, alas, it was too late to turn back.
>
> On The Day That Atlanta Disappeared the suburbs became doomed to a slow but sure death. The umbilical cord which had fed them had been severed.
>
> Oh, if only those people had understood that the fortunes of their city and their suburbs were inextricably tied together. Opie L. Shelton, "The Day That Atlanta Disappeared," Reprinted from *Atlanta Magazine,* April 1966.

30 The recent NBC comedy show, "Laugh In," has made this facility one of the best known in the nation.

31 For a definition of urbanized area, see Burea of the Census, *United States Census of Popula tion: 1960* (Washington, D.C., United State Government Printing Office, 1961). Definition are in the introduction.

32 The Audit Bureau of Circulation publishes cir culation data — numbers and places — for news papers subscribing to the service. These dat may be obtained at the newspaper genera office.

:ying phenomena. They provide a vital : between city and hinterland, one that rates in both directions but which is cen- d on the city.

:ertainly, telephone communication, re- dless of the location of equipment or :es, has fostered urban growth. It is pos- y the most important factor permitting entralization of population, business, in- try, and virtually all major urban activi- [33]

nsportation

most visible and critical communications dium is transportation. Our homes are rounded by streets which, in turn, lead to iways and terminal type facilities. Indeed, this means of circulation associated with automobile that has formed our cities and ropolitan areas.[34] These surface pheno- ia are most obvious to us, and the recent ation of a Department of Transportation the recent federal requirements for trans- tation studies underline their importance nderstanding and planning the city (real).

Rail and rapid transit — the important -automobile surface lines — have played tively minor roles in recent years. How- r, as the need for parking facilities increa- as congestion becomes more severe, and air pollution becomes a greater problem, find renewed interest in this mode of vel. Although the Japanese have led the recently, the American cities are now awakening. In the Boston area the Massachusetts Bay Transit Authority has begun functioning during this decade and has returned rail commuter service to the South Shore and added it to the western suburbs. The Bay Area Rapid Transit System has been having difficulties, but is proceeding with development of a new system for the San Francisco-Oakland Metropolitan area. Los Angeles held a referendum on November 5, 1968 regarding a beginning on a rapid transit system. That these lines operate both in and out of the city needs no elaboration, but the major function is to get the hinterland dwellers into the central city for work and play, and then return them home. This type of development, together with the development of the Northeast Corridor rail commuter service will make the hinterlands even more attractive to the city escaper.[35]

Intercity, transcontinental, and intercontinental travel have created problems that the hinterland must solve for the city (and the metropolitan area as well). Air travel and the development of bigger and better planes has created traffic jams as dramatic as the "freeway rush." New York's commercial air traffic must be handled by three airports — two in the city proper and the other in Newark, New Jersey. The need for a fourth facility already exists. Surely, if political considerations can be sublimated, it will be planned in a hinterland location. Whether the airport represents a city exclave or not, new facilities must be placed farther from its center.

Although Chicago is now considering construction of a new air terminal in Lake Michigan, it long ago replaced Midway with

Note that telephone call data are considered by the Bureau of the Census in its definition of tandard metropolitan statistical areas.

For a recent discussion of Ekistics, see C. A. Doxiadis, "Man's Movement and His City," *Science*, Vol. CLXII, No. 3851 (October 18, 968), pp. 326-34.

[35] The Department of Transportation has taken the lead in developing high speed rail transportation between Boston, New York, and Washington, D.C.

O'Hare.[36] Dallas and Fort Worth have finally decided to cooperate in operating a joint airport between them. Kansas City is in the process of replacing a small "downtown" airport with a modern supersonic-capability airport in the northern part of the standard metropolitan statistical area. Dulles Airport, 50 miles from Washington, D.C., is finally receiving some of the traffic for which it was designed. Surely it will replace National as the major airport for Washington, D.C. Here, too, there was experiment which was perhaps too early. Friendship Airport between Washington and Baltimore has not become for D.C. what might have been expected; rather, it is Baltimore's airport.

Airports and airport problems are not restricted to any part of the country or the world. It appears that we may look forward to a decentralization of the large jet ports which, in turn, may be connected to the central city by smaller local airports serviced by helicopters or STOL craft.[37] The hinterlar is already performing a "break of bulk" fun tion for the city and is likely to increase th particular function.

CONCLUSION

City-hinterland relationships cover the who range of economic and social functions withi the metropolitan complex. Whether or whe a line is drawn creates the initial and maj problem in delineating inside vs. outside fun tions and relationships. The major focus this chapter has been on the administrati boundary line. This focus is on intra-metr politan phenomena. Had the line been draw at the extreme of the urbanized area, a ne set of problems would arise, but they wou still fit the major categories discussed here.

There are no self-sufficient cities or metr politan areas. Neither are there complete static situations. Cities are in a dynam equilibrium with their hinterlands and the total environments. Analysts may stop th action momentarily to view relationships, b they must recognize that such relationshi are ever changing. These are entities whos limits are indeterminate.

[36] After virtually abandoning Midway for commercial flights to Chicago, a number of airlines have recently re-established service there. This change has been due to the traffic problems at O'Hare, and has been fostered by the development of jet aircraft that can safely use Midway's shorter landing strips.

[37] Great emphasis has been placed on developme of Short Take Off and Landing aircraft, bo conventional and vertical. These craft have be very busy on commuter runs to and from L Angeles International Airport, supplementi and possibly supplanting helicopter service some Los Angeles metropolitan area airports.

26

rban Expansion— /ill it Ever Stop?

ason Gaffney

ıen you walk down Main Street in any ge city, each step takes you past several usand dollars' worth of frontage. Frontage common measure of city land, and it goes the foot, like a precious commodity. A nt foot is a foot along the sidewalk with a p behind it 100-150 feet to the rear of the . A foot on the right street is worth whole ms.

Among the dearest is State Street in icago, where some frontage goes for 30 usand dollars a foot. At that rate an acre uld bring 13 million dollars. Market Street San Francisco runs up to 10 thousand dol- s a foot. A foot on Fayetteville Street in leigh, N.C., is worth about 4 thousand lars.

Why do these strips of otherwise common t command such prices? The answer lies in forces of urban centralization.

Urban land, which serves a region much the farmstead serves a farm, is a central rage base for collecting and distributing puts and inputs and for sorting, process- , and reassembling them.

It is a center that affords easy, reliable access to enough volume and variety of resources to supply complex, specialized, continuous, and large-scale operations, and enough markets to absorb their outputs and byproducts.

It is a reservoir of goods and labor whose abundance gives the slack to allow flexibility of operations, meet emergency needs, and afford the innovator endless possible combinations of skills and resources to experiment with.

The city is a convenient gathering place where buyers can rely on finding sellers, and sellers buyers — a place to inspect, compare, and exchange goods and render and receive services. Its large local market attracts a variety of specialized goods and services. Its compactness permits cheap distribution, which in turn facilitates savings from large-scale central operations.

It is a central store of information and ideas — a place to confer and arbitrate face to face, to plan and administer, to do research and educate. It is a place where many minds

ɔrinted from *Land, The Yearbook of Agriculture* (1958), pp. 503-22 by permission of the author and United States Department of Agriculture.

can associate freely to stimulate, evaluate, and diffuse new techniques and ideas: in all, the brain, control, and power center of society.

Urban land commands a premium, too, as a place to reside. For living, as for business, its advantage is access to a wide selection of opportunities and associations.

Although it need not be fertile, or flat, or even dry, good urban land is scarce. The value of land for urban functions depends on its location relative to transportation, resources, and markets. Large-scale producers attach a special premium to the best lands, as they require access to the widest markets for economical operations. Being large, they also require large areas, so that competition for the best land is extremely keen.

The entire network of location factors defies simple analysis. But the greatest cities develop at strategic central locations, where they assemble and process many resources for many markets. Junctions and hubs of transportation have obvious merits, as do heads of navigation and other load-breaking points.

Good location is not enough to fit land for urban functions. Access, the basic urban resource, is partly man-made. The city enhances its natural advantages by pushing out routes to tap wider territories, but that is only a start. To realize its full potential, the city develops a network of local transportation — a system of general access through which its lifeblood moves.

So vital is transportation that most cities devote more than half their developed land to it. In 53 central cities — "central" meaning the major downtown city of a metropolitan region, excluding suburbs and satellites — which were studied by Harland Bartholomew for his book, *Land Uses in American Cities,* streets and alleys alone occupied 2 percent of the developed area.

Autos are voracious off-street land con sumers, too. One parking space, with acces lanes and a little to spare to allow for huma weakness, preempts more than 300 squar feet. The driveway and garage on a resider tial lot occupy about as much surface as th house. Many factories occupy less space tha their own parking lots and loading and del very aprons. The modern, auto-oriente shopping center allows 4 or 5 square feet c parking for each square foot of floor area Filling stations are almost entirely ope space.

Other forms of transportation are less de manding, but still they take a good deal c land. Railroads took 5 percent of the citie studied by Mr. Bartholomew, including muc very costly land near downtown. Conside able space is devoted also to docks, bu terminals, airports, and easements for pipe and wires to transport water, gas, and elec tricity. Halls, elevators, and stairs take spac inside buildings.

Most of this spacious network of publi and semipublic lands dedicated to free move ment yields little direct income, but the cit can ill afford not to devote generous space to these corridors, which allow full release c the enormous productive forces inherent i specialization and exchange and give the pr vate lands their value.

The final essential for productive urba land is the improvement of adjoining lanc One lonely storehouse no more makes a cit than one smoldering stick makes a fire. As sembled buildings compete for customer suppliers, and use of public spaces, bu generally they also complement each other s as to enhance enormously their overall proc uctive value.

For the essence of urban value is acces

d every resource the city adds increases the lume and variety of resources accessible to . Each new seller is a magnet for more yers.

Each buyer is a magnet for sellers, pulling de from farther away, attracting more nsportation routes and scheduled runs, and ping establish the city as the place to rely finding what you want, selling your wares d services, and, in a dynamic, competitive rld, keeping touch with the latest products, ormation, techniques, and ideas.

Each addition to the local market helps o to spread the overhead of more spec-ized and larger operations. Each new tax-yer shares the burden of large public works d improves the city's credit. Each new oducer helps diversify the city's economic se and insure its stability. Each new seller ds either to bring in outside money or luce leakages of money to outside sellers, d thus he creates new demand for local vices.

A growing city therefore may enjoy a long ge of increasing returns, when growth gets more growth. Thus the one best loca-n in a region has a decisive advantage over second best, and the earliest development s a commanding lead over later comers. e largest urban nucleus tends to snowball, ile others shrivel.

The European scholar Georges Widmer s provided an interesting demonstration of reasing returns in urban growth. Widmer rked with Swiss census data, and published results in the Revue Economique for arch 1953. He found a direct relationship tween size of city and several measures of r capita economic activity, such as wages d tax revenues.

A limit to increasing returns is the cost of nsportation. The larger the city grows, the ther it has to range for markets and materials. And many cities are stopped short of this limit by the city fathers' fears of spoiling their markets, lowering rents, risking money on public works, raising wages and taxes, admitting outsiders, spoiling the fishing, or losing control of city hall.

But a number of metropolitan titans have burst these bonds to accumulate a large share of the population, capital, and the land value of the country. New York City (excluding suburbs and satellites) in 1955 had about 7.8 million people (4.8 percent of our population), and its annual real-estate taxes were 746 million dollars, 7 percent of the national levy.

The gravitational pull of a city does not stop at its fringes. The center of gravity, the downtown district of maximum access, draws the whole city in upon itself, story on story. In this focusing of demand, the city finds further increasing returns from large-scale building.

The most economical layout to interconnect given space users is in three dimensions, in which central heating and other utilities can be distributed over shorter conduits than in two dimensions and each room has quicker access to most of the others. One roof and one foundation serve many stories. Inner partitions need not be weatherproof; the outer surface of a cube increases in less proportion than the space it encloses. So a large, multistory building provides given space, services, and access more cheaply than several small buildings.

There are limits to the economical height of buildings and to the amount of crowding people will endure, of course, and everyone knows that a conspicuous centrifugal surge started some years ago. But nevertheless a city keeps its basic cohesive tendencies, which are its reason for being.

Just how large an area cities occupy no

one knows, for no one can say where a city ends. The United States Census defines "urbanized areas" roughly as those in and around cities of at least 50 thousand inhabitants. That was about 8 million acres in 1950, evenly divided between the central cities and their urbanized fringes. Eight million acres equals the area of Maryland and Delaware, 0.42 percent of the continental United States, and a little less than the 9 million acres in farmsteads. It seems a modest space requirement for its 70 million residents, particularly the 50 million in central cities.

The census has been conservative in its definition, for the area enclosed inside far-flung urban outposts would be much greater. Eight million acres is the area of a circle with a 63-mile radius, or two circles with 45-mile radii, and stray bits from any one of our metropolitan giants may be found that far from its center.

But even the census' limited area is urbanized only in a loose sense. Despite the advantages of compact land use, central cities themselves are surprisingly patchy. In Mr. Bartholomew's 53 central cities, the undeveloped portion was about 29 percent. Although his surveys are not all up to date, many local planning surveys show comparable figures after 1955.

His developed urban land was about 0.06 acre per capita, or 5 yards on a football gridiron. At that density, the 50 million inhabitants of central cities of more than 50 thousand use nearer 3 million acres than 4 million.

Even some of that 3 million acres they "use" only in a poetic sense. It is mostly open space. The area actually covered by buildings is probably less than 400 thousand acres, less than some western ranches and less than 15 percent of the developed area of the central cities.

No one expects that every building should occupy 100 percent of its site, but just how big a yard and grounds should be so as to be designated as developed by a building somewhere on it is a puzzle. Some urban buildings do occupy their entire sites, and by contrast such other sites as the 75 acres around Ford's new administration building in Dearborn seem nearer akin to undeveloped lands. No one can say exactly how we are to designate such lands, but some sort of allowance would certainly reduce the central cities' land "use" appreciably below 3 million acres.

If the central city is a little patchy, its outskirts are in shreds. Here, to be sure, are big users of land like golf courses, dumps, drive-ins, and airports, serving the central city. But it would be hard to define any segment of this nebulous territory that was not largely in weeds. Probably less than half the 4 million acres of urban fringe cited in the census deserves to be called "developed."

For cities under 50 thousand, our data are progressively less detailed.

Hugh H. Wooten and James R. Anderson of the Department of Agriculture, estimated that all cities of more than a thousand inhabitants in 1954 occupied 18.6 million acres — about the area of South Carolina and 1 percent of the continental United States. Smaller communities may occupy another 10 million acres. But all these figures include empty spaces, which make up larger portions of the smaller cities.

As to urban values, they are prodigious. It is easy to underestimate them because of the comparatively modest space requirements of cities. There is nothing modest about the prices of urban land, however.

Residential lots in respectable established neighborhoods sell for 50 dollars to 250 dollars a foot and for more than 500 dollars

›ot along a few gold coasts. Apartment ; average higher, going above 1 thousand ars along Lake Shore Drive in Chicago. n sites are often held at fancy prices be-se of an expectation of future industrial, ımercial, or public demand. Some sub-ary shopping districts sell for 1 thousand ars a foot. The best industrial sites in e central cities command well over 100 ısand dollars an acre.

'rices of land out from the center are :h lower, but still impressive, especially r the multifold increases since 1950. Un-eloped residential or industrial land along superhighways was bringing several ısand dollars an acre in 1957, and more ınd New York City. Industrial acreage r Eastshore Freeway, Oakland, averaged 500 dollars as early as 1953. Potential s of shopping centers brought 10 thousand i0 thousand dollars an acre, as did motel s near the better interchanges of the new ıpikes and thruways.

ιirspace above the golden ground of the also carries high price tags. An option on over the Pennsylvania Railroad tracks in v York specified more than 3 million dol-an acre in 1955. A Times Square bill-rd brings 15 thousand a year.

ιt such prices, it does not take many :s to outvalue all the farms in whole :es, and in most States one or a few of largest cities do. New York City real te in 1955 was worth some unknown but e amount over its assessed valuation of 20 on dollars, which was the current market ıe of all the farm real estate in New York :e and 19 other Eastern States. For the ›le country, urban values exceed farm ıes several times over.

t may even be that urban values exceed n values per capita. One cannot be certain. ıd prices swing violently and rapidly, yet the only general source of data on urban values is from moss-covered tax assessments. Urban assessments are more obsolete than rural assessments — if that is possible.

But we do know how much taxes property pays. It may surprise some farmers to learn that farm property taxes are less per capita than nonfarm property taxes — roughly 54 dollars, compared to 72 dollars in 1956. Of some 11.7 billion dollars levied that year, farm property bore only 1.2 billion dollars.

The higher urban levies might reflect higher urban tax rates, rather than per capita values. The average rate on farm real estate, as reported in 1957 by the Agricultural Finance Review, was about 1 percent of market value. There is a general impression that urban real rates average higher — and some evidence to back it up. David Rowlands, of the University of Pennsylvania, in a report on the Property Tax in Atlanta and Other Large Cities, estimated effective tax rates in 20 large cities for 1956. Only 2 of them fall under 1 percent, and a few exceed 2 percent.

On the other hand, a study published in the Review of Economics and Statistics for February 1957 found otherwise. Scott Maynes and James Morgan, analyzing voluminous questionnaire data from the University of Michigan Survey of Consumer Finances and the United States Census Residential Financing Survey, found the real rate of property taxation on owner-occupied urban residences in 1953 to be nearly 1 percent.

They did not check the possibility that respondents may have tended to understate their taxes. Nor did they discover to what extent the low tax rates on owner-occupied residences resulted from homestead exemption, which would not apply to other classes of real estate. Still, it is other classes of real estate, especially rented slum and vacant land, that are most frequently found to be underassessed.

One might reason that city tax rates must be higher because city property pays city taxes on top of county taxes — although, of course, the most urbanized counties might have lower overall rates than predominantly rural counties. City people get more local governmental services, it is true, but they get them cheaper because they live closer together. They also have more non-property-tax sources of revenue.

Then, too, a census study under Allen Manvel found farm real estate overassessed — hence overtaxed by the counties — relative to urban real estate in 101 counties of downstate Illinois in 1946. Arthur Walrath found the same in several counties around Milwaukee in 1955. Remember, too, that an appreciable share of urban real estate is tax-exempt institutional ground.

None of these studies provides a solid basis for estimating urban real-estate values. The United States Census of Governments planned to release in 1958 what should be a definitive study of tax assessment ratios. Even that omits tax-exempt real estate from consideration, and also it omits suburban acreage, but still it may provide the first firm estimate of urban real-estate values in the United States.

Meanwhile, we have reasonable grounds for putting the real rate of urban property taxation between 1 and 2 percent, which means the aggregate value of urban real estate is of the order of seven or eight times greater than farm real estate. It is entirely possible that a 100-percent comprehensive reckoning, including tax-exempt holdings and suburban acreage, would reach as high as 10 times farm values, or 1 trillion dollars.

Other indirect evidences of real-estate values are the mortgages they carry. As of September 1957, the farm mortgage debt w 10 billion dollars, compared to 143 billic dollars on nonfarm residential and cor mercial real estate. Nonfarm real estate 1957 probably carried a higher ratio of de to value — it is impossible to say for certa because most real estate is unmortgaged. C the other hand, however, nonfarm mortga figures do not include the debt on industri rail, or utility holdings, or on institutional a public real estate.

Several studies also indicate that urb families occupy dwellings valued at two three times their annual incomes. This su gests that urban residences alone are wor more than 500 billion dollars.

These last two lines of reasoning yield definite numerical estimates of urban value but they do confirm the belief that they dwa the value of farm real estate.

Real estate is more than land, of cours and conceivably urban real-estate values i here largely in the buildings — we hear good deal about the declining importance land in an urban society. That may be a mi conception, however.

Builders putting new single-family hom on cheap outlying land reckon the site at on sixth or one-fifth of the total cost. But n many urbanites live in new homes on che outlying land. Even in 1957, after 12 yea of record-smashing construction, 75 perce of all urban dwelling units were built befo 1945 and most of them before 1929. The are almost no new residences in older ce tral cities. A study by the Real Estate Boa of New York in 1953 found that 80 perce of Manhattan's apartments were more th 50 years old.

In fringe areas, where new buildings outvalue their own sites, a large share of t sites have no buildings. Around Clevelan for example, 57 percent of the Cuyaho unty Planning Commission's "suburban g" and 84 percent of the "rural ring" were cant in 1954. In commercial districts, with ir majestic frontage prices, it takes a new d substantial structure to match the site ue.

All in all, from the limited information ailable, there is no reason to dismiss land ue as a minor part of urban real-estate ue, especially if we include vacant lands their current market prices. It may even the larger share. And, interestingly ough, the ratio of land to building values ds to be highest in the centers of large, nsely populated, and built-up cities, where onomic life is supposed to have lost touch h the land most completely.

A striking aspect of today's cities is their oid outward thrust. Urban values being at they are, cities gobble up farmland at l. There is no accurate survey of the wide d ragged urban frontier, but various esti- tes suggest it has been advancing recently out 400 thousand acres a year into the art of America's farmlands.

Is this in the farmers' interest? Many ughtful observers are raising voices in rm for the future. The most vocal of them m to think the city should be contained. ere is another side to the question, though. The city serves the farmer and buys his oducts. It is the farmer's interest that cities ve ample land to serve him well. He would ly suffer if he were to confine the city into ottleneck between the barn and the table. In fact, the city is all too likely to become ottleneck, anyway, with no help from the mer — but much to his detriment.

Because of increasing returns in urban owth, many cities in strategic spots have a easure of monopoly power over parts of eir trade territories. Without the spur of mpetition, they are easily tempted to settle back comfortably and take their customers' money without the costs and bother of offering very adequate or modern service. Their strong position lets them do this simply by vegetating quietly without necessarily having any active monopoly motive. Because downtown sites are favorite investments for absentees and heiresses, too, a high proportion of them fall into ownerships that tend to resist progressive management and risky improvements.

There is competition within each city, of course, but the city fathers who are so inclined can minimize it by restrictive policies. They may lay out streets so as to limit the business frontage; maintain obsolete traffic patterns to protect vested investments; discourage new buildings by overassessing them relative to old — a practice that has become especially common since the war — and assessing undeveloped land at next to nothing; zoning out new developments; limiting the height of buildings; winking at tax-delinquent land speculators and selling off foreclosed properties only slowly; fostering obstructive building codes; endowing tax-free institutions with grounds vastly beyond their needs; neglecting essential public works and services; and refusing to act decisively against obsolescence and blight.

Whether by design, apathy, or sincere devotion to an obsolete tradition, probably most cities contrive to remain inadequately developed to serve fully the demands on them.

To protect themselves, the farmers' best assurance of adequate, modern, and competitive urban services may be to release lands for new development around stagnant central cities. With all its faults, such expansion does introduce new competition for farm trade.

The urban expansion bears critical watching, however.

Are efficient cities evolving — cities that distribute goods with minimum time, motion, and cost?

Are cities swallowing much more farmland than they need?

Above all, does the present pattern of urban expansion contain the same elements of instability that have brought most previous land booms to collapse?

To answer these questions, it is necessary to analyze the process of urban expansion more closely.

Like the eager suitor who leaped onto his horse and dashed madly off in all directions, the city moves out hither and yon with little apparent consistency or reason. Here is Washington, D.C., growing out from the back door of the Capitol, in defiance of its planner's best-laid schemes. There is the shopping district gravitating toward a high-income residential area, but radiating influences that create slums in its van and erode away the attracting force. Here are sewers without-houses, while out beyond arise new houses without sewers. There is hardly any predicting where the construction crews will turn up next.

What are the builders seeking?

More space? There is considerable unused space in the central city itself.

Lower taxes? Fringe residents, scattered broadcast with more school-children per capita and without the downtown commerce and industry to share tax burdens, in general must pay more taxes to finance given municipal services.

Surveys in 1955 by Amos H. Hawley and Basil G. Zimmer, of the University of Michigan, found fringe residents around Flint, Mich., actually more willing than residents of the central city to assume higher taxes. And it is evident that many people flee central cities in search of better schools and other costly public services that the cit fathers are to parsimonious to finance.

Freedom from traffic? The farther on lives from jobs and markets the more traffi he must buck in between.

Freedom from restrictive policies? Ofte so — yet many suburban enclaves becom more restrictive than the central city.

Of the many, many things that urba refugees are seeking, most are to be foun in the central city. The refugees want munic pal services, access to social and economi opportunities, and other urban advantage — but not at any price. To oversimplify complex politico-socio-economic phenome non, urban outmigrants, like the westwar pioneers before them, are seeking cheap lan The very advantages of the city prove it major liability when they promote askin prices so high as to drive builders out c town.

The quest of cheap land leads the city n just to expand, but to disintegrate. Th quest turns very much on the individu seller. Asking prices for comparable land vary widely with the seller's finances, ta position, information, sentiments, or jus plain cussedness. Jack Lessinger, of the Un versity of California's Real Estate Researc Program, has found tentatively that in th Santa Clara Valley, around San Jose, it i the smaller farmers who succumb earliest t the city, and larger landholders who hold ou longest. The French geographers, M. Phlip poneau, J. Tricart, and C. Precheur, describ the same tendency around Nancy and Pari Buyers find a bargain here, another yonde and build accordingly, so that developmer proceeds in patches and freckles.

State highway builders can stretch fund much further where the right-of-way is cheap Besides, holders of cheap land are less likel to band into militant "Property Owners otective Leagues" and the like to block new uways; and railroads are just as happy to highway funds diverted to routes not ralleling their own. New highways, like lroads before them, often tend to bypass ngested areas and develop earliest and st fully in less settled territory. They open de new areas to hunt-and-peck develop-nt and establish new urban nuclei where y converge.

These outlying nuclei are bases from which en farther flung developments are nched. Especially along trunk routes, they alesce into gangling, diffuse urban com-xes that some writers, fancy running free, describing as "polynucleated urbs," "co-bations," "cities as long as highways," omic megalopolises," and "scrambled gs" and hailing, with enthusiasm or re-nation, as forerunners of a new era.

Our first question was, "Are they efficient ies?" By any ideal standard they are not. Transportation and utility lines to join the ttered pieces cost billions. The result at st is a poorly coordinated tangle. Com-rce bypasses old bottlenecks but meets an stacle course that consumes untold time d motion and can hardly avoid reflecting it-f, among other ways, in a wider farm-rket spread.

Such coherent patterns as do emerge are ometrically imperfect. Some variation on inear theme, strung out miles along a rail-y, waterway, or highway, is commonest. t why go 20 miles west when there is open d 5 miles north? It takes three-dimensional velopment to afford maximum access at nimum cost among given users of space. near developments do not even use two nensions, but force all traffic along one ng, congested line. That, often as not, was ilt originally for through traffic.

One can probably understand how linear patterns develop. Cities fail to provide adequate two-dimensional street networks; and interurban trunk lines, financed by the State or National Government, offer ready-built, open-ended avenues of escape to cheap, accessible land. Landholders along existing routes can subdivide without dedicating 25 percent of their land for streets and without submitting to central controls over subdivision plans. But to explain is not to justify.

Our second question was: "Do cities need to swallow so much good farmland?"

We should probably concede the city first choice over the best land, even the most fertile, just as farmers concede corn first choice of the best wheatland. It may not make much sense to farm steep slopes in the Ozarks, but it would make less sense to put St. Louis there, to put Minneapolis in the north woods, and so on. But this hardly settles the question.

Cities, even central cities, are not using nearly the land they already contain. These undigested pieces are of negative value to the city itself. Cities exist to bring people together. Vacant and underdeveloped lands keep them apart and thus destroy part of the city's basic resources: cheap distribution and easy access. Even if land had no alternative use in farming, it would pay many a city to draw itself together.

Dispersion also forces heavier reliance on those hungry land gobblers, automobiles and trucks. Their demands for highway, turning, and parking space displace tens of thousands of dwelling units a year, scatter the city out farther, and consume more farmland. Dispersion requires that each plant, far from the storehouses and services of the central city, be more self-sufficient, which of course increases its space requirements.

It is especially out from the center, though, that cities preempt vast lands they do not use and may never use. Little urban fragments, prospering busily among fields and orchards, excite speculative hopes for land sales around and between them until urban price influence extends millions of acres beyond the city limits.

Urban prices have a baleful influence on farming. The dirt farmer has struggle enough financing title to lands priced by their anticipated income from agriculture alone. Urban prices push him out of the market completely. Landholders near cities must be speculators as well as farmers.

Often they are not farmers at all. High-priced lands in areas with urban possibilities tend to gravitate to those who have the financial power to wait.

Urban financial power is something few working farmers can match.

Federal income-tax laws tend to aggravate the dirt farmer's disadvantage, for they make speculative gains especially attractive to those in higher tax brackets. To begin, any interest and local taxes are fully deductible. Then the speculator may qualify for "capital gains" treatment — that is, for excluding 50 percent of any realized increment from taxable income, with a maximum tax rate of 25 percent on the increment. That is of great value to the man in an 80-percent tax bracket and tends to make him a high bidder in the market for appreciating suburban lands.

To qualify for capital gains treatment, the speculator must establish that he is not "in the real-estate business," but is a passive "investor," neither improving land for sale nor soliciting buyers. Or he may establish that he is "using the land in his trade or business" (other than real estate).

Should he lose on one sale he can offset the loss against other capital gains. Better yet, if he establishes that he is using the land in his trade or business, he can offset losses against ordinary income, even though an gains would not be taxed as such.

Still better, if it is his residence that h sells, and he puts the proceeds into a nev residence within the year, the entire gain i tax free — and with a little effort a commute may learn to "reside" over a considerabl investment.

Best of all, one who buys land years ahea of his own needs never pays a tax on the ris of value so long as he does not sell — some thing many large corporations, with hug reserves "for expansion," have little expecta tion of doing. Wilbur Steger, writing in th National Tax Journal for September 1957 estimates that 90 percent of all capital gain were thus left tax free from 1901 to 1949.

The result of all this is a virtual scorched earth policy for many lands around cities Why risk any improvement or overt sale effort that might land you "in the real-estat business" and thus disqualify your increment from "capital gains" treatment? Why nc hoard up vast industrial estates for "futur expansion"? Should your alleged need actu ally eventuate and if the value of the land ha gone up in the meantime, you will hav achieved a kind of tax-free income. Shoul you sell, you can probably get capital-gain treatment for increments and ordinary offse for any losses.

For lands that do remain farmed, the influ ence of urban prices often means a wastin away of farm fertility and capital.

Dr. Lessinger has documented this pheno menon in his dissertation, *The Determinatio of Land Use in Rural Urban Transition Area* (Berkeley, Calif., 1956). Around expandin San Jose, Calif., prune and apricot orchard are deteriorating as the city infiltrates th Santa Clara Valley. He analyzes the age dis tribution and bearing condition of orchard in different zones around the city and find rioration of orchards closely related to cipations of urban demand, as reflected in prices.

hus the city takes land from the farm before actually putting it to urban use. a degree this is economical: farm im- ements are wasted on lands marked for ediate urbanization. But Dr. Lessinger's ies indicate that urban prices, with their hting influence on agriculture, already nd over an area of the Santa Clara Valley beyond any likely urban demand. Is this eneral condition throughout the United es?

uppose we allow the entire nonfarm popu- on of the United States the luxury space dards of Winnetka, a Chicago suburb. h a golf course, spacious parklands, play- ls, beaches, wide, tree-lined streets, two road rights-of-way, large lots and yards, ate driveways and two-car garages, estate ricts, and almost no apartments, Win- ka has 0.16 acre of developed land per dent — far more than the 0.06 acre in 53 central cities that Mr. Bartholomew veyed.

t the Winnetka standard, an urban popu- on of 150 million would require 24 million s — about the area of Indiana — which can safely take as beyond any foreseeable nand.

he "regional cities" that enthusiasts are isioning and promoters are touting along Atlantic, Pacific, and gulf coasts, the at Lakes, dozens of State freeways and npikes, resurgent inland waterways, and cipated Federal-program superhighways ng with a more conventional accretion und established cities) by the simplest nt exceed that 24 million acres by a wide rgin.

wenty-four million acres would be con- ed in 6 circles with 45-mile radii; or 24 circles with 22-mile radii; or 120 circles with 10-mile radii. As small a city as Eugene, Oreg., extends its price influence more than 10 miles from the center (not around a full circle), but there are 340 cities in the country larger than Eugene and the price influence of some of them radiates more than 50 miles. If that were not enough, there are thousands of smaller towns. A careful survey would probably show at least 100 million acres — the area of California — under the influence of urban prices.

The answer to our second question, then, is that cities are taking and leaving undeveloped more farmland than they need.

This raises the third question: "Can urban expansion continue?" Or have the onrushing urban armies overextended their lines and lost themselves in agriculture's defense in depth?

Many writers since 1955 have been projecting trends of the past 10 years forward another 20 years or so and viewing with alarm the startling inroads on farmland. History warrants few things less than it does projecting land booms far into the future. Cities typically have expanded in waves.

May we expect the present wave to break and recede?

This also is a prospect to view with alarm. The enormous financial impact of urban expansion is a vital element of our prosperity. New construction, excluding farm and military construction, has been running around some 40 billion dollars annually. That is nearly 12 percent of the national income. It consists mainly of residential, commercial, industrial, highway, and public-utility building. Most of it is tied closely to urban expansion.

The role of construction in sustaining the flow of spending is greater than its volume alone would suggest. A good deal of purchasing power in most years leaks out of the circular flow of spending into savings and allowances for depreciation. The leakages must be offset each year by new investment to avoid a multiple decline in national income.

A decline of annual investment under most conditions will produce a multiple decline in national income because consumption spending, which declines when income declines, is also a creator of money income. Lower investment means lower income. Lower income means lower consumption. That in turn means still lower income — and so on through several stages.

Autonomous declines in consumption would have similar multiple effects, but consumption usually is a relatively passive factor, which economists are inclined to treat as primarily a function of income itself. Investment is more independent and temperamental a variable, and probably most economists would agree that maintaining national income is in large part a problem of maintaining investment spending.

Of the investment on which so much hinges, 40 billion dollars of construction spending is a large share. It is also the most independent share. Other private investment is mostly in less durable goods — machinery, equipment, and inventories. Replacement and turnover of these are passive functions of time and income to some extent. Other public spending is mostly relatively rigidly committed.

The importance of the third critical question is equaled by the difficulty of answering it.

On one hand, cities have rarely expanded rapidly without tragedy — neither, for that matter, has agriculture. We have experienced land development booms along wagon roads, canals, steamboat channels, plank roads, steam railways, horse railways, cable carlines, trolleys, subways, elevated railroads, and motor highways, with townsites and subdivisions proliferating on every hand. Most of th booms busted.

The disasters of 1819, 1836, 1857, 187 1893, and 1929 greet the tourist through hi tory like bones bleaching by the trailsid Will future historians shake their heads sad over the "second automobile bubble," today they do over the first, and over th "canal fever," "plank-road delirium," ar "railroad mania" of the past?

Perhaps — but, on the other hand, histo is under no iron necessity to repeat itsel Optimists who seem to believe that collap is unlikely today cite several reasons: i creasing population; strengthened moneta and banking regulation and insurance; Fe eral willingness and ability to spend; long term, fully amortizable mortgages; more pr dent subdividing practices; large private hol ings of liquid assets; and other reassurir phenomena.

These are not completely tranquilizin however, in light of the cocksure optimis that has preceded and even accompanied – yes, even followed — great crashes of th past. It is worthwhile questioning mo closely the stability of forces that lead citi to preempt lands beyond their needs.

The dynamic process of overexpansic seems to be a complex urban variation on familiar problem of agricultural land settl ment.

The process in simplest outline is this: ne demand raises land prices; supply respon slowly but massively; high prices over th long period of response ultimately stimula more new supply than the demand ca absorb.

Supply responds very slowly to deman because the process of converting land urban use involves many steps by sever slowly moving, poorly coordinated, fr quently reluctant and sometimes downrig obstructive public and private agents and b ıse it usually takes land speculators a long .e to release or develop most of the sites actual service.

Say a new State-financed freeway begins process of bringing farmland into an ›an market. Besides transportation, the d needs water, storm and sanitary sewers, ›phone, gas, electric power, schools, fire d police protection, and sidewalks, to name ne elementary items.

Not only are many services needed. Several ps must be taken to extend most of them m trunklines out through forks and ınches to the ultimate distributive tracery ıt finally brings service to each parcel of ıd. Governments and utilities must decide extend their lines and networks to indi- ual parcels. Landholders must decide it is ıe to receive them — that usually means ›dividing, dedicating lands for streets and ements for utilities, often paying for part the utility extensions and street improve- nts, and perhaps being annexed and sad- d with municipal taxes.

It would be nice for each party involved if the others would commit themselves to velopment before he did — or at least en he does. Then he need only pluck the e fruit from the tree, instead of undergoing ırs of risk, interest, depreciation, and obso- cence while he waits for complementary estments to help his own pay out. The ıation lends itself to a long impasse of 'ter-you-my-dear-Alphonse." At every ge, there is inertia, nostalgia, fear, and long gaining and jockeying.

The final step — actual building on pre- red lots — may be as slow as the others, there are still the lot speculators to wait t. Even when all utilities are in, there is a ther rise to speculate on as homes, stores, ırches, and so on make a community.

We are also witnessing a sort of municipal ıd speculation on a grand scale. Many metropolitan suburbs have incorporated un- developed land, which they proceed to over- zone out of reach of the middle-class market. That is done in hopes that its exclusive tone will one day attract upper crust residents who will pay high taxes, handsomely support local merchants, and send their few children away to school. Many communities are ready to wait a long time for such profitable fellow citizens, even when chances of success are slim.

Ralph Barnes and George Raymond, New York planning consultants, warn in the Jour- nal of the American Institute of Planners for spring 1955, that such municipal policies have become more restrictive than even the communities' parochial self-interests would dictate. New Canaan and Greenwich, Conn., New York suburbs, have actually increased the minimum size of building lot to 4 acres in some sections, in the most congested metro- politan area in the United States. Mountain Lakes, N.J., has gone so far as to buy up a large share of its land to forestall building.

Now scarcity breeds substitution, and while supply is thus developing so dilatorily in areas most logically destined for urban growth, the impatient demand probes out- ward. It finds a warm welcome in many outly- ing communities that have urban aspirations. Some of them even offer subsidies, tax favors, and sites to woo industries.

Moreover, a large share of building is out- side any incorporated area. The Sacramento housing market is an extreme instance. An unpublished report of the Federal Housing Administration, dated April 1957, states that 80 percent of all private dwelling units authorized there from 1954 through 1956 were outside incorporated areas.

These latter-day pioneers demand utilities, which often are willing to come if the custom- ers are there first, especially if rival sellers are within striking distance and if regulatory commissions let them balance any losses with higher rates charged to all their customers. The newcomers also demand public services, which usually come where there are votes and a tax base.

Thus the scattering of urban settlement leads the basic urbanizing distributive networks and services to proliferate over wider territories than the ultimate demand can absorb.

Just how wide and how empty these territories are is startling to discover. The New York engineering firm of Parsons, Brinckerhoff, Hall & MacDonald surveyed land uses and potentialities in connection with its 1953-1955 report to the San Francisco Bay Area Rapid Transit Council. It found ample suitable acreage in the Bay area for the entire projected 1990 population of the whole State of California: 22 million to 31 million people — 7 to 10 times the Bay area's population of 3 million in 1953-1955. This is allowing ample areas for recreation and industry.

The California State Water Resources Board surveyed the area independently in 1955, using aerial photographs, and published the findings in its Bulletin No. 2. For the 10-county Bay area metropolitan region, only 15 percent of the suitable urban land, or 10 percent of the gross land area, was actually developed for urban use in 1955.

In the crowded city of San Francisco itself, the Water Resources Board survey showed 23 percent of the usable land was undeveloped in 1955. Along the Bay side of San Mateo County (the "Peninsula"), which is often hastily described as having become "a solid mass of suburbs," 75 percent was undeveloped. On the Bay side of Alameda County, which includes Oakland and Berkeley, the survey reported 62 percent was undeveloped.

In the Santa Clara Valley (around San Jose), whose "total urbanization" is ofte forecast as imminent, 86 percent of the sui able land was undeveloped for urban use i 1955. The total suitable urban land in thi valley, 155 thousand acres net of street exceeds the area used in 1955 in the entir Bay area (129 thousand acres, also net c streets). The developed portions, howeve are scattered over the valley floor. By on estimate, 7 square miles of postwar subdiv sions in 1954 were scattered over 200 squar miles of Santa Clara County, with at least on subdivision in each square mile. Transporta tion and utility networks are or must some day be extended to most of these urban islet and thereby to the lands among them.

The California Water Resources Boar bulletin said that 65 percent of the suitab land was undeveloped for urban use in th Los Angeles hydrographic unit — that is, i the city of Los Angeles, the immediately su rounding cities, and the more or less urban ized unincorporated lands.

Another 1955 survey, Bulletin 87 of th Regional Planning Association of Ne Jersey, New York, and Connecticut, reporte the following percentages of suitable lan undeveloped in some of the counties of metro politan New York: Bronx, 9 percent; King (Brooklyn), 44 percent; Richmond, 32 pe cent; Hudson, 21 percent; Bergen, 54 pe cent; Westchester, 63 percent; Fairfield, 8 percent. (They counted estates of 2 acre and more as "undeveloped.") For the entir 22-county, tristate metropolitan regior dotted from end to end with fragments c New York City and laced with transportatio and utility lines, only 21 percent of the sui able land, or 16 percent of the gross lan area, was developed for urban use.

To occupy these vast territories calls nc only for transportation and utility network but also for enormous private investments i

:os, trucks, service stations, and the whole nplex of individualized transportation uipment. This mobilizes consumers to ng their demand to every nook and cranny undeveloped territory. Scattered stores, ools, factories, churches, and other basic ators of urban land value also shed their luence on the included undeveloped lands. The unfilled demand pushes upward, too. e high price of land stimulates more in- sive vertical building (and generally closer onomy of land) on a few sites than demand n begin to absorb over the entire area sub- t to urban influence.

Here are the makings of a cycle of over- pansion that should come to light when eculators holding the better lands try to d markets. But a great deal remains un- ar.

Perhaps some land developers do plunge ead under the sole stimulus of current ces, but it seems doubtful whether most estors would commit themselves for long ms without an eye to the future.

How shall we explain the tenacity of the eculators who confidently hold for a rise d the dauntless optimism of developers, ilders, home buyers, utilities, municipali- s, and still more speculators who invest in owing areas in contempt of mounting ards of half-urbanized land within the mar- t sphere?

One reason for surplus development is that al districts and cities race for position. cing differs from economic competition, as ually conceived, in that races end. Where w population and transportation are open- g and promising to open new urban poten- lities, the fixed layout of routes becomes nporarily fluid. During the developmental riod of uncertainty, several contestants vie thusiastically for prized positions in the w pattern before it freezes.

Because of increasing returns in urban development, these positions, once established, are quite secure and should appreciate in value as outsiders flock to them. So it makes sense for each contestant to risk great resources in a race which most of them must lose.

Cities and districts race by improving themselves to attract trade, routes, and investments. They push out their own routes to capture undeveloped trade territory from rivals, just as some cities push out aqueducts to stake out scarce waters well ahead of need. Because the motive is to secure territory and position quickly before it is too late, extension of trunklines may proceed when the fever is high without much thought for immediately foreseeable demands.

Trade racing also helps explain the behavior of land speculators. Should a district win its race, it is primarily the land that would appreciate, buildings being duplicable. But should it lose, any buildings, being immobile and fairly specialized, would stand a good chance of finding themselves obsolete. The rational gambler therefore may often prefer to bet on the race from the sidelines by holding unimproved land, postponing building until the uncertainties of racing have been resolved.

He thus lessens his district's chances of victory by retarding its development, of course, but one individual is not likely to think his influence is great.

The irrational gambler also is a factor — a major one — to consider. With several contestants running for the same prize, the average chances of success obviously are not good. Yet land prices in each contending district often seem to run higher than the statistical probability of success would warrant, and the sum of the prices over entire developing areas seem to exceed considerably what would reasonably be justified by income from the land.

Just why this should happen is a mystery social scientists are only beginning to probe. Milton Friedman, of the University of Chicago, and G. L. S. Shackle, of Cambridge, England, have developed some interesting hypotheses about it. The fact that it does happen is well established, however. Economists of several generations have observed, with Alfred Marshall, a renowned Victorian economist, that ". . . if an occupation offers a few extremely high prizes, its attractiveness is increased out of all proportion to their aggregate value." Certainly the urban land market is of that description — frontage prices in some areas increase 100 times within a few blocks.

Just as gamblers who love gambling for its own sake will bet against a wheel they know is fixed, land gamblers bid up land prices higher and over more area than the possibilities of urban income can justify.

Perhaps the most powerful stimulant to demand for land is the emergence of a Malthusian climate of opinion. Opinion is a powerful agent in the land market because land prices are based on opinions of the future and because there is so little factual information to go on.

Try to find a simple statistic, like the number of lots subdivided annually in the United States or, indeed, in any region. Few jurisdictions compile even this information, and few of those include entire metropolitan areas.

Urban outskirts especially are beyond the ken of established centers of information — and it is in these far reaches that the greatest excesses have occurred in the past. There might be enough land prepared and preparing for urban use to swamp a metropolitan market for 20 years, and it is doubtful if more than a few real-estate men, who are not given to broadcasting such gloom, would be aware Not until June 1957 has there been any semblance of an inventory for the Nation. That compiled as part of the study of urban ta assessments by the Census of Governments does not purport to tell anything about th lots other than that they are "of record."

We have no systematic data at all on mor difficult but equally important questions, suc as the trend of land prices, the number o unrecorded and illegally subdivided urba sites, the areas in various stages of partia urbanization, plans for impending redevelop ment, and so on.

Land developers must grope to decision primarily by the present feel of the marke without factual basis for the longer sighte analysis that is so essential to an activit whose product is as nearly permanent as any thing produced by man.

And so, lacking information, the marke relies on opinions, which always are in lon supply. Some of these are based on carefu inference. Others are sheer folklore or gli platitudes circulated by professionally opti mistic salesfolk.

Many students of past booms have com mented on the propensity of contemporar opinion, unsoundly based, to underestimat the emerging supply of urbanized land an overestimate the demand for it. It is pos sible to trace out several primrose paths b which opinion falls into these errors.

One is the plausible presumption that con struction tends to exhaust the supply of urba land. The sight of childhood haunts covere with fresh masonry seems especially to sti deep Malthusian anxieties that find their wa into poignant articles, indignant editorials goading investment counsel, and finall urgent land hoarding that transcends prosai computations of supply and demand.

Yet construction urbanizes as much lan

t consumes, or more. Even if a city grew compact circle, the ring around its widen- circumference would grow ever larger, ghly with the square of its radius. And be- se cities scatter out all over the landscape, ding (especially of roads and utility net- ks) brings wide supplies of new land into urban market.

Another primrose path is the equally isible presumption that skyrocketing land es reflect an acute scarcity of urban land. this is to reckon without the vast sup- s held in cold storage by speculators and douts of one kind and another. The eco- nist's nightmare of inflation without full ployment of resources has characterized d markets toward the close of every boom iod.

There also seems to be a tendency to erestimate the regenerative power and orptive capacity of downtown.

There is no denying that autos and trucks, ound by central terminals and fixed tes, have made it more feasible to bypass vntown and thus have drastically weak- d its central position. The big swing has n toward expansive, cheap-land, single- ey development. But many persons in r enthusiasm tend to write off downtown d as though it had become as obsolete as buildings on it, without due account of nan factors like inertia, monopolistic king, absentee ownership, speculative d pricing, and restrictive policies.

Others seem to have accepted too uncriti- y part of the thesis of the late Harvard nomist, Joseph A. Schumpeter, and ers, that capitalists require security from npetition before they will risk funds in e investments like buildings.

But the sleeping giant downtown once used by the sting of effective competition running scared is still no mean competi- tor itself. Decentralization has tended to de- flate speculative anticipations that buoy up downtown land prices and thus has made the most expensive land in the world a bargain relative to outlying sites whose asking prices have multiplied since 1950. Downtown can rebuild and finally has begun to do so.

When downtown rebuilds, it still has the primary advantage of location that made it downtown in the first place — why run around end when you can step through cen- ter? And a few skyscraping hotels, office buildings, department stores, and apartments — as only downtown has the focused demand to support — can do the work of square miles of sprawl outside the city limits. 3-D develop- ment can work wonders with very little sur- face. In Philadelphia, for example, just one building, No. 3 Penn Center, increased by 4 percent the city's rental office space when it opened in 1955.

There has been a widespread idea that downtown building space is saturated. Yet the editors of Architectural Forum noted in March 1957 that the architect, Victor Gruen, retained to replan downtown Fort Worth, found that "the underused or derelict reser- voir was large enough to provide space for a belt highway, parking garages for 60,000 cars, greenbelts, a 300 percent increase in office space, 80 percent in hotel space, and new civic, cultural, and convention centers. . . . Fort Worth is not a special case. . . ."

The urban economic geographers, R. E. Murphy, J. E. Vance, and B. J. Epstein, dis- covered from a close study of eight central business districts that six of them were so de- cayed at the core that building heights in the zone of peak land values averaged much less than in the central business district as a whole. Large parts of the districts were taken up with what they considered "noncentral business district" uses, especially in the older eastern cities. Central business districts occupied well under 1 percent of the areas of their cities and thus had ample room to expand. The authors published their work in Economic Geography, January 1955.

In the downtown of downtowns, Manhattan's accelerating office boom accounts for much more than half of the postwar office space in the country. The postwar increase alone exceeds the total space in any other city in the United States. It is augmenting Manhattan's office space by 40 percent over 1946, yet — far from exhausting the land supply of that tiny island — it is contained in a mere 84 new buildings. And these are focused on two narrow districts, the financial and commercial centers, which are already most congested.

Homer Hoyt, an urban planning consultant, in his monumental *100 Years of Land Values in Chicago*, has shown how the percentage of Chicago land values contained in the Loop has risen and fallen many times in the short span of Chicago's lifetime from 1833 to 1933. Decentralization has not been a continuing process. In the development of American cities, both centralizing and decentralizing forces have worked. Now one dominates; tomorrow it may be the other.

Opinion often seems to stray, too, in interpreting the effect of a few skyscrapers and other intensive developments on future land values. Their advent convinces many landholders that high land prices can be met.

But multistorey buildings are substitutes—enormously effective ones — for land. A few of them can pay high land prices, but to do it they drain demand from blocks around. To be sure, they are also magnets pulling trade to the city from miles away. But when cities all over the country are racing to the sky, outside competition tends to offset this benefit.

High buildings are symptoms of high land prices. But to let a symptom be a cause is run a danger of circular reasoning.

If land prices are prematurely high to beg — higher than long-run supply-demand ba ance warrants — intensive vertical develo ment must ultimately deflate the price ba loon. The longer this deflation is delayed, t more the error compounds, and the mo violent must be the reaction.

The same general lines of reasoning app to horizontal urban expansion. This is lar substitution, too, destined ultimately cheapen urban land. Yet the psychologic impact may be to create a feeling of centr position that leads to higher asking price more horizontal extension, and a rude awa ening some day.

Along with those underestimates of supp there are overestimates of demand.

A prominent cause is exaggerated relian on population forecasts. These have be notoriously unreliable in the past. Techniqu have improved, but there is little warrant f the utter confidence with which forecasts a often repeated. But this is not the main poir

Population forecasts, if accurate, tell something about the volume of "need," b not so much about effective demand, which another animal, and the one whose pow makes the economic world go round.

Some half of the postwar building boo has been to produce more space per persc —that is, greater spending per capita h been as much a factor as greater populatio Undoubling of families, which was one el ment in this trend, has now virtually halte — the average number of persons per hous hold has leveled off at about 3.3 since 195 The recent and immediately forecast swellir of population is in the relatively unproducti age groups under 18 and over 65. But neith babies nor aged dependents increase one income or borrowing power.

Supporting them does tend to reduce ·adwinners' savings. Many analysts trans- ɜ this into increased effective demand. It y increase demand for toys and TV, but factor that increases the urgency of pres- over future needs is likely to increase the estment demand for a long-term, deferred- ome asset like title to land, especially un- eloped land. Reduced saving, higher erest rates, and lower land prices follow in ical sequence. More schoolchildren also an higher real-estate taxes, which tend to luce the investment demand for land.

Then there are two sources of demand that nost by necessity are only temporary but t operators on the field of action may be able to distinguish from more permanent ırces of demand.

One is demand premised on anticipations rising land prices. High prices themselves, ce realized, tend to depress demand, of ırse, but expectations of rising prices have opposite effect. They increase demand not y from avowed speculators but to some ex- t from all land buyers, including builders l owner-occupants, who are as glad as any- ɜ to board the price elevator on the ground or.

This demand is inherently very unstable. the way up, it helps fulfill its own expec- ons, in the familiar pattern of speculative rkets wherein expectations of rising prices ke prices rise. Eventually, however, even if her prices fail to dampen expectations of ther rises, they certainly increase carrying ts and dampen the basic demands of ulti- te consumers.

Once prices stop rising, this unreliable ele- nt of demand is likely to collapse. If it is a ge share of the total demand, its desertion l then let prices sink. Stability is next to possible in such a market. Prices either ntinue up or turn down.

A second unstable element of demand is that generated by investment in construction.

Construction is largely a migratory indus- try, which creates temporary demands on local facilities in areas of growth. This poses no difficult forecasting question around fly- by-night construction camps. But elsewhere it is all too easy to confuse temporary demand from construction spending with demand from more permanent sources. They are hard to distinguish in a complex, interdependent, growing urban economy.

A small confusion of this sort may be mul- tiplied into a large error because of the lever- age effect of outside money on the develop- ment of a region.

Because growth areas are capital-hungry as a rule, construction usually is financed largely from outside. Outside money flow- ing into an area serves as part of its economic "base" — that is, it sets up demand for local services and sustains it by offsetting the in- evitable cash outflows.

Because local services account for roughly half of the incomes of most cities, each dollar of income financed from outside serves as "base" for another dollar or so of income from services sold locally. Then there are many market-oriented or camp-following in- dustries, which move to an area largely be- cause consumers are there ahead of them. When we consider them, a dollar of outside money may exert several dollars' leverage on local income, depending on the locale.

Because these local sellers also require buildings and urbanized land with utilities for working and living, they set up demands for more construction, which means more out- side money — and so on. Such a sequence, once started wrong, can send development veering off course like a sliced golf ball. We have seen this happen in the midst of our postwar prosperity around the atomic boomtowns of Portsmouth, Ohio, Paducah, Ky., and Aiken, S.C. with full foreknowledge that construction payrolls were temporary, these three communities contrived to overbuild anyway, and each suffered its depression-in-a-teapot when the crews left town.

Expansion of local banking often adds to the possibility of error. Outside money flowing in increases the reserves of local banks and encourages them to lend. Under our banking system, they can expand their loans by more than the increase of reserves. This expansion would generally lead to drains on reserves that would stop it short. But it need not happen immediately, especially in a booming district, where much of the banking system's new loans come back to it in new deposits. The expanding loans of local banks meanwhile, serve like outside money, as part of the economic "base."

The situation may be complicated once more where outside money flows in, not simply to finance construction or buy land, but to speculate in the extreme sense of the word — to buy and sell and buy again. It is well known that New York banks have large deposits held to speculate in Wall Street. When a city or district catches the imagination of the more colorful part of the investment community, funds pour into its banks for similar purposes. Homer Vanderblue, then of Harvard University, found that bank deposits tripled in14 months of 1924 and 1925 in the Florida land boom, only to flow out rapidly with the crash.

The wisdom of investors, or at least their conservatism, might seem proof against this sort of folly. But investors in boom times have been notoriously susceptible to fads and stampedes.

Homer Hoyt laid down as a general rule: "In each successive land boom there is a speculative exaggeration of the trend of the period. . . ."

And as long as outsiders are ready t finance it, there is nothing to stop a new dis trict or town from prospering while the resi dents, exporting little but mortgages, deposi slips, and land titles, simply build the plac and take in each other's washing.

Outside investors are not going to do thi knowingly. Jacob Stockfisch, economist a the University of Wisconsin, maintains tha individuals can foresee tolerably well th complex interactions of their investmen with those of others and trim their sails s as to achieve an orderly integrated economi development. But history leaves little doub that this ideal behavior presupposes a fore sight and exchange of information which fa lible, suspicious man seldom achieves.

We return to our third critical question can urban expansion be a stable process

A pattern of expansion that stimulates va oversupplies of urbanized land to meet a de mand that is partly collapsible obvious presents some danger of instability. Th United States Census of Governments, in i Advance Release No. 3 for 1957, reporte the number of vacant lots of record in th United States at nearly 13 million (not coun ing parking lots). That is 21 percent of a city lots, and about 13 times the annual con sumption in new construction.

The census figure does not purport to b more than an aggregation of local records and some of the "lots" recorded are no doub that in name only. On the other hand, som actual lots never find their way into loca records. And the figure is especially strikin in light of the universal observation that sub dividing land for sale of lots to avowed specu lators has been at a minimum during the post war building boom, with its emphasis o mass-produced suburban developments fron which lots are sold only underneath houses

The larger part of the land hanging ove urban markets is acreage not yet subdivide o lots, but with ready access to farflung ɔan transportation and utility networks.

A study of Greensboro, N.C., in 1956 by orge Esser, Jr., of the Institute of Govern- nt of the University of North Carolina, ınd 125,000 persons scattered over a asi-urbanized area big enough for all the ɛds of 600,000. We have no reason to be- ve that that is anything but typical of nerican cities.

Will private and public developers add lefinitely to so swollen an inventory?

Will speculators and holdouts want to con- ue meeting the rising carrying costs on just present supply?

Will lenders continue to extend credit on ch hazardous collateral? With 143 billion llars in nonfarm residential and commer- l mortgages (in September 1957), could credit system stand a real-estate collapse?

No one knows for certain. History puts the rden of proof on the affirmative. Cities ve rarely expanded other than in crashing ves, and today one sees several portents niniscent of previous crests.

Some of these portents are:

The rapid, manyfold rise of land prices ɔund growing cities since 1950;

the sharp rise of construction costs;

the wildfire spread of municipal zoning and gulations very hostile to mass-market ilding;

the decline of residential construction since rly 1955, coupled with an increase of land- bstitutive construction in extensions of ads and utilities, and multistorey buildings;

the disproportionate increase of transpor- ion costs and utility rates since 1950;

the disproportionate increase and high el of residential and commercial debt. (Its erage annual increase has been 9.5 billion llars from 1945-1956, and its annual per- ntage growth rate 14.4 percent over the 1946 base. That compares to 2.2 billions, and 9.4 percent, for the period 1920-1930. In September 1957, it reached 143 billions, 48 percent of disposable personal income. That compares to 37 billions, and 45 percent, in 1929.);

the general deterioration in the quality of credit, as noted by Geoffrey Moore, of the National Bureau of Economic Research, and others, and as exemplified by the growth of second-mortgage financing;

the high level of interest rates;

the almost universal confidence that grow- ing population and living standards are press- ing on the land supply and insure a continual rise of land prices.

The result of these combined causes will depend largely on human response, private and public, which few would be so bold as to forecast.

Past mistakes, if that is what they are, have not trapped us in any dilemma beyond the power of informed, intelligent action to resolve.

It is heartening to see so much concern quickening today over problems of urban ex- pansion. There is hope that today's more lit- erate and prudent American public can avert the disasters that beset the past.

But whatever the immediate outcome, the public and its representatives, including farm- dominated State legislatures, would probably serve themselves well to attend closely to the compelling problems of harnessing urban land. This resource holds economic forces of titanic power for welfare or destruction. Har- nessed, these forces could serve the public commensurately with their unrivaled market values. Untamed, unpredictable, and irre- sponsible, they could figure in a national calamity.

Indeed, they have already done so in a measure. The disintegration of our cities could be described conservatively as a national calamity of some proportions, whose mischievous consequences only wait to be recognized. To forestall more of the same, the reasoning of this chapter suggests that policymakers might do well to take steps to lower the prices asked for urban lands.

The thesis of this chapter is that urban land prices are uneconomically high — that the "scarcity" of urban land is an artificial one, maintained by the holdout of vastly underestimated supplies in anticipation of vastly overestimated future demands. I think this uneconomical price level imposes a correspondingly uneconomical growth pattern on expanding cities. High land prices discourage building on vacant lands best situated for new development and divert resources to building highways, utility networks, and whole new complexes of urban amenities so as to provide and serve substitute urban lands further out — substitutes for something that is already in long supply. Not only is this pattern wasteful of time, steel, cement, gasoline, and good farmland; it founds national prosperity on the film of a land bubble.

And so it would seem wise for policymakers to set about lowering asking prices for urban land. But here they meet a dilemma. What stimulates building is not falling prices, but the end result of the fall — low prices. Falling prices themselves tend to depress building. Few there are who want to invest money on the foundation of a sinkin land market.

Policymakers are tempted to put off th day of reckoning, to tolerate and, in fac actively support high land prices. But th irony of such policies is that they stimulat development of still more substitute urba lands, and set the stage for more drastic ultimate collapse.

There seems one obvious escape from thi dilemma. As it must be done, do it quickl Bring land prices down fast, and get it ove with.

If this is a desirable policy, however, history offers little comfort that it will be enacte without painful changes in established attitudes. Squeezing the water from speculativ land prices has usually been a slow process c attrition, with public agencies often bendin their efforts toward delaying the inevitable a long as possible, while building stagnated.

But whatever policies are desirable, I believe there certainly is urgent need for public-minded citizens to agree on what those ar now, before an emergency strikes. For th suburban land boom shows many evidence of evolving along the same lines as its notorious predecessors, which have confronted u with several of the most trying crises i American history. We can ill afford to mee one today as indecisively and ineffectively a in the past.

rban Sprawl and peculation in rban Land

27

arion Clawson

: rapid spread of suburbs across the pre-
usly rural landscape is a common pheno-
non in the United States today. Even the
st casual observer cannot but be impressed
n the magnitude of the changes. There has
n much criticism, on aesthetic and other
unds, as to the kind of suburbs being built;
y have also had their defenders, or at least
se who say the results cannot be hopelessly
l because people still move in great num-
s to the new suburbs. This article will not
empt a general critique or appraisal of
dern suburbanization but rather will con-
r only one phase of it.

One feature of postwar suburbanization
been its tendency to discontinuity — large
sely settled areas intermingled haphaz-
ly with unused areas. This intermixture of
n and developed areas is largely indepen-
t of the density of the settlement within
developed areas; the question of the ideal
sity of settled suburban areas is another
ie, which we shall not explore. The lack
continuity in expansion has been given the
criptive designation of "sprawl," which
l connotes its hit-or-miss character.

"Sprawl" has been widely criticized as leading to unnecessarily high costs of social services and of private transportation, as well as for the frequent lack of publicly available open areas. It is also responsible for, or associated with, much wastage of land, since the intervening unused areas are mostly not used at all. Others have tended to minimize these deficiencies, arguing that they are but part of a growth process, not too serious in nature. Whatever may be the verdict on sprawl, it is clear that suburbanization has been the result of a relatively aimless process. It seems highly doubtful if any participant in suburban growth, or any observer, actually chose the pattern which has resulted. Possibly no one objects violently enough to exert the force required to change it but neither will anyone defend it as ideal. One aspect of this picture has been large-scale speculation in land, with consequent high costs to the actual settler and with large areas priced out of any market except urban usage, but the latter not yet taken over. Although nearly everyone seems aware of this process, and although most are critical of the results, yet it appears there is a serious

rinted from *Land Economics* by permission of the Regents of the University of Wisconsin.

lack of understanding as to just what is going on.

The purpose of this article is to explore the economic process in suburbanization — why some areas are developed, why intermingled ones are not, why land speculation invariably accompanies the process, and the like. The economic forces will be described, as far as possible, and some judgment offered as to which are manipulatable and which are not, and how. A basic premise is that no significant progress can be made in developing better suburbanization until the present processes are better understood.

Role of Agriculture

Perhaps the place to start is by eliminating one possible major causative factor of suburban sprawl — agriculture. Urban growth and urban demand have a major effect upon agricultural land use as a whole; in fact, as one surveys the history of agricultural development in the United States, one concludes that urban demand has been the main causative factor in agricultural development.[1] But the differential or locational effect of agriculture upon suburban land values has been very small. For one thing, some of the physical qualities which make land valuable for agriculture also make it suitable for urban use.

Locational theory as applied to agriculture, from von Thünen downward to the present, has emphasized the effect of the urban market on agricultural land use and land value but has also stressed the effect of transportation costs, as well as such differential factors as land fertility.[2] Under conditions of primitive transportation methods and high tran
port costs, agricultural production may b
highly stratified, with bulky, low valu
perishable products near the market, ar
those with higher value in relation to weig
and with less perishability produced farth
away. The width of the zones in any mod
depends upon transportation costs to a lar
extent; and the sharpness of the boundari
between zones depends largely upon natur
production conditions and upon intra-far
economies, such as the need to grow feed f
draft animals.

Today, the chief agricultural commodi
with a clear orientation to the nearby urba
market is fluid milk and it is held the
largely by "health" regulations which mak
its importation from more distant areas in
possible. Although there are some adva
tages in producing fruits and vegetables ne
the market, yet as a matter of fact the gre
bulk of urban supplies of these commoditi
comes from a considerable distance — fro
across the continent in many cases. Toda
this nation has a combination of good, rel
tively cheap transport, and technology –
such as refrigeration which makes long di
tance transport of perishable commoditi
possible. Even the fluid milk zone is mo
than 50 miles in radius for our larger citie
and within this zone there is almost no loc
advantage to agriculture. The widest arc
the suburban spread is far less than the nea
est edge of the zone within which agricultu
might have any differential effect upon loc
land values.

Some farm or rural land near cities wi
indeed come to have relatively high values
country estates or as a certain type of gentl
man farming. But in this case the value of th
land arises from the urban settlement n
from agricultural production. Although suc
estates may lie outside of the usually define
suburban area, yet in fact the same value

[1] Marion Clawson, R. Burnell Held and C. H. Stoddard, *Land for the Future* (Baltimore, Johns Hopkins Press, 1960), see esp. p. 247.

[2] For a clear recent statement of locational theory as applied to agriculture, see Raleigh Barlowe, *Land Resource Economics: the Political Economy of Rural and Urban Land Resources Use* (New York, Prentice-Hall, 1958).

.king processes are at work. It is the city as •lace of residence and of work which gives ue to such estates, not their agricultural :put.

Farmers in some areas, notably in Cali-nia, have tried to protect their farm dis-:ts from encroachment. In general, such orts have not been conspicuously success-, in part because such farmers are ambi-ent: they want their land left in farms but y also want a chance to sell at the best ssible price. It seems highly doubtful that iculture can perfect an institutional barrier ainst urban expansion; at the most, it may p guide the direction and nature of the ourbs which develop. If we are to explain suburban growth and land speculation ocesses, we must therefore look to forces er than agricultural land use and output.

aracteristics of the Market for Raw burban Land

e market for raw, undeveloped suburban d has several peculiar characteristics. First all, land for suburban development is not omogenous commodity, any more than is d for any other possible use. While differ-:es in soil texture and fertility may be less oortant, as compared to these same quali-; for agriculture, they are not neglible. pe of land may be highly important, as ecting building costs. The risks of flood nage differ greatly from area to area. In se and in other ways, the native or natural alities of potential suburban land may dif-greatly.

The history of land ownership usually re-ts in a present ownership pattern of vari-le size tracts of land owned by different ners. Some pieces are large, others small. me owners have one objective, others an-er. A potential new owner must deal with at he finds, buying as he can. He will find it impossible to buy exactly as he wishes but must deal with discrete tracts in different ownerships. Subdivision of large tracts often creates a "plottage" value, which is at its peak when the size of tract coincides with the tract best suited to the use for which the land is intended. Tracts either larger or smaller than the optimum have lower value. The passage of time may change the use of land and hence the optimum size of tract. It is significant, we think, that since the war the major railroads of the country have purchased potential industrial sites along or near their tracks when they could, largely to prevent subdivision which would spoil the larger tracts for industrial development.

The owner of a discrete tract often must sell it all, or a major part, if he wishes to sell any. Suburban land, equally with or perhaps more than other land is not, perhaps cannot be, sold in incremental pieces, but rather in relatively large chunks — chunks not necessarily adjusted to the needs of the buyer or seller.

Society, acting through government at some level, has given suburban land further special characteristics. Location with respect to transportation, to water supply, to sewerage, and to other services vitally affects potentiality of land for suburban development. These qualities were given the land without action by the landowner, except as far as he was able to influence the public action which resulted in these services. Individuals may buy and sell land to take advantage of the services provided by group action but they are not responsible for the services.

Society has affected the value of suburban land in other ways — by taxes, by zoning and building codes, and the like. If master plans, zoning, and building codes were explicit, firm, enforceable, and enforced, and if there were confidence they would remain so, they would greatly limit if not completely determine land values in many areas. In fact, zoning in particular and others to some degree can be changed under political and other pressures. Even the courts do not always accept values consistent with zoning regulations, when private land is condemned for public use. Public action through zoning and other related measures affects land values; but the major effect may be through the uncertainty created. While some of these services or action by society affect land over large areas more or less equally, yet some have a highly local effect.

Suburban land also differs greatly in accessibility, especially to major highways and sometimes to rail lines. The quality of accessibility may affect its price and its saleability greatly. Accessibility is generally not provided by the individual landowner but rather through the public, as in the case of highways, or by large private undertakings, as in the case of rail lines.

The market for suburban land is a derived one, dependent upon the market for the dwellings, shopping centers, or industrial plants erected on it. As such, it is subject to the uncertainties of market for the final product, compounded by the uncertainties of the conversion process. The market for suburban housing is a fragmented and not wholly consistent one, often variable in short distances or over brief times. Differences in price for houses are often reflected back into differences in price for undeveloped land, but in varying degree.

Lastly, the market for suburban land is usually very thin. There are very few buyers and very few sellers at any one time. Annual turnover in relation to total area is small. For almost any commodity there is a liquidation value at forced sale; a normal value between willing seller and willing buyer; and a forced purchase price when for some reason the buyer must buy almost regardless of price. For suburban land these prices might well stand in the ratio of 50 or less, to 100, to 200 or more, respectively. The time required to make a sale of land may be considerable, and directly related to the price obtainable. Part of these variations may be due to lack of information on the part of buyers and sellers but much is probably due to the character of the commodity itself. One need only contrast these characteristics of the market for suburban land with the market for wheat or even for autos. For these latter and for many other commodities there are many buyers and sellers; and forced sale, normal sale, and forced purchase prices stand in much closer relationship to one another. Some of these characteristics we have described for urban land do apply to all kinds of land for any purpose. Although empirical studies are lacking we hazard the judgment that these factors are more serious for urban than for other land.

Value-Making Process for Undeveloped Suburban Land

Undeveloped suburban land, not yet in use for urban purposes but already taken out of other land uses, obviously must derive its value from the expectation of its later development as urban land. As we have noted agriculture does not contribute in any important way to the value of potential suburban land, especially when the land is no longer actually used for agriculture. Most land has value based on its agricultural productivity of less than $400 per acre, although of course there are exceptions; much suburban building land at the time of development sells by the lot at prices equivalent to $4,000 per acre or more, with modest subdivision improvements, or at least $2,000 per acre as com-

:tely raw potential suburban building sites, d often at much higher figures. The potent- subdivision value depends on many factors t the least of which is the popular estimate the kind of suburban district it will ulti- tely be, which in turn depends somewhat on neighboring districts but also somewhat on the prices the subdivider puts on his s: that is, to a degree, to put a high price suburban lots gives them a high value. The nversion value of the raw potential subur- n land into actual developed suburban land somewhat uncertain at any date, depend- ; in part upon the action of the community a whole, and in part upon the skill of the bdivider and developer himself.

The date at which there will be an active mand for the raw suburban land for actual velopment is to a large extent uncertain. some instances a piece of land may lie se to areas developed within the past few ars and toward which the tide of develop- :nt is flowing. Under such circumstances its esent value is moderately forecastable on basis of estimated probable future con- rsion date and value. In other cases, land y lie at greater distance or in directions ere future development is less certain; then th its conversion date and its conversion lue are more uncertain. The timing of velopment of a particular piece of subur- n land is partly outside of his control. He y obviously withhold it for later develop- :nt, if he thinks a greater net income can be tained thereby — he is less able to speed its development. The large, well-financed, llful developer can bring about the devel- ment of a particular tract more nearly on terms than can a smaller developer; but ch operates within the general market ucture.

An expected future income or value can discounted back to a present worth or value. An interest or discount rate is required to do so. The discount rate may be thought of as having two parts; a more or less normal interest rate based upon alternative sources of investment or alternative sources of funds in competitive money markets; plus an uncertainty factor. The latter relates not only to the date of future conversion from raw to developed status for the land — and even as to "whether" as well as "when" — and the value at that date, but probably should include a large allowance for illiquidity as well. As we have noted, suburban land can be sold quickly or at forced sale only at prices substantially below its normal value when ample time is available to negotiate a sale. In practice a single discounting figure will be used, large enough to include all these and perhaps other factors as well.

The appropriate interest rate in land speculation depends to a large extent upon the situation of the particular individual.[3] A man with ample investment funds, perhaps faced with a high marginal income tax rate and hence eager to secure caiptal gains on which a lower rate is paid, could afford to speculate on land at interest rates perhaps no higher than 2%. A farmer, short of capital and hence forced to ration his scarce capital among various potentially profitable farm enterprises, or forced to borrow at 6% or more, would necessarily use a much higher rate — perhaps 6, 8, or even 10%. A real estate developer, perhaps short of capital and eager to use his available capital in enterprises where the turnover was rapid, would be in a position similar to that of the farmer. These differences among individuals would logically lead to greatly different positions in land

[3] For a very stimulating discussion of this point, see Mason Gaffney, "The Unwieldy Time-Dimension of Space," *Am. Jour. of Economics and Sociology* (October, 1961).

speculation, but we shall not explore them in more detail here.

In addition to delays and uncertainties as to time and value of suburban land for conversion to development, there are some holding costs to be taken into account. Taxes over a period of years may be considerable even at low assessments and low rates. Occasionally charges other than taxes must be met annually. One cost of holding is interest on the value of the land if sold, but of course the discounting formula includes this factor.

One could easily construct or adopt formulae to show these relationships, or give illustrative tables of different time periods, different final conversion values, different discount rates, and different holding charges. The best guess as to land values 10 years from now will justify present values well under half of that level; and the best guess as to values 20 years from now will justify present values much less than a fourth as high. It is altogether possible that normal or free market values may be higher than this because of widespread optimism over ultimate values, time of conversion, costs of holding, uncertainties, and the like.

The ownership of any suburban land for a rise in value is a speculative undertaking. Profits, when all factors are taken into account, are by no means assured nor large on the average. Everyone knows, or at least has heard, of others who have made substantial gains from holding suburban land for a rise in price. This type of common knowledge nearly always is ignorant of or ignores the cases, perhaps more numerous, when increases in value were much less or even negative. The chance for profit in holding suburban land for development arises entirely out of error in consensus or out of individual judgments more astute than the consensus. If there was complete knowledge as to the time of future conversion, as to value at that time as to holding costs and as to discount rate then obviously everyone would be in complete accord as to present worth. There would be no opportunity for speculative gain, because all future value would have been fully and accurately discounted into present value. It is altogether possible that at times the consensus on these matters is in error — everyone is sure of something which later history proves not to be true. Under such circumstances, a sounder judge with a minority view may reap a profit. At other times a consensus may be lacking but one view may prove in time to have been closer to the fact than any other; if the person who held it acted upon his convictions, he may have profited.

As long as the price of land ripe for conversion from undeveloped to developed status is relatively high, then the price of land less ripe for development will be somewhat lower until at the margin the prospects for conversion into developed status are so uncertain or so remote that even the most optimistic will not bid up the value of this land. As long as we have free markets in suburban land and as long as the total effect of the various factors in the formula promise some present value above alternative use value, and given imperfections of knowledge and incomplete consensus, then we can reasonably expect speculative bidding up of suburban land values. Viewed in this way, land speculation in and beyond the suburbs is not only normal but inevitable. The possibilities of its control will be explored later.

Forces Leading to Development of Particular Suburban Tracts

Given the nature of the market for raw suburban land and given the value-making process such land, what are the forces leading to development of particular tracts of such 1? How can we account for the fact that a tively few of the many possible suburban ts are developed in a particular year and can we predict which ones will be devel- d and which left for a later future?

)ne basic factor is the over-all market and for urban land for the whole urban concerned. Some cities or metropolitan as are growing rapidly, others at a more lest pace, and some are essentially stag- t. The amount of new land needed for an purposes annually will obviously vary atly among cities, depending upon this or. At some times the real estate and ding market is much more active than at ers, depending in large part upon credit ilability as well as upon general economic nand. When the demand for new urban 1 is high, not only is more land needed but profitability of conversion is probably ater. This means not only greater profits to lowners, on the average, but also that e tracts or types of development which ld be marginal in other circumstances will be promoted — it is the time for the long nce, for the unusual deal.

'he extension of essential public services articular areas or districts will bring land in such areas or districts closer to the it of actual development or building. Pro- on of new roads, schools, water supply, erage, and other services, or marked im- vement in them, add greatly to the impetus development. The possibility of alterna- devices, such as septic tanks instead of k sewer lines, may have the same effect. wing the subdivision developments which ally take place, one can hardly say that se public services, which he is tempted to essential, are in fact either essential or essary to building development on specific sites — one sees too many areas that get built up, at least to a degree, without them, or at least without satisfactory services. Yet the provision of new services undoubtedly gives a fillip toward development. On the other hand, it is unlikely, of itself alone, to be sufficient. That is, mere extension of one public service, or even of a group, to an area previously lacking them, may not lead to much actual building. Other factors — above all, over-all demand — must be present.

Though empirical data are lacking, at least to this author, yet one cannot but suspect that the personal desires, projections, and preferences of present landowners must be a major factor responsible for some tracts developing while other intermingled ones do not. Institutional factors, such as estate holdings, trusts, defective titles, covenants, and others, may affect marketability of particular tracts, especially in the shortrun. Some present landholders may be optimistic about future increases in value of their land, others more cautious; some may have ample capital for which they seek investment outlets, others may have pressing need of any capital they can raise by sale of their land; and in other ways landowners may differ considerably. It seems wholly probable that owners of identical land (if one can imagine such a thing) might react quite differently to exactly the same offers for their land. Moreover, the differences between individual landowners may well be so great that a small increase in offered price, such as another year or two might bring, will be insufficient to move the man who wants to hold for later profit. Anyone familiar with urban real estate knows of many tracts remaining vacant for many years while all around them development proceeds apace. Surely one major factor must be the characteristics of the landowner himself.

As we have noted above, raw suburban

land differs greatly in physical suitability for development and also in size of parcel which each owner possesses. A residential builder may wish a moderate size tract; some will appear too small for his needs, others larger than he needs, but not available in part. An industrial development is likely to need a relatively large tract, as well as one of specific locational and other qualities and thus many smaller tracts are practically unavailable or nonexistent to him.

When all of these factors are combined one should expect a rather hit-or-miss type of suburban development as normal; it will not normally be incremental, even regular. Instead, some tracts will be developed, other nearby ones remain vacant for long periods, relatively more distant ones developed sooner than some nearby ones, and so on. One should, in fact, anticipate exactly what we have experienced: sprawl! The frontier of urban land use or building will not move slowly and regularly, taking in all land as it goes; instead, development will leap ahead to more distant tracts, passing over nearby ones, taking in some large and some small tracts, and leaving others of assorted sizes. While there has been much criticism of sprawl, and even a little wonder at why it looks as it does, in fact, given the institutional and economic forces we have described, one should have expected exactly the same kind of sprawl we have experienced in such a large way since the war. Those who are surprised at it have even ignored history for this is exactly the way the farm frontier passed across the nation a century or so ago. Canada has succeeded in some provinces in requiring a more uniform filling-up of the frontier areas before additional areas are opened for settlement but this has been difficult to enforce even there.

Effect of Speculative Land Prices and of Suburban Sprawl on Use of Intermingled Land

It is a matter of fairly common knowled that the land within the suburban zone sprawl for the most part is not used for a economic output until it is actually develop for urban usage. Vacant lots, larger vaca leap-frogged areas, and surrounding vaca lands characterize the suburban scene. W should there be so much idle land, hopeful "ripening" for later transfer to urban use?

The processes we have described bid the price of this land far beyond its value f agriculture, forestry, or other rural land us This alone need not render the land idle f these purposes. It is true that the farmer w formerly farmed it is likely to prefer to ta his gain, go elsewhere, and buy a bigg and/or better farm with his enhanced capit It is also true that the new buyer, particular the land speculator, may not know how farm, or perhaps care to try. Yet it is possib that he might lease the land to an actu farmer; the gains, while small, perhaps wou nevertheless meet the annual cash holdi costs of the land and possibly more, th facilitating in some degree the holding of th land speculatively.

When land comes within the zone of subu ban influence, or possible later developmer its taxes often rise. Until new public servic are extended to the area, the increases taxes may be small, often less than the ri in land values. Land speculators and t "Court House gang" are sometimes the sa people, or at least not unknown to each othe But special services in the form of mo roads, better or bigger schools, water line sewer lines, and the like often are extend the potential urban area; and this is almost e to lead to higher taxes on the land. The es may indeed rise so high that they exceed y possible return from land used for farm-. But this alone would not necessarily take h land out of farming. High as the taxes ght be, some net income from farming uld seem to be preferable to none at all m idle land. High taxes mean high annual rying costs to the land speculator and thus ier depress present values (future values counted back to the present) or provide an entive for early sale, especially by the downer who either cannot meet these costs is pessimistic about future increases in ue. Thus, a farmer might be even more ling to take a rise in land prices and trans- his capital elsewhere where farming was re profitable. But again, presumably the d could be rented for farming and at least ne income obtained. Taxes in excess of ome attributable to land make continued d use of any type impossible but not neces- ily short-term use of this type.

A more serious fact is that in the suburban ie the planning horizon has shortened stically and uncertainty greatly increased any land user. The farmer now does not ow when he may one day receive an offer his land so high he simply cannot resist the speculative landholder is faced with a ilar situation. Each knows that such gen- us offers come at most irregular intervals d to forego this one does not mean that other equally good one will come along n. A tenant farmer under these circum- nces will have no assurance of continued eration. A generation ago, agricultural nomists pointed to the depressing effect good farming of the uncertainty in the ical Southern share crop farm. The crop- r never knew from year to year where he would be the following year; hence he made no investment nor plan for more than the current year. Under that circumstance, however, the landlord knew that the farm would be operated by someone in successive years and at least some expectation of continuity existed on his part. The farmer, whether owner or tenant, in suburban zones has no such expectation of continuity. If he has high fixed investment in land improvements such as an orchard it will pay him to operate it as long as he can and recover as much of the sunk investment as he can. If he has high movable investment that might be jeopardized by loss of the land, such as a herd of high-producing dairy cows, his move will be accelerated. Not only is the actual farm operator affected by this shortened planning horizon and increased uncertainty but so also are the innumerable marketing and supply services which are indispensable to modern agriculture. As farming declines, some of these move out also, further hampering successful farming within the zone.

At any rate, land within the suburban zone, not actually used for urban purposes, typically is not used at all. Our best estimate is that there is about as much idled land in and around cities as there is land used (in any meaningful sense) for urban purposes. In the suburbs, the idled land is an even larger proportion. While this is a waste, we think it is inevitable, given the economic and institutional structure we have described. That is, land speculation, sprawl, and intermingled idle land are all natural outgrowths of economic and institutional forces, not perversions of them. Instead of surprise and shock that these situations exist, we should expect them. Perhaps we regard the result as socially undesirable; if so, we should examine wherein the economic and institutional base might be modified. We should look for causes, not moan over or try to treat results.

Possibilities of Controlling Sprawl and Suburban Land Speculation

The following criticisms have frequently been levied at sprawl. Others denied them or at least argued that in practice the situation is not as bad as pictured. Our personal conviction is that sprawl deserves many of these criticisms. But, regardless of the reasons, society (acting through government at some level) might decide to reduce or eliminate suburban sprawl and speculation in raw suburban land. The common criticisms of sprawl are: (1) A sprawled or discontinuous suburban development is more costly and less efficient than a more compact one, each of the same density within settled areas. Many costs depend on maximum distance or maximum area; if these were reduced, costs would be lower per capita or per family served. (2) Sprawl is unaesthetic and unattractive. (3) Sprawl is wasteful of land since the intervening lands are typically not used for any purpose. (4) Land speculation is unproductive, absorbing capital, manpower, and entrepreneurial skill without commensurate public gains. It destroys or impairs economic calculations that ideally lead to maximum general welfare. (5) It is inequitable to allow a system in which the new land occupier is required to shoulder such a heavy burden of capital charges or debt merely for site costs — costs which in large part are unnecessary and avoidable.

That is, we may accept urban sprawl and speculation in raw suburban land as the natural consequences of the economic and social processes we have described and at the same time we may seek to change one or more stages or bases of those processes because we dislike their final outcome if unchecke Where might society intervene, and how? number of possibilities seems to exist. Th following suggestions are largely complemen tary; most would be effective alone but jointl they would be more so.

First of all, effective market reporting c transactions in suburban land would be hel ful. If numbers of parcels, total area, locatio of parcels, prices paid, and other terms c sale were widely reported and generally pub licized, this would provide a solid factu basis now lacking or at least not generall known. Such market reporting for unim proved urban and suburban land should b supplemented by similar reporting for subu ban developed real estate. We have in min something like the market news reporting fo agricultural produce markets for other prim ary materials such as metals and lumber, o even stock market reports. With the low turn over in the real estate market perhaps montl ly reports would be frequent enough. Obvi ously, such reports must be city by city to b really useful. But broad regional and nation totals and averages would be helpful, als Such reports might well be limited to infor mation of public record such as recorde deeds or transfer tax receipts unless buyer and sellers could be induced voluntarily t report unrecorded sales. If limited to publi information some may doubt the advantag of such reports. However, even if everythin in them was known to the alert land specula tor, such reports, if widely distributed, woul bring useful information to many who other wise would be uninformed. This type of mar ket information might be provided by federa state, or municipal government or conceiv ably by real estate boards.

Secondly, this type of reporting on trans actions made could be supplemented by de mand and outlook studies of the type lon

:ablished in agriculture. Given the best pos- ›le forecasts of population growth in a city metropolitan area, how much land will be eded annually, and over the next 10 years? ›w does the amount required compare with ; area presently available? Several past ıdies have shown platted and subdivided ıd adequate to accommodate 20 or more ars anticipated growth in a city.[4] The ratio land available to average area developed s varied greatly from time to time though rhaps nearly always far in excess of a :ionally optimum area. Under these circum- ınces a few astute speculators may make bstantial profits; but all speculators as a oup will lose unless present prices are lower an one-fourth to one-tenth of sale price ıen actual development occurs. Informa- ›n of this type would at least help actual velopers and builders to avoid some specu- :ive traps and excesses and should exert me stabilizing effect on speculation.

Thirdly, urban planning and the subdivi- ›n controls and zoning which make it effec- e might be made into a stabilizing force ther than the unsettling one we have sug- sted it now is. This assumes that some eans could be found which would make the sults of urban and suburban planning more nerally known and more widely accepted that the necessary public and political sup- rt would be forthcoming to secure adher- ce to the plans in the face of aggrieved oup or sectional interest. As we have noted, ning controls and similar regulations are nply not taken seriously in the land valua- ›n process; it is assumed they can be chang- upon a political or interest group demand. If planning, zoning, and subdivision were m — enforceable and enforced — then the area available at any one time for each kind of use could bear some reasonable relation to the need for land for this use. That is, area classified for different purposes could be consciously manipulated or determined in relation to market need. Sufficient area for each purpose, including enough area to provide some competition among sellers and some choice among buyers, should be zoned or classified for development; *but no more*. By careful choice of the areas concerned sprawl could be reduced, perhaps largely eliminated. Forcing relatively full development of each zone before opening up the next zone to settlement would put landowners in a very strong position to exploit buyers. This could be dealt with in a different way, discussed below.

Fourthly, local real estate taxes could be made into a conscious instrument to implement plans. This could be done by gearing taxes more closely to land values as the latter are affected or established by zoning and subdivision regulations. Taxes should be sharply raised in most suburbs on land zoned and classified for reasonably early development. They could be put high enough to bring severe pressure on landholders, forcing or inducing them to sell relatively soon. High taxes in the zones classed for early development would increase the cost of speculative holding of land and thus make early sale more attractive. At the same time, taxes might well be lowered on lands not classed as ready for early development. This would remove one incentive for early development. It would also lower costs of holding land and thus would encourage speculative holding and higher prices. This could be dealt with as explained below. Keeping taxes lower but at the same time putting the land in a class for deferred development might encourage use of intermingled and adjacent land for other

For a summary of better known studies of this kind, see Clawson, Held and Stoddard, *op. cit.*, pp. 70-74.

purposes, at least for a few years more in each case.

The lands not classed for early development might have part of the tax deferred. The part payable annually could be adjusted to a reasonable level for other land use; the deferred part would reflect value for later development. The deferred part would accumulate from year to year and would be a lien on the land. The deferred part might come due when the owner sold for actual development or it might come due when the public planning body classified the area as ready for development. The former would encourage longer holding for speculative gain and hence probably more urban sprawl. Many more owners would prefer to gamble on higher future land prices. Making deferred taxes due when the planning body classified the areas as ready for development would have the opposite effect: now pressure would be exerted for early sale and hence more nearly solid development encouraged.

Assessment and taxation have not generally been used deliberately to modify land use but they have nevertheless exerted great influence in this direction. Some may question the wisdom or the legality of taxes based on land use plans or the conscious use of taxes to implement plans. But, to the extent the plans are backed up by vigorous land zoning and subdivision controls, they do in fact vitally affect if not determine land values.

Fifth, the public, acting through government at some level, should acquire as much of the vacant lands as it needs for public purposes. By-passed or leap-frogged areas are often suitable for parks or other public purposes. Owners of such areas are often willing to sell. Others will prefer to hold for later possible gain. The area actually required for public use is often small compared to the total vacant area. But its early public action would have two effects: (1) the parks, schools, and other public uses would considerably affect private land use; and (2) offer by public agencies, or asking prices by such landowners, establish the market price of such land. If local tax assessments for such land could quickly be adjusted to the customary ratio of assessed to market price, then the profitability of continued speculative holding would be sharply decreased and urban sprawl correspondingly lessened. Close cooperation between school, park, and other bodies interested in acquiring land for public purposes on the one hand, and tax assessment bodies on the other, could be most effective.

Sixth, a more purposeful and coordinated use of public services such as roads, water lines, and trunk sewers could greatly affect urban sprawl. By refusing to extend any of these or other services to more distant areas until most of the intervening area was filled up, urban sprawl could be substantially reduced. The wisdom to plan public improvements in this way and the courage to enforce such plans would require a substantially higher level of performance than urban and metropolitan public service agencies typically now have. Such a program by public agencies should be accompanied by an educational program so that the general public would understand how and why such services were used for this purpose. Unless accompanied by some of the measures previously described for bringing pressure on closer-in landholders to sell, this too would give them monopoly power and large gains in land prices.

Underlying all these suggestions is the idea that government at some level possesses great powers for influencing, if not controlling, the future form of the city and metropolis. To achieve positive goals, suburban sprawl and speculation in raw suburban land must be greatly reduced or eliminated. The net effec ıese various measures would be to greatly nge general expectations of future land es and dates of maximum net gains. Some ıction in land prices at time of develop-t might be achieved; timing of develop-t and hence of gain in land prices based t would be more predictable. Hence, some he basis for land speculation would be e. Users of land for other purposes would e a longer and more certain planning izon.

'hese suggestions assume that these varied vers of government can be marshalled to h a coordinated program. This may be ealistic. Most of these programs are for ıl government. Local government is no-ously fragmented and uncoordinated — ch more so, really, than federal govern-ıt about whose deficiencies we hear so ch. If a really coordinated and effective ck is to be made on urban sprawl and culation in raw suburban land then per-s we shall have to use the Suburban De-opment District which I have proposed where.[5] Under that proposal, various ıl governmental and private interests, sub-: to some regulation by the state, would empowered to form special districts, with y wide powers over all aspects of the sub-anization process. Such powers would e a limited time duration and the districts ıld pass out of existence once an area were sonably well settled.

The ultimate in public control over land settlement and land speculation is achieved only when a public agency first acquires all the land from present owners and then sells it to new owners. Experience in the United States and abroad with forced land reform, land colonization, or other land use arrangements where the public objective diverges significantly from private objectives has shown rather clearly that anything less can somehow be evaded in some way. However, it seems to this author most unrealistic to think of wholesale public acquisition of land in potential suburbs with its subsequent sale to actual occupiers and developers. We have, it is true, gone about this with the seriously decadent slums in the older parts of our cities; one reason we were willing to do so there was that the process required a major infusion of public funds. Public acquisition, possibly public development, certainly sale to private users of land in new suburbs would, on the other hand, be highly profitable; for that reason, if none other, private interests will bitterly oppose it. On the governmental level we are not willing to take strong measures to prevent a possible or probable future disaster or difficulty; we wait until it is upon us. Thus, while for logical completeness one should include wholesale public acquisition and subdivision of suburban land as a means of achieving better cities, through reducing sprawl and speculation, as a practical matter it probably is not a real alternative.

Iarion Clawson, "Suburban Development Dis-ricts: A Proposal for Better Urban Growth," *ournal of the Am. Institute of Planners* (May, 960).

The Effects of Urbanization

Section C

Scholarly critics seldom agree on the correct description of our urban society. Some argue that the achievements outweigh the defects, while others see only problems and the ultimate destruction of the total environment. Too often the detrimental effects of urban growth are the most publicized by our news media. But, if the picture of our cities depicts much that is wrong, why do they continue to grow and attract more people? Analysis of some aspects of the problem may lead to at least partial answers.

Jane Jacobs' *The Kind of Problem a City Is* presents an interesting concept of a city as a process. The dictionary defines process as 'a continuing development involving many changes'. If the process is the essence of a city, then man is the catalyst of that process and cities may contain the seeds of their own regeneration.

In his discussion of Riverside, the Olmsted designed suburb of Chicago, Eaton, in *The American Suburb: Dream and Nightmare,* criticizes the modern suburb as a monument to the greed of the developer. The parkways, spacious lawns and open spaces of Riverside have been forgotten in the search for wealth. *Why Planners Fail* by Stanford presents an explanation for the decline of the American suburb. He maintains that the fault lies with the planners themselves who do not see their work as an integral part of the administration and planning of a city but as something above all the red tape. Stanford suggests that planners should enter the lifeblood of a city. The failure of planners to adequately consider rurban lands is the main issue of Kinsel's *A Concept of Rural-Urban Regions.* He presents an argument for more study of those agricultural lands that have become enclosed or surrounded by an expanding urban region. Thorsell, *Open Space for the Urban Region,* gives us an alternate look at open areas. Nc only agricultural lands but parks, green-belt and forested areas must be preserved an maintained in the city.

Spelt, in *The Development of the Toront Conurbation*, opens to the student a discussion of the growth of a large portion of th Golden Horseshoe (Mississaga) with som thoughts to the future. Following a literal us of the title *A Tale of Two Cities*, Richardso discusses the development of two planne cities.

The average citizen is usually not awar of the political and administrative red tap needed to put urban renewal into effect. Tanner in *Who Does What in Urban Renewal* and Fountain in *Zoning Administration i Vancouver* have clarified many aspects c this vexing problem. Tanner's explanation c the general manner in which renewal i achieved is backed by detailed steps in planning by zoning.

Urban areas have presented unique problems that are often contradictory in term of development and solution. Stokes i *A Theory of Slums* presents a sociologica explanation for the existence of slums in citie of a very wealthy nation. Rooney, *The Urba Snow Hazard in the United States,* attempt to explain why the northern cities, which ar the best prepared, are often the worst disrupted by heavy snowfalls.

It is only fitting to terminate the book witl a plea for better cities. Beecroft in *Let U Make Our Cities Efficient* points out tha better access routes might encourage mor people to stay in the city rather than mov out to the suburbs. Shopping plazas are popular because they are efficient. Wilson, *Cumbernauld New Town*, describes an efficien but pleasant urban development, suitable fo man, now and in the future.

'he Kind of Problem
City Is

28

ane Jacobs

ıinking has its strategies and tactics too, ıch as other forms of action have. Merely think about cities and get somewhere, one the main things to know is what *kind* of oblem cities pose, for all problems cannot thought about in the same way. Which enues of thinking are apt to be useful and help yield the truth depends not on how might prefer to think about a subject, but ther on the inherent nature of the subject elf.

Among the many revolutionary changes of is century, perhaps those that go deepest the changes in the mental methods we can for probing the world. I do not mean new chanical brains, but methods of analysis d discovery that have gotten into human ains: new strategies for thinking. These ve developed mainly as methods of science. t the mental awakenings and intellectual ring they represent are gradually beginning affect other kinds of inquiry too. Puzzles at once appeared unanalyzable become ore susceptible to attack. What is more, the very nature of some puzzles are no longer what they once seemed.

To understand what these changes in strategies of thought have to do with cities, it is necessary to understand a little about the history of scientific thought. A splendid summary and interpretation of this history is included in an essay on science and complexity in the 1958 *Annual Report of the Rockefeller Foundation*, written by Dr. Warren Weaver upon his retirement as the foundation's Vice-President for the Natural and Medical Sciences. I shall quote from this essay at some length, because what Dr. Weaver says has direct pertinence to thought about cities. His remarks sum up, in an oblique way, virtually the intellectual history of city planning.

Dr. Weaver lists three stages of development in the history of scientific thought: (1) ability to deal with problems of simplicity; (2) ability to deal with problems of disorganized complexity; and (3) ability to deal with problems of organized complexity.

printed from *Death and Life of Great American Cities* (New York, Random House, 1963) pp. 428-, by permission of the author and publishers.

Problems of simplicity are problems that contain two factors which are directly related to each other in their behavior — two variables — and these problems of simplicity, Dr. Weaver points out, where the first *kinds* of problems that science learned to attack:

Speaking roughly, one may say that the seventeenth, eighteenth and nineteenth centuries formed the period in which physical science learned how to analyze two-variable problems. During that three hundred years, science developed the experimental and analytical techniques for handling problems in which one quantity — say a gas pressure — depends primarily upon a second quantity — say, the volume of the gas. The essential character of these problems rests in the fact that . . . the behavior of the first quantity can be described with a useful degree of accuracy by taking into account only its dependence upon the second quantity and by neglecting the minor influence of other factors.

These two-variable problems are essentially simple in structure . . . and simplicity was a necessary condition for progress at that stage of development of science.

It turned out, moreover, that vast progress could be made in the physical sciences by theories and experiments of this essentially simple character. . . . It was this kind of two-variable science which laid, over the period up to 1900, the foundations for our theories of light, of sound, of heat, and of electricity . . . which brought us the telephone and the radio, the automobile and the airplane, the phonograph and the moving pictures, the turbine and the Diesel engine and the modern hydroelectric power plant . . .

It was not until after 1900 that a second method of analyzing problems was developed by the physical sciences.

Some imaginative minds [Dr. Weaver continues] rather than studying problems which involved two variables or at most three o four, went to the other extreme, and said "Let us develop analytical methods which ca deal with two billion variables." That is t say, the physical scientists (with the mathe maticians often in the vanguard) develope powerful techniques of probability theor and of statistical mechanics which can dea with what we may call problems of *disor ganized complexity* . . . .

Consider first a simple illustration in orde to get the flavor of the idea. The classica dynamics of the nineteenth century were we suited for analyzing and predicting the mo tion of a single ivory ball as it moves about o a billiard table. . . . One can, but with sur prising increase in difficulty, analyze the mo tion of two or even three balls on a billiar table. . . . But as soon as one tries to analyz the motion of ten or fifteen balls on the tabl at once, as in pool, the problem become unmanageable, not because there is an theoretical difficulty, but just because th actual labor of dealing in specific detail wit so many variables turns out to be impractical

Imagine, however, a large billiard tabl with millions of balls flying about on its sur face. . . . The great surprise is that the prob lem now becomes easier: the methods o statistical mechanics are now applicable. On cannot trace the detailed history of on special ball, to be sure; but there can b answered with useful precision such import ant questions as: On the average how man balls per second hit a given stretch of rail On the average how far does a ball mov before it is hit by some other ball? . . .

. . . The word "disorganized" [applies] t the large billiard table with the many ball . . . because the balls are distributed, in thei positions and motions, in a helter-skelte way. . . . But in spite of this helter-skelter o unknown behavior of all the individual vari

:s, the system as a whole possesses ain orderly and analyzable average pro- :ies . . . .

wide range of experience comes under label of disorganized complexity . . . . It lies with entirely useful precision to the erience of a large telephone exchange, licting the average frequency of calls, the bability of overlapping calls of the same ber, etc. It makes possible the financial oility of a life insurance company . . . . The ions of the atoms which form all matter, vell as the motions of the stars which form universe, all come under the range of e new techniques. The fundamental laws eredity are analyzed by them. The laws hermodynamics, which describe basic and itable tendencies of all physical systems, derived from statistical considerations. whole structure of modern physics . . . s on these statistical concepts. Indeed, the le question of evidence, and the way in ch knowledge can be inferred from evi- ce, is now recognized to depend on these e ideas . . . . We have also come to realize communication theory and information ory are similarly based upon statistical s. One is thus bound to say that prob- ity notions are essential to any theory of wledge itself.

Iowever, by no means all problems could robed by this method of analysis. The life nces, such as biology and medicine, ld not be, as Dr. Weaver points out. se sciences, too, had been making ad- ces, but on the whole they were still con- ed with what Dr. Weaver calls prelimin- stages for application of analysis; they e concerned with collection, description, sification, and observation of apparently elated effects. During this preparatory e, among the many useful things that e learned was that the life sciences were neither problems of simplicity nor problems of disorganized complexity; they inherently posed still a different kind of problem, a kind of problem for which methods of attack were still very backward as recently as 1932, says Dr. Weaver.

Describing this gap, he writes:

One is tempted to oversimplify and say that scientific methodology went from one extreme to the other . . . and left untouched a great middle region. The importance of this middle region, moreover, does not depend primarily on the fact that the number of variables involved is moderate — large compared to two, but small compared to the number of atoms in a pinch of salt . . . . Much more important than the mere number of variables is the fact that these variables are all interrelated . . . . These problems, as constrasted with the disorganized situations with which statistics can cope, *show the essential feature of organization*. We will therefore refer to this group of problems as those of *organized complexity*.

What makes an evening primrose open when it does? Why does salt water fail to satisfy thirst? . . . What is the description of aging in biochemical terms? . . . What is a gene, and how does the original genetic constitution of a living organism express itself in the developed characteristics of the adult? . . .

All these are certainly complex problems. But they are not problems of disorganized complexity, to which statistical methods hold the key. They are all problems which involve dealing simultaneously with a *sizable number of factors which are interrelated into an organic whole*.

In 1932, when the life sciences were just at the threshold of developing effective analytical methods for handling organized complexity, it was speculated, Dr. Weaver tells us, that if the life sciences could make significant progress in such problems, "then there might be opportunities to extend these new techniques, if only by helpful analogy, into vast areas of the behavioral and social sciences."

In the quarter-century since that time, the life sciences have indeed made immense and brilliant progress. They have accumulated, with extraordinary swiftness, an extraordinary quantity of hitherto hidden knowledge. They have also acquired vastly improved bodies of theory and procedure — enough to open up great new questions, and to show that only a start has been made on what there is to know.

But this progress has been possible only because the life sciences were recognized to be problems in organized complexity, and were thought of and attacked in ways suitable of understanding that *kind* of problem.

The recent progress of the life sciences tells us something tremendously important about other problems of organized complexity. It tells us that problems of this *kind* can be analyzed — that it is only sensible to regard them as capable of being understood, instead of considering them, as Dr. Weaver puts it, to be "in some dark and foreboding way, irrational."

Now let us see what this has to do with cities.

Cities happen to be problems in organized complexity, like the life sciences. They present "situations in which a half-dozen or even several dozen quantities are all varying simultaneously *and in subtly interconnected ways*." Cities, again like the life sciences, do not exhibit *one* problem in organized complexity, which if understood explains all. They can be analyzed into many such problems or segments which, as in the case of the life sciences, are also related with one another. The variables are many, but they a not helter-skelter; they are "interrelated in an organic whole."

Consider again, as an illustration, th problem of a city neighborhood park. An single factor about the park is slippery as a eel; it can potentially mean any number o things, depending on how it is acted upon b other factors and how it reacts to them. Ho much the park is used depends, in part, upo the park's own design. But even this parti influence of the park's design upon the park use depends, in turn, on who is around use the park, and when, and this in turn de pends on uses of the city outside the par itself. Furthermore, the influence of the uses on the park is only partly a manner how each affects the park independently the others; it is also partly a matter of ho they affect the park in combination with o another, for certain combinations stimula the degree of influence from one anoth among their components. In turn, these ci uses near the park and their combinations de pend on still other factors, such as the mi ture of age in buildings, the size of blocks the vicinity, and so on, including the presen of the park itself as a common and unifyin use in its context. Increase the park's si considerably, or else change its design in suc a way that it severs and disperses users fro the streets about it, instead of uniting an mixing them, and all bets are off. New se of influence come into play, both in the par and in its surroundings. This is a far cry fro the simple problem of ratios of open space ratios for population; but there is no use wish ing it were a simpler problem or trying make it a simpler problem, because in real li it is not a simpler problem. No matter wh you try to do to it, a city park *behaves* like problem in organized complexity, and that at it is. The same is true of all other parts features of cities. Although the inter-relans of their many factors are complex, there nothing accidental or irrational about the ys in which these factors affect each other. Moreover, in parts of cities which are rking well in some respects and badly in ers (as is often the case), we cannot even alyze the virtues and the faults, diagnose trouble or consider helpful changes, witht going at them as problems of organized mplexity. To take a few simplified illustrans, a street may be functioning excellently the supervision of children and at produc- a casual and trustful public life, but be ing miserably at solving all other problems cause it has failed at knitting itself with effective larger community, which in turn y or may not exist because of still other s of factors. Or a street may have, in itself, cellent physical material for generating ersity and an admirable physical design casual surveillance of public spaces, and t because of its proximity to a dead border, may be so empty of life as to be shunned d feared even by its own residents. Or a eet may have little foundation for workaity on its own merits, yet geographically in so admirably with a district that is workle and vital that this circumstance is enough sustain its attraction and give it use and fficient workability. We may wish for ier, all-purpose analyses, and for simpler, gical, all-purpose cures, but wishing cant change these problems into simpler mats than organized complexity, no matter w much we try to evade the realities and to ndle them as something different.

Why have cities not, long since, been ntified, understood and treated as probns of organized complexity? If the people ncerned with the life sciences were able to identify their difficult problems as problems of organized complexity, why have people professionally concerned with cities not identified the *kind* of problem they had?

The history of modern thought about cities is unfortunately very different from the history of modern thought about the life sciences. The theorists of conventional modern city planning have consistently mistaken cities as problems of simplicity and of disorganized complexity, and have tried to analyze and treat them thus. No doubt this imitation of the physical sciences was hardly conscious. It was probably derived, as the assumptions behind most thinking are, from the general floating fund of intellectual spores around at the time. However, I think these misapplications could hardly have occurred, and certainly would not have been perpetuated as they have been, without great disrespect for the subject matter itself — cities. These misapplications stand in our way; they have to be hauled out in the light, recognized as inapplicable strategies of thought, and discarded.

Garden City planning theory had its beginnings in the late nineteenth century, and Ebenezer Howard attacked the problem of town planning much as if he were a nineteenth-century physical scientist analyzing a two-variable problem of simplicity. The two major variables in the Garden City concept of planning were the quantity of housing (or population) and the number of jobs. These two were conceived of as simply and directly related to each other, in the form of relatively closed systems. In turn, the housing had its subsidiary variables, related to it in equally direct, simple, mutually independent form: playgrounds, open space, schools, community center, standardized supplies and services. The town as a whole was conceived of, again, as one of the two variables in a direct, simple, town-greenbelt relationship. As a system of order, that is about all there was to it. And on this simple base of two-variable relationships were created an entire theory of self-contained towns as a means of redistributing the population of cities and (hopefully) achieving regional planning.

Whatever may be said of this scheme for isolated towns, any such simple systems of two-variable relationships cannot possibly be discerned in great cities — and never could be. Such systems cannot be discerned in a town either, the day after the town becomes encompassed in a metropolitan orbit with its multiplicity of choices and complexities of cross-use. But in spite of this fact, planning theory has persistently applied this two-variable *system of thinking and analyzing* to big cities; and to this day city planners and housers believe they hold a precious nugget of truth about the *kind* of problem to be dealt with when they attempt to shape or reshape big-city neighborhoods into versions of two-variable systems, with ratios of one thing (as open space) depending directly and simply upon an immediate ratio of something else (as population).

To be sure while planners were assuming that cities were properly problems of simplicity, planning theorists and planners could not avoid seeing that real cities were not so in fact. But they took care of this in the traditional way that the incurious (or the disrespectful) have always regarded problems of organized complexity: as if these puzzles were, in Dr. Weaver's words, "in some dark and foreboding way, irrational."*

Beginning in the late 1920's in Europe, and in the 1930's here, city planning theory began to assimilate the newer ideas on probability theory developed by physical science. Planners began to imitate and apply thes analyses precisely as if cities were problem in disorganized complexity, understandabl purely by statistical analysis, predictable b the application of probability mathematic manageable by conversion into groups c averages.

This conception of the city as a collectio of separate file drawers, in effect, was suite very well by the Radiant City vision of L Corbusier, that vertical and more centralize version of the two-variable Garden City. Al though Le Corbusier himself made no mor than a gesture toward statistical analysis, hi scheme assumed the statistical reordering of system of disorganized complexity, solvabl mathematically; his towers in the park wer a celebration, in art, of the potency c statistics and the triumph of the mathe matical average.

The new probability techniques, and th assumptions about the kind of problem tha underlay the way they have been used in cit planning, did not supplant the base idea c the two-variable reformed city. Rather thes new ideas were added. Simple, two-variabl systems of order were still the aim. But thes could be organized even more "rationally" now, from out of a supposed existing systen of disorganized complexity. In short, the nev probability and statistical methods gav more "accuracy," more scope, made possibl a more Olympian view and treatment of th supposed problem of the city.

With the probability techniques, an ol aim — stores "properly" related to im mediate housing or to a preordained popula tion — became seemingly feasible; ther arose techniques for planning standardize shopping "scientifically"; although it wa early realized by such planning theorists a Stein and Bauer that preplanned shoppin centers within cities must also be mono-

* *e.g.* "a chaotic accident," "solidified chaos," etc.

listic or semimonopolistic, or else the tistics would not predict, and the city uld go on behaving with dark and fore- ding irrationality.

With these techniques, it also became fea- le to analyze statistically, by income ups and family sizes, a given quantity of ople uprooted by acts of planning, to com- ne these with probability statistics on rmal housing turnover, and to estimate curately the gap. Thus arose the supposed sibility of large-scale relocation of citizens. the form of statistics, these citizens were longer components of any unit except the nily, and could be dealt with intellectually e grains of sand, or electrons or billiard lls. The larger the number of uprooted, the re easily they could be planned for on the sis of mathematical averages. On this basis was actually intellectually easy and sane contemplate clearance of all slums and re- rting of people in ten years and not much rder to contemplate it as a twenty-year job.

By carrying to logical conclusions the sis that the city, as it exists, is a problem disorganized complexity, housers and plan- rs reached — apparently with straight faces the idea that almost any specific mal- nctioning could be corrected by opening d filling a new file drawer. Thus we get ch political party policy statements as this: 'he Housing Act of 1959 . . . should be sup- mented to include . . . a program of hous- for moderate-income families whose in- nes are too high for admission to public using, but too low to enable them to obtain cent shelter in the private market."

With statistical and probability techniques, also became possible to create formidable d impressive planning surveys for cities — rveys that come out with fanfare, are read practically nobody, and then drop quietly to oblivion, as well they might, being nothing more nor less than routine exercises in statistical mechanics for systems of disorganized complexity. It became possible also to map out master plans for the statistical city, and people take these more seriously, for we are all accustomed to believe that maps and reality are necessarily related, or that if they are not, we can make them so by altering reality.

With these techniques, it was possible not only to conceive of people, their incomes, their spending money and their housing as fundamentally problems in disorganized complexity, susceptible to conversion into problems of simplicity once ranges and averages were worked out, but also to conceive of city traffic, industry, parks, and even cultural facilities as components of disorganized complexity, convertible into problems of simplicity.

Furthermore, it was no intellectual disadvantage to contemplate "coordinated" schemes of city planning embracing ever greater territories. The greater the territory, as well as the larger the population, the more rationally and easily could both be dealt with as problems of disorganized complexity viewed from an Olympian vantage point. The wry remark that "A Region is an area safely larger than the last one to whose problems we found no solution" is not a wry remark in these terms. It is a simple statement of a basic fact about disorganized complexity; it is much like saying that a large insurance company is better equipped to average out risks than a small insurance company.

However, while city planning has thus mired itself in deep misunderstandings about the very nature of the problem with which it is dealing, the life sciences, unburdened with this mistake, and moving ahead very rapidly, have been providing some of the concepts that city planning needs: along with providing the basic strategy of recognizing problems of organized complexity, they have provided hints about analyzing and handling this *kind* of problem. These advances have, of course, filtered from the life sciences into general knowledge; they have become part of the intellectual fund of our times. And so a growing number of people have begun, gradually, to think of cities as problems in organized complexity — organisms that are replete with unexamined, but obviously intricately interconnected, and surely understandable, relationships.

This is a point of view which has little currency yet among planners themselves, among architectural city designers, or among the businessmen and legislators who learn their planning lessons, naturally, from what is established and long accepted by planning "experts". Nor is this a point of view that has much appreciable currency in schools of planning (perhaps there least of all).

City planning, as a field, has stagnated. It bustles but it does not advance. Today's plans show little if any perceptible progress in comparison with plans devised a generation ago. In transportation, either regional or local, nothing is offered which was not already offered and popularized in 1938 in the General Motors diorama at the New York World's Fair, and before that by Le Corbusier. In some respects, there is outright retrogression. None of today's pallid imitations of Rockefeller Center is as good as the original, which was built a quarter of a century ago. Even in conventional planning's *own given terms*, today's housing projects are no improvement, and usually a retrogression, in comparison with those of the 1930's.

As long as city planners, and the businessmen, lenders, and legislators who have learned from planners, cling to the unexamined assumptions that they are dealing with a problem in the physical sciences, city planning cannot possibly progress. Of course it stagnates. It lacks the first requisite for a body of practical and progressing thought: recognition of the kind of problem at issue. Lacking this, it has found the shortest distance to a dead end.

Because the life sciences and cities happen to pose the same *kinds* of problems does not mean they are the *same* problems. The organizations of living protoplasm and the organizations of living people and enterprises cannot go under the same microscopes.

However, the tactics for understanding both are similar in the sense that both depend on the microscopic or detailed view, so to speak, rather than on the less detailed, naked-eye view suitable for viewing problems of simplicity or the remote telescopic view suitable for viewing problems of disorganized complexity.

In the life sciences, organized complexity is handled by identifying a specific factor or quantity — say an enzyme — and then painstakingly learning its intricate relationships and interconnections with other factors or quantities. All this is observed in terms of the behavior (not mere presence) of other specific (not generalized) factors or quantities. To be sure, the techniques of two-variable and disorganized-complexity analysis are used too, but only as subsidiary tactics.

In principle, these are much the same tactics as those that have to be used to understand and to help cities. In the case of understanding cities, I think the most important habits of thought are these:

1. To think about processes;
2. To work inductively, reasoning from particulars to the general, rather than the reverse;
3. To seek for "unaverage" clues involving very small quantities, which reveal the way larger and more "average" quantities are operating.

f you have gotten this far you do not d much explanation of these tactics. How- r, I shall sum them up, to bring out points erwise left only as implications.

Vhy think about processes? Objects in es — whether they are buildings, streets, ks, districts, landmarks, or anything else can have radically differing effects, de- ding upon the circumstances and contexts hich they exist. Thus, for instance, almost ning useful can be understood or can be e about improving city dwellings if these considered in the abstract as "housing." dwellings — either existing or potential are *specific* and particularized buildings *ays involved in differing, specific processes* h as unslumming, slumming, generation of rsity, self-destruction of diversity.*

or cities, processes are of the essence. thermore, once one thinks about city pro- ses, it follows that one *must* think of lysts of these processes, and this too is he essence.

he processes that occur in cities are not ane, capable of being understood only by erts. They can be understood by almost body. Many ordinary people already erstand them; they simply have not given se processes names, or considered that by erstanding these ordinary arrangements ause and effect, we can also direct them e want to.

Vhy reason inductively? Because to rea- , instead, from generalizations ultimately es us into absurdities — as in the case of Boston planner who knew (against all the -life evidence he had) that the North End to be a slum because the generalizations make him as expert say it is.

This is an obvious pitfall because the generalizations on which the planner was depending are themselves so nonsensical. However, inductive reasoning is just as important for identifying, understanding and constructively using the forces and processes that actually are relevant to cities, and therefore are not nonsensical. I have generalized about these forces and processes considerably, but let no one be misled into believing that these generalizations can be used routinely to declare what the particulars, in this or that place, *ought* to mean. City processes in real life are too complex to be routine, too particularized for application as abstractions. They are always made up of interactions among unique combinations of particulars, and there is no substitute for knowing the particulars.

Inductive reasoning of this kind is, again, something that can be engaged in by ordinary, interested citizens, and again they have the advantage over planners. Planners have been trained and disciplined in *deductive* thinking, like the Boston planner who learned his lessons only too well. Possibly because of this bad training, planners frequently seem to be less well equipped intellectually for respecting and understanding particulars than ordinary people, untrained in expertise, who are attached to a neighborhood, accustomed to using it, and so are not accustomed to thinking of it in generalized or abstract fashion.

Why seek "unaverage" clues, involving small quantities? Comprehensive statistical studies, to be sure, can *sometimes* be useful abstracted measurements of the sizes, ranges, averages and medians of this and that. Gathered from time to time, statistics can tell too what has been happening to these figures. However, they tell almost nothing about how the quantities are working in systems of organized complexity.

To learn how things are working, we need

Because this is so, "housers," narrowly special- g in "housing" expertise, are a vocational rdity. Such a profession makes sense only if assumed that "housing" *per se* has important eralized effects and qualities. It does not.

pinpoint clues. For instance, all the statistical studies possible about the downtown of Brooklyn, N.Y., cannot tell us as much about the problem of that downtown and its cause as is told in five short lines of type in a single newspaper advertisement. This advertisement, which is for Marboro, a chain of bookstores, gives the business hours of the chain's five stores. Three of them (one near Carnegie Hall in Manhattan, one near the Public Library and not far from Times Square, one in Greenwich Village) stay open until midnight. A fourth, close to Fifth Avenue and Fifty-ninth Street, stays open until 10 P.M. The fifth, in downtown Brooklyn, stays open until 8 P.M. Here is a management which keeps its stores open late, if there is business to be had. The advertisement tells us that Brooklyn's downtown is too dead by 8 P.M., as indeed it is. No surveys (and certainly no mindless, mechanical predictions projected forward in time from statistical surveys, a boondoggle that today frequently passes for "planning") can tell us anything so relevant to the composition and to the need of Brooklyn's downtown as this small, but specific and precisely accurate, clue to the *workings* of that downtown.

It takes large quantities of the "average" to produce the "unaverage" in cities. But as was pointed out in Chapter Seven, in the discussion on the generators of diversity, the mere presence of large quantities — whether people, uses, structures, jobs, parks, streets or anything else — does not guarantee much generation of city diversity. These quantities can be working as factors in inert, low-energy systems, merely maintaining themselves, if that. Or they can make up interacting, high-energy systems, producing by-products of the "unaverage."

The "unaverage" can be physical, as in the case of eye-catchers which are small elements in much larger, more "average" visual scene They can be economic, as in the case of on of-a-kind stores, or cultural, as in the case an unusual school or out-of-the-ordina theater. They can be social, as in the case public characters, loitering places, or res dents or users who are financially, vocatio ally, racially or culturally unaverage.

Quantities of the "unaverage," which a bound to be relatively small, are indispe sable to vital cities. However, in the sen that I am speaking of them here, "unaverag quantities are also important as analytic means — as clues. They are often the on announcers of the way various large quan ties are behaving, or failing to behave, combination with each other. As a roug analogy, we may think of quantitative minute vitamins in protoplasmic systems, trace elements in pasture plants. These thin are necessary for proper functioning of th systems of which they are a part; howeve their usefulness does not end there, becau they can and do also serve as vital clues *what* is happening in the systems of whic they are a part.

This awareness of "unaverage" clues - or awareness of their lack — is, again, som thing any citizen can practice. City dweller indeed, are commonly great informal exper in precisely this subject. Ordinary people i cities have an awareness of "unaverage quantities which is quite consonant with th importance of these relatively small quant ties. And again, planners are the ones at th disadvantage. They have inevitably come regard "unaverage" quantities as relativel inconsequential, because these are *statistical* inconsequential. They have been trained dicount what is most vital.

Now we must dig a little deeper into th bog of intellectual misconceptions about citi in which orthodox reformers and planne ve mired themselves (and the rest of us). derlying the city planners' deep disrespect their subject matter, underlying the jejune ief in the "dark and foreboding" irration- y or chaos of cities, lies a long-established sconception about the relationship of cities and indeed of men — with the rest of ure.

Human beings are, of course, a part of ture, as much so as grizzly bears or bees whales or sorghum cane. The cities of man beings are as natural, being a prod- of one form of nature, as are the colonies prairie dogs or the beds of oysters. The tanist Edgar Anderson has written wittily d sensitively in *Landscape* magazine from e to time about cities as a form of nature. ver much of the world," he comments, an has been accepted as a city-loving crea- e." Nature watching, he points out, "is ite as easy in the city as in the country; all e has to do is accept Man as a part of ture. Remember that as a specimen of *mo sapiens* you are far and away most ely to find that species an effective guide to eper understanding of natural history."

A curious but understandable thing hap- ned in the eighteenth century. By then, the es of Europeans had done well enough by m, mediating between them and many sh aspects of nature, so that something be- ne popularly possible which previously had n a rarity — sentimentalization of nature, at any rate, sentimentalization of a rustic a barbarian relationship with nature. rie Antoinette playing milkmaid was an ression of this sentimentality on one plane. e romantic idea of the "noble savage" was even sillier one, on another plane. So, in s country, was Jefferson's intellectual ection of cities of free artisans and me- anics, and his dream of an ideal republic of f-reliant rural yeomen — a pathetic dream for a good and great man whose land was tilled by slaves.

In real life, barbarians (and peasants) are the least free of men — bound by tradition, ridden by caste, fettered by superstitions, riddled by suspicion and foreboding of whatever is strange. "City air makes free," was the medieval saying, when city air literally did make free the runaway serf. City air still makes free the runaways from company towns, from plantations, from factory-farms, from subsistence farms, from migrant picker routes, from mining villages, from one-class suburbs.

Owing to the mediation of cities, it became popularly possible to regard "nature" as benign, ennobling and pure, and by extension to regard "natural man" (take your pick of how "natural") as so too. Opposed to all this fictionalized purity, nobility and beneficence, cities, not being fictions, could be considered as seats of malignancy and — obviously — the enemies of nature. And once people begin looking at nature as if it were a nice big St. Bernard dog for children, what could be more natural than the desire to bring this sentimental pet into the city too, so the city might get some nobility, purity and beneficence by association?

There are dangers in sentimentalizing nature. Most sentimental ideas imply, at bottom, a deep if unacknowledged disrespect. It is no accident that we Americans, probably the world's champion sentimentalizers about nature, are at one and the same time probably the world's most voracious and disrespectful destroyers of wild and rural countryside.

It is neither love for nature nor respect for nature that leads to this schizophrenic attitude. Instead, it is a sentimental desire to toy, rather patronizingly, with some insipid, standardized, suburbanized shadow of nature —

apparently in sheer disbelief that we and our cities, just by virtue of being, are a legitimate part of nature too, and involved with it in much deeper and more inescapable ways than grass trimming, sunbathing, and contemplative uplift. And so, each day, several thousand more acres of our countryside are eaten by the bulldozers, covered by pavement, dotted with suburbanites who have killed the thing they thought they came to find. Our irreplaceable heritage of Grade I agricultural land (a rare treasure of nature on this earth) is sacrificed for highways or supermarket parking lots as ruthlessly and unthinkingly as the trees in the woodlands are uprooted, the streams and rivers polluted and the air itself filled with the gasoline exhausts (products of eons of nature's manufacturing) required in this great national effort to cozy up with a fictionalized nature and flee the "unnaturalness" of the city.

The semisuburbanized and suburbanized messes we create in this way become despised by their own inhabitants tomorrow. These thin dispersions lack any reasonable degree of innate vitality, staying power, or inherent usefulness as settlements. Few of them, and these only the most expensive as a rule, hold their attraction much longer than a generation; then they begin to decay in the pattern of city gray areas. Indeed, an immense amount of today's city gray belts was yesterday's dispersion closer to "nature." Of the buildings on the thirty thousand acres of already blighted or already fast-blighting residential areas in northern New Jersey, for example, half are less than forty years old. Thirty years from now, we shall have accumulated new problems of blight and decay over acreages so immense that in comparison the present problems of the great cities' gray belts will look piddling. Nor, however destructive, is this something which happens accidentally or without the use of will. This is exactly what we, as a society, have willed to happen.

Nature, sentimentalized and considered as the antithesis of cities, is apparently assumed to consist of grass, fresh air and little else, and this ludicrous disrespect results in the devastation of nature even formally and publicly preserved in the form of a pet.

For example, up the Hudson River, north of New York City, is a state park at Croton Point, a place for picnicking, ballplaying and looking at the lordly (polluted) Hudson. At the Point itself is — or was — a geological curiosity: a stretch of beach about fifteen yards long where the blue-grey clay, glacially deposited there, and the action of the river currents and the sun combined to manufacture clay dogs. These are natural sculptures, compacted almost to the density of stone, and baked, and they are of a most curious variety, from breathtakingly subtle and simple curving forms to fantastic concoctions of more than Oriental splendor. There are only a few places in the entire world where clay dogs may be found.

Generations of New York City geology students, along with picnickers, tired ballplayers and delighted children, treasure hunted among the clay dogs and carried their favorites home. And always, the clay, the river and the sun made more, and more, and more, inexhaustibly, no two alike.

Occasionally through the years, having been introduced to the clay dogs long ago by a geology teacher, I would go back to treasure hunt among them. A few summers ago my husband and I took our children to the Point so they might find some and also so they might see how they are made.

But we were a season behind improvers of nature. The slope of muddy clay that formed the little stretch of unique beach had been

molished. In its place was a rustic retaining
ll and an extension of the park's lawns.
'he park had been augmented — statisti-
lly.) Digging beneath the new lawn here
d there — for we can desecrate the next
n's desecrations as well as anyone — we
ınd broken bits of clay dogs, mashed by the
lldozers, the last evidence of a natural pro-
ss that may well have been halted here
'ever.

Who would prefer this vapid suburbaniza-
n to timeless wonders? What kind of park
pervisor would permit such vandalism of
ture? An all too familiar kind of mind is
viously at work here: a mind seeing only
order where a most intricate and unique
der exists; the same kind of mind that sees
ly disorder in the life of city streets, and
hes to erase it, standardize it, suburbanize

The two responses are connected: Cities, as
ated or used by city-loving creatures are
respected by such simple minds because
y are not bland shadows of cities subur-
nized. Other aspects of nature are equally
respected because they are not bland
ıdows of nature suburbanized. Sentiment-
ty about nature denatures everything it
ches.

Big cities and countrysides can get along
ll together. Big cities need real countryside
se by. And countryside — from man's
int of view — needs big cities, with all
ir diverse, opportunities and productivity,
human beings can be in a position to ap-
ciate the rest of the natural world instead
to curse it.

Being human is itself difficult, and there-
e all kinds of settlements (except dream
es) have problems. Big cities have diffi-
ties in abundance, because they have people in abundance. But vital cities are not helpless to combat even the most difficult of problems. They are not passive victims of chains of circumstances, any more than they are the malignant opposite of nature.

Vital cities have marvelous innate abilities for understanding, communicating, contriving and inventing what is required to combat their difficulties. Perhaps the most striking example of this ability is the effect that big cities have had on disease. Cities were once the most helpless and devastated victims of disease, but they became great disease conquerors. All the apparatus of surgery, hygiene, microbiology, chemistry, telecommunications, public health measures, teaching and research hospitals, ambulances and the like, which people not only in cities but also outside them depend upon for the unending war against premature mortality, are fundamentally products of big cities and would be inconceivable without big cities. The surplus wealth, the productivity, the close-grained juxtaposition of talents that permit society to support advances such as these are themselves products of our organization into cities, and especially into big and dense cities.

It may be romantic to search for the salves of society's ills in slow-moving rustic surroundings, or among innocent, unspoiled provincials, if such exist, but it is a waste of time. Does anyone suppose that, in real life, answers to any of the great questions that worry us today are going to come out of homogeneous settlements?

Dull, inert cities, it is true, do contain the seeds of their own destruction and little else. But lively, diverse, intense cities contain the seeds of their own regeneration, with energy enough to carry over for problems and needs outside themselves.

29

The American Suburb Dream and Nightmare

Leonard K. Eato

With American cities expanding at an unprecedented rate, the time has arrived for a reconsideration of the traditional role of the suburb in the urban picture. While the history of the suburb goes back into medieval times, its most characteristic development in this country took place in the 19th Century. The typical suburb was an attempt to escape from the crowded conditions and ugly industrialism of the central city. Its virtues were openness and greenery. Invariably it was connected to the metropolis by railroad, and most of the male inhabitants were commuters. At the end of a hard session in the office or factory, the businessman could enjoy some of the advantages of country living, while his wife and children might have them during the entire day.

Among American practitioners of the difficult art of suburban design, the name of Frederick Law Olmsted, Sr. stands out most clearly. No less an authority than Lewis Mumford has commented extensively on Olmsted's achievement, laying particular emphasis on Riverside, a Chicago suburb, and Roland Park, just outside Baltimore. Because the plan has been so well preserved, th example of Riverside is particularly interest ing. What is the story behind the planning c this remarkable community, which even to day offers unusual amenities?

In 1868 a group of affluent Eastern busi nessmen formed a company to build wha they hoped would be a necessary adjunct t the rapidly growing city of Chicago. The recognized the need for a suburb easily acces sible to the city, a place affording home site where the families of businessmen could en joy some of the benefits traditionally associ ated with rural life. After a long search the settled on a sixteen-hundred-acre tract o land west of the city and bounded on tw sides by dense woods and the Des Plaine River. The area already had some reputatio as a resort center, since it possessed a goo country hotel where fashionable Chicagoan were accustomed to take their families fo Sunday dinner; moreover, it had been con nected to the heart of the city by a railroa in 1862. In 1869 the company showed bot good business judgment and esthetic sense b employing Olmsted, at that time the mos

Reprinted from *Landscape*, Vol. 13, no. 2 (Winter, 1963-64), pp. 12-16 by permission.

nent landscape architect in the United
es, to lay out the plan.

)lmsted, at this time age 47, was some-
g of a public figure. He had written what
e universally admitted to be the best social
economic commentaries on the pre-war
th; with Çalvert Vaux had won the comtion for Central Park in New York; and
, in fact, developed a national reputation
n expert in park design. At the same time
ısted was planning Riverside, he was also
ng Delaware Park in Buffalo and Prospect
k in Brooklyn, two of his finest efforts. It
nteresting to note that although the Gen-
Grant era was characterized by wide-
ead political corruption, it was also the
iod when American cities began to set
le substantial quantities of land for public
eation. While Chicago owes its magnifi-
t lake front to events following the World's
r of 1893, the land for its equally impor-
West Park System was purchased as
y as 1871. The park commissioners re-
ed William LeBaron Jenney, the noted
scraper architect, to do the first landscape
k.

)lmsted's writings reveal that his object in
design of Riverside was a combination of
an convenience and rural charm. In com-
n with many of his contemporaries, Olm-
l had an extremely romantic attitude
ard nature. He held contact with it was a
itive good; while he did not share the
theistic views of Emerson and Thoreau,
believed that experience of nature had a
nite moral value. In studying the area he
therefore delighted to find that it was
vily wooded with elm, oak, hickory and
nut trees and that a river bounded one
e of the property. His design took full
antage of this favorable natural situation.
h gently rolling land and a curving river
his major terrain features, he sought to
avoid the harsh regularity of a rectangular
plan. Instead of the squares and rectangles so
common elsewhere in 19th Century planning,
the pattern becomes a series of ovals and
circles of various sizes. Where the ovals and
circles meet, spaces for small parks and village greens are developed. It is not surprising
to find Olmsted writing to his clients, "The
ordinary directness of line in town streets,
with its resultant regularity of plan would
suggest eagerness to press forward without
looking to the right hand or the left; we should
recommend the general adoption, in the design of your roads, of gracefully curved lines,
generous spaces and the absence of sharp
corners, the idea being to suggest and imply
leisure, contemplativeness and happy tran-
quility." Today the residents of Riverside
sometimes boast that not one of their streets
follows a straight line, and Olmsted's small
parks are much used by children for ball
games, tree climbing, or just sitting in the
grass and talking.

Not neglecting the problem of access from
Chicago, Olmsted observed that the railroad
was not a sufficient means of transportation
and that a road should be constructed from
the center of the town. This approach road
was another romantic scheme. The plan involved a separation of traffic by green strips
planted with trees, watering places and
benches at intervals for rest. The entire idea
was to induce a feeling of serenity and rest-
fulness as one traveled along the road. "We
see no reason why," wrote Olmsted, "if this
suggestion is carried out liberally, it should
not provide, or at least begin to provide, another pressing desideratum of the city of Chicago, namely a general promenade ground."
In other words, Olmsted proposed a formal
parkway with naturalistic planting. His route
today is heavily traveled and is called Long
Common Road because of the large strip of

grass and trees between the two directions of traffic. Here is one of the earliest and most effective American usages of the parkway idea, later employed with excellent affect in the Burnham plan of 1909.

Olmsted's thinking on the design of individual dwellings, while advanced for its time, did not contemplate strict architectural controls. "We cannot judiciously attempt to control the form of the houses which men shall build," he remarked, "we can only, at most, take care that if they build very ugly and inappropriate houses, they shall not be allowed to force them disagreeably upon our attention when we desire to pass along the road upon which they stand. We can require that no house shall be built within a certain number of feet of the highway, and we can insist that each householder shall maintain one or two living trees between his house and the highway." In truth, Riverside is not a community made up of architectural gems from the past, although it does possess distinction in Frank Lloyd Wright's Coonley House, one of the masterpieces of his early period, as well as several other structures of the same vintage. The town's other showplace was Louis Sullivan's Babson House (1909), now torn down to make way for a "development." Most of the dwellings are medium to large size clapboard houses, set back from 30 to 70 feet from the roadway. They are placed at various angles to the street, thus contributing to the planned irregularity of the town. Many of the older homes were designed by Olmsted and Jenney.

While Riverside's commercial district possesses little architectural quality, the town's center itself has a strongly defined form. This is partly a matter of scale and partly a matter of vertical accent. Borrowing a concept from Kevin Lynch, we may say that the water tower is the major landmark. Somewhat difficult to see in summer because of the hea planting, it is easily viewed in winter wh the leaves have fallen from the trees. Rive side is, in fact, a highly imageable communit The river is a strong, natural edge, and t small parks act as nodes. As might be e pected, the entire area has a park-like qualit This impression is reinforced by one of Ol sted's master strokes, the suggestion that t river bank be reserved for a public pla ground. In winter it is heavily used for sli ing, and the river itself becomes a skati rink. All year round, Riverside provid ample opportunity for both active and pa sive recreation.

For the visitor to Riverside the domina impressions are the winding street patte and the luxuriance of the landscaping. T constantly changing direction of vision whi strolling or driving on these streets is a sour of pure delight to pedestrian and motori alike. In this respect the town has an almo medieval quality; it recalls the twisting stree of Siena and Perugia, though it lacks t changes in elevation so characteristic of tho cities. Because of this unusual street patter there is very little through traffic and aut mobile accidents are few. The landscapi makes an overwhelming impression. Some the trees which are today so stately we planted by Olmsted himself; others, such the lilacs, are carefully tended replacemen The thinning out process is apparently he to a minimum. In fact, many of the residen seem quite conscious of their heritage a are intent on preserving it insofar as is po sible. For example, they insist on retaini the charming (but antiquated) street lamp A typical comment in an interview was, "M husband and I have lived here forty years a raised our children here. They live in Ca fornia now, and we go there to visit them, b we always have a yen to get back to Rive e." One is reminded of the famous remark D. H. Burnham "... a noble, logical dia-m once recorded will never die, but long er we are gone will be a living thing, assert-; itself with ever-growing consistency."

In short, Riverside exemplifies the virtues the 19th Century upper middle class burb. It has a strong and compact core, a cinating street pattern and an ample allow-ce of light, air and greenery for everyone. landscaping, which differentiates it im-diately from the surrounding communities, unique.

By way of contrast, let us examine Allen rk, a 20th Century suburb of Detroit, med after the land speculator Lewis Allen. orporated as a village on April 4, 1927, h a population of 664, it has subsequently ng steadfastly to the gridiron pattern im-sed at its founding. The name Allen Park of course, an ironical indication of the nantic landscape atmosphere which was ight. By 1927 Detroit was well developed ustrially, and the town purported to offer customary suburban luxuries of fresh air d green space to those who wished to ape from the industrial atmosphere of the tral city. For a few years it must indeed ve been a refuge. In 1940 the federal sus counted 37,393 inhabitants; Riverside contrast has remained small — slightly r nine thousand in 1960. The cause of the narkable growth of Allen Park was the t-World War II boom in the Detroit auto-bile plants. The chief beneficiaries were real estate speculators who had laid out town in a completely mechanical pattern. eoretically, it is a bedroom community, in cept much like Riverside. Every morning male population arises an hour before job e to make the long, frustrating trip into troit over expressways which are continu-y more crowded.

The inhabitants are mostly middle class with an average income of about $8,000 per year. The average house costs anywhere from $15,000 to $25,000 and is provided with the customary city services such as paved roads, utilities, snow removal and fire and police protection. The city has numerous commercial establishments, one hundred acres of parks and playgrounds, six churches of various denominations, four hospitals, a public library and a community building. A close observer remarks that most of the citizens of Allen Park have moved there for one of three reasons: (1) It is "a better place to raise a family." (2) A house in Allen Park is a status symbol, and most of these people are upwardly mobile. (3) It is a retreat from the hustle and bustle of Detroit. This inventory of institutions and sociological analysis indicates nothing whatever about the visual quality of this environment. A photographic survey of Allen Park shows that the suburban dream has become a nightmare.

The center of the town has become a tawdry commercial area filled with small businesses, most of them struggling to meet the competition of neighboring shopping centers. Striking landmarks, such as the water tower in Riverside, are altogether lacking. Even more appalling is the carefully engineered monotony of the residential districts. For this hideous dullness the city building code is partly responsible. Even on Allen Road, which is the strategic backbone of the town, the setback is uniformly established at sixty feet from the center line, while on the three secondary roads it is seventeen feet. An attempt has been made at the suburban atmosphere of open greenness through the regulation of fences, which are restricted to side and back yards only and must be of non-solid construction with a maximum height of four feet. These regulations would tax the ingenuity of a Leonardo Da Vinci and contrast dramatically with the lack of restrictions in Riverside.

Equally important in the development of Allen Park has been the avarice of the speculative builder. The residences are completely standardized; almost all are brick veneer with white trim. There is a maximum residential building height of thirty-five feet or two and one-half stories, but the majority are a storey and a half. These dwellings are placed in a minimum lot of 5,000 square feet, and the structure must cover not more than 30 percent of the total lot area. Lots must be at least fifty feet wide with a front yard minimum of twenty feet, an interval of three feet on one side and eight feet on the other, and a backyard depth of thirty-five feet. The land has been developed as intensively as possible. The result is a series of ludicrously small spaces between the houses; even the bicycles are crowded together. Population density is relatively high and privacy is almost impossible to obtain.

Are there any elements of interest in this community? Quite clearly, the most striking buildings are the schools. The somewhat conventional but neatly detailed South Junior High School is typical of the structures recently put up by a harassed school board, which must struggle to educate an increasing number of children on a decreasing tax base. Such green space as the community possesses tends to be concentrated around the schools. Here little league baseball games go on, while in the background the monotonous rows of identical houses extend in straight lines far into the distance.

The primacy of the school is symbolic. Along with the church, the bridge club and the family circle itself, the school is a major center of social activity. As Lewis Mumford has pointed out American suburbs are turning more and more away from metropolita centers for their social interests. Allen Par is hardly conscious of the museums, theatres concerts and general intellectual stimulu offered by Detroit. In contrast, Riverside precisely because it is bound closely to Chicago by the Burlington Railroad, has a muc more organic connection with its mother city One wonders about the children growing u in Allen Park. Will they be forever oriente to suburbia and ignorant of the potentialitie of the city? It is a terrifying possibility.

Like most other suburban communities i the Detroit area, Allen Park has two sever economic problems. They are: insufficien parking area to accommodate the invasion o the automobile, and a rising public deman for city services accompanied by an unwillingness to pay for these services through personal taxation. The response to these problems will determine its future.

Because the housewife ordinarily insists o using the automobile to go to the corner market, she has difficulty finding a parking place Consequently, she patronizes a shoppin center in a neighboring town, leaving the individual commercial establishment bereft o customers. In order to save some of the business of the small merchant, the town is nov forced to cover large quantities of valuabl land with asphalt or concrete. Frequentl residences have had to be torn down, an sometimes park area has had to be taken fo parking lots. This radical measure, of course removes valuable land from the tax rolls s that Allen Park is truly caught in a viciou circle. In small compass it has the same problem as downtown Los Angeles, where 60 percent of the total land area is now donated t super highways, parking structures and parking lots.

At present Allen Park has an excellen array of city services, but, because of the unlingness of its citizens to pay for them, it s been forced to allocate a portion of its ıd to industry. Since most of its area has eady been built up, this allocation is rather ited in extent; nonetheless, five major instrial concerns have been attracted: The rd Motor Company, B. F. Goodrich Co., a ttern works, Wolverine Tube division of lumet and Hecla, Inc., and a Montgomery ard regional office. The land zoned for instrial use is near a branch of the Detroit, ledo and Ironton R.R. (an important ight carrier, not a commuter's railway), d in addition, super highways carry the oducts of Allen Park to midwestern marts such as Detroit, Chicago and Cleveland. l of this development scarcely squares with the classic picture of the 19th Century suburb, which we can still observe in Riverside. Olmsted's suburb offers the very real advantages of close contact with the city, less emphasis on commercial expansion and more places for informal sociability. Its townscape is a living monument to the wisdom of Olmsted and the foresight of its citizens. Allen Park, by way of contrast, is a monument to the greed of the real estate speculator, though it was supposedly founded to secure for its people most of the same benefits as Riverside. When there is such a great discrepancy between ideology and reality, it is usually wise to admit that a concept is outmoded and to begin thinking anew.

30

Why Planners Fai

John H. Stanfor

One need only sample the literature of planning to sense the pervasive underlying emotion among planners these days: frustration. In their more frankly written essays, particularly those aimed at enlisting sympathetic support, the cities are pretty clearly assessed as going from bad to worse. Then men-with-the-plans who are supposed to reverse what they generally see as distressing urban deterioration seem discouraged, and irritable. Despite more recognition, more planning offices, bigger budgets and a general sense of respectability that contrasts sharply with the pre-World-War-II attitude toward planning, the sum total of specific accomplishment seems disappointing, even to the planners themselves.

Let us recognize that some of the invective against "the city" is deliberately heightened for the purpose of influencing public attitudes and opinions, as, for example, to sell the need for more urban planning. The natural sales campaign of any specialist group which seeks increased status is to paint the blackest possible picture of the present and an even more ominous vision of what will happen if thei particular prescriptions are not accepted.

Beyond all this, however remains a gener ous measure of frustration, of even self-doub among planners. Looking at the grand scop of the mission which planning theory assign to them, and at the impressive amounts o time, talent, money and hope invested, on may wonder why the return is so small. Wh do so many plans remain unimplemented Why do planners feel so remote from th centers of effective decision-making, so alien ated almost from the city? Some planners i their frustration seem well along toward hostile rejection of modern urban life itself.

My hypothesis is a modest one: I believ that a major cause of unproductivity in plan ning is the general failure of planners t understand how to do effective staff work Moreover, most planners do not even sens the causes of this failure because they do no conceive of their work as an integral part o the administrative processes of government.

Concentrating on the physical descriptio of the plan he would like to achieve, the tradi

Reprinted from *Landscape,* Vol. 13, no. 2 (Winter, 1963-64), pp. 8-11 by permission.

al planner is naïve in his concept of how et things done. The traditional approach tains the causes of its own frustration.

et us look at some of the traditional roaches of the planner and the assump-s behind them. Certainly not all planners them all of the time, but they are very tential in planner behavior.

'irst is a belief that fact-finding, objectivity application of planning principles will (or east, should) lead to general acceptance the resulting plans. The preparation of s is considered essentially an intellectual cess, consisting of the application by ex-s of generally accepted standards to the s at hand and the drawing therefrom of clusions and recommendations. Or (and is really not quite the same thing) the ner thinks of himself as already knowing ain "right" answers, and the plans based hese should therefore be accepted. While e is considerable attention to "selling", the itional planner does not seek much real icipation by non-specialists until after the clusions have been pretty clearly estab-ed. The dominant idea is that planning ıld be turned over to planners, who can get on with it if not excessively inter-d with by others.

econd in the list of traditional planning roaches is the faith in the "master plan." future patterns of desired physical devel-ent are portrayed on a master plan dia-n or in a model, and this is supposed to le future development. Supplemental de-s are used — zoning, descriptive reports pamphlets, lists of future public works ects, statistical projections of basic data. because the planner's approach tends to basically architectural, the "master plan" print remains central among his devices. hough it is customary to describe it as flexible and subject to revision, the planner really would like it to be the civic command-ment by which the right-thinking citizen and public official will pattern their own actions.

Third, the planner does not expect to have to be very articulate about the "how" of plan-ning. How, for example, he derives a specific master plan from his basic data is seldom described in much detail. One does not hear much about the value judgments involved, the alternatives considered, or the whole complex process by which such a plan is evolved. Nor does the planner have much to say about how he intends people to live their daily lives in this future city. One sees the idealized physical layout, but hears little of the "pro-gram" for the occupants. Will life be better there? The planner tends to assume that phy-sical environment controls all else, and into that physical design he tends to project his personal goals and values as universal.

Similarly, the "how" of implementing the master plan is fairly casually treated. The troublesome specifics of public law, finance, organization and administration and the com-plex issues of public and private ways and means are mostly left for future consideration. At worst, the planner seems to say, "I've told you what to do; now it's up to somebody else to figure out how to do it." Too often, of course, nobody does.

A fourth tenet comes into play when the planner *does* decide that virtue alone is not enough protection for his plans. At this point, he tries either to get power enough vested in the planner himself to force others to comply or attaches himself as closely as possible at the ear of top executive or legislative authority so that the voice of that authority will silence opposition. Such opposition or even question-ing of plans is likely to be labeled as mere ignorance or sabotage. The possibility that resistance to plans reflects real shortcomings in the plans themselves is a possibility scarcely to be considered.

What are the defects of this approach to planning? It is easier to answer this question now than it was twenty years ago or even ten. Other staff specialists in the managerial system have relied on these same approaches, found them wanting and are evolving other, newer, more effective techniques. In recent years, the concepts and techniques of management have become somewhat more knowledgeable and sophisticated. New insights from the behavioral sciences make us much less cocksure about the scientific objectivity of our approaches and much more aware of the importance of the processes by which people work together to evolve and to achieve common goals. We still don't know a great deal but we do know more than we used to.

During these developments planners seem mostly not to have been listening. With fragmentation of knowledge and the increasing specialization of specialists this is not so unexpected, but still it is curious that planners in their frustration seem not to have recognized their kinship with others who try to bring the impact of special skills and values to bear upon public management. I do not doubt that individual on-the-job planners who have achieved results have learned new and nontraditional approaches, whether by pragmatic common sense, by bitter experience or by borrowing from other specialties. It still seems fair to say, however, that the central body of planning literature and of planners has remained remote from the behavioral approach and from the general field of public administration.

The defects of the traditional approach in planning or in any of the other staff specialties are (1) an assumed "right" to take over very broad and important areas of public policy as belonging only to one particular speci discipline — an assumption which generat conflict and resistance from others who interests extend to these same areas; (2) tendency to repeat over and over the san litany of proposals, with little energetic se criticism and great reluctance to go deep enough into specific real situations to evol specific and creative solutions; (3) an ove simplified view of the administrative proce and of society and particularly a reluctan to accept and understand the processes power, status, tradition, human feelings, cor munication and persuasion which are intens ly "real" elements in that society; (4) ivory-tower tendency to prefer the "big" pi ture and the "big" plans, and to scorn t drudgery (and the discipline) of carrying down to detailed ways and means of imp menting broad goals and of stimulating t processes of evolution along desired line (5) an implicit faith in centralized power administration (since the peak of the pyram of hierarchy is where the planner prefers reside) and a reluctance to experiment wi planning approaches where the planner hir self might not be dominant; (6) a tenden to ignore or to depreciate all planning n done by planners.

This partial catalog also suggests why pla ners who use traditional approaches are fru trated by resistance and meager results. T experience of specialists using traditional a proaches in other fields of administration the same, whether one looks at organizatic analysts, systems specialists, manageme trainers, personnel experts, fiscal specialists any of the other burgeoning fields of st endeavor.

The traditional approach in any of the fields tends to be pseudo-scientific, authorita ian, stereotyped, and a largely closed syste concentrating upon emphasizing limit

.ues central to the specialty. Such ap-
)aches, however well-intentioned, generate
istance and opposition which limit effec-
eness and frustrate achievement. Only in
riods of crisis will large systems yield pas-
ely to drastic surgery by such traditional
cialists; but unfortunately it is the business
planners to prevent crises, and the very
ure of their work focuses on the long-range
ure rather than the immediate solution of
lay's problem.

If the traditional approach to planning is
ective in this regard, what alternative is
ilable? Again, planners might look over
fence to other staff specialties for an
swer. The alternative is for planners to
cept and to learn to perform a *consultive*
ff role in the administrative processes of
vernment.

The essential elements of this approach are
plied in the criticisms which have been
.de of traditional methods. Most funda-
ntally, the staff role means accepting that
s is a multi-valued world and an extremely
nplex one. It means giving up the drive
exclusive power or control by a specialist-
ented group, and undertaking instead the
re complex task of gaining acceptance for
as, of working with and through older peo-
, and of evolving *specific* solutions to probns. It means seeking for solutions which
egrate a broad spectrum of conflicting
ces into new and mutually acceptable
irses of action. It means emphasizing the
nning process as much as the plan or more,
d aiming not so much at a single static
aster plan" solution as at a continuing
olutionary change. It means learning in
oth the real community and its people in all
complex variations. It means a basic con-
ot of obtaining results through others by
mulating, by communicating, by educating,
"making things happen." In short, the consultive role means that the planner gives up the effort to dictate the shape of the future by "divine right" and accepts instead the tasks of catalyst and counselor.

What, specifically, would such a planner do differently? He would work as part of the management processes of the government in which he is located, considering his work inseparable from other problems of public policy and administration.

He would seek to heighten the ability to plan effectively throughout the government, and to launch systems by which planning considerations will be more recognized and given weight in day-to-day operations. He would seek to accomplish his objectives through building planning into the entire activities of government, rather than through the activities of his own office alone. He would try to be a stimulant, an expediter, balancing his objectivity with an informed empathy toward those with whom he works.

He would seek to build confidence in planning and greater acceptance of its worth by seeking out first those problems which others have with which he could be helpful, and working hard on their solution.

He would accept decentralization and evolutionary change as normal parts of the administrative processes of large organizations and would seek a proper pace and balance in both.

He would try to develop throughout the government an enlightened understanding of planning values and a broad consensus on goals, so that gradually the whole activities of the governmental entity and the full exercise of its powers would be contributing toward the achievement of these goals.

He would try to remain free of bureaucratic routines and paper approvals which, however helpful in justifying staff increases and in calming bureaucratic feelings of insecurity, steal precious time from more basic activities. He would try to keep his staff small, highly trained and mobile — seeking always to place continuing operating work elsewhere so as to remain free to deal with emerging problems.

In relation to the broader urban environment in which he works, the planner would accept the fact that no absolute power can be found, no "owner" of the city to be client for his blueprints. He would accept the necessity of working within a framework of cooperation and complexity rather than seeking to achieve simplicity through power. He would approach his planning much more in terms of people and their relationships rather than in terms of architectural arrangement of urban spaces.

The same general techniques can be applied in the community at large as within the government itself, but the setting is much more complex. The planner must understand in detail how the city really works, in the full social economic and political sense, and how people live in it. He must accept the role of leadership without power on planning issues. He must evolve effective planning processes which work in that specific setting and wit that particular combination of people an forces. And, he must have a willing enthusiasm for processes of citizen participation education and communication that immers the planner deeply in the realities of his community.

The planner as staff consultant to government and to community alike must earn th right to influence the future. This right i earned not merely by being a planner but b entering into the life of the city and seekin out every possible opportunity to influenc and to assist the people and the processe which guide its development.

Perhaps all this seems too demanding task. Is it reasonable to expect the planner t obtain results without power? There is n question that it is difficult; but who eve thought that planning could be simple? Th fault, if fault there be, lies rather with over simplified aproaches which have promise more than they can deliver and which hav frustrated and disappointed the planner themselves.

31

‸ Concept of ural-Urban Regions

ohn Kinsel

e are becoming urbanized in a hurry. Of the o million new noses counted in Canada's 56 Census, 92 percent were in cities or vns of 1,000 population and up. Two-thirds all Canadians now live in these urban mmunities.

By and large, this postwar move to the ies caught us unprepared. And now we're nning to keep from slipping backward. ater supplies, sewage disposal systems, using developments, recreation facilities, hools and many other services have had to expanded enormously to accommodate w thousands each year. Many urban muni- alities have solved one emergency only to d themselves faced with a new one. Costs ve compounded faster than new resources uld be uncovered, and capital borrowing s increased rapidly.

any Problems are Rural-Urban

is not surprising, then, that we have tended regard the need for planning as something culiarly associated with urban expansion. e obvious differences between rural and urban problems, the separation between rural and urban local government jurisdictions, and the urgency of extending urban facilities have succeeded in splitting our vision. Preoccupied with our special problems, we have lost sight of the many common interests which are shared by the residents of an urban centre and by those who live in its rural area of influence.

What are some of today's problems where this community of interest is evident?

First of all, consider the physical expansion of urban centres. Urban dispersal may take one or more of several forms: gradual encroachment on the surrounding rural land, encirclement of non-urban territory, radial penetration along main highways, or "leap-frogging". Industrial decentralization, which is becoming steadily more prominent, contributes to the dispersal of population and creates special problems of its own. In a few cases, urban dispersal may be systematically planned as a green belt development.

There can be little argument that cities and towns must have room to expand. But in our

printed from *Community Planning Review* (September, 1957), by courtesy of *Community Planning view*.

approach to the planning problem we have tended to overlook the fact that "rurban" transition involves rural as well as urban adjustments. For example, one technique which is extensively applied in "rurban" areas is agricultural zoning. But agricultural zoning has been used almost exclusively as a tool for urban land use — a means of controlling undesired urban development. A parcel so zoned may or may not be an economic farm unit. The zoning may or may not be permanent enough to justify certain kinds of intensive agricultural development. As pointed out by Ernest A. Engelbert, Land-Use Planning for 'Rurban' Areas, (Farm Policy Forum, Winter 1957), the role of agriculture in "*rurban*" land use deserves considerably more study than it has received thus far.

Countless problems encountered in urban expansion require joint rural-urban planning. The question of zoning outside urban limits is one such problem. The location of industries on rural land well outside the jurisdiction of the neighbouring town or city where workers live is another. "Leapfrogging" suburban communities create demands for access routes of high standard through non-urban territory — demands which rural jurisdictions are not prepared to meet. Nearly every aspect of the problem of "urban spread" affects rural as well as urban interests.

No urban centre can be self-contained. It must seek its water supply outside the city gates. Usually it must dispose of its waste products in the rural area. Certainly its people must be fed from farms, near or far. Many of its people earn their living as handlers or processors of farm products or as suppliers of farm production needs. Similarly, the farmer is directly dependent on the urban centre to market his produce, to provide and service his equipment and to supply his many needs as producer and consumer. In this sense, urban and rural people are closely inter dependent.

The need for joint rural-urban planning ha not gone unrecognized. Nearly every expand ing urban centre of any size has attempted t set up some kind of machinery in conjunctio with the surrounding rural jurisdiction to dea with mutual planning problems. But suc efforts have not been uniformly successful. I the first place, such arrangements are almos invariably and necessarily informal, with fina action depending upon ratification by th separate jurisdictions. In the second place, th relationship is frequently one-sided, since ver few units of rural local government have plan ning resources of their own. Finally, it fre quently occurs that the urban centre must dea with not one but two or more rural jurisdic tions to embrace the area necessary for plan ning purposes.

And, while the need for urban planning i receiving much attention, coordinated plan ning in our rural areas is no less urgent. Cit dwellers have no corner on problems. Th same forces which are concentrating ou people in urban areas are throwing life out o joint in farming communities.

Tendency Toward An Urban-Centred Life

For rural people (in the Canadian West at an rate) the past two decades have been year of continual social and economic adjustmen A rapidly advancing farm technology ha meant substantial increases in the output pe unit of farm labour. Farm size has increase rapidly. As a result, some farm families hav higher incomes; many others are forced t leave farming and migrate to cities and towns In many areas traditional rural neighbour hoods have disappeared entirely.

At the same time, farming is becoming les of a distinct way of life and assuming more o attributes of other commercial enterprises. ater commercialization means less self- iciency. The impact of higher incomes, s communication, and the automobile has a distinct urbanizing effect on farm values. habits and attitudes of farm families are oming less distinguishable from those of in families. The social and economic focus 'arm living is becoming centred on the rby urban community.

'arm people are less content than they e to accept second class services: one room ools, impassable roads, inaccessible doc- and hospitals. At the same time, sparser ulation means higher per capita costs for rly all rural services. A substantial number arm families have sought individual solu- s to these problems by taking up residence rban centres near their farms. Where this occurred, remaining farm residences are more isolated. Even initially, the prairie n settlement pattern was one of extreme ersal. Depopulation has made that dis- al doubly extreme. The costs per farm of viding essential services such as roads and cation have risen sharply because of this or alone. Reinforcing this rise has been teadily-growing demand for services of er quality.

he need for intelligent rural planning is ed urgent. Planning is needed to ease the sition of rural living to a new and larger n-centred community. Planning is needed nprove and extend services to rural people services comparable to those available in ern urban centres. Planning is needed on oordinated rural-urban basis to contend the growing list of mutual rural-urban blems.

k of Workable Planning Areas

e province of Saskatchewan is typical, we are ill-prepared to meet these planning needs. The Royal Commission on Agriculture and Rural Life has recently completed a four-year study of a broad array of rural problems in that province. Its findings and recommendations, comprising over 3,000 pages in 14 volumes, cover subjects ranging from farm credit to the rural family. But throughout its investigation, the lack of facilities for rural planning, the lack of awareness of the need, and the lack of appropriate rural planning areas were matters of recurring concern. It devoted special attention to the problem of defining appropriate regions for rural planning and administration. And the regional concept which the Commission developed has important implications, not only for rural areas, but for urban centres as well.

The Commission's concern with defining rural-urban regions grew out of its appraisal of rural local government in Saskatchewan. It found a near-chaotic situation, characterized by:

(1) A basic local government unit in inadequate size. Although there is some variation, the typical rural municipal unit in Saskatchewan contains nine townships in an 18-mile square. This uniform small size creates some serious inefficiencies in planning and administering services. Small size means limited fiscal capacity, small population, and the absence of a meaningful planning area. In the provision of roads, for example, tax resources in most municipalities do not permit the purchase of modern efficient machinery; the small area precludes its economic use. Qualified supervisory personnel cannot be hired. Few local planning facilities exist and, in any case, road use is often oriented to points outside the municipality's jurisdiction.

(2) A multiplicity of units. In addition to almost 300 rural municipalities and some 500 incorporated urban centres, Saskatchewan has

literally thousands of special purpose districts and quasi-governmental jurisdictions. These include school districts, consolidated school districts, larger school units, union hospital districts, health regions, municipal doctor plans, agricultural representative districts, and many, many more which fulfil some kind of local government function. Including them all, there is one such unit for every 130 people in the province.

(3) Overlapping jurisdictions. Most of the special purpose districts, which have been superimposed on the small rural municipal units, are larger and fail to conform to the municipal boundary lines. The resulting welter of overlapping boundaries makes integrated planning and administration practically impossible.

The historical explanation for this confusing situation is simple enough. Initially, rural municipalities were logically designed to provide very limited services in a horse and buggy age: dirt roads, minimum health and welfare services, and some agricultural services. School districts needed to be large enough only to supply population for a one-room school with a single teacher. What happened? Population began to thin out and at the same time people began to demand more and better services. Each new or expanded service failed to fit into the established rural municipal pattern. Rather than reorganize the basic structure to fit changing conditions, succeeding governments added new jurisdictions to meet each new situation. The resulting unwieldy structure resembles a building in which each floor has been constructed according to a different design and unique specifications.

What is a Rational Planning Area

It was obvious to the Commission that the situation demanded rather drastic reorganization. As a first step the Commission was fac with the fundamental task of dividing t province into regions of appropriate size wi appropriate boundaries. Certain criteria establishing size were readily apparent: fisc capacity, population, administrative effic ency, and so on. But where to draw the bou dary lines? The settled portion of Saskatch wan is a reasonably homogeneous farmi area with few natural boundaries and wi few impediments to transportation, other th the condition of roads. What was the co mon denominator whereby the interrelat services of local government could be coor nated? What constituted a rational planni area for Saskatchewan's sparse and scatter rural population?

No single set of boundaries, of cours would be suitable to define all local gover ment services and regional administrati functions. Some involve the administratio and use of natural resources; here, boundari and size are determined largely by the occu rence and use of the resource. Water use districts would be one example; forest administration another. Then there is a gro of services which are oriented to consume with dispersed consumption — such things rural power, telephones and police protectic The location of boundaries and administrati centres for this kind of service are large matters of technical necessity and administr tive convenience.

In the third category are those servic which are supplied more or less universal and which require consumption at a centr point. These include some of the more impo tant services typically assigned to local go ernment agencies — such things as healt education and recreation. Roads become vital adjunct to this group of services becau of the mobility required for central consum tion — bus routes for schools, access ctors and hospitals and to municipal offices the payment of taxes, and so on. These re the services which most concerned the mmission. They were services vital to the ieral welfare of people and they also exhib-d the clearest need for coordination and egration.

e Trading Area

e search for means to define an area of atural association" which possessed a focal itre led the Commission to examine the ban centre and its trading area. The idea a trading area is, of course, a familiar one. has always been a preoccupation of retail erchants and is part of the stock-in-trade market analysis. It defines a region, not in ms of geographic characteristics, cultural its or typical economic activity, but rather terms of economic interdependence. The ding area describes a pattern of association ilt on years of trial and error in the ex-inge of goods and services essential to our onomic mode of life. If a farmer goes to ntre "X" to repair his equipment, buy his ts and play golf, why should he not go to same centre to get hospital care, pay his tes and educate his children?

The logic appeared inexorable. But to termine whether the trading area did in fact ve the qualities necessary to define a mean-ful region, the Commission conducted two ld studies in widely separated areas of the ovince. In each of these field studies, far-ers selected from a cross-section of location the trading area of a medium-sized urban itre were interviewed. Detailed information is obtained on the economic and social rela-nship between the farm family and all the ban centres, large and small, which it vis-d. From these surveys, the Commission was le to construct the patterns of association for each area. Moreover it was able to gauge the effectiveness of the trading area as an organizing principle; to measure its economic and social meaning to the rural residents of the area. The result of these surveys offered strong confirmation to the validity of the Commission's concept.

The Service Centre Principle

Although encouraged to proceed, the Commission soon faced a number of problems in attempting to apply the service centre principle:

(1) It was apparent that any given farm family was "attached" to several urban centres rather than one. Thus, the farmer markets his grain or buys his groceries in the hamlet closest to his farm. For other needs he travels farther to the village, the town or the city. The small centre performs certain functions within its small trading area. But the larger centre performs additional functions for residents of a wider area — including the residents of the smaller centres within its orbit. Out of well over a thousand centres of varying size in Saskatchewan, how was the Commission to determine which were the suitable centres and areas to define regions appropriate to the given public services?

(2) Regional boundaries must exhaust the area of the province. Did trading areas offer a reasonable basis for dividing the entire populated area?

(3) What about regional subdivisions? For some purposes sub-areas were necessary. Was there a basis in the organization of service centres for a rational subdivision of larger regional planning and administrative areas?

(4) The accepted method of defining an urban trading area — the market analysis technique — involved costly local surveys.

Was there any alternative practical method for delineating trading areas?

The Commission pursued its analysis of these questions in a report on Service Centres. The study set out to do three things: (1) establish a basis for classifying service centres according to function; (2) examine the principles governing the location of service centres in an agricultural economy; and (3) test the applicability of the trade-centred community to the definition of regions through an actual analysis of service centers in a portion of the province.

Without attempting to deal here with some of the more complex and theoretical aspects of the Service Centres report, it is useful to examine some of the Commission's conclusions and their practical application.

First of all, the Commission demonstrated that it is relatively easy and inexpensive to classify urban centres in the order of the functions they perform for the rural population. For its purposes, the Commission adopted a classification of six levels of centres, ranging from the crossroads hamlet to the provincial city. As a measure, the Commission counted the number of services — both commercial and public — which farmers and their families used in each centre. Centres with 2-10 services were designated Hamlets; those with 11-25 were called Villages; and so on. Additional classifications were Towns, Greater Towns, Cities and Provincial Cities. The range of services available was the key to the classification system.

In the course of this classification, it became apparent that each rank of centre was marked by certain characteristic services. Services for which the demand was universal and the required scale of operation small were found in centres of all sizes. In Saskatchewan's farming area, for example, every centre has at least a grain elevator and a general store. These are the minimum services characteristic of a hamlet, although most hamlets also have one or more of the following: post office, railway depot, one-room school, church.

The next higher rank of centre — the village — typically offers all the services available in hamlets. In addition, it has a new range of services such as lumber yard, fuel dealer, municipal office and telephone exchange. These services require a larger market

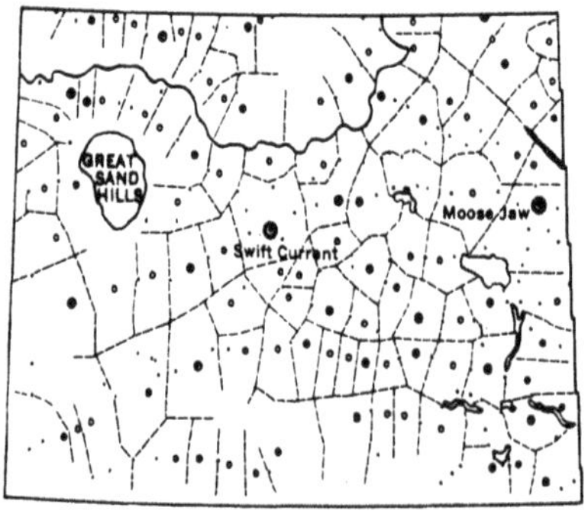

Figure 1. First approximation of boundaries of Village-centred Service Areas in Southwestern Saskatchewan.

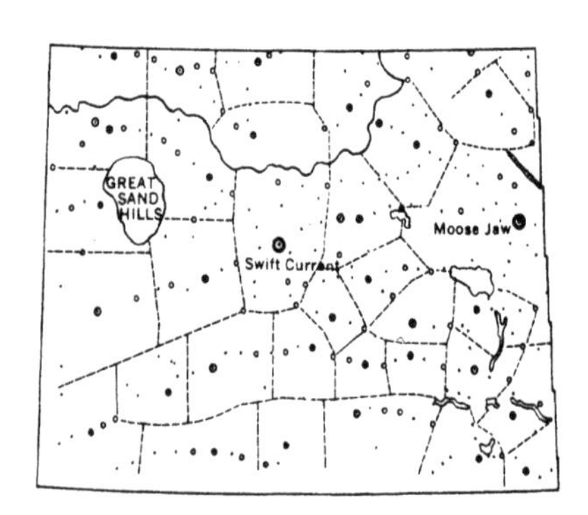

Figure 2. First approximation of boundaries of Town-centred Service Areas in Southwestern Saskatchewan.

: economic operation than hamlet services. ıe trading area of the village, therefore, is rrespondingly larger and includes not only : farm population but the population of the mlets which surround it.

For each succeeding rank, the progression similar. The number of centres in the higher ık is smaller and the trading areas are ger. And each rank has its characteristic ıge of services which are seldom found in nters of lower rank.

With respect to the location of centres, the mmission found evidence to indicate that ntres are distributed with reasonable uni- mity throughout any populated agricultural ea.* In Saskatchewan, the adherence of ntres to rail lines disturbs the uniformity of tribution to a degree. The populated area, wever, has a relatively dense network of l lines; as a result, the "gaps" between ding areas are not of serious proportions. nerally speaking, the trading areas for any rticular class of centre pretty well blanket : populated area.

It was also found that, by utilizing the butary areas of two ranks of service centers,)-regions could be delineated, although ne compromises in boundaries were neces- y to keep subsidiary regions wholly within : major region. Because of certain charac- istics in centre location, it proved best to ect alternate ranks of centres to define major regions and their subdivisions. The trading area of a Greater Town, for example, is most satisfactorily subdivided by Village trading areas rather than those of the intermediate Town. The reason for this is that Towns tend to straddle the border separating the trading areas of Greater Towns, while Villages tend to mark the limits of Town influence, and so on.

This latter characteristic proved extremely useful in developing a method for preliminary mapping of trading areas without resorting to costly local surveys. Between two adjacent Towns, for example, one will usually find a Village; between two Cities, a Greater Town. The Commission found evidence to support the thesis that trading area boundaries between two centres of a given rank are marked by the occurrence of a single centre of the next lower rank. By locating all such boundary markers, the general outline of trading areas can be derived.

In general, the Commission found that the trading area concept fulfilled its initial hopes as a sound method of approach to defining meaningful regions. In a detailed analysis of centres in Southwest Saskatchewan, it was

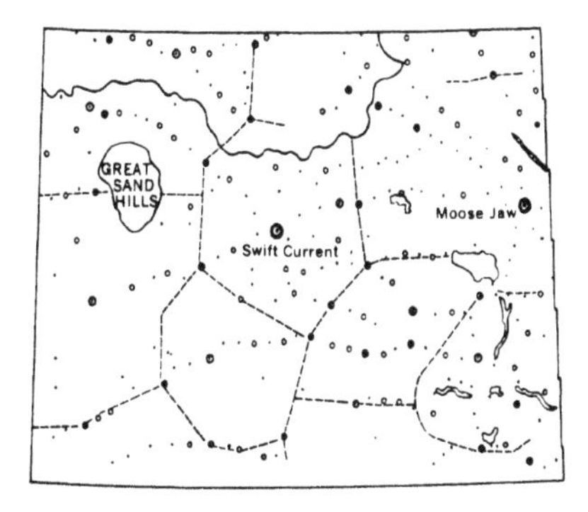

Figure 3. First approximation of boundaries of Greater Town-centred Service Areas in Southwestern Saskatchewan.

*The concept of centre location adopted by the mmission depends largely on theories advanced Walter Christaller in explaining the location of ıtral places in southern Germany. (See Die Zen- len Orte in Suddeutschland, Jena, Gustav Fischer rlag, 1933). The development and application of ristaller's theories in the Saskatchewan environ- nt is the subject of a monograph: P. Woroby, nctional Relationships Between Service Centres l the Farm Population, unpublished thesis, iversity of Manitoba, 1957.

able to rough out the approximate boundaries of trading areas for various ranks of centres without recourse to actual field surveys (see maps).

At the same time, the Commission makes no claims that boundaries located in this fashion can be considered final in any respect. They are only first approximations. In the first place, boundary lines in the above figures were drawn without detailed knowledge of local factors which might cause residents in the boundary areas to gravitate to one center or the other. In any event, any final boundaries for administrative or planning purposes would need to be altered so as not to pass through smaller centres, and their local trading areas. In the second place, the Commission points out that other factors such as fiscal capacity, population, administrative loads, and problems of coordination must be considered in establishing boundaries. The trading area is proposed only as a rational starting point in defining regions.

Advantages of the Trading Area

So much for the method developed by the Commission. What are some of the advantages of the trading area as a planning unit?

(1) It defines an existing pattern of association of people for certain economic and social purposes. It therefore lends itself to efficient organization for some of the more important public services. It defines boundaries (albeit in general terms) and locates the logical administrative centre for greatest convenience to the population being served. It provides the only sound orientation for planning local roads and contains the proper area for planning a local road network.

(2) Because of the functional differences between ranks of service centres, trading areas provide the basis for integral tiers of majo and minor planning areas. Different levels o services and different levels of administratio can be accommodated in an integrated system Health services provide a good example here General hospital care can be provided econ omically to a relatively small population, an minimum facilities need to be distribute widely to give adequate service. More special ized hospital and diagnostic services, how ever, require a much larger population uni for economic operation. A two-tier arrange ment composed of smaller trading areas con tained within the larger area tributary to a City, provides the logical base for integratin two levels of service.

(3) Because the trading area is oriented t its urban centre, it delineates the most suitabl area for integrated rural-urban planning. I attacking the problems of urban expansion the trading area is the natural unit. It define and includes satellite communities whic attract suburban movement. It includes al the access routes to the major centre. It in cludes the rural population dependent on th centre for commercial services, public service and recreational and social activities. At th same time, it includes those nearby rural area which the urban population uses for recrea tional activities. While it certainly does no solve all rural-urban planning problems, a unit based on the trading area nonetheles provides a framework with a minimum o structural handicaps.

Other Considerations Affecting Regional Planning

The Commission, of course, was primaril interested in defining regions appropriate t the requirements of rural local government Its recommendations for the reorganizatio ocal government in Saskatchewan em- ed additional considerations, a number of h have implications for regional planning. was the Commission's conclusion that l government — in the sense of a single l authority with comprehensive responsi- y — had practically ceased to exist in atchewan. In its place was a series of rate and sometimes conflicting jurisdic- s, each serving some segment of local ls. The rural municipal unit, once the nstay of local government, was essentially with two residual functions: construction maintenance of roads and collection of s. Health services were administered by ital districts and health regions. Schools administered by larger school units and ol districts. Agricultural services were inistered in a variety of ways: a few ugh municipal councils, some through ial districts and others through relatively rmal local arrangements. Other functions, locally administered, were now in pro- ial hands. Jurisdiction had become so nented that to identify any given rural area a single responsible unit of local govern- t was impossible.

he implications of this situation were far- hing, in the Commission's view. The in- ity of citizens to fix local responsibility contributing to an obvious decline in tical participation and to a growth of thy in rural areas. The lack of any com- nensive budgetary control over local rural enditures was making long-term plans in allocation of resources virtually impos- . The accurate determination of local tax l and tax carrying capacity was also out he question. This, plus the very number axing authorities, made debenture financ- difficult and costly. In addition, the ficiencies and added costs involved in providing related services through unrelated jurisdictions were obviously high.

Aims of the Royal Commission

The Commission's recommendations for a fundamental and sweeping reorganization of rural local government in Saskatchewan are directed towards the achievement of the following:

(1) Coterminous planning and administrative areas of adequate size for a maximum number of local government functions. First priority is given to the functions of education and public works (roads).

(2) Establishment of boundaries on the basis of trade-centered communities. This involves matching units of optimum size to trading areas of the appropriate rank of service centres.

(3) Integration of the maximum number of functions under a single authority within the reorganized areas. The Commission favours the county form of administration, with standing committees assigned to individual local government functions: education, public works, agriculture, social welfare, area planning, etc.

While the Commission made no firm recommendation in the matter, it also called attention to the need for integrated rural-urban jurisdiction over a greater number of functions. Particularly is this true in the case of Saskatchewan's small and middle-sized urban municipalities. The trend in education is toward integration of rural and urban school systems. Health services on an area basis often include both rural and urban residents. Certainly the joint rural-urban planning problems cited earlier would be immensely simplified if jurisdictional gaps were somehow bridged.

The Commission recognized that a number of difficulties stand in the way of achieving unified rural-urban jurisdiction. Land assessment as a tax base is not fully comparable in rural and urban areas. In the matter of political representation, rural residents hold some fears of urban dominance. And, despite the growing community of interest, certain aspects of local government are exclusively urban, others exclusively rural. Above all there are age-old prejudices and traditions to be broken down.

Nevertheless, the Commission proposed that careful study be made of incorporating villages and towns into the county system in Saskatchewan. The obstacles to this final step in integration may be more apparent th real.

In any event, the core of the Commissior approach to reorganizing and unifying loc government in Saskatchewan is its conce of the rural-urban region — a region bas on the patterns of association which peop have built up to satisfy their day-to-day ec nomic and social needs. How generally th concept may be applied remains to be see It appears to be particularly suited to Sa katchewan's problem: the definition of mea ingful regions in an area characterized relatively uniform agricultural developme and by a system of service centres whi evolved primarily to serve the rur population.

pen Space for the rban Region

32

mes W. Thorsell

e onto them that join house to house, that lay
d to field, till there be no place that they may
placed alone in the midst of the earth.
Isaiah V, viii.

e unto Urban Man who in the twentieth tury is fast approximating the condition phesized by Isaiah 2200 years ago. Today anization has become the way of life of a jority of our population. Social factors such the need to "get out of the crowds into the ntry" and the growing cultural sanction for ensive travel made possible by improving nomic factors, are making leisure and rec- tion great problems and challenging oppor- ities in developing our "Great Society".

Herein lies an obvious paradox: outdoor reational resources are becoming more and re remote from the urban dweller at a time en improved social and economic condi- ns have increased his need to enjoy this edom. Perloff and Wingo sum up the pres- problem:

> suddenly we find ourselves face to face with ajor urban need unsatisfied and commanding ention. For two generations we have accumu- lated a backlog of recreational needs; now wealth and leisure have converted these needs into an active economic demand and a pressing political force (Perloff and Wingo, 1962, p. 83).

This need for open space stems from the relatively recent massive urbanization and the technological revolution which gave it birth. Canada's urban population set at 62 per cent in 1966 should rise to 80 per cent by 1980. Reinforcing the effects of the pull of the cities on the need for recreational open space is the effect of a 50 per cent reduction in the work week in the last hundred years. Three-sevenths of the year will be available for leisure time pursuits in 1980.

While the number of potential recreation seekers is increasing as a result of these factors, the developing suburbs are ironically making recreational escape from the city more difficult. Haphazard fringe development has spread a thin veneer of low-density dwellings

printed from *Ontario Geography*, no. 7 (1967) by permission.

over large tracts of potential recreational open space that is needed by recreation-hungry residents of the central city. Moreover, only rarely have planners provided for sufficient public open space within the new communities.

What are the specific problems involved in the complex task of providing open space for urban man? The remainder of this paper will offer a brief survey of the literature on the subject and attempt to analyze some of the problems of demand for and supply of open space.

What is Open Space?

Tankel defines open space as "... all land and water in and around urban areas which is not covered by buildings ... the space and light above as well" (Tankel, 1963, p. 57). More specifically, open space includes all non-asphalted private and public land that confronts the urban resident in the different spatial frameworks through which he moves from his backyard to a far distant wilderness preserve. In this manner a hierarchy of open space is conceptualized which proceeds from street to community to county to regional levels.

The form and function of open space can vary considerably. More than just parks and playgrounds, open space consists of all open "wanderable" land such as may be found on college campuses, institutional lands, or agricultural areas. Basically open space then exists to provide relief from the urban social, commercial and industrial jungle by offering free open land with greenery and flowers to replace the artificial frame of everyday city life.

In their stress on recognition of open space as a valid urban land use, Tunnard and Pushkarev (1963) identify four functions of open space: protective, productive, ornamental, and recreational. The first two can be groupe as open space for structure and the latter tw as open space for service. Similarly Tanke distinguishes between open space that peopl can see and open space they are probably no aware of. Open space that can be seen is tha which is *used* for recreation, *viewed* fro vantage points or *felt* in that it offers land scape relief. That which is not necessarily per ceived either does urban work or helps t shape the development pattern of the city (water supply area or an airport buffer zone).

This latter use of open space as an urba form determinant is a major weapon in plan ning procedure recommended in the ope space plan of the Chicago metropolitan are One of the primary uses of open spaces ther is:

> ... to give structure, shape and form to the cit — separating clusters, preserving wedges, divic ing and giving identity to urban communitie and maintaining a balance between urban an rural land uses (North-eastern Illinois Metr Area Planning Commission, 1963, p. 1).

With the often haphazard character o metropolitan development underlying th need for this controlling and directing func tion, open space thus assumes a multi-purpos role. No longer is open space planning onl concerned with what to do with the leftove areas and "sloips".* It has now become a integral part of the master plan as an urba land use, as important as other types of lan use in the city.

Metropolitan Structure and Open Space Distribution

Much has been written on the per capita an per acre areas of open space that are neede

* A colloquial term meaning "space left over i planning".

: a city region. Reference standards, how-
er, are only helpful guidelines. The land set
de for urban and suburban recreation space
s been reported as low as 4 acres per 1,000
ɔan population (the present goal in Lon-
n, England) and as high as 12.5 acres per
)00 population (the present Chicago ratio).
ndon, Ontario has 4.5 acres per 1,000
pulation, well below an arbitrary desirable
ndard of 19 acres per 1,000 population. As
apin points out:

. the amount of land required for open space
open ended for as yet we have no basis for
antifying the need for breathing space in cities
hapin, 1964, p. 419).

However, the emphasis should perhaps not
on the amount but on the proper distribu-
n of open space. The conspicuous need for
en space occurs where the man-land con-
t is most intense — in and near large cities.
en the most urbanized areas have some
en spaces which are potential recreational
es, such as ravines, river-valleys, woods,
d wetlands. Furthermore the acquisition
sts for these marginal and commercially
developable sites are usually lower than for
surrounding land.

In examining the problems of urban open
ace it is helpful to view the region in three
atial contexts: the central city, the outer
g, and the urban region.

ntral City Open Space

e residents of urban cores of large cities
y rarely have adequate space. Along with
keen competition for, and intense use of
d one of the reasons for this deprivation
s been the obliteration of the remaining
protected open space by the expansion
m the city core to the urban fringe. Except for some of the larger, more outstanding urban parks such as are found in the cities of Vancouver, New York and Auckland, the central core of cities is too often devoid of adequate areas of greenery, silence, and "stretching room". Even in the above exceptions the existing facilities are being taxed to their limits.

Different socioeconomic characteristics of the central city resident add further to the awkwardness of his open space situation. As the United States Outdoor Recreation Resources Review Commission pointed out, the average central city resident has below median income, lesser mobility (53 percent of New York City families have no car), and lives in an area with a much higher residential density than a resident of the suburbs. Moreover, it is here that the senior citizens and racial minority groups congregate in an environment with fewer possibilities to engage in outdoor recreation activities. Even the high income resident of the central city to some extent shares with the lower income groups the disadvantages of the crowded core. The Commission concludes:

. . . the people on whom the environment of the city with its close-packed living, its constant pressure on the concrete, brick and asphalt environment, and its lack of pleasant surroundings which is most oppressive have less than average access to the out of doors (Outdoor Recreation Resources Review Commission, 1962, p. 10).

Outer Ring Open Space and the Greenbelt

Deficiencies in recreational amenities and blight in the central city were factors hastening the flight to the suburbs which began after World War II. Physical deterioration, appropriation for other uses of remaining open space in the core, as well as the desire to return to a form of life more akin to country living were reasons for this suburbanization. But the initial ready access of the older suburbs to recreation space declined as the city, in its outward growth, absorbed more and more of the surrounding countryside. The subdivisions of yesterday became the slums of today. Indeed, the suburban residents of large metropolitan areas found themselves crowded out of parks and off the highways that lead to them by the sheer weight of numbers of neighbours from the core. Thus, the urgent planning problem is to find methods to preserve suburban open space by controlling the gradual sprawl of the suburbs.

There is one technique which can be used in the outer ring that helps to alleviate this problem: encircle the urban perimeter by a greenbelt. This idea is not new. In 1898 Ebeneezer Howard provided Letchworth, England, his model garden city, with a greenbelt (Figure 1) to prevent the town from overspilling its natural boundaries, to act as a link between town and country, and to prevent less skillfully planned adjoining neighbourhoods from invading the Letchworth territory (Howard, 1898). But setting a fixed limit to th growth of the city often presents too many ad ditional problems to the modern explodin metropolis. The original greenbelt concep has been adapted to present urban condition by town planners and often takes more th form of green wedges, ribbons, or interna blocks of open land (Figure 1). These "fin gers" are frequently scattered throughout th city thus allowing more flexibility and bette access for central residents.

The Conservation Authorities of Ontari are unique in considering river valleys an ravines as elements in greenbelt formation Anderson's work on the importance of day use river valley recreation in the London Ontario region is an outstanding example o open space research using this modern multi purpose greenbelt approach (Anderson 1962).

Metropolitan Toronto offers an interestin example of the use of drainage corridors a the main structural elements in the provisio of urban open space (Figure 2). Here, as i such other large cities as Winnipeg and Ed monton, extensive river flood plains an

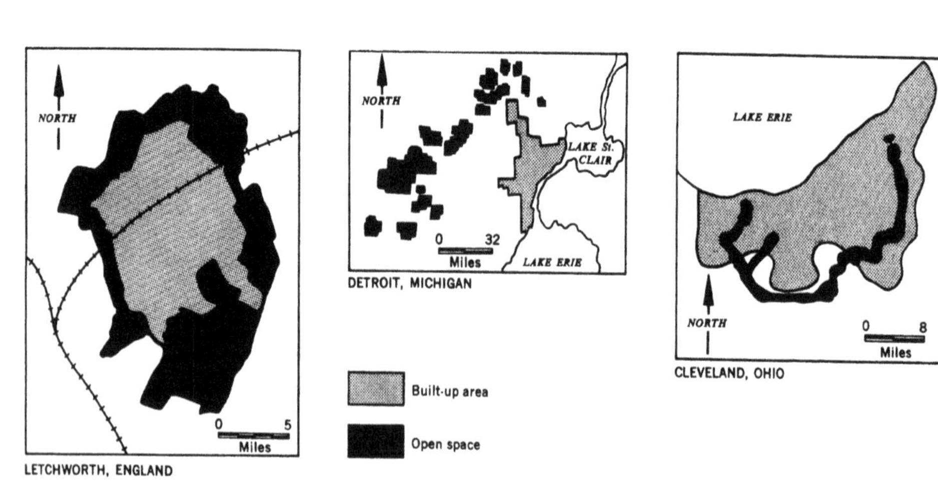

Figure 1. Open space pattern in Letchworth, England, Cleveland, Ohio and Detroit, Michigan.

ded ravines are presently, or can readily developed as open space reserves. At the e time they can function as storage areas flood waters and serve as open space ers and wedges to relieve the monotony rban development.

ional Open Space

critical problems appear in the provision rge scale regional areas of accessible and inuous countryside. Perloff and Wingo for a more urban-oriented open space cy where emphasis is on nearby, user-nted space rather than on distant re-ce-based preserves:

> he evolving need is not so much for more emites, Grand Canyons, and Okefinokees is for millions of acres of just plain open e endowed in many cases with only perhaps modest landscape and topographical interest but richly and imaginatively developed — this is the basic shift in approach that is called for by the logic of the present day situation (1962, p. 62).

The critical shortage in this country is underlined by Brooks:

> No large population centre in Canada can be said to have even a passable system of close-in regional parks to meet the day-to-day and week-end outdoor recreational demands . . . (Brooks, 1965, p. 6).

Even though provision of extensive stretches of national Resource-oriented space such as National Parks and Recreation Reserves are required for preservation of natural environments offering the ultimate in urban contrast, the problems of such provision are less immediate. An ecological, holistic view

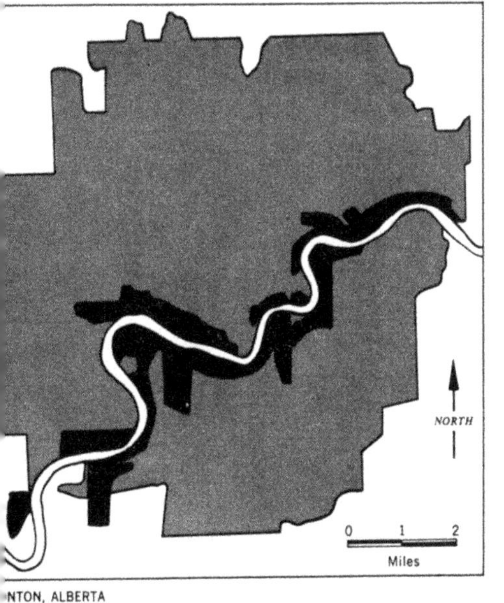

NTON, ALBERTA

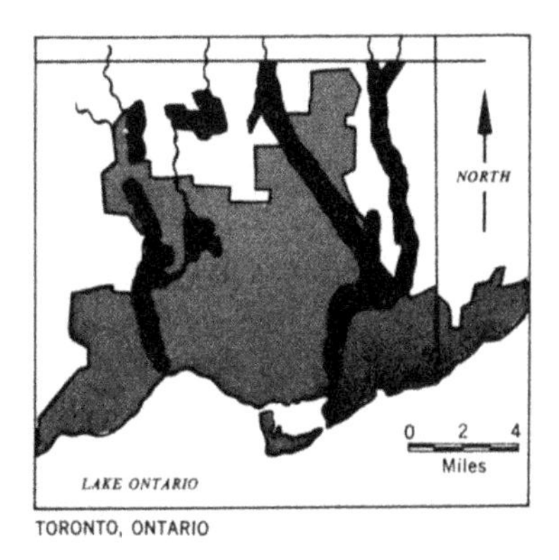

TORONTO, ONTARIO

Built-up area

Open space

re 2. Open space pattern in Edmonton, Alberta and Toronto, Ontario.

does, however, demonstrate that these large regional tracts must be provided in order to achieve a viable "system". In fact, the problems of incompatible uses presently being forced into Canada's National Parks can be seen as a type of negative feedback which originated in the deficient and over-crowded conditions that exist at lower open space levels.

Planning a Metropolitan Open Space System

It is evident that the provision of adequate urban open space involves an area greater than the local area. The very nature of providing adequate open space calls for the design of broad regional recreational systems which can best exploit and merge the available resources and human demands. For example, the city of Denver, Colorado has gone far beyond its urban boundaries in purchasing several mountain peaks for the use of its residents. Likewise the Lincoln County, Ontario lakeshore study requested co-operation as well as financial aid from surrounding municipalities in developing the public shoreline. Concluding that "the lakeshore does not and cannot exist independently of the active hinterland", the report stressed the need for a regional outlook to solve a regional problem (Niagara Regional Development Association, 1959, p. 9).

Because of present conditions, however, long range planning for recreation space is usually extraordinarily difficult. In transcending city boundaries there are seldom conventional government structures or analytical frameworks available to deal with the broad environment planning required. Striving for a new open space outlook, Perloff and Wingo suggest there is great merit and potential in using the systems approach to bring into better perspective the recreational behavior of the urban population. The systems techniq would identify such key elements as popul tion, activities and facilities and determine t interactions and casual relationships that ex among them. Invaluable and relevant too available for analysis are the gravity mod and linear programming. Dr. Clawson h vividly advocated the necessity of syste planning by sketching what happens if it is n used (Clawson, 1963). The birds-eye vie of the "recreation shed" of the urban regi appears to be a most fruitful method f planning and integrating open space into t environment framework of the region.

It is useful to look at an example of t methodology that could be used in planni for the provision of open space in an urb region. A suggested general procedural ou line based on the North Eastern Illinois stu is:

1. *Inventory* the existing regional op space and evaluate its adequacy.
2. *Estimate* the future open space r quirements for recreation and other purpose
3. *Examine* the possible use of op space as an urban form determinant.
4. *Survey* the open space resources identify the areas of greatest adaptability f open space use.
5. *Review* the methods available for pr serving and acquiring this space.

The Outlook

Action is needed as the hunger for more pu lic open land becomes acute. One of the fir major forward steps was the enactment Ontario's Conservation Authority Act whic suggested the provision of open space watersheds as a major multi-purpose b product of other conservation measures. T United States Outdoor Recreation Resourc Review Commission, the ARDA Recreation nd Inventory, and the recent formation of National and Provincial Parks Association Canada are some important federal and vate examples of progress which has been de.

On the local level positive action has been en in setting up regional planning commisns in large metropolitan areas. Prominent ong these are the California State Outdoor creation Planning Committee, the North stern Illinois Metropolitan Area Planning mmission, and the Metropolitan Toronto d Regional Conservation Authority. In te of the obvious value of these regional ards much work is needed in establishing m. In stressing the need for a comprehene planning approach because of the interpendence of all open space uses, an Israeli hitect concludes:

> In looking for a solution to the recreational blem, our main concern must be with regional velopment and regional design. We cannot urn to past conditions, and we are therefore npelled to turn our thoughts and energies to comprehensively planned reconstruction of vn and landscape as well as to the change of tude towards environment (Glikson, 1965, p. 4).

ikson, by broadening the scope of the open ace and recreational land problem, includit as part of the general problem of natural ources conservation. It is also in this wider ntext of total environmental reconstruction that open space plays an important role in the evolution of the civilized landscape.

In this discussion of the need for open space, two themes have been apparent. First, open space for urban needs is a pluralistic concept. It involves, at the same time, recreation, conservation of natural resources and wildlife, flood control, aesthetic satisfaction, and control over the direction of urban development. Secondly, there is need for regional planning in systems terms to effectively meet and manage the myriad problems and opportunities of developing urban areas. As a prodigious user of land and water resources, the city challenges our ability to use our outdoor resources wisely. Urbanization leaves its indelible imprint on the natural landscape but, unfortunately, not always in a planned and efficient way. In addition to the effects of relentless metropolitan growth, the accumulated backlog of inadequate past solutions have tended to be "specific rather than general, limited rather than extensive, piecemeal rather than comprehensive, and finally . . . site rather than market oriented" (Outdoor Recreation Resources Review Commission, 1962, p. 40). In preventing future open space shortages, we must act now to allocate and reserve open spaces for all required purposes. The ultimatum issued by the United States Department of the Interior sums it up: "we must blend beauty with utility or sacrifice quality to chaos" (U.S. Department of the Interior, 1965, p. 6).

ferences

Anderson, D., *River valley recreation — London day-trip zone,* Unpublished M.A. Thesis, Department of Geography, University of Western Ontario (1962), 201 pp.

Brooks, L., *Perspectives on Canada's parklands,* Paper presented to the Parks and Leisure Seminar, University of British Columbia, Vancouver (1965), mimeo, 19 pp.

Chapin, Jr., F. S., *Urban land use planning* (Urbana, University of Illinois, 1964), 479 pp.

Clawson, M., "Planning and managing a system of parks for a nation," *Proceedings of the federal-provincial parks conference* (Ottawa, 1963), pp. 28-35.

Glikson, A., "Recreational land use," *Man's Role in Changing the Face of the Earth,*

Thomas, W., *ed.* (Chicago, University of Chicago Press, 1956), p. 1193.

Howard, F., *Garden Cities of Tomorrow* (London, Faber and Faber, 1898), 168 pp.

Niagara Regional Development Association, *Lincoln's lakefront problems* (Niagara Falls, Ontario, 1959), 30 pp. North Eastern Illinois Metro Area Planning Commission, *Open space in north-eastern Illinois* (Chicago, 1963), 107 pp.

Outdoor Recreation Resources Review Commission. *The future of outdoor recreation in metropolitan regions of the United States*, Study Report No. 21, Vol. I (Washington, D.C., 1962), 161 pp.

Perloff, H. and Wingo, L., "Urban growth an the planning of outdoor recreation," *Outdoo Recreation Resources Review Commissio* Study Report No. 22 (Washington, D.C., 1962) pp. 81-100.

Tankel, S. B., "The importance of open space i the urban pattern," *Outdoor Recreation Re sources Review Commission*, Study Report N 22 (Washington, D.C., 1962), pp. 57-71.

Tunnard, C. and Pushkarev, B., *Man-mad America: chaos or control* (New Haven, Conn Yale University Press, 1963), 479 pp.

United States Department of the Interior, *Que for quality* (Washington, D.C., 1965), 45 pp.

'he Development of the 'oronto Conurbation

33

acob Spelt

owards a New Municipality

ie 19th century development had created me serious problems making a further effi- ent functioning of the urban complex ex- eding uncertain. Thus the citizens and their vernment were forced to make a critical -appraisal of their environment, a process iich led to the gradual emergence of plan- ng as a branch of government administra- n and this in turn helped to bring about creation of a new administrative frame- ork.

The first attempts at improvement, with exception of the reconstruction of the ter supply and sewage system, were almost tirely aimed at beautification of the city, order to discard some of the more ugly pects of the 19th century legacy. It was a ase which more or less would last until the d of the 1920's and which culminated in University Avenue project. There were o attempts to find a solution for the traffic oblems and the deficiencies of the street ttern. Suggestions for improvement in- ded the building of an impressive system of boulevards, also the construction of a large viaduct in Bloor Street across the Don Valley, and the northward extension of Bay Street from Queen to beyond Bloor. To en- hance the beauty of the City and to give per- sonality to its heart, it was proposed in the years before the First World War to create a civic centre between the City Hall of 1899 and Osgoode Hall. It even was suggested to open a new, wide avenue through the middle of the long blocks between York and Bay Streets to connect the proposed centre with a new Union Station. The square was eventu- ally acquired by 1947 and became the site for the new city hall.

The problems of street lay-out and traffic conditions remained major concerns until the 1930's, when attention also began to be focused on the deteriorating housing condi- tions. A subsequent report noted the inade- quacies of city planning and recommended among others the immediate establishment of a city planning commission. This together with a concern about post war reconstruction and planning led to the appointment of the

is contribution is in essence a revised part of an article originally published in the Buffalo Law Review, l. 13, No. 3, Spring 1964.

Toronto City Planning Board in 1942. Although this board was only an advisory body, its work had far-reaching effects in the late forties and fifties. Its report, rather gingerly received by the City Council, was the first comprehensive plan for the city in that it dealt with all aspects of land use, from greenbelts to a civic square, from substandard housing to a flight of industries to the suburbs. In dealing with the greenbelt plans and policies for industrial land use the Board became increasingly aware of the need for co-ordination between the municipalities of the conurbation and the creation of a planning authority more regional in scope. The result was the formation in 1947 of the Toronto and York Planning Board, York being the county which contains the Toronto conurbation.

In 1947, of the 13 municipalities which later would enter into a federation, only three, North York, Etobicoke and the City had appointed planning boards. Three of the remaining municipalities had only planning committees and the remaining seven nothing. This illustrates clearly to what extent the Toronto built-up area, which contained a total population of more than 1.1 million in 1951, had grown under a system allowing a minimum of official planning. It is not surprising, therefore, that problems had arisen of such magnitude that they no longer could be solved within the existing administrative framework.

Through lack of funds, several municipalities were unable to make the necessary investment in public services. A 1947 planning report stated that the taxes obtainable from a six-room house could not hope to meet the financing of public services, including education, unless additional revenue could be obtained from commerce and industry. This was especially true for East York, North York and Scarborough. The suburban municipalities therefore attempted to promote industrial development by laying out properly serviced industrial areas, but not all municipalities succeeded in balancing their assessment in this manner.

Most serious were the problems in connection with water supply and sewerage facilities. The city acted as a barrier between Lake Ontario and some of the inland municipalities, a condition reminiscent of Yorkville many decades earlier. The lack of proper sewage facilities had become a menace to public health. Dwellings had been crowded on small lots with soils entirely unsuitable for septic tank disposal systems. To the extent that local municipalities did build sewage treatment plants, they tended to locate them on the Don and Humber Rivers so that the discharge of these soon overloaded installations created other problems farther downstream.

By the late forties, serious water shortages had developed and enforced curtailment of the use of water became a normal feature of suburban living. Especially in the Township of North York the situation became critical. However, through lack of foresight, Etobicoke Township although fronting on Lake Ontario suffered almost as badly.

The highway network and public transportation in the conurbation were poorly integrated. In 1949, fully 30 per cent of the population found itself outside the limits of the universal fare system; co-ordination of transportation services was most urgently needed. A plan for the preservation of open space could not be implemented, because of lack of co-operation among the municipalities concerned, each viewing the problem from a purely local point of view. Thus chao

rapidly developing and in its 1949 report, Toronto and York Planning Board was e outspoken in pointing out the underg causes: ". . . constructive progress is in y case barred by the difficulties of securmunicipal co-operation in the developt and extension of public services. . . ." of the solutions suggested by the Board a unification of groups of municipalities. recommendations of the report were ensed by the City Council and an application made for amalgamation. Eventually this lted in 1953 in the creation of a new mupality in the form of a federation of the and the 12 suburbs; it was called Metrotan Toronto. Soon the morphology of the began to reveal the effects of the new inistrative organization. Some of these, as expressways, housing developments, improvements, better water supply and age disposal facilities profoundly innced the further development of the builtrea.

he further growth of the Toronto urban plex, however, also meant continued nges in population distribution and assessts patterns. In 1953 Toronto accounted 59 per cent of the total metropolitan ulation. By 1966 this had declined to 37 cent. Industrial expansion improved conrably the taxation base of the suburb. n changed circumstances and also the inity of the city to solve urban renewal and sing problems led to a reorganization of metropolitan administrative structure. 13 original municipalities were reduced x. The Metropolitan council was enlarged 2 seats with a distribution more in accorce with the new population and assesst patterns. A slightly enlarged City of onto obtained 12 seats compared with seats for the other five municipalities or boroughs as they are now called. The city, however, retained half of the seats on the Executive Committee. The newer organization became effective January 1968.

THE PRESENT-DAY LAND USE PATTERN

The most important factor in the area differentation within cities undoubtedly is the transportation system. After Los Angeles and Detroit, Metropolitan Toronto has the highest motor vehicle ratio on the continent in terms of registrations per 1,000 of population. Gradually it became impossible to handle the volume of traffic, especially in the downtown part of the city where the great majority of streets originally were 66 feet or less in width. Until quite recently, the only through highway in Toronto was the Lakeshore Boulevard which had emerged from a partial reorganization of the waterfront. It linked the Queen Elizabeth Way with Highway 2 on the east side of the city. Even so, access to the road was not controlled and its traverse of an amusement park in the west and the harbour zone in the downtown area reduced its capacity to handle through traffic.

At present, an extensive plan for expressways, proposed in broad outline as early as 1943, is being implemented. The broad scheme is to enclose Metropolitan Toronto within a triangle of expressways. Two of these, Highway 27 to the west and Highway 401 to the north are provincial roads, while the third side of the triangle, the F. G. Gardiner Expressway, mainly an elevated structure, has been completed from the West as far as the Don River and the Don Valley Parkway. The Province is engaged in widening Highway 401 over a distance of 17 miles to a minimum of 12 lanes.

Within this triangle, the crosstown arteries

will be supplemented by a system of widened streets and expressways, giving access to the centre of the city. The construction of one of these, the Don Valley Parkway, is completed, and a beginning has been made with the building of the Spadina Expressway which will run in a northwesterly direction. These two highways combined with the Gardiner Expressway and possibly another east-west expressway south of the Iroquois shoreline may eventually form a ring around the inner city. Nevertheless, the system of expressways, extensive though it may be, is not expected to be able to meet all transportation needs; in the field of public transportation it is being supplemented with a network of subways.

The traditional importance of Yonge Street to the central business district is reflected in the building of the first subway. Completed in 1954, the Yonge Street subway contributed greatly towards improving the connections between the centre of the city and the areas to the north. It reinforced certain functions of the central district, but paradoxically it also stimulated some decentralization from downtown and favoured the expansion of office buildings near the stations. This together with new apartment complexes, accounts in part for the new traffic generated by the Yonge subway. At present the line, which runs over a distance of 4.6 miles from Union Station to Eglinton Avenue, operates at capacity during rush hours. The original system was extended with a line along Bloor-Danforth and under University Avenue. The Yonge line is being extended into North York. The total system is at present (1970) 21 miles long.

The railway pattern in the metropolitan area was established in the 19th century and is focused for the most part on the downtown area, not far from the waterfront. Access from the city to the waterfront was improv with the completion of a new viaduct in 193 Seven new underpasses between the D River and Spadina Avenue replaced the tim consuming crossing of myriads of tracks. the same time all the railway stations we eliminated by the new Union Station, open in 1927.

The extensive marshalling yards and a sociated facilities in the downtown part of t city will be replaced by new establishmer now under construction by the two railw companies outside the metropolitan area. I deed, this may provide the city with a ne opportunity to re-appraise its links with t waterfront, since the downtown yards longer will be needed. It may be possible fulfill the aspirations of the founders of t city and to create an organic link with t lake, giving new dimensions to a confined a crowded city core. Attempts along these lin are also made in the area of the central wate front. The reduction of freight trains movi through the Union Station area has made t latter more accessible to commuter trains, t first of which has begun to serve the lakesho communities between Hamilton and Picke ing.

The pattern of residential districts Metropolitan Toronto is similar to that other large North American cities. Around inner core with only a small number permanent residents, extends a zone of hou ing largely built in the 19th century wi densities varying from 101 to over 150 pe sons per net residential acre. Beyond this, t net residential densities decrease to 16 pe sons per acre along the fringes of the bui up area. About 60 per cent of the develop area of Metropolitan Toronto is in residenti uses. The Metropolitan Toronto Plan aims an increase in population densities by e couraging a mixture of varying house type luding apartment buildings in the low isity districts. It is expected that the gross idential population density will be in-ased to between 20 and 29 persons per e in most of Etobicoke, North York and rborough. A main advantage of increased isities would be the possibility of providing equate public services, in particular public nsportation.

In the last decade, suburban development urs with more imagination and serious empts are made to create attractive com-nities. This trend began with the building Don Mills, a community with a variety of ise types, including multiple dwelling uctures; it is built on an irregular street tern, protected against the inflow of out-e traffic, and arranged around a modern ll-type shopping and service centre. It nds in sharp contrast with the earlier amor-ous subdivision plans devoid of any per-iality or identity. Swedish concepts are ng incorporated in other projects.

One of the most striking changes in the dscape of Metropolitan Toronto over the t decade has been the great increase in the mber of apartment buildings. Until the late 40's, Toronto was predominantly a city of ached and semi-detached homes. Even as e as 1953, there were fewer than 30,000 artment units in Metropolitan Toronto, or out ten per cent of the total number of using units. Apartment complexes have sen in widely scattered areas, but especially locations with easy access to subway sta-ns. The widely assumed identification of artment living with residence close to the ntre of the city is not valid for Metropoli-Toronto. Indeed substantial complexes e found at the very edge of the built-up ea. Apartment developments tend to arid eas suitable for redevelopment or urban newal, but they prefer good residential areas, especially when these also have large lots.

A second significant change has taken place within the residential population itself. Throughout the 19th century and well into the 20th, Toronto was almost entirely a city with Anglo-Saxon population. The city was not only Anglo-Saxon in composition, but perhaps even more so in outlook — *plus royaliste que le roi*. Even as late as 1951, the British element accounted still for about 70 per cent of the population. Ten years later, however, 46 per cent of the city's population came from countries other than the British Isles. The largest non-British elements are the Italian, German, French (largely French Canadian), Polish and Ukranian groups. The newcomers are almost entirely concentrated in the city proper, especially in its 19th century parts. Here some districts have experienced tremendous changes in the ethnic composition of their population. Once solidly British in character, now the language spoken in the streets, the style of clothing, the stores with their signs, and their displays of strange merchandise, remind one of cities and towns in Europe. It is to the credit of the city, that these profound changes in the make-up of its population have taken place without ethnic strife.

As noted previously, the site was endowed with a rich potential for the development of park land. Yet over one-third of the inhabitants of Metropolitan Toronto live in areas which are deficient in parks. Since the early fifties, vigorous programs initiated by the city and after 1954 augmented by the Metropolitan Parks Commission have resurrected old parks and developed several new ones. The Island, just opposite the city's most crowded residential areas, is being transformed into a multifunctional park. It involves the removal of some 650 residential and commercial structures and eventually some 575 acres will become available for recreational purposes.

Approximately 10,000 acres, or 11.7 per cent of the developed land in Metropolitan Toronto has been appropriated for manufacturing, wholesale and warehouse complexes. The first industries and warehousing arose on the waterfront. There were no water power sites which could have attracted major industrial concentrations. Instead, manufacturing tended to move to open land just outside the built-up area at points of good access. The heavy dependence of manufacturers on railway transportation before World War II is reflected in most pre-war industrial districts — long tentacles of industry extending out from the centre of the city, mainly to the west and northwest, along the lines of the C.P.R. and the C.N.R. In recent years, the rise of the trucking industry has reduced the dependence of manufacturing on the railways and many of the industries in new suburban developments have no access to rail sidings. Nor are there any rail facilities in large parts of the inner city where industry also is based entirely on truck transport.

The modern metropolitan area therefore, contains several areas of conflux in the pattern of the daily journey to work. The city proper has some seven major concentrations of manufacturing and warehousing within its limits giving employment to a total of over 120,000 workers. Outside the city, ten additional concentrations are well distributed throughout the old and new suburbs, with a total employment of more than 85,000.

By 1950, industrial employment in the city had reached a peak of 160,000, after which time it has declined steadily mainly due to the migration of firms from old and overcrowded quarters to the suburbs. The newer suburbs showed remarkable increases in manufacturing employment. Etobicoke, North York, Scarborough, East York and York saw their industrial employment rise from 6,000 in 1950 to 66,700 in 1960. This has led to the formation of a much more solid assessment base and consequently a desire for greater influence in metropolitan matters. The city contributes 42 per cent of the total taxable assessments in Metropolitan Toronto.

Toronto's central business district offers employment to some 145,000 people, or about a fifth of the total employment in Metropolitan Toronto. Within its confines, it reveals a great deal of diversity both in form and activity. According to the 1951 census of retailing, 34 per cent of all sales in Metropolitan Toronto were concentrated in downtown. By 1961, however, the proportion had declined to 23.5 per cent, a loss the more striking when viewed against the background of population increase in the total metropolitan area. Nevertheless, downtown is still by far the leading retail concentration between Montreal to the east and Winnipeg to the northwest.

The retail centre migrated from its early location on the market along King Street westward to the intersection of King and Yonge Streets which by the end of the 1870's had become the city's leading shopping area. Eventually, however, Yonge Street was to assume the part played by King Street. This re-orientation of the retail trade began with the establishment of the department stores at the intersection of Yonge and Queen Streets around 1870. The success of these companies stimulated an expansion of retailing along Yonge Street which accelerated in the 20th century and at present by far the bulk of downtown retailing is concentrated in a strip along Yonge Street, to the north of King Street. The confinement of downtown retailing to a single street is a unique feature ong the large North American cities, ere retailing of this type generally encom- sses several blocks.

The department stores hold complete sway er downtown retailing, leaving little room competitors. A significant concentration high-class stores which effectively could mpete with them does not exist in down- vn Toronto, but is located farther north, on oor Street between Yonge and Avenue ad. In spite of the improvements in trans- rtation and the very substantial growth of population in Metropolitan Toronto and rounding areas, the downtown stores have t been able to expand accordingly. It is ite obvious that competition from other siness centres, such as the Bloor area and large suburban plazas, have contributed the relative decline of downtown retailing. e department stores and other downtown ops have responded by opening branches the major suburban plazas, such as York- le.

In the late 1920's, the low buildings along y Street were demolished and tall office ildings arose in their place. By that time, connotation of Bay Street as a leading ancial street had been established. The ilding of the Toronto Stock Exchange gave it further definition. In recent years a remarkable expansion of office buildings has made University Avenue an integral part of the downtown office core. In contrast to retailing, the financial and general office concentration appears to be the expanding cell of the downtown business core.

Of some 32 million square feet of office space in Metropolitan Toronto, about 11.9 million square feet is found in the downtown area. Increasingly, office complexes are being erected outside the central business district. Some of the subway stations and major highway intersections have attracted office concentration.

A distinct part of the general office zone is the complex of municipal and other administrative buildings such as the city hall, the city registry office and Osgoode Hall, the seat for the administration of justice in the province. The completion of the new city hall and civic square have brought about the realization of plans formulated many years ago. The commanding style of the new structure seems a fitting symbol for the throbbing metropolis which has arisen since the Second World War and has at present a population of more than 2.3 million.

34

A Tale of Two Cities

N. H. Richardson

Between the Pacific Ocean and the 4,000-foot crest of British Columbia's Coast Mountains, the westernmost mainland range of the Canadian Cordillera, is a rugged, forest-covered land of rivers and fiords. Among the mountains, away from the sea, its summers are cool and its winters bitterly cold; but on the lower levels, along the inlets and the coast, the climate is milder, with mean monthly temperatures between 55° and 65° in July and between 20° and 35° in January. Precipitation is high, reaching 160″ or more annually in some areas, and rain or snow fall throughout the year; except on the sea-coast the land is blanketed by thick snow during the winter months.

Until the latter part of the nineteenth century the region was virtually unknown to the white man, though scattered along the fiords and the rivers were the villages of the Tsimshian, Bella Coola and Kwakiutl, dependent on fish for their livelihood as most of their descendants are today. The first systematic exploration was carried out in 1859 and 1860 by Major William Downie, who was impressed by the potentialities of the Skeena Valley as a transportation route. Five year later the "Collins Overland Telegraph" which was to link North America with Europ across the Bering Straits and Siberia, reache the Skeena, but in 1866 came the news tha a cable had been successfully laid across th Atlantic, and the project was abandoned But many of the workers stayed to look fo gold, which in due course was found on th upper Skeena, and by the seventies the rus was on. The riches of the Skeena, however did not compare with those of the Yukon an the Cariboo, and, as elsewhere, frustrate prospectors turned to less colourful occupa tions: logging, farming or fishing, for the mos part. By the 1890's fish canneries were bein established at several points along the coas and the rivers. White settlement was firml established by the end of the century.

Surveys for a transcontinental railway wer carried out as early as the 1870's, and i seems that serious consideration was give to the adoption of a northern route throug the Yellowhead Pass and along the Skeen Valley. But the CPR chose the souther route instead, and Vancouver rapidly blos

Reprinted from *Plan*, Vol. 4 no. 3 (1963), pp. 111-25 by permission.

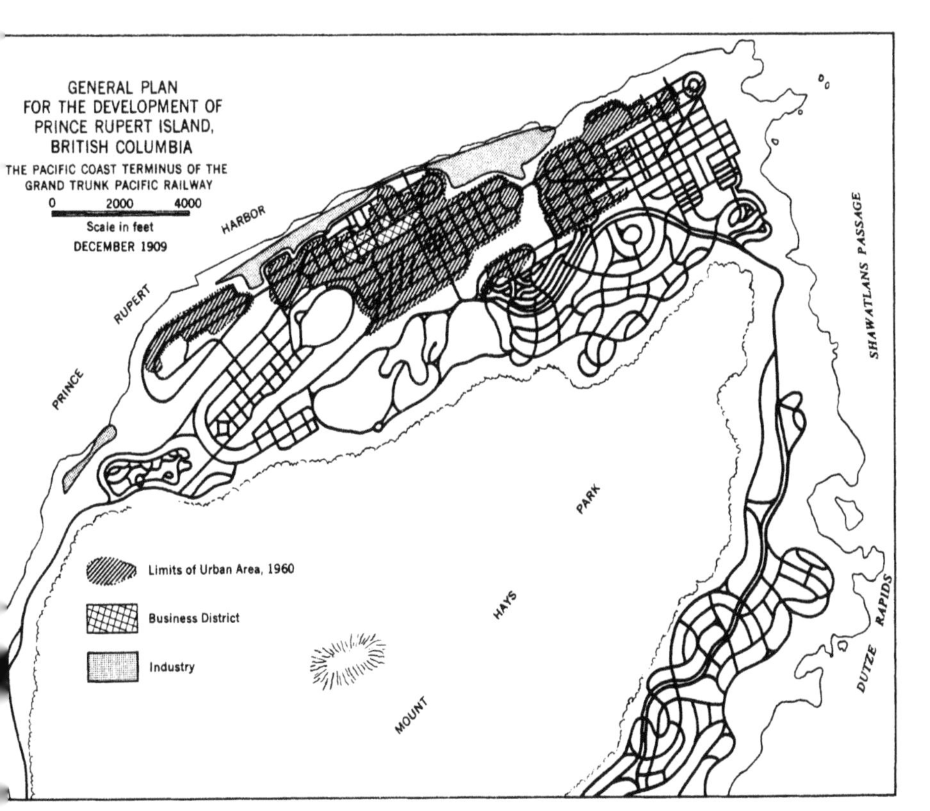

ed from a cluster of shacks along the re to British Columbia's chief city. How- r, the possibilities of the northern route e not forgotten. In 1903, with the support ir Wilfrid Laurier, the Grand Trunk Paci- Railway Company was incorporated. The nd Trunk Pacific was the brainchild of M. Hays, General Manager of the Grand nk Railway, a line beset by chronic finan- difficulties, and it was Hays' scheme to d a second transcontinental railway to npete with the CPR and, in fact, by bring- the railhead five hundred miles closer to ental ports, to outstrip the CPR in the race carry the rich trade that was expected to elop with the Far East. Construction of the eastward from the coast was started in)7; in 1914 the route was completed, from rth Bay, Ontario, to the Pacific.

Prince Rupert: New Town, 1904

Meanwhile, even before any track was laid, a site had been chosen on Kaien Island, some ten miles north of the mouth of the Skeena, for the city that was to be the line's western terminus. The company bought 10,000 acres of land from the provincial government at $1 per acre, on condition that a quarter of the townsite and a quarter of the waterfront would be reserved for government use. It then took a step of considerable significance in the history of Canadian town planning by engaging a prominent U.S. firm of landscape architects, Messrs. Brett and Hall of Boston, to design a townsite layout for the new city of Prince Rupert.

The planners were faced with a twofold objective. First, they had to plan for a city of 100,000 people, which might before very long have a population considerably exceeding that figure. If that now seems a trifle ridiculous, one must recall that in 1904 Vancouver, which was barely a village at the time it was reached by the CPR, less than twenty years previously, already had a population nearing 50,000 and in fact destined to reach 100,000 only seven years later, and a spectactular rate of population increase was no more than consistent with the rosy prospects that were thought to lie before Vancouver's rival.

Second, Brett and Hall were expected to design a city of beauty and dignity, according to the formal civic aesthetics of the day, on a topographically difficult site consisting largely of rock and muskeg and covered by dense forest. Considering the circumstances, their layout was remarkably sucessful. A series of monumental thoroughfares, terminating in natural elevations or sites for public buildings, was skilfully adapted to the form of the land. The streets connecting them were laid out broadly on a grid, but where topography demanded it the planners were prepared to modify the grid or even depart from it altogether. In fact, the care and deftness with which a formal street layout is fitted to a rough and broken site commands great respect, particularly when the Prince Rupert plan is compared with the infinities of rectangles which were being imposed elsewhere upon the land of British Columbia with brutal disregard for hill, stream or coastline. The two main axes which dominate the plan, with their related systems of avenues, circles, crescents and public sites, were in fact intended as the commercial and institutional foci of a great city. The residential areas which were to house a hundred thousand people were to be more freely and informally laid out, their streets curving along the sides of the hil that rise behind the city center.[1]

In anticipation of the tremendous boo that was to occur when the railway w opened to traffic and the docks of Prin Rupert were crowded with transoceanic shi ping, and in an attempt to regain some of t money that it had spent, the Grand Tru Pacific soon began to sell lots. And in t euphoric atmosphere of the time, and wi the rosey image of a Canadian San Francis before their eyes, many people bought the Streets, sidewalks and utilities were laid the forest. Within a couple of years of esta lishment of the town, Prince Rupert had population of over four thousand, a Counc an apparently firm financial base, electricit water supply and a telephone system.

Prince Rupert: Development After 1912

Then the bubble burst. In 1912 Hays we down in the *Titanic*: the new docks were a most unused; the real estate boom was ove the City was in serious financial difficultie In Prince Rupert's hinterland there we riches of minerals and timber indeed, but t time was not yet ripe for them to be tappe Farming was scattered and marginal. In 19 Prince Rupert was stagnant, and the comi of war destroyed what hope there might ha been that the opening of the railway wou revive the golden prospects of a few yea before. Eight yearst later the Grand Tru Pacific was absorbed into the Canadi National Railways. In 1929 most of the la purchased by speculators twenty years b

[1] For a contemporary account of the planning Prince Rupert, see George D. Wall, "The Futu Prince Rupert as conceived by Landscape Arc tects," *Architectural Record*, Vol. XXVI, no. (August, 1909).

'e began to revert to the City for tax de-quency. In 1933 the City itself went nkrupt.

For a quarter of a century Prince Rupert's pulation remained almost unchanged at out 6,500 people engaged mainly in fishing d fish processing, but also in government d other services, logging and sawmilling. e Second World War brought a temporary om (during which the population, includ- servicemen, rose to some 20,000), plus more lasting benefits of enlarged docks, a plane base, and, most important of all, a d which for the first time linked Prince pert to the provincial highways system. A lp mill was established near the City in 51, a new airport has recently been built, d a ferry service to Alaskan ports will soon started. The City's population, which opped to 8,500 after the war, reached ,000 by 1961.

Physically, the City's growth has taken ce almost entirely within the area origin-y intended for commercial, institutional d governmental use. Ironically, however, street which was intended as the princi- boulevard of the City is now not even the in business street, chiefly because specula-n raised lot prices so high that businessmen ferred to establish themselves on a paral- street a block away, which is now in effect nce Rupert's "Main Street" — a lesson not thout contemporary relevance. Prince pert is still not a wealthy place; shabby me houses and weed-grown vacant lots join the wooden sidewalks of streets built er gullies or blasted through rock, streets t were to have been lined with tall and tely buildings. Comparing the vision of 05 and the plan which grew out of it with reality of a half-century later, one is re-nded of an impoverished family squatting in the ballroom of a palace abandoned long before it was completed.

But from the viewpoint of physical development Prince Rupert has two great assets. Almost all its unused land is in public ownership, amounting in total to over half the City's land area; and, partly as a consequence of this, the built-up area is fairly well-defined and compact, with a minimum of "sprawl". There is, however, a housing shortage, and much existing housing is in poor condition.

Today, the future of Prince Rupert looks brighter, perhaps, than it has since 1912. The old dreams are unlikely ever to be realized, but new transportation links to the south and north, development of timber and other resources, and tourism, hold out hopes of better times to come. At the same time, the silver lining has a cloud in the form of indications of a decline in the City's staple industry, fishing. In the face of these prospects, the City Council recently engaged a Vancouver firm to carry out an economic survey and a study of housing conditions, and a report on planning.[2] These were completed early in 1963, and may provide a sounder basis for future development than the febrile optimism in which Prince Rupert was born.[3]

The Origin of Kitimat

From the beginning of the First World War to

[2] Associated Engineering Services Ltd., *Prince Rupert: Economic Prospects and Future Development* (Vancouver, B.C., January, 1963).

[3] For general accounts of the geographical background, early history, and planning of Prince Rupert, see A. D. Crerar, *Prince Rupert, B.C.: The Study of a Port and its Hinterland*, Unpublished M.A. Thesis, University of British Columbia (1951); and P. D. McGovern, "Prince Rupert, British Columbia," *Town and Country Planning*, Vol. XXVIII, nos. 4-5 (April-May, 1960).

the end of the Second, as mining was almost completely abandoned in the face of adverse economic conditions, the North Coast — Skeena region as a whole depended almost entirely on fish and lumber for its livelihood. By 1951 it had only 20,000 people, about a third of them Indian. 8,500 people lived in Prince Rupert and the remainder mainly in small towns strung along the railway — now part of the Canadian National Railways system — or in isolated logging communities scattered along the coast and in most cases accessible only by sea or air, still the only direct means of communication between the region and the main population centres of the province. But in addition to fish and forests the north coast had an almost untapped reserve of hydroelectric power from the drainage of the Nechako Plateau between the Coast Range and the Rockies, estimated at three million kilowatts. It was this potential supply of cheap power which in 1951 led the Aluminum Company of Canada ("Alcan") to decide on the establishment of a major aluminum smelting plant; for while the raw materials of aluminum reduction can be shipped cheaply in bulk by sea, the process requires enormous quantities of power — 20,000 kilowatt-hours for every ton of ingots produced. Smelters are thus tied to sites near sources of ample power and with deepwater docking facilities as well as road and rail transportation, requirements which were met at the head of the Kitimat Arm, some fifty miles inland and connected with the railway and the Prince Rupert-Prince George highway by a convenient valley.

To provide the power, Alcan undertook several major engineering projects, including the reversal of the drainage of an entire lake system. The physical and economic problems were thus overcome; but the human problem remained. The plant, located in almost virgin wilderness forty miles from the nearest town needed a workforce of thousands, and to get and keep a stable and contented workforce meant providing living conditions that would compare favourably with those obtainable in the cities and towns to the south. Thus Kitimat was born, the region's second planned "new town", only eighty miles as the crow flies from its first, making them next-door neighbours in the terms of that vast country.

The Plan for Kitimat

Alcan, like the north coast, had had previous experience with new towns, and in the Company's case, like the region's, the experience had not been entirely happy. It is likely that the problems created by Arvida, established under similar circumstances on the Saguenay in Quebec in 1926, had a good deal to do with the care devoted to the physical and administrative forms of the new venture. To shape the general conception and principles of the town plan for Kitimat, the Company engaged Clarence Stein, the author of the Radburn plan, working in collaboration with the New York firm of Mayer and Whittlesey; while a team of experts prepared proposals on the administrative, fiscal and social aspects of the scheme. The result was a remarkably comprehensive series of detailed studies and plans embracing every important consideration

[4] For general accounts of the Kemano-Kitimat project, see: Aluminum Company of Canada Ltd., *Kitimat-Kemano: Five Years of Operation 1954-1959*; Paul Clark, "Kitimat — A Saga of Canada," *Canadian Geographical Journal*, Vol. XLIX, no. 4 (October, 1954); L. G. Ecroyd "Start-Up at Kitimat," *Western Business and Industry*, Vol. XXVIII, no. 7 (July, 1954); "The Nechako-Kemano-Kitimat Development," *The Engineering Journal* (April, 1953 and November, 1954); "Kitimat: A New City," *Architectural Forum* (July and August, 1954); Pixie Meldrum, *Kitimat — The First Five Years*, Corporation of the District of Kitimat (1958).

ıich could be expected to apply to the ilding and operation of a town of ten thou- ıd or more people in a virgin wilderness, ·ty miles from its nearest neighbour and ır hundred from a large city.[5]

Twelve years have now passed since the cision was made, and a decade since Kiti- ıt was formally incorporated as a Munici- l District by a special Act of the British ılumbia Legislature. The town is today ıerally recognized as one of the most sig- icant of contemporary North American ıtributions to planning principles and prac- e, and one of the few ventures on the conti- ıt to compare in scale and character with : British New Towns. A first assessment of : true extent of its achievement can now be empted.

In doing so, however, one must bear in nd the physical circumstances and the ndicaps which were imposed upon the ınners thereby. When Alcan came, the ıntry was largely unexplored and unsur- yed. Since it was covered by dense timber, stakes were made even after aerial photog- ›hy, mapping, ground surveys and test llings. Ravines were discovered that had t been suspected, and the site of the City ntre was found, after construction had ·ually started, to consist in part of muskeg, d the intended arrangement of the Centre d to be drastically altered.

These were the circumstances under which ·in had to devise his plan. There were two ırses which he could have adopted: to apt to geography and climate or to adapt to ›ple and culture. The former would have ·ant probably a tight design, a huddling together of people and buildings about a compact, perhaps multilevelled centre, a solution such as has been adopted in northern Sweden and has been proposed in other parts of Canada. Instead Stein chose the latter. He assumed that the people of Kitimat would be the same sort of people as those of Vancouver or Winnipeg or Toronto, living as far as possible in the same way; and he adapted to the rugged landscape the "Radburn idea", the principle of the loose-knit community based upon but not subjected to the use of the automobile. He designed a pattern of trafficways linking the City Centre to neighbourhood centres, and, connected to the trafficways, a series of loops and cul-de-sacs around which the houses are built. Among the buildings run pedestrian walks and park strips by means of which supermarket, drugstore, movie theatre, church and school can be reached on foot, in complete safety from vehicular traffic.

All this presupposed a car-owning society, and although the car is kept to its place, serving but not dominating the houses and excluded entirely from the inner ways which link homes to community facilities, the mobility which it gives and the demands which it makes have clearly been the chief determinants of the pattern of the town to the almost complete exclusion of any deferring to the exigencies of land and climate. Kitimat has as a result been described derisively as "Radburn revisited" and "A suburb without a city", and indeed it seems at first a little strange to find amid the forests and rugged hills of British Columbia a town which would seem quite at home in the suburban fringe of any North American city. But social values and habits are facts to be faced as much as are snow and muskeg, and less easily overcome; and this being so Stein's application in Kitimat of the principles which he had

See Clarence S. Stein et al., *Kitimat Townsite Report*, edited and reissued by the Corporation of the District of Kitimat (December, 1960).

pioneered in Radburn to permit peaceful co-existence between man and machine was not only sensible but perhaps inevitable.

Kitimat Assessed

In fact, if any criticism can be levelled at Stein's general conception it is that he did not fully appreciate the extent of automobile domination; while he allowed in effect for two separate but equally complete systems of circulation, one for motor traffic and the other for bicycles and pedestrians, it has proved in practice that much of the latter has been neglected and has served little useful purpose. While the internal "greenway" or park belt is well used, the connecting footpaths running between the rows of houses have mainly degenerated into mud tracks littered with garbage cans and junk, and they are likely to be abandoned in the design of new neighbourhoods. It could be argued that for a town set in the midst of a wilderness of tremendous potential for hunting, fishing, skiing and other forms of outdoor recreation, with fifteen feet of snow in the winter and much rain the rest of the year, the whole system of parks and greenways is unrealistically generous; that for its real value the cost of landscaping and maintenance will be too high. The present municipal administration would not agree, but as time passes and the consciousness of and pride in being part of a new venture fades, future municipal councils may well look on all this open space flowing through the town with more jaundiced eyes.

While the quality of the two-dimensional plan of Kitimat can be considered worthy of the tremendous enterprise of which the town is a part, the same unfortunately cannot be said of its physical form. Once the decision was made to go ahead with the project, time was of the essence; production had to be started and houses for the workers had to be built as quickly as possible. For several years shortage of accommodation was a chronic condition, and quality of design and construction was hardly considered. Alcan in any case left the provision of housing to others as far as possible; where necessary it had houses and apartment buildings erected to rent to its workers, but for the most part it sold land to private developers, mainly from the Vancouver area, and left them to build as they saw fit. This meant in practice that the standards of construction were those imposed under the National Housing Act as a condition of N.H.A. financing, while the standards of design were, aesthetically speaking, nil. In terms of space, Kitimat houses and apartments are quite acceptable; in terms of solidity of construction they are no better than N.H.A. minimum, which is not high; in terms of appearance they are at best undistinguished and at worst quite bad. One looks in vain for any evidence of care for civic design. Among other things, the visitor to Kitimat quickly notices that the builder, accustomed to the standard Vancouver suburban grid, was clearly puzzled by the Stein concept of the inward-facing house turning its back upon the street, which is quite contrary to the North American convention. It is obvious that the same applies to many of the residents; the consequence is that some houses are oriented toward the street, in the usual manner, with the orderly, well trimmed "front yard" located accordingly and the washing hung out at the back; while others are reversed according to the Stein conception to face the pedestrian ways or greenways.[6] As a consequence, the vista whethe

[6] After this was written the author was informed that this was in fact a deliberate attempt to introduce "variety". If so, it succeeded only in introducing visual chaos.

:he street or of the interior frontage is ɜrally scruffy.

'o such criticisms Alcan's reply would be its business is to produce aluminum, not uild towns; and that its policy, based on Arvida experience and pursued vigorously means of second mortgages at N.H.A. s, home-ownership bonuses and repur- se guarantees, is to encourage its em- vees to own their own homes, to avoid "company town" stigma, and generally to ourage Kitimat to become in every pos- e way a "normal" Canadian community. se principles are sound enough, and have oubtedly greatly helped the town to be- ne, as it now is, a vigorous community ely free of company paternalism. Never- ess, it is a matter for regret that Kitimat's sical manifestation is so far removed from high aspirations of Stein's original con- tion; and it is particularly regrettable that attempt has been made in the building of town to take account of its geographical umstances. The same buildings might ally well have been erected in Vancouver Halifax or for that matter in Chicago or ton; in conception they are as much mass- duced as cars or coffeepots. As a single mple, the canopy which surrounds the n block of the City Centre — the com- rcial heart of the town — is perhaps six wide, a token shelter which is of little ue in a place with heavy rain or snow ing most of the year. This failure to re- ct, even to consider, the nature of the ce is unfortunately completely character- of the building of Kitimat; no attempt atever has been made to try out ideas and hods geared to the circumstances of the rather than to custom and convention. n apart from the lack of experimentation general level of design is low. There is not building in the town with any claim to architectural distinction, though there are several examples of uninspired but respect- able contemporary design. The only obvious examples, in fact, of imaginative design lie in a few such details as the street-name signs, which are colourful, legible and handsome (and made of aluminum).

There has in fact been a general lack of concern with the quality of execution, all the more striking and disappointing when con- trasted with the care and skill devoted to the preparation of physical, administrative, fiscal and social plans. There has been a similar lack of concern, the job once having been done, with the success of these plans or with finding out how they have worked, despite the largely experimental nature of the project and the enormous potential value of such an experiment for the planning not only of future ventures of a comparable nature but of new communities generally. In particular, it is impossible to know how successful the con- ception and exception of the scheme has been for the people most closely concerned, those who live in the town, and it is impossible also to know what relationship exists between physical conditions, planned and unplanned, and their way of life and the degree of satis- faction which it provides.

This failure to consider Kitimat as any- thing but an expedient design, however competently, to meet the exigencies of an immediate practical problem has in fact been apparent from its very inception. Consider the circumstances; here was a town estab- lished in a region with a great and almost untouched store of natural resources, a town confidently expected to be the largest in hun- dreds of thousands of square miles, yet next door to another which had been founded only a few decades earlier amid even rosier hopes only to become a shabby fishing community. But almost no thought was given to its prospects as a regional centre, almost none, in fact to the eventual settlement pattern of the region or to the appropriateness of Kitimat's location from this point of view, or even from the point of view of the eventual development of resources other than hydroelectric power. Here was a site that met the requirements of the company for the production of aluminum; that was sufficient. Its suitability in terms of the future economic development or human settlement of the region was simply not a relevant consideration.

To sum up, Alcan must be given credit for the care which was taken in the planning, physical and otherwise, of Kitimat. The site plan is not beyond criticism but it is unquestionably distinguished. But a townsite had to be planned and from a purely businesslike point of view it made more sense to do the job well than to do it poorly. Good industrial and a stable labour force were important to Alcan, and anything necessary to aid in maintaining them made sense. This attitude was hardly novel. Alcan is in the direct line of descent in this respect from the New England mill-owners of the early nineteenth century, from Salt, Lever, Cadbury, Pullman and other industrialists who, over the last century and more, have provided their workers with better living conditions for the sake not only of social ideals but of hard cash. Alcan probably had its eyes fixed more firmly on the latter rather than the former than had some of its distinguished predecessors. As evidence, one may point to the three main areas of criticism which have been discussed: the lack of interest in the quality of building design, particularly from the aesthetic viewpoint; the failure to make any but the most tentative efforts to learn from so valuable an experiment in town planning, social organization and local administration; and the lack of regard for Kitimat's place in the economic, geographical and social evolution of tl region as a whole. Kitimat was, in short, see in exactly the same light as was the Kenn Dam, the Kemano powerhouse, the plant i self, or any of the other components of single tremendous industrial project: as a jo to be designed by the best men available, one end only — the production of aluminur

The Lessons of Prince Rupert and Kitimat

A comparison of the histories of Prin Rupert and Kitimat produces some striki parallels. Both towns were created (almc exactly half a century apart) as integral par of great commercial enterprises intended tap the vast resources of the north. Bo towns received generous allotments of la from the provincial government but bo were developed entirely by private enterpris with little or no government interferenc Both towns were planned by American firn according to the most up-to-date principl of their time.

But the most significant similarity lies the fact that neither town has achieved wh was to be its appointed destiny. In the ca of Prince Rupert, the economic base upc which a great city was to grow proved to an illusion; the second Canadian transcon nental railway was a failure; and the ne town, almost stillborn, had to struggle alo on a single industry which had hardly, if all, entered the calculations of its founder The economic planning which produced Ki mat was more thorough and more realisti The ingots flow from the Alcan plant and w probably continue to do so — but they a not flowing at the rate that was origina expected. In the face of unexpected adver conditions in the world markets, productic has levelled off and has even been cut bac instead of rising steadily as was anticipate en operations were started. Furthermore, Company and the municipality have so found little success in their search for new lustries to broaden the community's eco-mic base and lessen its dependence on can. Consequently, the town's population ms to have stabilized at a figure between)00 and 8,500, having dropped from a ak of ten or eleven thousand accounted for construction workers. This may be com-red with the estimate of 35,000 to 50,000 ed by Stein as the basis of his plan.

Today the attention of Canadians is turn- more and more to the north — to the last d frontier, with what we are assured is its st storehouse of riches. Where there are hes men will certainly go; new communi-s will come into being; thousands of people eventually, perhaps, millions — will call north their home. What lessons do the ries of Prince Rupert and Kitimat hold for as we contemplate this prospect?

That depends on the answer to another estion. When we talk about northern evelopment" (a word which, as every nner knows, covers a multitude of sins), we mean *resource* development or do we an *regional* development? Because, in actice, these are very different things in-lving very different philosophies. Resource evelopment" is too often mere exploitation natural riches for financial gain, without gard to the long-run fate of the region ich supplies them or of the communities which the process gives birth. Canada can pply many examples; the fact that no con-eration other than the requirements of minum production entered into the oice of Kitimat's site is only one illus-tion. It may well be that in this partic-r instance no other location would have rved, but in view of earlier experience in her parts of Canada — experience in which Alcan has been involved — it seems most regrettable that so little thought was given to future regional development. The case of the Saguenay-Lac St. Jean region may be cited: this part of Quebec offered a variety of natural resources, including, like British Columbia's north coast, minerals, timber and water-power; but each of these resources was exploited independently without reference to the development of the others or to the future of the region as a whole. The result today is a settlement pattern consisting of a scattering of small or medium-sized one-industry towns (including Alcan's Arvida) within a relatively small area, in some cases only a few miles apart, none of which can aspire to the size or functions of a genuine regional centre. The economic dangers, human deficiencies and general inefficiency of such an arrangement both for the individual community and for the region as a whole are obvious enough; yet the same pattern seems to be emerging around the Skeena Valley.

The best known example of the alternative approach, of course is TVA. Certainly TVA was concerned with making use of resources, but always as part of the process of developing, improving and helping a *region*. TVA's achievement lies not in the number of dams it has built or the number of kilowatt-hours of energy it produces, but in the fact that the Tennessee Valley is a better place, its economy stronger and more diversified, its communities more prosperous, the lives of its people fuller, healthier and more secure, than they were thirty years ago. This is regional development in the true sense of the words, and it is to be hoped that the future settlement of Canada's north and the utilization of its resources will be carried out in this spirit. On this assumption, let us return to the question: what lessons can be learned from the experiences of Prince Rupert and Kitimat?

First, a single industry is not a satisfactory basis for a permanent community. This is perhaps a truism; certainly after the experience of Elliot Lake, to name only the most famous recent example, the point is unlikely to be disputed. But even if an industry is soundly based, stable and prosperous, it is not a healthy thing for a community to be almost solely dependent upon it for employment, for tax revenue, sometimes even for welfare assistance, entertainment and social facilities. Sometimes it may be unavoidable; but when a new town is to be created, every effort should be made to find a site with the best possible chance of attracting industries other than the one which called it into being; and every effort should be made to encourage such diversification within the community and the concurrent development (in the true sense) and settlement of the area around it. In other words, the enterprise should not be seen as a matter of a single industry and a community serving and subordinate to it, but as the initial phase in a coherent regional development programme whose object will be not only the exploitation of the region's natural resources but also the establishment of a sound and broadly based local economy and of a full and secure way of life for the people of the region. Really successful new town planning, in short, is inextricably bound up with regional planning and economic planning, and regional planning and economic planning are but two sides of the same coin.

The second lesson to be drawn from the history of the Skeena-Coast region is that regional development planning of this coherent and comprehensive kind cannot be expected from private enterprise. This is no criticism of the Grand Trunk Pacific or of the Aluminum Company of Canada or of any other private firm. Their function is to run a railway or to produce aluminum or whatever the case may be, and to pay dividends to their shareholders, not to design towns or to develop regions. It is to the credit of both the GTPR and Alcan that, the nature of their commercial activities requiring the creation of new communities, they went to considerable pains and expense to create good ones at least to the extent that enlightened self interest dictated. They could not be expected to do more. It follows, therefore, that if future regional development based on the exploitation of the natural resources of the north is not to be the hit-or-miss process that it has been in the past; if the tapping of natural riches is to provide the basis for permanent and stable settlement providing the best possible way of life for the entire regional community, then public participation will be required in very considerable measure.

New Approaches to Building Frontier Regions

Let us consider briefly where models of such participation might be found.

Several Canadian provinces, notably, perhaps, Alberta and British Columbia, have legislation providing for the establishment of regional planning agencies. But these are clearly envisaged as essentially *intermunicipal* bodies operating in settled areas. Their function, broadly, is to prepare plans to regulate private development and to guide the works programmes of other public agencies. Several provinces also have statutes providing for special forms of administration for new communities.[7] But again, this legislation is in general terms designed to provide a special form of local government to apply

[7] e.g., Local Improvement Districts in Newfoundland and Manitoba; Mining Towns and Mining Villages in Quebec; Improvement Districts in Ontario; New Towns in Alberta; Local Administrative Districts in the North-West Territories

ring the life of a temporary settlement or nding the establishment of normal munici- l administration. In neither case does it pear to be envisaged that the special ency, either regional or local, will play a ry positive role in the development of the ea under its jurisdiction. On the other hand, ere is a number of agencies established by leral statute — the Prairie Farm Rehabili- ion Administration, the Agricultural Re- bilitation and Development Administration d the Atlantic Provinces Development ouncil, for example — whose purpose is ite explicitly to play an active and con- uctive part in the development or revital- tion of the economy of some part of the untry, including making changes in the customed manner of land use. None of ese, however, is concerned primarily, if at , with the opening up of largely undevel- ed and uninhabited areas, but rather with proving the situation of regions where velopment has been unsoundly carried out has not kept up with the more prosperous rts of the country. (One is tempted to sug- st an analogy with urban growth, wherein e profitable building of new houses at the ge of the city is left to private firms while vernment assumes the cost of tearing down e old and decayed areas at the centre and placing them with something better.)

Thus there exist in this country at least the eds of three relevant ideas: regional plan- ng, special administrative arrangements for w towns, and government action to aid and mulate the economies of particular regions. t in pursuing possible avenues of approach Canadian regional development it is also orthwhile to consider two cases outside anada, neither of them new: the British ew Towns Act of 1946 and the Tennessee alley Authority itself, born in 1933. Both e too familiar to Canadian planners to warrant detailed description, but it is useful to recall briefly the features of the two schemes which are particularly relevant.

The New Towns Act, based upon the Scott and Barlow Reports and upon Abercrombie's Greater London Plan, had as its chief object the dispersal of population from Greater London. What is of interest here, however, is not the purposes of the New Towns but the manner in which they are established, financed and operated. A New Town, under the Act, is the responsibility of a Development Corporation appointed by the Government with the power to acquire land, to build, to construct utilities, to provide community facilities, and ". . . generally to do anything necessary or expedient for the purposes of the new town or for purposes incidental thereto." (the words of the Act). Apart from the fact that they are financed by Government loans and grants, the role of the Development Corporations in relation to the New Towns has in fact been very much the same as the role of the Grand Trunk Pacific or Alcan in relation to their respective offspring — with the very important exception that the former are instruments of national policy rather than private investment, and that consequently the first consideration has been the establishment of sound, viable and permanent communities, and each New Town (apart from special cases such as Corby) has been carefully sited and designed with a view to attracting a wide range of industries and not just to serve the needs of one.

The special relevance of TVA lies in the fact that the utilisation of a resource was regarded not as an end in itself but as a vehicle for the economic and social enrichment of an entire region, and that this enrichment took many forms and was effected in many ways — direct action by the federal government being the exception rather than the rule.

TVA's way was to educate, to demonstrate, to assist, to encourage, not to step in and do the job itself, beyond its immediate statutory task of harnessing the Tennessee River for power production, navigation and flood control; but it did these things to such good effect that a poor and primitive region achieved prosperity and vitality.

From all these — from present Canadian regional planning and special local administration legislation, from PFRA, ARDA and APDC, from the British New Towns and the American TVA there are valuable lessons to be learned, and it is to be hoped that they will be noted and applied. For the experience of Prince Rupert and Kitimat shows unmistakably that good town planning under the aegis of private firms is not enough. Each of the two towns is a fine example of the best *town* planning thought and skill of its day, and the differences in approach, planning and execution provide a fair measure of the progress that has been made in this field in half a century. But in origin and concept, in the kind of consideration that led to and conditioned the establishment of the two communities, there was no significant change at all. In each case the first consideration was economic exploitation, not the creation of a good environment for living or the future wellbeing of a region. It is here that progress in thinking is long overdue; it must be understood that the use of the resources of the north means the building of new human communities urban and regional; that the creation of such communities demands an enlightened and comprehensive economic development and regional planning policy; and that this in turn requires a large measure of public participation, public iniative and public responsibility.

While the manner in which these responsibilities could best be exercised is obviously a matter for careful and expert study, the requirements of the job as well as experience both in Canada and elsewhere indicate the need for some special form of regional planning and development agency, adequately financed and with the authority and capacity to undertake a variety of functions, including resource surveys, provision of roads, harbours, airports and other elements of the economic "substructure", industrial promotion and investment, town planning and building, and perhaps also local administration. There is no exact model to be slavishly followed in meeting the special conditions of the Canadian north, but there are many examples to learn from, not only those already mentioned but others in many parts of the world. By such means, not only could the resources of the north be brought into use more rapidly and more efficiently, but settlement and urban development could take place in a stable and orderly fashion to the lasting benefit of the people who are, in the final analysis, what really matter.

35

’ho Does What in
rban Renewal?

den Tanner

across the country, more and more people waking up to the fact that their cities are rouble, and that it is up to them to do ething about it. But what? Here is what e hard-headed businessmen, profession- and good citizens have done, each con- uting his special talents and resources to common problem.

y now most Americans know that their s need help. More and more people ize that slums are dangerous cancers cting everyone, and that a crowded, de- ng downtown somehow must be reshaped new and better principles. But few know to go about it — what practical moves individual in the community can make et the vital process moving.

there is a lesson that can be drawn from hundreds of examples of redevelopment oss the country, it is this: broad-scale ewal of real consequence can never get he ground unless it 1) gets strong, work- support from the city’s business leaders, 2) gains wide understanding and help ugh the city’s communications network its newspapers, associations, clubs, rches and neighborhood groups right down to the store-owner, the home-owner, the voter. A city planner, or a mayor, or a chamber of commerce can not get it done alone.

The great surge of interest in renewal in the U.S. is not just a sudden coincidence of civic virtue. It is growing out of the business facts of life in every lagging community: the mounting distribution costs of inefficient traffic patterns, the loss of trade to the suburbs and to newer cities, the headaches of manufacturers, the bad living conditions of employees. These were the reasons behind such classic transformations at Pittsburgh, where the Mellons and other bankers and businessmen formed the Allegheny Conference to pitch in and clean up and persuade business to stay. These were the reasons behind St. Louis’ equally famous Civic Progress, Inc., a smaller spearhead of 21 leaders who joined with the mayor and patiently dramatized the issues of renewal for two years before letting $100 million in bonds be put to a vote.

What businessmen can do

The experience of Pittsburgh and of St. Louis

rinted from *Architectural Forum* (November, 1956), pp. 128-34 by permission.

and of Cleveland, Kansas City and other alert cities suggests some practical steps for any smart businessman interested in his own future and the future of his town, large or small:

• Get into the parts of government whose proper functioning depends on the part-time service of responsible citizens: the planning commission, the zoning board of appeals, the housing or redevelopment authority, the board of education, the board of health.

• Publicly support those local officials who are doing a good job for the community. These men are often under attack from special interest groups, and sometimes their worthwhile projects are politically unpopular. Unless community leaders back good officials, government becomes tired and timid; its projects become those which will bring the least amount of criticism.

• In particular, get acquainted with the people in the city planning office, find out what the city looks like from their trained, over-all point of view, where blight exists and where it will strike next, what movements are underway or expected. Eventually, this is where renewal must fit into a comprehensive plan.

• Work on Chamber of Commerce city planning committees, use the National Chamber of Commerce's growing fund of advice and literature on urban renewal.

• Help set up a citizen's urban renewal committee that can enlist and coordinate the myriad organizations and interests of the city through a representative cross-section of its leaders.

• Join with fellow businessmen in creating a revolving fund and a redevelopment corporation. Without these financial tools, renewal can never get started (see Cleveland, below).

• Pay top architects and planners to draw up rebuilding schemes that not only will work, but ones that are capable of firing the publ
imagination.

• Publicize to reach all elements of the com
munity. Enlist the thinking and support
newspapers, and of those trained in adverti
ing and promotion. *Do it in the beginnin*
rather than coming in later with a "story
and expecting it to be worthy of printin
Donate advertising space, billboards, sp
announcements on radio and TV. Ma
short, hard-hitting films on renewal and ur
theaters and TV stations to show them
their contribution to the citizen effort.

• Focus first on one outstanding issue th
can unite the whole community and create
climate in which the rest of the issues can
brought up and carried out. In Pittsburg
this theme was soot; in St. Louis, soot ar
blight; in St. Louis and Chicago, rat-bitt
children; in Cleveland, the slow and painf
loss of standing among the first rank of U.
cities. Things like Los Angeles' famed sm
could become a useful banner for a broad
battle.

• Get other citizens out to renewal exhibi
and meetings. Some 90 cities are now focu
ing attention on big citizen gatherings to s
a wide-screen slide and movie show made
LIFE magazine and now touring the count
as a contribution to ACTION (the we
publicized American Council to Improve O
Neighborhoods, which is rapidly moving
its own thinking from "fix-up" to incluc
over-all city planning redevelopment ar
rebuilding).

Some examples:

In Detroit, a "Detroit Tomorrow Commi
tee" of 100 business and professional men
giving Mayor Cobo invaluable advice ar
support on slum clearance, civic center r
development, park programs. Businessm
have raised a $16 million Metropolit
Detroit Building Fund to finance capital in

ɔvements for 43 health, welfare and recrea-
nal agencies. The nonprofit Citizens development Corp., conceived by Mort-ge Banker Walter Gesell and other com-nity leaders (including the UAW's Walter uther), raised an initial $400,000 from lustry and labor, helped a big developer rt on 53 acres of the long-delayed Gratiot-leans clearance and rebuilding project .F, April '56).

Fort Worth's famed downtown plan by chitect Victor Gruen (AF, May '56) was nmissioned by a utilities man, J. B. omas of Texas Electric Service Co., who s understandably concerned about the ıg-term growth of the whole area in which power lines and lights were permanently ilt in.

In Indianapolis, a group of downtown sinessmen have formed the Civic Progress sn. aimed at rehabilitation and erection of w office buildings, a new civic auditorium, molition of blighted areas, new walk-to-rk apartments developed with private ɔital.

In Cleveland, the new Garden Valley ınned community offers a close-up study how renewal can get started and the roles ıt various business and professional men ı play. Four years ago, a *Plain Dealer* iter assigned full time to cover urban re-wal reported to his Editor Paul Bellamy ıt redevelopment would never get moving less it had strong support from the City's lustry and commerce. At his suggestion, llamy called a meeting of prominent citi-ıs, including the mayor, the Chamber of mmerce president and Utilities President ner Lindseth of Cleveland Electric Illumi-ting Co. Lindseth proposed a development mmittee, suggested it be headed by John C. rden, then a lighting fixtures manufacturer d board chairman of Cleveland's Federal Reserve Bank. When City Planning Director James Lister came up with 247 acres of industrial wasteland owned by Republic Steel as the only available site for a needed housing project, the "Virden Committee" turned itself into the Cleveland Development Foundation, a nonprofit group of industrialists who put up a $2 million revolving fund for this and other renewal projects. Working with the Chamber and with five company presidents, Virden campaigned to raise the money from some 100 companies. Republic Steel's Thomas Patton got his company to sell the land to the Foundation at cost and fill in a deep gully traversing it. (The Foundation in turn has sold the land to the city and will get its money back as it is purchased by private developers and the Metropolitan Housing Authority.) Construction company President A. M. Higley donated engineering studies and construction advice estimated at $25,000. Lawyer Seth Taft worked at cost as the Foundation's counsel, supported by Republic's Patton, also a lawyer. Through Lister the city planning department did the site planning. Board Chairman A. A. Stambaugh of Standard Oil of Ohio released his assistant, Upshur Evans, to become executive director of the Foundation. So concerned was Cleveland's business community about the need for redevelopment all over the city that Newspaper Editor Louis Selzer tried to get Planner Ernest Bohn to run for mayor on a renewal platform. (Bohn insisted he could serve the community better in his role as planning and housing expert.)

In Oakland, Calif., the Henry J. Kaiser Co. helped get renewal back on the track by heeding the call of an embryo citizens' committee, lending the full-time administrative help of its vice president in charge of community relations. Under Norris Nash, the committee has coordinated the city offices concerned with renewal, obtained a budget for its official coordinator, got Oakland designated as a pilot city by the federal government and is now making a sample survey of a 78-block slum area with the help of the University of California. Nash sees the work of his committee as "priming the pumps so that private enterprise can go ahead and do the job."

In neighboring Vallejo, Bank of America Branch Manager Leon Coleman picked up an earlier redevelopment survey, and working through Vallejo's downtown association got a redevelopment agency appointed, the first big step toward rebuilding downtown. In nearby San Leandro Coleman's opposite number, A. J. Oliveira, helped bring together 52 community leaders aimed at forming a redevelopment agency.

Across the bay in San Francisco, Paper Manufacturer J. D. Zellerbach got into the fray after looking downtown for an office building site and finding the foot of Market St. almost entirely blighted. After talking to Mayor George Christopher, Zellerbach and Investment Broker Charles Blyth formed a group of 11 businessmen which has advised on the problems of a product market and an Embarcadero Freeway, and helped persuade Architects Skidmore, Owings & Merrill to contribute a study of freeway costs to the city for a nominal fee.

In East Chicago, Ind., Inland Steel, Cities Service, du Pont, Socony Vacuum, Standard of Indiana and Youngstown Sheet and Tube joined Purdue University in a $1 million foundation for slum clearance, neighborhood rehabilitation and conservation.

What big companies can do

Companies with a nation-wide interest in healthy communities and their buying power are adopting urban renewal as a major policy. Recently some 60 General Electric branc plant executives met in New York wit ACTION executive vice president Jame Lash and his staff for a second worksho session to swap ideas on how GE could par ticipate in the dozens of communities i which it is involved.

Some of the reports:

In Cicero, Ill., Hotpoint Executive R. H Thomas heads a citizens' action committe which has torn down 65 shacks in variou parts of town, turned a municipal dump int a drive-in theater, persuaded the railroad t demolish 30 railroad workers' shacks an find them decent housing elsewhere, organ ized high school students into block-by-bloc surveys to photograph substandard houses fo later evidence, encouraged Boy Scouts t clean up vacant lots, cooperated with th Cicero *Light* on a series of renewal article made renewal films available to other loca organizations. Next steps: a Cicero ACTIO information center and staff, a professiona planner for the city's staff, renewal films an lobby displays in local movie houses.

In Brockport, N.Y., a town of about 5,00 GE Appliance Executive Joseph Orbin ha taken the lead in the new Chamber of Com merce, spoken on urban renewal to a handfu of community organizations, helped th Chamber beat the drum for a new plannin commission. Orbin faced the typical attitude of old, tightly bound communities, had t work carefully and quickly, in part to diver the idea that it was all a conspiracy betwee big business and local realtors. He found hi civic job easier after a blue-sky planner fro out of town had aroused the citizens' ire b proposing that Main St. be turned into a hug pedestrian mall. The day after that appeare in the news, the paper was full of letters to th editor and the next Chamber of Commerc meeting was packed to the rafters with fearfu d angry property owners. Brockport had rted to think about its future. Says Orbin his civic work: "We're trying to show a all town that we're not in here as a carpet-gger, but that industry and community ed each other."

In Bloomington, Ill., GE Executive Rich-d Ehrman is chairman of the 50-member tizens Advisory Committee and a member the City Planning Commission. The CAC mbership was made purposely large to get many people as possible interested in dying the new city plan report by Planner rland Bartholomew. Especially effective in mulating public support and understanding s a series of 15 four-page special tabloids blished by Bloomington *Daily Pantagraph* hose publisher is also the chairman of the y planning commission). These profusely strated special editions, once a week, amined each aspect of Bloomington at se range. Sponsors of the tabloids, along th the *Pantagraph*, were 105 local busi-sses, ranging from the big GE plant down the smallest local drug store. The CAC also sed $50,000 in subscriptions to remodel YMCA, got local labor organizations to nate free labor, local suppliers to give free terials. Every major lumber company in vn now has a free advisory service which ludes plans for new houses or remodelings.

Sears, Roebuck, with 700-odd stores ound the country, has its own Director of ban Renewal, Harry N. Osgood, encour-es its local representatives to take an active rsonal part in local programs. Sears, like E, held an ACTION workshop to acquaint field men with techniques, has provided em with such tools as urban renewal glos-ies and legislation manuals. The effects are eady being felt in such towns as Columbia, C., Greenboro, N.C., and near Sears' big icago plant, where the company has backed the Greater Lawndale Conservation Commission by sponsoring a contest for home improvement. Says Osgood: "We can help arrest a million houses a year from sliding into slums. And in bringing about better communities, Sears will make money too."

What merchants can do

In any city, store owners obviously have a big stake in a healthy community, especially in its downtown business district. Some of the contributions merchants have made are mainly self-interested, such as the promotion of downtown expressways and parking. Others, showing a broader awareness of civic problems, have:

- Contacted their own national associations for renewal information. The National Retail Dry Goods Association, like its counterparts in the real estate and building fields, has accumulated considerable data, and the chain druggists and others are on the verge of their own programs.
- Stirred up whatever local bodies they be-long to or can join.
- Campaigned for *peripheral* parking near downtown to keep traffic from snarling the very heart of the city.
- Put on a bright front by remodeling their stores and encouraging other owners to follow their example.
- Initiated area competitions among archi-tects and planners, with the help of city plan-ning departments and local AIA chapters.
- Given over display windows and inside store space to these and to other city planning and architectural exhibits.
- Donated some of their large budgets for newspaper advertising space to periodic backing of renewal projects.
- Encouraged editors to give editorial sup-port to these projects.

In Chicago, Marshall Field, which has a special vice president in charge of civic affairs (Earl Kribben), backed the Ft. Dearborn renewal project. Carson, Pirie, Scott & Co. celebrated its 100th birthday by sponsoring a $32,500 competition to redesign the Loop area for the coming 100 years; the contest drew 106 professional entries and provided the city with invaluable data and ideas (AF, Nov. '54).

In Detroit, the J. L. Hudson Co., working with the citizens' committee and the city planning department, donated its display staff and auditorium to stage a big "Detroit Tomorrow" exhibit.

In Denver, renewal got started when Developer William Zeckendorf was prodded into making good his intentions of major development in Mile High Center and following projects. Another step forward came when Joseph Ross, president of Daniels & Fisher department store, became president of the city's new Urban Renewal Commission as well as director of the executive committee of the Downtown Denver Improvement Association. Ross, who had headed a big Dallas rehabilitation project while at Niemann-Marcus, is now looking to a ten-year program for Denver that includes three pilot projects in renewal, a downtown plan for blight elimination and a new expressway, two more major slum clearance projects. Says Ross: "The first important thing in urban renewal is a good housing code, enforced humanely, flexibly, and patiently. You must prepare neighborhoods for it by working with local ministers, school principals and other leaders and helping them form neighborhood councils to advise the central citizen group." Ross' citizens' commission is a good cross-section of community skills: a merchant, a banker, a private developer, an insurance man, a lawyer, a professional planner, a labor leader, a councilman and a newspaper publisher.

What bankers can do

As a financial expert a banker, mortgag banker or insurance executive can contribut valuable experience to a citizens' committe in setting up redevelopment corporations t deal with federal, state and city government and with private developers. He can also se aside a portion of regular funds for low-cos home-improvement loans, always a desperat need in a tight money market. For example:

In Kansas City, Mo., Banker Jame Kemper helped form and head up the Down town Redevelopment Corp., a group of 9 businessmen which bought part of the city' "skid row" prepared under Title I, turned i into the 1,850-car Northside Parking Projec just inside Kansas City's ring-road redevelop ment (AF, No. '55).

In Cleveland, five banks backed up "Oper ation Demonstrate" by extending home improvement loans from three to eight years offering to consolidate any old mortgage with new ones to reduce monthly payments.

In Memphis, Mortgage Banker Willian Galbreath, besides serving on the city plan ning commission, has worked to get othe bankers interested in Title I loans for rehab ilitation.

What realtors can do

Any citizens' group needs real estate men t point out where blight is coming, to advise o land values, trends, taxes, costs. When re newal gets underway, they can also set up relocation bureau to help small businesses i the path of redevelopment find new space o building sites elsewhere in town. Other con tributions:

n Norristown, Pa., a committee of the l real estate board surveyed every prop- in town, notified owners of unsightly ditions, got them to bring 500 units up to dard and demolish 50 others.

n Memphis, Realtor-Developer Russel kinson helped his local real estate board blish a loan and advisory group to help ne owners finance improvements.

n San Francisco, Realtor Lloyd Hanford arheaded the formation of a Citizens' ticipation Committee which has set off hot healthy public debates over San Fran- o's freeways, produce market and other evelopment projects. Hanford's committee circulates among property owners, show- them how things are financed, what it will t them to do and what the committee ks they must get to do eventually, sketch- out for them how to work with city artments, architects, contractors.

n Atlanta, Real Estate Developer Frank Etheridge helped prod the Georgia legis- re into urban renewal laws, pushed for a n redevelopment program for a downtown a near the Capitol. The real estate board agreed not to lease property designated as ns by the city building inspector, and rance men have ceased to write slum fire rance.

at architects can do

the trained professionals most needed to pe actual redevelopment, too many archi- s are still busy deploring wrong solutions ong themselves instead of getting out and ting right ones. On the lists of various ups sponsoring urban renewal — in big ns and small, at the state house and on pitol Hill — the number of architects' nes is distressingly small. Not enough of them know their public officials, inform themselves on prospective redevelopment jobs or go out and find developers to promote and build their ideas. Few are able to present a clear, dramatic picture of urban design and renewal to the general public. Some notable exceptions:

- Nathaniel Owings of Chicago, former planning director for the city and now a participant in several large projects for its renewal.
- Oskar Stonorov of Philadelphia, who helped set up a remarkable city planning exhibit at Gimbel's store and has worked endlessly for rehabilitation projects sponsored by the Friends' society.
- Harry Weese, also of Chicago, who has helped on plans for both the north and south sides of town and has stimulated Chicago's vision of what it might do in the middle — create islands off the downtown lakefront.
- Edmund Bacon of Philadelphia, a planning director with a broad architectural background and an active liaison with the city's leaders.
- Kenneth C. Welch and Victor Gruen, two architects who trained themselves in planning and merchandising, have not only built outlying shopping centers but have developed challenging schemes for whole commercial districts downtown (Gruen at Fort Worth, Welch at Grand Rapids).

Other recent examples:

In San Francisco, Wayne Hertzka led the local AIA chapter in a successful attempt to get a coordinated plan for civic center expansion, something the chapter had been trying to achieve for seven years. Guided by the architects, the center will now enjoy a new exhibit hall, and underground garage topped by a park with properly laid out fountains and floral displays.

In Tulsa, members of the Architectural League went to the city with a proposal to design, at cost, a brand new civic center of public buildings linked by plazas and underground parking. Developed step by step with the practical advice of a mayor's committee of leading citizens and a six-man staff paid for by the city, the $28 million scheme has been well publicized and made part of the city's master plan (AF, Feb. '56).

In Springfield, Mass., Planner Reginald Isaacs used the town as a case study for his Harvard students, got the citizens aroused enough to consolidate groups and go to work on urban renewal problems (AF, July, '56).

In Elyria, Ohio, 31-year-old Architect Richard Miller came back from a Harvard conference on urban planning, wrote a dramatic series of ten sharply worded, well-illustrated articles showing how other cities were handling their problems and how Lorain County's new planning commission and citizens' groups could work together to get action at home. The windup: a full-page, four-color sketch plan, complete with ring roads and radials, parks and parking, for an Elyria of Tomorrow.

What newsmen can do

As Publisher Otto Schoepfle of the *Chronicle-Telegram* did with Miller's series in Elyria, other newspapermen have led their cities in the first steps toward urban renewal. In Washington, both the *Post* and the *Star* have not only given full-time coverage in news, features and editorials, but have in fact assembled so much valuable data that they are continually sought out for vital information — and for speakers. The famed St. Louis *Post-Dispatch* series "Progress — or Decay?" was invaluable in getting broad-gauge support. In addition to editorial support, the Detroit *News* actually sponsored a pan study of the city's problems by local leade and an Urban Land Institute team of expert In Cleveland, both the *Plain Dealer* and th *Press* got into renewal in the early, talkir stages, have full-time reporters and occasio ally help coordinate action. (Said one Ci cinnati leader: "If we had the kind newspaper support Cleveland is getting, we be twice as far along in our program.")

In Fresno, Calif., the *Bee* ran 18 articl under a "Community Crisis" headline, lat compiled the pieces in a special tabloid ec tion that went like hot cakes. Since then volunteer committee has drawn up a ne building code, others are improving coun zoning and studying metropolitan-area go ernment. Says the author of the series, B reporter Gordon Nelson: "In political car paigns around here today, being for planni is as necessary as being for motherhood."

In Bloomington, Ill., the *Pantagraph* g local businessmen to support a series special supplements. In St. Louis and oth cities radio and TV stations have broadca interviews with slum-fighters, renewal film and cartoons, round tables, spot commercia

What the professions can do

• *Educators* are bringing the story of slum renewal and planning to school childre through children to their parents, and parents directly. Schoolbooks like Chicagc famed old *Wacker Manual*, a child's prim on community planning, have appeared Atlanta and other cities. High school civi teachers are sitting in on briefing sessio with citizen committees, taking current pr grams back to class.

In Memphis, Dr. Laurence Kinney Southwestern University turned adult-educ tion studies into a highly successful series ıe television programs called "The City is ›u," with the help of a small grant from the ventieth Century Fund. Timed with the ›sentation to the city of a new master plan ›ort, the series showed how each citizen ıld get into the urban renewal act, ended th a picture of what Memphis could be in 84.

In Sioux City, Iowa, John Schmitt of the ›al adult education center obtained a grant ›m the Fund for Adult Education, used TV d other media to mobilize citizens behind mmunity improvement.

In Newark, N.J., the city's education de-rtment followed up a citizen's rehabilita-n program with neighborhood studies by ;h school juniors and seniors, has included ım prevention in its adult education pro-ım as well.

On New York's west side, other schools ; following the lead of Joan of Arc High hool in teaching housing problems and m prevention.

Doctors and health departments have pro-led urban renewal groups with data on ıitary facilities, health figures and hospital ›ds and have given valuable appraisals of ; group health and mental health aspects of ›an programs.

Lawyers have proved indispensable mem-rs of citizen groups in studying existing vs and codes at city, state and federal els, recommending enabling legislation, isions and new laws to implement urban ıewal. Lawyer-volunteers representing the izens can serve as a valuable check on the y attorney's office, can also help set up a ›akers bureau to explain renewal, a pool free or low-cost legal advice for property ners and tenants in the path of redevelop-nt projects.

Ministers have been invaluable on the ard of citizens' organizations, both as ad-visers and as spreaders of the word to their flocks. Not only do neighborhood people have more confidence in projects affecting them when ministers are involved, but the backing of church groups makes it difficult for special interests to attack real movements for civic progress. The church sometimes starts the movement itself:

In Cleveland, two Catholic priests have been the leaders in winning the city's interest in their area and getting officials to apply for a federal grant. Rev. William McMahon and Rev. Stephen Radecky repeatedly stressed neighborhood improvements from the pulpit, insisted that parishioners attend meetings of the West Side Civic Council to talk about them. They met with residents, street by street, to explain proposed zoning changes. Result: when public hearings on the zoning changes were held before the City Planning Commission, the people were prepared and the proposals moved along smoothly. Mc-Mahon also wrote to over 100 cities and agencies all over the country to collect available literature on urban renewal for distribution among West Side Civic Council members.

Another minister who has played a vital role in Cleveland's renewal is Rev. Robert L. Fuller, Negro pastor of Mt. Hermon Baptist Church. When Fuller learned that his church was to be demolished in clearing a central slum area, he came in to discuss the redevelopment plan with Cleveland's urban renewal chief, James Yeilding. He studied it and became convinced that it would work to the advantage of his church and his parish. Then he went about interpreting and selling the plan to his people, dispelling the fears that could have built up to solid resistance of the program.

Still another spark plug for renewal just getting underway in downtown Cleveland is

Reverend John Bruere, pastor of Calvary Presbyterian Church and leading member of the Hough Area Council. In this area, where fine old residences have been allowed to run down as absentee landlords cut them into suites and crowded them with low-income minority groups, Bruere has worked doggedly with street clubs comprising the Council to get members to do voluntary rehabilitation. After ten years of hounding city hall to enforce building codes and save the area, Bruere has finally been able to get the City Planning Dept. to make a comprehensive study of the area, concerned not only with which buildings should be rehabilitated and which demolished, but also with entire neighborhood problems of overcrowding, traffic parks, playgrounds and zoning. Volunteer are helping out by making house-to-hous surveys to get planners valuable informatio on size of families, income, number of autos etc.

Says Bruere, summing up the grass-root problem and, in a sense, the whole nature o urban renewal: "Individually, people fee they are alone and nothing can be done abou their problems. When they are brought to gether in a common cause, they find they ha been thinking along the same lines after al Things begin to happen. They take action – and city hall listens."

36

.oning Administration
ı Vancouver

. F. Fountain

any planners know zoning in its relation to
preparation of a zoning by-law but beyond
at point administration of the rules laid
wn in the by-law is left largely to others,
ually to the City's Building Inspector. Thus
close contact of the planner with the appli-
tion of his zoning by-law is often missing.

There is a school of thought that asserts
planner should stick to planning and leave
implementation of his plans to others. In
s way the planner can sit back and take a
tached view and with half-closed eyes and
t up on the table can best produce that
ich is best for the City.

The other school of thought is that the
anner should live with his plans after they
produced and by being in daily contact
th the problems of land use and its develop-
ent he will then more readily appreciate the
oblems that arise in making zoning a prac-
al tool of Government.

The latter is the method adopted in the City
Vancouver. Land use administration is the
sponsibility of the Director of Planning who
s under him a division of the Planning De-
rtment whose duty it is to deal with zoning in all its aspects and with subdivision control, the other major factor in land use administration.

This paper will not deal with the theory of zoning, nor with other methods of zoning administration, but will describe the way in which the zoning problem is tackled in Vancouver.

Vancouver Planning Department Establishment

The Vancouver Planning Department under the Director and the Deputy Director, divides into three main divisions, Administration, Plan Production and Redevelopment.

The main activities of the Administration Division are Development Permit Control, Rezoning Procedure, Zoning Appeal staff work and Subdivision Approval. The Planning Department has close contact with three independent boards, the Technical Planning Board, the Town Planning Commission and the Zoning Board of Appeal.

The technical planning board consists of eleven senior officials, representing all those

printed from *Plan*, Vol. 2, no. 3 (1961), pp. 115-24 by permission.

civic departments whose responsibilities include the execution of public projects contained in the unfolding development plan for the City. This Board has broad planning powers delegated to it by the City Council. It acts as a co-ordinator in technical and administrative matters bearing on the development of the City. It acts in an advisory capacity to the Council on planning matters, and has power to approve of development permit applications and to grant relaxation from specific compliance with some phases of the zoning regulations.

The Vancouver Town Planning Commission consisting of 15 members is now basically an advisory body to the City Council on planning. Prior to the creation of the City Planning Department in 1952, the Commission had a modest but efficient staff of its own under the late Mr. J. Alex Walker, as its Executive Secretary, and it carried out the planning function of the City. The Commission has operated continuously for 35 years under able leadership and is held in high regard by the City Council. The early planning in Vancouver was done by the consulting firm of Harland Bartholomew and Associates, engaged by the Commission. Now the Commission confines itself to consideration and advice on rezoning applications, to certain land use applications such as churches, hospitals and community centres, to advice on the unfolding Plan for Vancouver, and, by special legislation, to approval of subdivisions in a high-class residential area of the City known as Shaughnessy Heights.

The Zoning Board of Appeal is a statutory body of five members, two appointed by the Provincial Government, two by the City Council and the fifth, being the Chairman, by the other four. The term of office is three years, with the members receiving a modest annual stipend for their services. The Board meets every two weeks to hear appeals agains decisions on questions of zoning. Basicall the Board decides cases concerning the issu or refusal of development permits or matter concerning relaxation of the provisions of th Zoning By-law. The work is onerous but im portant in ensuring equity in zoning admini stration.

Legislative Authority for Planning in Vancouver

Vancouver operates under its own Charte from the B.C. Legislature, whereas all th other municipalities in the Province operat under the B.C. Municipal Act.

Insofar as Vancouver is concerned thi arrangement, though having its good and ba features, is generally satisfactory. It allow Vancouver to do things, if it can persuade th Legislature to enact them, without first havin to persuade the other municipalities to co operate. Thus there are various ideas in th Vancouver Charter, particularly related t planning, which are not yet available to othe B.C. municipalities. For instance, we hav through our own Charter, secured the righ to control development by the developmen permit procedure and to introduce condition uses and to delegate Council powers to Technical Planning Board and to relax th provisions of a zoning by-law in certai instances.

The Legislative authority granted the Cit is then in turn translated by the City Counci into enabling by-laws, the most important o these by-laws insofar as this discussion is con cerned, are the Zoning and Development By law, the Zoning Board of Appeal By-law, an the Subdivision Control By-law.

The Zoning and Development By-law

The present Zoning and Development By-la

ıe into force in 1956, superseding the ner zoning by-law which originated back 928. The new by-law is more positive in racter than its predecessor. It tells you it uses and regulations are allowable in the ous zoning districts, rather than what may be allowed. It adopts a uniform style of entation in each zoning district, and for first time it sets out full standards for cing and loading requirements applicable e various types of land use irrespective of particular district. It provides for both ight and conditional uses in each zoning rict, and it brings into force a development nit procedure.

'he number of zoning districts is steadily easing in Vancouver. We seem to add one ost every year: so that at the present time have 22 zoning districts. These include One-Family Dwelling districts, two Two-ily, four Multiple Dwelling, seven Com-cial, two Industrial districts and three rs covering special categories.

ı each zoning district the permitted uses divided into two categories, namely, those ch are allowed, as of right, provided the ication meets all the specified regulations cerning height, yards, site area, floor space ɔ, parking and loading, etc. These are wn as outright uses. The other category the uses which are allowed subject to r approval of the Technical Planning rd. These are known as conditional uses, g in the main, borderline cases which d be allowed provided they do not ad-ely affect the neighbourhood.

'or instance, an automotive repair shop is outright use in the industrial districts, but conditional use in (C-2) Suburban Com-cial District, and is not allowed at all in (C-1) Local Commercial District. Some he conditional uses, such as a hospital in ne Family Dwelling District, also require prior consultation with the Town Planning Commission, before the Technical Planning Board may act.

The Development Permit System

Probably the most interesting feature of our new Zoning and Development By-law is the introduction of the development permit system.

Most zoning by-laws couple administration and enforcement to the issuance of the building permit, and prescribe that the building permit shall not be issued until the applicable zoning provisions have been met. Under the development permit procedure the building permit (with minor exceptions) is not issued until the applicant first produces a development permit.

The development permit relates solely to zoning matters, leaving the building permit to take care of the structural features. There are a number of land uses where no building permit is involved, such as a storage yard or parking lot. These in a proper case would be issued a development permit. Conversely there are a number of situations arising where a building permit is necessary to ensure structural safety, but a development permit is unnecessary. For instance a development permit would not be required for internal structural alterations to a building where no change of use is involved.

For administrative purposes, and as a convenience to the public, all applications for a development permit are made to the Building Inspector's Department and not to the Planning Department. More often than not the applicant applies for both his development permit and his building permit at the same time. But he does not need to do so. If he wishes assurance that his proposed development will be acceptable in regard to zoning prior to spending time and effort on detailed drawings, he applies for his development permit first, on the basis of preliminary sketches. If then he secures his development permit he has 12 months in which to obtain his building permit.

When the development permit application is received by the Building Inspector it is identified in relation to its zoning district and to the use within the district. If the use happens to be an outright one, the application is processed by the Building Inspector. If it is a conditional use, it is forwarded to the Planning Department for processing and for presentation to the Technical Planning Board.

In both outright and conditional uses the application is referred to the City Engineer for a clearance on any public works matters such as street widening, lane opening, and availability of sewer and water services; also with regard to such matters as drainage, soil stability and air pollution.

In like manner all development permit applications are checked against a comprehensive Index Map (or series of maps on 100 ft. scale) kept up-to-date in the Planning Department. The Index Map shows all former development permit applications, all zoning appeals, all subdivision applications and similar matters affecting the site. In addition, by a series of transparent overlays all planning proposals which are under consideration or which have been approved are shown. Thus it is possible, on short order, to identify all the features of public knowledge which might have a bearing on the development permit application.

If perchance the proposed development is found to be in conflict with a public project, such as, for instance, the extension of a park site, issuance of the permit is deferred until the case can be reported to the City Council. The Technical Planning Board refers the application to Council, for instructions either negotiate for the purchase of the site to allow the application to proceed. Ord narily no such complications arise and in th case of an outright use, which meets all th prescribed regulations, issue of the develo ment permit is made by the Building Inspe tor in about 48 hours.

In the case of the conditional uses the tin is considerably longer as the application h to be processed through the Technical Pla ning Board.

It will be appreciated that the members the Technical Planning Board, who are th top officials of the City, cannot spend tc much time away from their own departmen discussing development permits. Instead th processing work is done in the Planning D partment and the application is then sent to Sub-Committee of the Board on which th Planning, Building, Health and Engineeri Departments are represented. The Su Committee meets each Wednesday mornir and makes a written report with recomme dations on each case in the form of minut of the meeting which go to the Board meeti the following Friday afternoon for approva Any difficult or contentious cases or any i volving new matters of policy or principle a specifically brought to the attention of th Board for decision and for future guidan of the Sub-Committee. Then the ratificatio of the minutes of the Sub-Committee, amended, becomes the official approval the Technical Planning Board to these co ditional uses.

The Civic Design Panel

The Vancouver Zoning and Developme By-law makes provision for control of th external appearance of new buildings by pr viding for the referral of any developme

:mit application to a Design Panel for ad-
e on the building design prior to issuance
the permit. This procedure has operated
:cessfully now for four years and already
direct and indirect effect is quite evident
the improved appearance of the new build-
;s in the City. The Design Panel is com-
sed of six members appointed by the City
uncil, consisting of the Director of Plan-
g as Chairman, the City Building Inspec-
, three local architects nominated by the
ncouver Chapter of the Architectural In-
:ute of B.C. and one professional engineer
minated by the Association of Professional
gineers of B.C. In addition four alternate
mbers are appointed for the architectural
d engineering representatives. The alter-
tes are permitted to attend all meetings of
Panel and participate in the discussions,
t they only have a vote when the regular
mber is absent.

The work of the Panel is purely advisory
the Technical Planning Board. All public
ildings, all apartment buildings and all un-
al buildings are submitted to the Panel for
vice. Ordinarily single family dwellings and
like are not dealt with by the Panel. In
ler to expedite the processing of the design
ect of the buildings, the plans are first
mitted to the Director of Planning. As
airman of the Design Panel he has a good
owledge of the views of the Panel, and the
nel has agreed in general terms on what
inarily it would be prepared to approve.
us by agreed arrangement the Chairman is
owed to clear those designs which obvi-
sly would meet the Design Panel approval
hout waiting for a meeting. However, if
re is any doubt the design is submitted to
Panel in the usual way.

The Design Panel does not seek to dictate
style of architecture. It is concerned only
h quality. In the case of poor design, it encourages improvement, quite often by meeting with the architect and showing him where improvement can be made. However, the Panel is careful not to impose its ideas on the designer and does not undertake to give gratuitous architectural service for the improvement of defective work.

The Design Panel reports to the Technical Planning Board by submission of minutes of its meetings, giving a report with recommendations on each case that comes before it. The Technical Planning Board considers these minutes and, of course, endorses any approved designs. If the report from the Design Panel is unfavourable, the Board then decides whether in its opinion the design is sufficiently bad that it would adversely affect the amenity of the neighborhood. If it is that bad, the Board has the power to reject the development permit application.

The Bonus System For Improving Exterior Appearance

Zoning by-laws ordinarily contain provision for regulating the size, and location of new buildings. Inevitably such provisions tend to dictate the shape and external appearance of the building. The classic example of this is the "wedding cake" type of tall buildings produced in New York City under their original zoning by-law. Different setbacks were required at different heights for the purpose of allowing light and air to percolate through into the streets below. The basic intention of the by-law was good, but the resulting appearance of the buildings left much to be desired.

In 1956, Vancouver introduced a fairly new idea into our apartment zoning regulations whereby the windows of any habitable rooms had to command an unobstructed view through a horizontal arc of 50 degrees, for a radius of 80 ft. In addition, in the denser zoning districts there was a vertical light angle control which to some extent achieved a similar result to the New York pattern of setbacks. We also had the normal yard requirements. At the same time, a floor-space-ratio was introduced as a control of density. These various controls were intended to be mutually in harmony, so that no one of them was more restrictive than another. In other words, the maximum amount of building that could be produced under the setback, yard, vertical and horizontal angle and parking controls would just about meet the maximum size of building permitted by the floor-space-ratio. However, this has tended to produce a stereotyped building with little room for flexibility in the design.

This is one of the reasons which has led us to introduce a bonus system in our apartment regulations. We still provide for the normal type of setback and light angle controls, and the same old type of shoe box design can be produced but if an owner wishes to do better he is allowed to erect a larger and taller building through earning an increase in the floor-space-ratio. He is allowed to score an increase in the floor-space-ratio for any one or all of three basic improvement factors. These three, in order of merit, are:

(1) Encouragement of more open space around the buildings.

(2) Increase in the proportion of required off-street parking placed below ground or within the building.

(3) The development of larger sites.

An interesting problem has been to set the bonus factors in such a way that they will, in fact, be attractive to developers. In Vancouver, the Building Code permits a two-storey building with a basement and a penthouse to be constructed in wood frame, but anything bigger must be constructed in masonry. Masonry construction is more expensive by 15 percent to 20 percent tha wood frame. To achieve a bonus for th extra space it is necessary for developers t build higher. Consequently the bonus systen encourages masonry buildings, but the bonu offered had to be sufficiently attractive to off set the extra 15 percent to 20 percent cos above that for wood-frame construction.

A different type of bonus system has jus been proposed to the City Council for th Vancouver downtown area. The three factor to be bonused are:

(1) Open area at ground level.

(2) Open volume up to a height of 80 ft

(3) Arcading.

The open volume factor gives the larges bonus, through the factors are adjusted t provide a different bonus for the three factor in different districts. For example, in the pro posed retail shopping district the arcade bonu has a higher factor than in other districts Again, in the center of the downtown are where we want to concentrate and stabiliz the maximum shopping and office develop ment, the bonus factors make it easier for developer to achieve a maximum floor spac ratio of 12.00 than in the fringe areas wher it is necessary to develop a larger site t achieve the same F.S.R. We have eliminate height and setback controls in the downtow area thus giving designers more flexibility.

In both the apartment areas and in th downtown core we hope to secure a bette type of building than in the past. Those archi tects who have tried the proposals repor favourably on them, which leads us to believ that we have found a solution through th bonus system of encouraging better exteric appearance of buildings.

Illegal Suites

Most of our larger cities have a problem i seeking to preserve the integrity of their singl

nily dwelling districts, and of the other resi-
ntial districts as well, against the encroach-
nt of illegal suites. Our problem in Van-
uver started in the West End of the City
the early '30's with the migration of the
ginal home owners to newer areas on the
tskirts of the City. Many of these old homes
re surreptitiously converted into multiple
cupancies of various kinds without the
cessary building, plumbing or electrical perts. Some were well done, but mostly the
ults were substandard accommodation
rely meeting minimum health requirements.
Then during the War years the blight
read to other parts of the City. With the
vent of P.C. Order No. 200 allowing the
aring of living accommodation as a War
ergency measure, the problem got out of
ntrol and the City could do nothing about
Rightly or wrongly Vancouver interpreted
Federal Order No. 200 to mean that struc-
ral alterations and increased plumbing
uld be allowed in dwelling houses to facili-
e the sharing of accommodation. In conse-
ence at the end of the War there were
usands of houses, both large and small,
w and old, good and bad throughout
City which had been converted to other
n single family use. The spread of illegal
nversions continued after the War during
period of housing shortage with no real
ion being taken to control it, only within
last half-dozen years have firm steps been
en to deal with the problem.

We now have a program whereby any ille-
suites found to have been created since
56 are required to be removed forthwith,
erwise the owner or operator faces prose-
tion. Any illegal suites created prior to
56 and continuously occupied since then
n be allowed to remain for a limited period
time depending upon the quality of the
commodation and the character of the dis-
ct. A large old dwelling can receive a permit to convert to other residential uses except in the best class of our One Family Dwelling districts. In our (RS-1) One Family Dwelling Districts no further conversions to multiple use are allowed and a program of progressive elimination of illegal suites is under way anticipating the complete restoration of such districts to one family dwelling use by December 31, 1970.

Originally we invited owners, by public notice, to seek validation of their illegal suites, with the promise that they would be allowed reasonable time to recoup their investment before restoration, but practically no one responded. We have to find the illegal suites through the various civic inspectional services.

The present policy then in other than the (RS-1) districts is to allow illegal suites to remain for a limited period of time, up to the life of the building, depending upon the quality of the accommodation, and the zoning district in which it is located. If the quality of the accommodation is rated very poor the owner is given up to three months to remove it. In all other cases the owner is encouraged to improved the accommodation and if it meets a rating of fair or good the owner can anticipate a renewal and continued renewals of his development permit during the ordinary life of the building.

In the (RS-1) One Family Dwelling districts a policy of concerted action to deal with illegal suites was put into operation at the beginning of January 1961. A systematic inspection of all houses in the district is being undertaken by a special housing inspection staff under the Building Inspector. When illegal accommodation is found the owner is told to restore the building to its lawful use or to apply for a development permit. If the owner elects to retain the accommodation, a further inspection is made to determine what work is required to bring the plumbing, gas and electrical installations up to by-law standard. The owner is told what is needed and the approximate cost. If the owner still wants to proceed and elects to bring the services up to standard he then applies for his development permit. Meanwhile the Building and Health Department staffs have rated the accommodation into one of three categories — good, fair or poor. Likewise the (RS-1) districts have been subdivided on a map, into three categories of good, fair or poor, depending upon the amenity of the neighbourhood. Now by matching the rating for the accommodation to the rating for the district the agreed expiry date approved by Council for that particular accommodation is readily ascertained. The poorest accommodation in the best district gets the shortest time, and the best accommodation in the poorest district the longest.

It is too early yet to make any predictions on this concerted effort to eradicate illegal accommodation in the better one family dwelling districts. To date about one house in ten seems to have some form of illegal accommodation and the owners appear to be willing to spend up to $12.00 for the privilege of retaining it for a few years longer.

In-law Suites

A related problem to that of illegal suites is that of in-law suites. Actually in-law suites are illegal suites being used for a special purpose.

The Vancouver City Council shows some sympathy towards those owners who wish to use or create additional accommodation in one-family dwellings for their parents. There is no problem in allowing relatives to live together as one family. The problem comes when one group of relatives wishes to establish separate living accommodation from the rest of the family in the same house. How do you distinguish relatives from non-relatives and how do you ensure that after the family use is discontinued that the premises can be restored to accommodate only one household? The problem which the City Council posed was made even more complicated by restricting the in-laws to parents or grandparents who through age and infirmity or financial dependency are arranging to live in the homes owned and occupied by their children.

Insofar as Vancouver is concerned there seems to be no straight-forward answer to this problem. Our Charter does not allow us in zoning to discriminate between one class of persons and another. Thus we are not able to allow additional separate accommodation to be provided in a home to be used by parents unless non-parents and non-relatives are also allowed the same privilege. Then there is the difficulty of administering the further restrictions whereby only aged parents of ill health or low income can occupy the in-law suites, and the added restriction that the home must be otherwise a one family dwelling owned and occupied by the children of the parents.

We are making some progress, but no real solution has been found short of introducing a modified occupancy certificate system. The most promising solution seems to be to seek a Charter amendment from the Legislature to allow the City to discriminate between parents and others in permitting more than one household to be established in a one family dwelling. Also for the City to seek Charter powers to compel the removal of additional sinks and cooking facilities once the in-law use has expired. In addition it may be necessary to secure authority to set up an independent Board of Referral composed of members outside of the civic service who ıld process applicants for in-law suites and . proper case issue an occupancy certifi- allowing designated persons to occupy gnated premises within a home. Then if occupancy, on subsequent inspection, re- 's different people in residence than al- ed under the occupancy certificate rem- ıl measures could be undertaken.

'he in-law suite problem serves to illus- how complicated a seemingly simple al problem can become when it comes to matter of its administration.

ing By-law Enforcement

ty years ago there was no zoning, as we w it, in any of our Canadian cities. The iest zoning by-laws were introduced about 5. Since then the impact of zoning on our s has been tremendous. Zoning is evident rever you go. If you walk around the resi- tial areas of Vancouver you can quickly those which were developed without zon- controls. If you look at the skyline of New k you can see the horrible examples of effect on architecture of the early zoning ack regulations.

/e tend to take it for granted that zoning ow an accepted way of municipal life, perhaps it is. The early struggles to get Councils to adopt the first zoning by- are a long way behind us. We tend to et the shaky start when there were no pre- ents to guide us and the Courts were none friendly to this new invasion of property ts. The reasOn zoning succeeded was that pioneer administrators were content to gress slowly, but securely, being never too ahead of public acceptance. Some of us become impatient and expect our ideas e accepted because they are designed to rove conditions. It comes as a distinct ppointment when City Councils do not always follow along as fast as we want them to go. Sometimes we fail to appreciate that progress must be geared to public acceptance.

Another thought to remember is that the more complex our zoning laws become, the more difficult they are to administer. The calculation of floor-space-ratio, horizontal and vertical light angles and similar features of zoning control take time and skill to process, much beyond what was required in the earlier simpler zoning by-laws. Here again, unless there is general public acceptance of the more elaborate zoning controls, criticism and complaints of bureaucracy and of unreasonable delay in processing development applications can lead to difficulties for the zoning administrator.

It, therefore, behooves the planner in devising more and better ways for controlling our mode of municipal development to take heed that his ideas are reasonable and practicable when it comes to administration. In fact, at times, one wonders whether some of our zoning controls can be administered effectively and equitably, or whether we are just leaning upon our inherent national characteristic of obedience to the law. If we cannot enforce our zoning controls in those instances where enforcement becomes necessary, then zoning administration quickly falls apart and becomes a mockery.

We have had our enforcement problems in Vancouver, and still have them, but fortunately many of our former difficulties are being resolved. In the past enforcement suffered from divided authority, inadequate or unenforceable provisions in the by-law and a somewhat apathetic attitude of the Courts towards all by-law enforcement.

First we amended our Zoning By-law, under advice from the City Prosecutor, so as to make it an offence to disobey an order from a zoning enforcement officer. Previously we were taking offenders into Court for, say, having an illegal suite. But if it had been there longer than six months apparently it was protected under the Summary Convictions Act. Now the owner is prosecuted for failure to comply with an order to eliminate the illegal suite.

Another difficulty with us is that the Vancouver Planning Department has no inspectional staff of its own and has to rely on other departments for this service. Unless there is very close harmony between the several departments concerned enforcement can fall apart. We are arranging for all zoning enforcement to be handled through a branch of the Building Inspector's Department. The staff will work in close contact with the City Prosecutor and will become specialists in handling cases needing Court action. In this way we hope to solve some of our past shortcomings in dealing with enforcement.

It is not suggested that we are quick to turn zoning offenders over to the Courts for prosecution. Far from it. Court action is a last resort. It is costly in staff, time and resources and the fines collected do not reflect the cost of this service. Unfortunately, there are some people who do not want to co-operate and prefer to ignore zoning controls. These are the people who when all else fails must be brought into line by Court action. Otherwise, your influence over the more co-operative section of society is undermined.

This review of land use administration has concerned itself with some of the features of zoning control as practised in the City of Vancouver. I do not pretend that our Vancouver system is perfect or that it is the precise pattern for other cities to follow. However, we have had zoning control in Va couver for well over 30 years and the prese procedures have emerged through a lo period of trial and error. They seem to su us fairly well; but we in the Vancouver Pla ning Department are well aware of imperfe tions in the system and are always on t alert for improvement, particularly if su improvement tends to simplify and to exp dite the procedures.

Ony passing reference to the subject subdivision control will be made at this tim Vancouver carried the torch for so ma years, seeking to convince the B.C. Gover ment of the importance of giving the munic palities full control over the subdivision land. We may look with critical eye upon t stereotyped subdivision pattern common many of our B.C. municipalities and we w probably fail to realize that it took 30 yea of effort before the Legislature vested t right of full control in the municipalities. Th was only achieved in 1954.

In conclusion, it may be worth recalli that the early pioneers in the field of planni — idealistic and visionary as they might times have been — nevertheless, have giv us in zoning a tool whereby the orderly d velopment of land for the benefit of t people can be achieved. Zoning is one of t most progressive ideas to emerge in the Twe tieth Century. Without it our cities, toda would be unimaginable areas of confusion. is our responsibility as professional planne to jealously guard this heritage and to ensu that our cities develop and redevelop alo sound rational lines in step with the advanc ment in knowledge and in techniques in oth spheres of human activity.

. Theory f Slums

37

narles J. Stokes

e paradox of slums is that despite the alth and the high level of economic development of the United States, they are as prevent in our cities as in many an overseas an area. Moreover, slums persist here pite attempts over the past three decades least to eliminate them. Is there a justification or an explanation for slums in this coun-?

Many explanations of slums do in fact st. They are used as a basis for the actions en to eliminate slums. One is tempted, ugh he ought not yield to such a temptan, to suggest that a possible reason why elimination of slums has failed is that the planations are inadequate. It is, however, rthwhile to suggest that in view of the very ge sums of money about to be spent on an redevelopment more careful attention ght well be paid to a complete theory of ms. That is what this article attempts.[1]

[1] The research upon which this article is based was in part financed by the Faculty Research Fund of the University of Bridgeport which awarded a grant.

Assuming that slums do have a function in the development of the city, we intend in this paper to see how this function evolves and show its direct relation to the growth of the city. But theory of slums presented here will seem a bit odd to the reader familar with the literature because it is not concerned with buildings or neighborhoods except tangentially. Rather what emerges is a socioeconomic analysis which is in fact a branch of the theory of the labor force.

We seek to find meaningful relations among the major variables assumed to be associated with slum situations and to derive a hypothesis about the rates of change of these variables in such a manner that predictions can be made about slum development. What follows arose out of a need to explain to Latin American audiences both the future of the slum problem so obvious in their large cities and the reasons why they persist in the United States. It evolved as the author wandered through the slums of Caracas, Lima, Buenos Aires, Guayaquil among other cities and made comparisons with what was fami-

printed from *Land Economics*, Vol. XXXVIII, no. 3 (August, 1962), pp. 187-97 by permission of the hor and the Regents of the University of Wisconsin.

liar to him in American cities. The theory is outlined here in the hope that unemotional thinking about slum formation may provide an ultimate solution.

What is slum in the city landscape is of spontaneous origin. This very spontaneity makes the definition of slums difficult. Slums appear to be planless and even antiplan. Slums, it is argued, do not yield themselves readily to rearrangement. Indeed, are they not like cancer? Cancer has its own growth process distinct and ultimately inimical to the human body. Is the presence and continued growth of slums destructive of the city? Suppose the answer were affirmative. Then one possible definition follows. Slums, it may be asserted, are those areas of the city in which housing and resulting social arrangements develop by processes so different from those by which the general growth of the city proceeds that they will destroy the city. All we have now to do is to look for these areas.

Whatever other merits such a definition has, it does suffer from two difficulties. It does not tell us what to look for. And there is a doubt about the analogy. Slums may be a necessary and even helpful phase of the ecological processes by which city growth can be described.

A good descriptive definition must be capable of fitting a wide range of apparently analogous situations. We must in the definition account for the community housing standard, the community's evaluation of its housing requirements, as well as the community's techniques for handling and absorbing the poor and the stranger. Indeed, the slum is the home of the poor and the stranger, if nothing else. These are the classes not (as yet) integrated into the life of the city. The poor are not integrated because of an ability barrier which tends to separate the city populace into those who will be fully utilized in the economic and social life of the city an those who will not be regarded as being o the required level of social development. Th strangers are not integrated by reason of "different" culture and the stage to whic their own acculturation has come.

Slums differ from the districts in which th lowest stratum of the integrated classes liv by failing to conform to the standards whic this stratum has set for itself. The distinctiv feature of slums is not appearance as such then, but the relation between the slum an its inhabitants and that neighborhood and it inhabitants which the city regards as havin met minimum livability standards. By thi kind of definition what is slum in Lima ma not be slum in Guayaquil. The function o the slum at any moment in city developmen is to house those classes which do not partici pate directly in the economic and social lif of the city.

To illustrate the complexity of slun formation and to attempt a theory of slun growth which correlates with a theory of cit growth, it will be useful to construct a simpl model.[2] This is done in Figure 1.

The model sorts out from a welter of vari ables two which are thought to be deter minants of slums.[3] One of these is the psy chological attitude toward the possibility o success in moving up through the class struc ture by assimilation or acculturation to fu

[2] The model which follows and which is elabo rated in the appendix grew out of an attemp to explain the observations of Professor Rober Lampman. See his *The Low Income Populatio and Economic Growth*, Study Paper number 12 Joint Economic Committee, United States Con gress, Washington, D.C., 1959. See also, *Makin Ends Meet on Less Than $2,000 A Yea* Senate Document number 112, Washingtor D.C., 1952.

[3] Lampman finds many more variables than these *op. cit.*, p. 6.

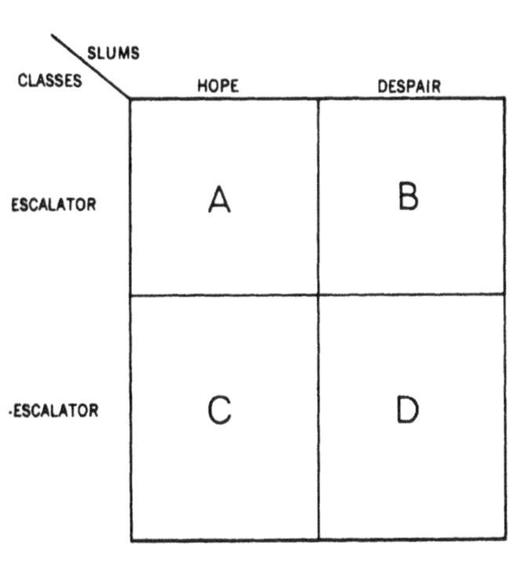

ure 1.

rticipation in the economic and social life the community. The other is a measure of cio-economic handicaps and barriers to ch movement.

Horizontally, we distinguish slums of ope" and slums of "despair," and vertic-y, escalator and non-escalator classes. The tinction between "hope" and "despair" is e which to some degree must be intuitive. "hope" is meant that quality of psy-ological response by the inhabitant of the m which indicates both his intention to etter" himself and his estimate of the prob-le outcome of such an effort. "Despair" by same token denotes either a lack of such ention or a negative estimate of the prob-le outcome of any attempt to change status.

An escalator class is a group of people who be expected, barring unusual circum-nces, to move up through the class struc-e. A non-escalator class is one which is nied in some way the privilege of escala-n.

The psychological distinction between ope" and "despair" may readily be con-rted into a distinction between the em-oyable and the non-employable. There may be some fuzziness about the exact location of the boundary line but its nature is clear. What is involved is some social measure of "ability" seen objectively as well as subjec-tively. At any moment of time, the city will have determined a minimum set of "ability" standards. For each job, for example, there may be age limits preventing those who are too old or too young from gaining employ-ment or it may prove difficult to find employ-ment for breadwinners whose education does not go beyond the 8th grade of grammar school.

The distinction between escalator and non-escalator classes implies the concept of caste. There are thus two categories of jobs one of which permits escalation and another which does not.[4] It is important to make a careful difference between caste and ability in this model. Caste membership need not mean residence in a slum provided adequate op-portunities for employment exist on this side of the caste line. This in no way denies that some form of racial or religious integration may be a more effective method than separate-but-equal job opportunities. Ade-quate job opportunities within the caste lines will, however, work to eliminate slums if permitted to do so.

The slums of "hope" have an additional characteristic of importance. They are the home of the stranger, the in-migrant, the recent arrival.[5] These strangers have been attracted to the city by the social or economic opportunities the city offers. Or they may

[4] It may be objected that a caste system implies inadequate as well as unequal job opportunities. Logically this need not follow. "Separate but equal" is a possible solution regardless of its desirability.

[5] In its *Primer About Slums* (1958), the New York State Division of Housing makes in-migration the first reason for slum formation. However, this reason is overstressed.

have been driven from their homes in the countryside or in other nations by economic, political or social upheaval. In this model, it has seemed best to emphasize the demand side postulating that an economic differential, more than likely wage levels, between the city and its hinterland is the motivating force leading to in-migration.

The strangers come to the city seeking "improvement" but, if they lack the language abilities, educational attainments and other necessary social and economic resources, they may find escalation difficult even if it remains possible. This is all the more true if as is likely to be the case the number of in-migrants in the early waves may exceed the actual new job opportunities made available. The result is that the strangers will tend to fill up the poor housing facilities and spill over into shanty towns.

Note that the slums of "hope" are so characterized because, given normal conditions, it can be expected that many of the strangers who inhabit them will be absorbed into the general employable population. They will learn the language. They will become acclimated and they will acquire the cultural resources necessary. So much the history of migration has taught us. At any moment of time prior to complete integration there will, however, be slums of "hope."

Not all the strangers, though, will be successful in moving up through the social class structure. Some will lack "ability." A group that we have called the "C" group will be debarred from making the same rate of progress as the "A" group by reason of skin color or religion. More important, a proportion of both the "A" and "C" groups will end up in slums of "despair" because they have found it impossible to get over the "ability" hurdles the city erects. Slums of "hope" disappear as migration slows down. Slums of "despair" do not disappear. For in the slums of "despair" live the poor.

"Poor people" in an underdeveloped or developing nation may be taken to be those whose standards of living are below some cut-off point. We may say, for example, that if we have made a careful study of necessary expenditures for normal subsistence, those who do not obtain this minimum amount are "poor". It is, of course, possible and even likely that in an underdeveloped society the per capita income may thus only permit a standard of living which must lie at or below the poverty level so defined. As society develops, in the sense that per capita income increases, there will be a widened gap between average incomes and the poverty level. But this is not the only observable change. Whereas, in underdeveloped countries it is very possible that the distribution of income will be positively skewed as contrasted with the distribution of abilities which is more than likely normal, with the growth in per capita income there is a tendency for the distribution of income to approximate more and more closely the distribution of abilities. What this means is that in advanced societies poverty and lack of ability become more and more correlative. Thus, in an advanced nation the "poor" — not only the poor in income terms but more significantly the poor in "ability" — will be found living in the slums of "despair." That is where those who cannot meet the society's minimum standards for full utilization and employability on any normal basis reside.

In short, we can say that slum formation depends on the rate of in-migration as well as on the rate of integration or absorption of the migrants. Obviously, too, slum formation depends on the existence of barriers to escalation as well as the distinction between income and "ability" classes.

laving set up the model, let us now see ' it fits actual cases. In what follows, four .nct city slum neighborhoods are pre-ed for analysis in terms of the model.

A

iuayaquil, Ecuador's largest city and sit-d near the mouth of the Guayas River, grown since World War II to have more 500,000 inhabitants. The fast rate of vth has been dependent upon in-migra-, a good deal of such migration coming the surrounding tropical countryside but aps an even larger proportion deriving the temperate and occasionally cold eys of the Sierra of the Andes. The "ser-os" have rather more Indian blood than "costeños."

rom Calle Los Rios west and Bulevar 9)ctubre south and especially along Aven-Gomez Rendon — the Boulevard of the r — a vast area of bamboo shacks ches to the salt water estuaries. Much of Barrio de los Pobres — city of the poor s below the tide, especially in the rainy ter." City planning engineers in Guaya-estimate that at least half the city's popu-n lives in such areas as this.

he Barrio de los Pobres is a Type A . It is inhabited by strangers who hope to and who see themselves held back by insurmountable barriers. Unlike Quito, ador's capital (where racial distinctions made much of) in Guayaquil, mestizos Indians who have been assimilated play ortant roles in the life of the city.

he institution of the "relleno" — the fill-in of the marshes — is a key indicator ne nature of the slum process in Guaya-. The slum dweller builds his bamboo k as high as the possible level of the "relleno," whenever he or the city can do the filling-in.

At the inner edges of this Barrio of the Poor and along its principal streets can be seen the active phase of the transition process. Each week finds the filled-in areas pushed farther out. With "relleno" completed, those who live in the marginal areas rapidly bring their neighborhood up to standard, particularly as the city paves the streets. Now more permanent buildings, largely of heavier wood than bamboo with facades of concrete, and in some cases of all concrete construction — the preferred building material — dot the reclaimed area. Out on the far edge of the slums the shack city extends itself farther west week by week as the squatters move in. Withal there is a sense of a victorious struggle against the elements.

For Guayaquil, slum formation is basic to the process of city growth.[6] The slum is a temporary home and a kind of a school house. If we classify the Barrio de los Pobres as a type A slum, we observe the paradox that the Type A slum is probably the least attractive of all slums because it is the most temporary. By the same token, it is the one most apt to "clean" itself up.

Admittedly this "cleaning up" process may not be apparent to the casual observer because it can take a number of forms depending upon the rate and nature of in- and out-migration. The slum families, for example, as they become adjusted to their new community and acquire the necessary social and

[6] Dr. Humberto Palacios, director of Economic Research Institute of the University of Guayaquil, argues that Guayaquil owes its growth entirely to this "pioneering" by the inmigrants. Not only is land for development provided but the process gives tone to the economic life of the city. Certainly Guayaquilenos are a very aggressive people.

economic resources may find it possible to leave their slum homes. If they do so and are followed in these same dwellings by other more recent arrivals with the same background, the dwellers have changed but the slums remain. On the other hand, if the growth of the city is rapid in a limited land area — as in the case of Guayaquil — the slum itself may disappear.

B

The first impression a casual visitor has of Boston's South End is one of rundown gentility. Along streets which often have parks down their center stand four- and five-storey brick houses with bowed fronts. There is a haunting charm to these elm-shaded streets. But the second impression is more lasting. Decay is everywhere. An air of hopelessness pervades the atmosphere.

This is a type B slum, an area in which the social residue live. While the South End does have Negro sections at its edges, it is largely a place for old, once well-to-do, poor, cast-off Bostonians. Here, too, live the shady characters, the prostitutes, the citizen at the margin of social respectability. Here and there occasional houses, churches and other public buildings indicate successful attempts to fight off the persistent down-grading. Except for the slum clearance project at Franklin Square, the cleaned out New York Streets area and the Hospital Zone, little if any new buildings have gone up in the South End for three generations.

The barriers to movement upward and outward for the South Ender are largely subjective. The poverty one sees in its streets appears to have some correlation with the large numbers of taverns. The South End does not expand physically. It slowly dies and with it what was once one of the best areas Boston.

The type B slum, the slum of "despair," frequently found in United States cities. Li the South End, they are often genteel in bac ground but are the present home of the cas off plus those who have been unable complete the process of acculturation. The are those for whom society has pity but h somehow been unable to help effectively. is unlikely that the major part of the i habitants of the slums of "despair" will ev leave them but there does appear to be steady movement into them.

It is difficult, however, to guess at wh rate these slums of despair increase or d crease. Notice that slums of type B are mo "attractive" than those of type A. They a much less livable; psychologically less pro uctive of progress.

C

Deceptively similar to Boston's South End Chicago's South Side — Bronzeville. Th area is a vast Negro slum which is continua pushing out from an area once bounded Roosevelt Boulevard on the north, the ra road tracks at 63rd Street on the south, Wen worth Avenue on the west, and the all-whi enclaves of Hyde Park and Kenwood on t east.

Social barriers against Negroes need detailing here, but that these barriers can overstressed is apparent in Bronzeville. Chi ago Negroes have had a reputation for aggre siveness. They have established newspaper magazines, insurance companies, factori and the like. In fact, "hope" is evident ever where.

Unlike Guayaquil, however, there is r easy transition in Bronzeville into bett ghborhoods. Bronzeville in its growth es little clue to the life and growth of the ;antic city of which it is a part. Forced develop their own subculture, Bronzeville's abitants are building their own city.

Type C slums, of which Bronzeville is an ample, are like type A with an important ference. While they are, to be sure, the m of "hope," homes of the stranger, their abitants have subjectively as well as ob- tively effective barriers to acculturation. ey do not belong to an escalator class. Yet se slum dwellers are "strivers." Evidenc- above average mental and social talents, they acquire a permanent foothold in the y into which they have come, they are more ely to reclaim their neighborhoods, if they a (as in Baltimore) or if they can't they push toward better neighborhoods. One aracteristic needs stressing, since type C m dwellers have substantial difficulties to ercome, they are likely to be militant along h their aggressiveness. Occasionally, this itance may manifest itself in a growth of nomic crimes — gambling, prostitution, otlegging, et cetera. Some of these activi- s are obviously wealth-producing and act pressures toward invasion of better neigh- rhoods.

D

Lima, Peru is a city of magnificence. But away from the charm of Avenida Are- pa, the impressive Miraflores and San dro districts as well the beauty of Plaza de mas, is the infamous Ciudad de Dios, an lian slum. Unlike Guayaquil where racial rriers are for all practical purposes non- stent, Lima in a subtle as well as direct y bars the Indian from participation in its nomic and social life. The Indian coming down from the Altiplano finds his home beyond the city limits in shacks built of what he can find. Though Lima grows, its growth neither benefits him nor affects him. Immigrants from Asia and Europe as well as elsewhere in Latin America, provided only that they are not Indian in culture, find more ready acceptance into Lima's life than he does. Not wanted, unaggressive, he sits in despair at the city's gates. Physically near, he is actually as far from social integration as he was in his mountain village.

The type D slum, of which Lima's City of God is an example, is even more a slum of "despair" than type B, for in addition to all the subjective disabilities characteristic of the dwellers of the type B slum, those of type D suffer because of their different color, religion or race. For them society has little sympathy. The community has made no provision for housing these people and does not intend to absorb them. There is literally no other place to which they can go. In the words of the Negro spiritual, "there is no hiding place down here."

Can the theory of slums we are outlining here be of help in setting forth a policy for slum elimination? We think so.

In underdeveloped and developing countries, slums of type A are quite evident as indeed they were in pioneer America. However disturbing these slums may be to the sensibilities of the casual observer, they are serving a necessary purpose. To be sure, better organization might reduce the need for this type of slum but it is not likely that developing nations will have either the resources or the governmental techniques which will permit a less disturbing way of handling the movements of laborers into the cities.

The advisor on economic development should be cautious in recommending that same form of solution for growing nations and their slums which he has seen tried at home. Perhaps a good example of the best form of slum elimination and even avoidance in an underdeveloped nation is that tried in the Puerto Rican city of Ponce, where basic sewage and water facilities along with sidewalks and macadamized pavement have been provided in slum areas as a part of the slum dwellers' own efforts to upgrade their neighborhoods. This implies a program of education, community organization as well as slum clearance.

South Africa's racial areas program, whatever may be said against it by those who quite rightly oppose residential segregation, is an attempt to find a rational solution to the helter-skelter growth of slums at the edge of the city. In Durban, Port Elizabeth and Johannesburg, among other cities, native "locations" have been bulldozed and have been replaced by carefully planned native townships. One of the world's most famous slums, Cato Manor, is now gone from the center of Durban.

Slums of types A and C (the slums of "hope") are self eliminating if the society has the time to wait. How fast they disappear depands upon the rate at which society absorbs the stranger to it. In West Germany we have an example of planning for housing of East Germans and of absorption which has prevented slum growth. However, the experience of the United States, Argentina and Canada suggests that the rate of absorption also depends upon continuing economic growth. The higher the demand for labor continues to be — assuming that there is no great self-impelled migration — the more rapid the integration of the stranger. But even if the self-impelled migration were large, for integration to proceed it would only be necessary that economic growth and absorption be faster.

While the slums of "hope" may pass away with economic growth, the same cannot be said for types B and D. To the extent that poverty and lack of "ability" become socially synonymous, the problem of the elimination of the slum of "despair" becomes more and more like that of a disease requiring a therapy which we have not yet worked out. Clearly evident in the United States is the fact that in many of our slum clearance projects we have tended to segregate types B and D people. If in any one generation there are the seeds of improvement of the next slum clearance — should it continue to produce the same sorts of results — may deter the elimination of slums. This is the paradox of today's efforts.

Conclusion

Some types of slums persist because they are an index of a paradox. Rising standards of living are accompanied by rising standards of ability and competence. In the United States poverty has become a term which describes the condition of a class more and more composed of the "incapable". These are the people who because of society's standards for entrance into job opportunities have not been integrated into full participation in the economic life of the community. How to provide for these unfortunates lest their presence yield a costly dividend of crime and disease remains the problem of highly developed society.

This slum problem is, however, distinct from the problem of earlier and less developed days and countries. In those distant days and lands, slums were and are an index of growth and of unabsorbed immigration to cities. Their presence is and was a sign of economic health though this need not mean that careful social organization might not ıke them largely unnecessary. Such a level organization implies a stage of develop-ɛnt not yet reached in most countries which ɛ yet developing.

Clearly outlined in this theory of slums is role of caste. That this caste membership ɛd not imply slum residence should be evi-ıt but it is one of the factors which provides a differential rate of absorption as be-een castes. These differential rates of ab-ption have led to lingering slums of "hope" ıong minority groups in the United States. ese, however, are disappearing. Ultimately problem of caste as a determinant of ms coalesces with that of "ability" and be-nes part of the same force at work in slum rsistence.

PENDIX

Migration and the Rate of Absorption

sume that at any moment of time the level population of a city of substantial size is ady. It is not increasing either from natural wth or from in-migration from the city's terland. Assume, moreover, that employ-nt is full. There is no excess demand for or.

In the hinterland, assume that there are no tors at work which would induce out-gration other than an excess demand for or in the city. To be sure, these assump-ns make in-migration a demand phenom-on. However, there is no neglect of the ply side. Rather supply enters into the alysis rather oddly. For one thing, if as is t of our tradition, out-migration depends on non-economic motivation, we can not oduce a normal supply function. For an-er, since our concern is with the impact of gration upon the formation of slums, we ɛd only to observe that migration will probably not be directed toward those cities where no economic opportunity exists.

An excess demand then for labor is the measure of existence of that economic opportunity and leads to in-migration. The excess demand leads at full employment to an increase in the city's going wage but, more important, given different rates of economic growth as between the city and its hinterland, to a widening of the differential between wages levels in the two market areas. It is in fact this growth in the differential which becomes the economic incentive directing migration toward the city.

Now we are ready to consider the nature of the migration. It is likely that, given the unsettled labor market in the hinterland plus the accumulation of non-economic motives, once the in-migration is induced to begin, it will exceed the actual initial demand. The city will be faced with absorbing the newcomers.

In addition, of course, to the excess supply of labor now created and the absorption problem thus inherent, there is the problem of the "quality" of the migration. It can be assumed that the migrants will be divided into those who can readily be absorbed into the labor force, provided such opportunity be available, and those who will not be readily absorbed because of the lack of "ability" of some of the migrants.

If there were no such division by reason of ability, then the rate of absorption would depend upon the rate of economic growth or what is the same thing, the rising level of average income. But if this rate of economic growth be slow — slow relative to the flow of in-migration — slums will appear. These will be slums of hope because in them will live a potentially utilizable labor force. These "temporary" slums will be an index of the rate of economic growth and the excess supply of labor.

The persistence of slums of hope will presumably depend upon the differential rates of growth of the cty and its hinterland as well as upon the internal rate of growth of the city alone. They will disappear as demand for labor moves toward equality with the supply. The slower the rate of absorption, the faster the rate of slum formation spreads blight across the city landscape. As growth in the hinterland — possibly a spillover from the city — proceeds, it will tend to counteract the non-economic motivation to out-migrate. The differential declines, in-migration slows down and absorption steps up. Internally, the city rate of growth insures that jobs are provided for the newcomers. Slums disappear.[7]

II. The Poor and Their Replacement

There are, however, ability hurdles which the in-migrants must surmount. A significant proportion of the newcomers will find or will react such a way as to indicate that they understand that little chance exists for full integration into the community. These are

			SLUM DWELLERS	
	P − ‖ CITY POP.	F + ‖ THE FULLY UTILIZED	H + ‖ SLUMS OF HOPE	E ‖ SLUMS OF DESPAIR
EXCALATOR CLASSES	P_a −	F_a +	A +	B
	+	+	+	+
CASTE BARRIER				
NON-ESCALATOR CLASSES	P_n −	F_n +	C +	D

Figure 2.

the "incapable," the un-utilized or underutilized.

To be sure, if the rate of growth is slow and if the proportion of "incapables" to the total flow of migration is low, then in the early period of in-migration, poverty arising from this source will be largely an economic phenomenon. But as absorption proceeds with ability barriers remaining at previous levels, a larger and larger proportion of slum dwellers (the "poor") will be characterized by a lack of "ability." In fact, if economic growth permits, absorption will finally have reached some maximum possible level and the "poor" and the "incapable" will be the same persons. Lacking any hope of integration, they inhabit the slums of despair.

It should be observed that some of the "incapable" may have come from the present population of the city. These are those who have "fallen" into poverty by reason of a lack of ability or who have been "cast off."

What can we say about the persistence of the slums of despair or about their possible rate of growth? In this case we can not rely upon the rate of economic growth, a rising level of income, or full employment to eliminate these slums.

To the extent that ability, however defined is normally distributed — a likely hypothesis — it may be expected that with a fixed ability

[7] In Figure 2 a diagram of the rate of absorption and its interaction with the formation of slums is presented. If we begin at OP_1 (equal to P_1F and to OF) where we have full utilization at a population equilibrium and observe what happens as population increases through migration from P_1 to P_2 to P_3 to P_4, we can trace out the genesis of slums. During period 2, in-migration occurs (equal to FA_2). B_2F is that proportion of the in-migration which is absorbed. The slope of $FB_2B_3B_4$ is the rate of absorption. By period 3 the rate of absorption has become parallel to the rate of full absorption indicating that a given absolute number of the in-migrants have found it impossible to get over the ability barrier. The gap A_2B_2, A_3B_3, A_4B_4 represents slums of despair which persist. The gap B_2F, B_3F represents slums of hope which finally disappear. This model assumes one "shot" of in-migration.

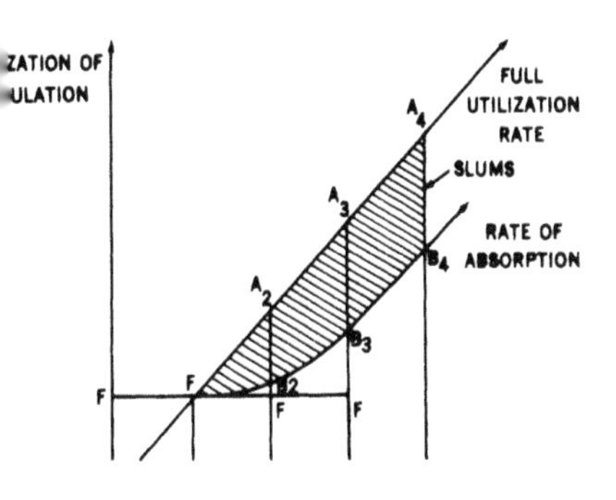

re 3.

dle, slums of despair will not only persist will grow at about the rate of growth of general population, assuming no further nigration. If in-migration continues it is sible that slums of despair may grow er than the general population. One of reasons for this likelihood is that society tend to impose a higher standard for ering than for remaining among the ized. It is also likely that the very growth he economy which makes possible absorp- brings with it a rising standard of com- nce — a level of ability — making a ger proportion of the in-migrants "in- able."

Income as a Determinant of Slum mation

night be argued that income should have n introduced at an earlier stage in the lysis but its consideration has been post- ed to now to permit a detailing of the role ability."

Assuming that ability is normally distri- ed is rather different from assuming that ome is normally distributed. Certainly in earlier stages of economic growth it is evi- t that income distribution is decidedly wed to the right. But there is also evi- ce that suggests a tendency for income distribution to move toward normality as income levels increase.

Just as there is an ability cutoff or barrier, so society has an income cutoff or barrier below which we have poverty. In earlier stages of economic growth, mean ability levels will lie considerably above mean income levels, so that the poor in income terms include many whose ability levels are above the ability cutoff. But as the mean income level shifts upward the income and ability cutoffs tend to become synonymous. To the extent that this becomes true, poverty and lack of ability become one and the same thing. This is another way of saying that a low level of income does lead to poverty and slums but that a rising income is limited in its ability to remove slums if ability barriers exist.

IV. Escalator and Non-escalator Classes

Up to now we have assuming that migration, ability and income growth are the factors which affect the rate of slum formation. So they do but it needs also to be pointed out that there may be differing rates of absorption within the city's population due to the presence of caste lines.

What is implied by caste is that, even if ability, however defined, did not act as a barrier to integration and full utilization, there would still be a third barrier. What is true is that the city will have divided its job opportunities into two categories. The one category permits those who retain the jobs to move up the social scale — escalation. The other category includes those jobs whose retention does not necessarily permit such escalation. This third hurdle or cutoff differs from the first two — ability and income — in being horizontal. It cuts across income and ability groupings.

V. The Basic Matrix.

We are in a position, now that we have the variables defined, to consider the basic matrix which underlies our model of slum growth. In Figure 3 we can distinguish two vertical barriers — income and ability. The income barrier separates the fully utilized from the poor and depends for its position upon the level of economic growth. The ability barrier divides the poor into two groups, those who reside in slums of hope and whose absorption and ultimate utilization depend upon the rate of economic growth and those who reside in slums of despair and who will not be fully utilized regardless of the rate of economic growth at present ability standards.

The horizontal barrier — the caste barrier — separates each grouping into two categories each. For each escalator group there is a parallel non-escalator group. For the non-escalator groups a different rate of absorptio applies, as well as a different level of utiliz tion and perhaps even a different and high cutoff for ability. Economic growth has weaker effect below the caste barrier in brin ing about the absorption of the capable. Th result is that poverty and slums of despa remain a more serious set of problems th for the escalator class.

It is, of course, possible to conceive of model in which the caste barrier is sufficient high that economic growth has practically n effect on the non-escalator class. *It is al* possible to conceive of a caste barrier whic is less effective at the upper end than at th lower. The one makes the model more near applicable to the Indian countries of Sou America, for example, and the other aids understanding better the present picture the United States.

he Urban Snow Hazard
n the United States

38

ohn F. Rooney, Jr.

aditionally the mid-latitude American city s coped with snow and ice in an organized t inefficient manner. Even in this age of loping technological advance, most of our rthern cities annually experience the crip-ng impact of at least one severe snow-rm. The brunt of the disruption, ironically, curs in areas that are "well prepared" to ndle any snow emergency. The continuing nd toward urban sprawl has been a major tor in accentuating the difficulties that m from an occasional snowstorm. Dis-nces separating urban dwellers from their eryday affairs and transactions have gthened, and dependence on both private d mass transportation facilities has in-ased. By introducing snow or ice into an ban setting with hypersensitive movement tterns, any form of chaos may be pre-itated.[1]

The present study is designed to assess the pact of snow in urban areas, using several lines of investigation. Snow's disruptive effects on man are analyzed, with an emphasis on the identification of the critical physical-environmental variables (amount and kind of snow, wind, temperature, terrain, and so on). Then the role of community adjustment and adaptation is examined, and, finally, attitudes concerning the snow hazard are probed, by means of interviews, to gain an understanding of the adaptations and adjustments that are characteristically made. The study focuses on seven selected cities, but the findings presumably have much wider application.

THE SNOW HAZARD

The snow hazard may be defined as comprising all the perils that snow and ice present, both in themselves and in association with other weather conditions. The hazard is also influenced by the nature of the terrain and the kinds of road-surfacing materials used in the area in question.

Most of man's activities are to some degree sensitive to weather. Certain of them are highly sensitive to cold (highway construction, outdoor painting), some to wind (struc-

Quite apart from snow and ice, consider the effect of an accident on an expressway in Chicago, New York, or Los Angeles, especially during rush-hour periods.

apted from *Geographical Review* (October, 1967) pp. 538-59. Copyrighted by the American Geographical
ciety of New York. Reprinted by permission.

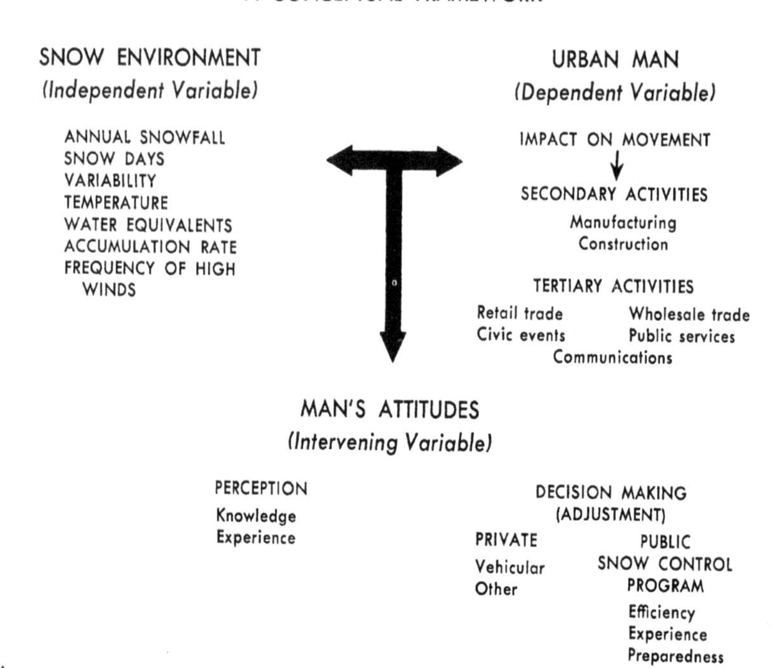

Figure 1.

tural-steel erection, roofing installation, recreation), and many others to precipitation, especially snow.[2] Transportation is the activity most critically affected by snow, but a host of others also suffer, including construction, merchandising, manufacturing, agriculture, power supply, communications, recreation, and public-health and safety services. The catalogue of problems caused by weather provides ample evidence of the disruptive impact of snow. It is noteworthy that snow cover is cited as a hindrance to transportation more often than any other meteorological phenomenon, and that surface icing and glaze are considered to be only slightly less severe.

The snow hazard has a number of important implications for urban geography. Perhaps the most notable is the impediment it poses to spatial interaction, both within the city and between the city and its tributary area. Normal movement — for example, the journey to work, shopping trips, and travel to participate in recreational activities — is often changed or curtailed. As many individuals rearrange their plans and patterns of action, the density and direction of traffic flow are also altered. In the snow hazard's most extreme form it may sever tributary connec-

[2] See, for example, J. A. Russo, Jr., K. Thouern-Trend, R. H. Ellis, and others, *The Operational and Economic Impact of Weather on the Construction Industry of the United States* (Hartford, Conn., Travelers Research Center, Inc., 1965), pp. 11-17.

[3] R. R. Rapp and R. E. Huschke, *Weather Information: Its Uses, Actual and Potential* (Santa Monica, Calif., The Rand Corporation, 1964), p. 31.

ns for extended periods, cutting off supply d distribution lines and thereby resulting emergency situations.

From an economic standpoint snow may luce heavy financial losses. The cost of mbating snow and ice, fixed capital and erhead costs for schools, factories, and res, and damage to property are some of direct losses. A diversion of expenditures also a common economic response; money at might have been spent on clothing or reation may be reallocated to snow-control uipment for homes and cars, for instance.

Some of the myriad relationships that exist tween "urban man" and his snow environnt are depicted in Figure 1. This diagram empts to portray the typical urban dweller nfronted by his snow environment, which ludes not only snow but the associated nditions (wind, air temperature, and other ms of precipitation) that often affect the gree of difficulty he experiences. Within s framework, man's accumulated contacts th snow hazards will influence the action takes. His decisions will be affected by the d of adjustments he has made to counter the hazard, both independently and as a rt of the total urban population. These adtments are influenced in turn by his attiles concerning snow and ice, which have en shaped by his perception of these enomena.

SESSMENT OF DISRUPTION

e investigation of the disruptions caused snow covers a period of ten years, 1953-63. This period is long enough to provide adequate sample of conditions and recent ough so that a reasonable amount of inmation was available.

Seven cities served as the basic laboratory which the effects of snow were examined. Cheyenne and Casper, Wyoming, and Rapid City, South Dakota, were the pilot sites. After testing for relationships between disruption and the snow environment in these cities, four more sites — Green Bay and Milwaukee, Wisconsin; Muskegon, Michigan; and Winona, Minnesota — were selected for study (Figure 2).[4] In addition to being in another section of the country, they represent different types of snow environments and public adjustments. Man's attitudes concerning the snow hazard and his subsequent adjustments to it were investigated in each city, but extensive interviewing was confined to Cheyenne, Casper, Rapid City, and Winona.

Six of the cities studied are of medium size, ranging in population from 25,000 to 65,000. One larger place, Milwaukee (*ca.* 750,000), was included so that the measures developed could be tested at a locale of authentic urban stature. Medium-size cities were selected for several reasons. It was felt that cities of this size would provide a less complicated setting than larger cities, and, in fact, they proved to be excellent laboratories, though their problems do not approach those of huge metropolitan areas such as New York, Chicago, and Boston. The seven sites chosen for study were large enough to possess a nearly complete span of urban functions, yet small enough to permit a careful analysis of snow-caused disturbances. At the same time, distances between sections of the cities were long enough to augment disruption when snow and ice were present.

METHODS OF INVESTIGATION

Information concerning disruption was taken

[4] Figure 2 also shows the location of ten other sites that were investigated more generally to test the significance of the moisture content of snow.

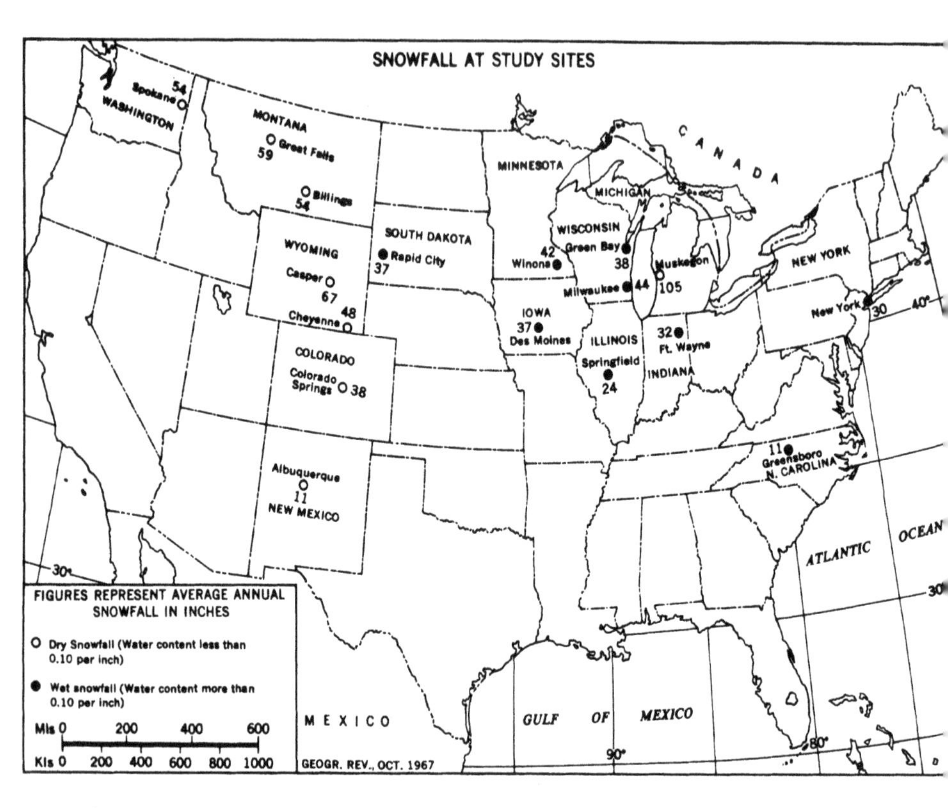

Figure 2.

chiefly from daily newspapers and public records.[5] From this material it was possible to classify the impact of all snow days[6] unde consideration. Each snow day (event) wa rated as to severity of impact within a hier archy or scale of disruption.

Some measure of data reliability wa thought desirable. The use of newspapers t identify trends and to form a basis for gen eralization falls under recognized researc techniques of "content analysis" — that is the categories of analysis used to classify th content are clearly and explicitly defined s that other individuals can apply them to th same content to verify the conclusions; th

[5] The major newspapers used were the *Wyoming Eagle* (Cheyenne), the *Casper Tribune-Herald*, the *Rapid City Daily Journal*, the *Winona Daily News*, the *Green Bay Press Gazette*, the *Milwaukee Journal*, and the *Muskegon Chronicle*. The public records that proved most helpful were police accident reports, parking-meter revenue summaries, and street maintenance reports. Telephone company files and information compiled by public utility firms were among the private sources consulted. The most dependable sources were newspapers, police and city records, and the data on power failures kept by public utility companies. School attendance records were too general for more than token use, and since the telephone company records were available only on out-of-city calls, they proved to be of limited value.

[6] A "Snow day" is one on which one inch o more of snow is recorded. In the case of storm that lasts more than one day, the entir snow period is considered as one snow day.

lyst is not free to select and report merely t strikes him as interesting, but must hodically classify all the relevant material his sample. By using this technique to lyze communications media, much of the oiguity is eliminated.

nterviews with private individuals and proprietors of commercial establish-its were used to probe attitudes and ad-ments concerning the snow hazard.[7] The stions were designed to reveal the re-ndents' attitudes toward snow and ice and letermine their ability to cope with the ard personally and as members of a group. chants were asked to comment about the ct of snow on their business, and to dis- the municipal snow-control program. rviewing was done on an area-sampling s. A total of 255 personal interviews were ducted, and 61 firms made up the com-cial sample.

TERNS OF SNOWFALL

ore analyzing the disruption attributable now, it is necessary to examine the snow erns of the seven cities. In terms of mean ual accumulation (ten-year averages),[8] skegon leads with 105 inches; Casper has Cheyenne, 48; Milwaukee, 44; Winona, Green Bay, 38; and Rapid City, 37. The western cities are characterized by a con-ration of snowfall from December through March, with January the peak month. In the Rocky Mountain area and adjacent plains, most of the fall comes later, and March and April are usually the snowiest months. In general, snow is wetter in the Midwest and in the East than it is in the Rocky Mountain region. The water equivalent of ten inches of snow at the Wisconsin, Minnesota, and South Dakota sites is approximately one inch, while at Casper and Cheyenne fifteen to sixteen inches of snow are required to produce one inch of water.

GENERAL IMPACT

Most of the difficulties caused by snow stem from disruption of transportation facilities. Even small accumulations may effectively curtail movement and contribute to accidents. Disruptions of retail trade, industrial production, school attendance, construction, civic events, and numerous other activities can be traced largely to the impairment of movement. In the disruption model developed here, highway transportation constituted the major link between the cities and their tributary areas. Although rail transportation is also vital, its functions are more important on a long-term basis; curtailment of rail facilities for a day or two does not generally produce the repercussions associated with disruption of highway transportation. The airlines that serve the study sites currently play only a minor role in the total transportation pattern.

To assess accurately the troublesome effects of snow in urban areas, some kind of categorization is desirable. The hierarchy of disruption presented in Table I rests on the assumption that snow unleashes its most damaging effects against transportation. The orders of disruption are ranked along a scale ranging from first order, the most severe, to fifth order, the minimal, which designates assumed but unvalidated inconvenience.

ne questionnaires used for both types of inter-ew appear in Appendix B of the writer's doc-ral dissertation, "The Urban Snow Hazard: n Analysis of the Disruptive Impact of Snow-ll in Ten Central and Western United States ties" (Worcester, Mass., Clark University, 66); also available from University Micro-ms, Ann Arbor, Mich.

ne ten-year averages were within 2 inches of e long-term averages at all sites except Muske-on, which was 33 inches in excess of the long-rm average of 72 inches.

TABLE I

Hierarchy of Disruptions: Internal and External Criteria

Activity	*1st Order (Paralyzing)*	*2nd Order (Crippling)*	*3rd Order (Inconvenience)*
INTERNAL			
Transportation	Few vehicles moving on city streets City agencies on emergency alert, Police and Fire Departments available for transportation of emergency cases	Accidents at least 200% above average Decline in number of vehicles in CBD Stalled vehicles	Accidents at least 100% above average Traffic movement slowed
Retail trade	Extensive closure of retail establishments	Major drop in number of shoppers in CBD Mention of decreased sales	Minor impact
Postponements	Civic events, cultural and athletic	Major and minor events Outdoor activities forced inside	Minor events
Manufacturing	Factory shutdowns Major cutbacks in production	Moderate worker absenteeism	Any absenteeism attributable to snowfall
Construction	Major impact on indoor and outdoor operations	Major impact on outdoor activity Moderate indoor cutbacks	Minor effect on outdoor activity
Communication	Wire breakage	Overloads	Overloads
Power facilities	Widespread failure	Moderate difficulties	Minor difficulties
Schools	Official closure of city schools Closure of rural schools	Closure of rural schools Major attendance drops in city schools	Attendance drops in city schools
EXTERNAL[a]			
Highway	Roads officially closed Vehicles stalled	Extreme-driving-condition warning from Highway Patrol Accidents attributed to snow and ice conditions	Hazardous-driving-condition warning from Highway Patrol Accidents attributed to snow and ice conditions
Rail	Cancellation or postponement of runs for 12 hours or more Stalled trains	Trains running 4 hours or more behind schedule	Trains behind schedule but less than 4 hours
Air	Airport closure	Commercial cancellations	Light plane cancellations Aircraft behind schedule owing to snow and ice conditions

[a]Warnings are the key to this classification. They provide excellent indicators because they are wide publicized.

TABLE I (*Cont.*)

4th Order (Nuisance)	5th Order (Minimal)
y mention	No press coverage
ffic movement lowed	
	No press coverage
casional	No press coverage
	No press coverage
y mention	No press coverage
y mention	No press coverage
y mention	No press coverage
	No press coverage
y mention or example, slippery in spots" warning	No press coverage
y mention	No press coverage
y mention	No press coverage

Urban snow disruption is of two kinds: internal, when interchange within the city itself is hampered; and external, when conditions affect the relationships between a city and its tributary area. First-order disruptions may occur in either or both of these situations. So far as internal activity is concerned, the complete restriction of mobility is normally the most serious problem that can be attributed to snowfall, since most functions characteristic of urban areas require movement from one section of the city to another (journey to work, shopping, appointments, and so on).

An inspection of the effects of snow situations on other forms of urban and tributary activity serves to confirm transportation curtailments. For example, if schools were dismissed and a number of business establishments closed, we have verification that traffic flow throughout the city was greatly impeded. In the same vein, the postponement of athletic contests between teams from the city schools and those from other institutions in or near the tributary area provides additional evidence. Another inconvenience associated with snow concerns electric power and communications. On rare occasions a wet clinging snow may result in widespread breakage of cables and wires. In some cases electricity and telephone service may have been disrupted for extended periods, in which event first-order categorization seems warranted.

The hierarchy of disruption is arranged largely on an economic basis. First-order disruptions generally, but not always, cost more than second-order ones, second-order more than third, and so on. However, cost is not always an ideal measure. Thirty accidents do not necessarily result in a greater monetary loss than twenty, and the infrequent fatality associated with a lower-order disruption may produce a more serious loss than might occur in some first-order situations, depending on the value placed on human life.

A combined internal-external first-order or "paralysis" disruption generally finds a community in a state that resembles suspended animation. Vehicular and pedestrian movement is at a standstill; most stores, schools, and offices are either empty or closed. Air, highway, and rail transport are severely hampered, and on occasion the city is completely cut off from its surrounding area.

Perhaps an example will best serve to illustrate both the conditions that may obtain in first-order disruptions and the operation of the classification system. In Cheyenne, snow began to accumulate at 7:00 p.m. on April 8, 1959, and continued throughout the following day, reaching a total of 8.4 inches.[9] Curtailment of internal and external movement was severe enough to merit a combined first-order disruption rating. The wet snow, accompanied by winds of 15 to 25 miles an hour, halted city bus service and brought private transportation to a standstill. Abandoned automobiles blocked streets throughout the city. Traffic in and out of Cheyenne was confined mainly to emergency vehicles, attempting to get aid to the more than three hundred motorists stranded on United States Highway 30 to the west of the city. All outbound air traffic was grounded, and no planes were able to land at the municipal airport. City and rural schools closed their doors, and retail trade was heavily curtailed. The ineptness of the Cheyenne snow-control program was all too evident on this occasion. Crews began the removal job the evening of the ninth, too late to combat effectively the impact of the storm. It was like beginning sandbag operations after the stream has crested.

[9] *Wyoming Eagle*, (April 9, 1959), pp. 1-3.

Classification as second- or third-orde disruption is based mainly on city motor vehicle-accident data and on reports on roa conditions in the tributary area filed by th State Highway Patrol. Other criteria includ the condition of rail and air transportation activity in the Central Business Distric traffic jams, postponements, and school attendance patterns. Disruption of the fourt order occurs when the impact of snow is no sufficient to cause a breakdown of the activi ties affected. A fifth-order classification des ignates an inconvenience so trivial as not t merit mention in the press.

ANALYSIS OF DISRUPTION

Disruption in the seven cities is summarize graphically in Figure 3. In general, lower order disturbances ("inconvenience," "nui sance," and "minimal") are the rule, an paralyzing disruptions occur only at intervals However, certain differences among the citie are conspicuous. The Midwestern communi ties tend to have a higher percentage of para lyzing and crippling situations than Caspe and Cheyenne do. Both Wyoming sites an Winona (for a different reason) demonstrat a greater clustering of disruptions at the lowe end of the scale. The profiles of the two larg est cities, Milwaukee and Muskegon, ar quite similar, and show greater concentra tions of first- and second-order situations Rapid City is also characterized by a highe proportion of crippling storms, though not o the paralysis variety.

Basically, frequency of disruption in creases with annual accumulation of snow (Table II). The relationship is not perfect but it is nevertheless significant. Muskegon which has recorded an annual average of 10. inches of snow, has experienced 206 higher order disruptions, or an average of 20.6

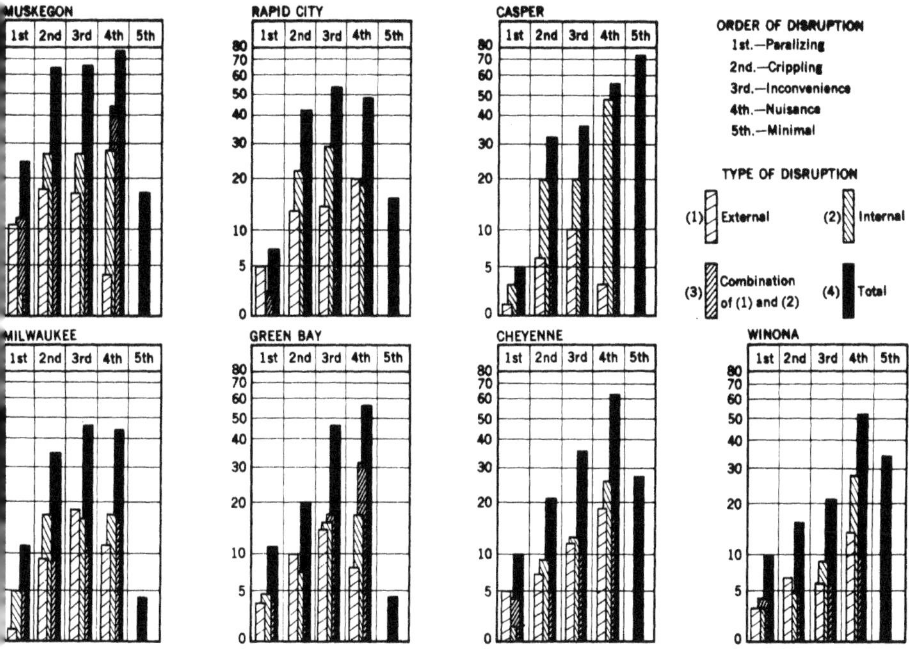

ure 3.

ar over the past ten years. Among the Mid-stern sites with snowfall averages ranging m 37 to 44 inches, the pattern of snow-ised inconvenience is similar, with Winona only exception.

The relationship between accumulation d disruption is less apparent when the stern sites are considered. Cheyenne (48 hes) averages 8.7 higher-order disruptions the normal year and Casper (67 inches) periences only 8.4. To search out the ises of this discrepancy, ten additional es were investigated, on a ten-year sampl- basis similar to that used for the original en sites. Western sites in the sample were okane, Great Falls, Billings, Colorado rings, and Albuquerque. To supplement Midwestern sites, New York City, Fort Wayne, Des Moines, Springfield (Illinois), and Greensboro (North Carolina) were selected.

The graphic comparison of disruption and the snow environment suggests that basic differences exist between the Western disruption patterns and those characteristic of the areas east of the High Plains (Fig. 4). Most of the Western cities experience fewer snow problems. This generalization applies regardless of whether magnitude or intensity measurements are used.

Why do these differences exist? It appears that the most important factor is the lower water content associated with the majority of snowfalls in the West. The drier snows generally present a less formidable barrier to movement and hence on the average tend to

TABLE II

Relationship Between Snow Environment and Disruptions of the First, Second, and Third Order

	Snow Environment		*Disruptions in 10-year Period*					
Sites	*Mean Annual Snowfall (in inches)*	*Snow Days per Year*	*No. of 1st, 2nd, and 3rd Order*	*Intensity per 10 Inches of Snow*	*No. of 1st and 2nd Order*	*Intensity per 10 Inches of Snow*	*No. of 1st Order*	*Intensity per 10 Inches of Snow*
	10-year averages							
Casper	67	18.5	84	1.25	36	.54	6	.09
Cheyenne	48	12.3	87	1.82	40	.83	14	.29
Spokane	54	13.5	129	2.39	70	1.30	25	.46
Great Falls	59	14.0	106	1.81	77	1.31	2	.03
Billings	54	13.2	99	1.83	55	1.02	15	.28
Albuquerque	11	3.3	34	3.27	19	1.83	5	.48
Colorado Springs	39	11.1	76	1.95	37	.95	11	.28
Muskegon	105	20.2	206	1.96	120	1.14	37	.35
Rapid City	37	11.2	119	3.21	56	1.51	9	.24
Milwaukee	44	10.2	118	2.68	55	1.25	16	.36
Green Bay	39	10.8	101	2.59	38	.97	15	.38
Winona	42	13.1	62	1.48	35	.83	14	.33
Springfield	24	9.1	76	3.18	44	1.83	16	.68
Des Moines	37	10.5	105	2.84	57	1.54	15	.41
Fort Wayne	32	9.4	119	3.72	72	2.25	16	.50
Greensboro	11	3.4	54	5.04	48	4.48	21	1.96
New York	30	7.2	96	3.20	71	2.36	42	1.39

cause less disruption.[10] However, this is not to say that in a given situation (for example, abundant snow buffeted by blizzard-force winds) dry snow is incapable of producing as much difficulty as the wetter variety.

A second factor that reduces disruption at the Western sites is the way in which the hazard is perceived. The view that regards snow as an element that must be coped with by the individual has produced a much more comprehensive range of personal adjustments than are commonly found farther east.[11] Suc adjustments are particularly effective at re ducing disruption of the second and thir orders. "Western perception," on the othe hand, has contributed to the pathetic inepti tude of public adjustment in the region. Thi inability of the public sector to react has cre

[10] Dry snow refers here to snow with a water content of less than one inch per ten inches of snow.

[11] A considerable percentage of those interviewe in the West carried emergency provisions suc as canned food and blankets in their aut mobiles. Also, many more Westerners tha Easterners equipped their cars with snow tir or chains, and made greater use of profession advice provided by the United States Weath Bureau and the State Highway Patrol.

TABLE III

Disruption in Relation to Physical Variables

Sites	Criteria (Averages)	*Order of Disruption* 1st	2nd	3rd	4th	5th
skegon	Depth (in.)	14.0	5.7	3.7	2.0	1.3
	Wind (mph)	15.2	12.1	11.2	11.7	13.8
	Water content[a]	16.1	16.2	16.2	12.1	15.2
	Number[b]	25	53	47	57	17
id City	Depth	8.8	2.3	2.1	1.5	1.4
	Wind	24.8	15.5	11.7	13.3	15.8
	Water content	10	9.1	10.1	9.6	10.1
	Number	7	33	35	22	11
vaukee	Depth	12.0	4.2	2.7	1.6	1.4
	Wind	23.5	16.6	13.3	14.3	7.5
	Water content	11.3	10.9	11.7	12.1	12.2
	Number	11	26	34	29	4
en Bay	Depth	6.9	2.9	2.7	1.8	1.4
	Wind	16.7	15.1	11.4	9.8	8.3
	Water content	10.5	11.7	10.6	10.3	10.2
	Number	11	14	33	42	4
ɔer	Depth	7.7	4.2	3.1	3.1	1.7
	Wind	15.9	12.1	11.8	12.9	12.2
	Water content	13.7	14.9	16.8	13.6	14.8
	Number	5	27	29	57	69
yenne	Depth	8.8	5.0	3.3	2.0	1.5
	Wind	25.6	15.6	14.3	13.7	13.6
	Water content	10.9	12.7	16.2	13.9	12.3
	Number	10	16	33	36	27
ona	Depth	10.7	5.8	3.6	2.8	1.7
	Wind					
	Water content	11.6	11.2	10.2	12.9	12.6
	Number	10	11	19	44	34

ount of snow equivalent to one inch of water.
resents only the highest-order disruption for any snow day. For example, if a snow day is rated 1st
er internal and 3rd order external, only the former is included in the sample. The fact that combina-
disruptions are not counted twice in these computations also decreases the number in the sample.

a greater vulnerability to severe snow
litions than exists in the East. As a re-
the Western cities are no better than their
ern counterparts in reducing first-order disruption, and under most circumstances their recovery period is considerably longer.

The graphs also suggest that disruption, though it increases with annual snowfall, does

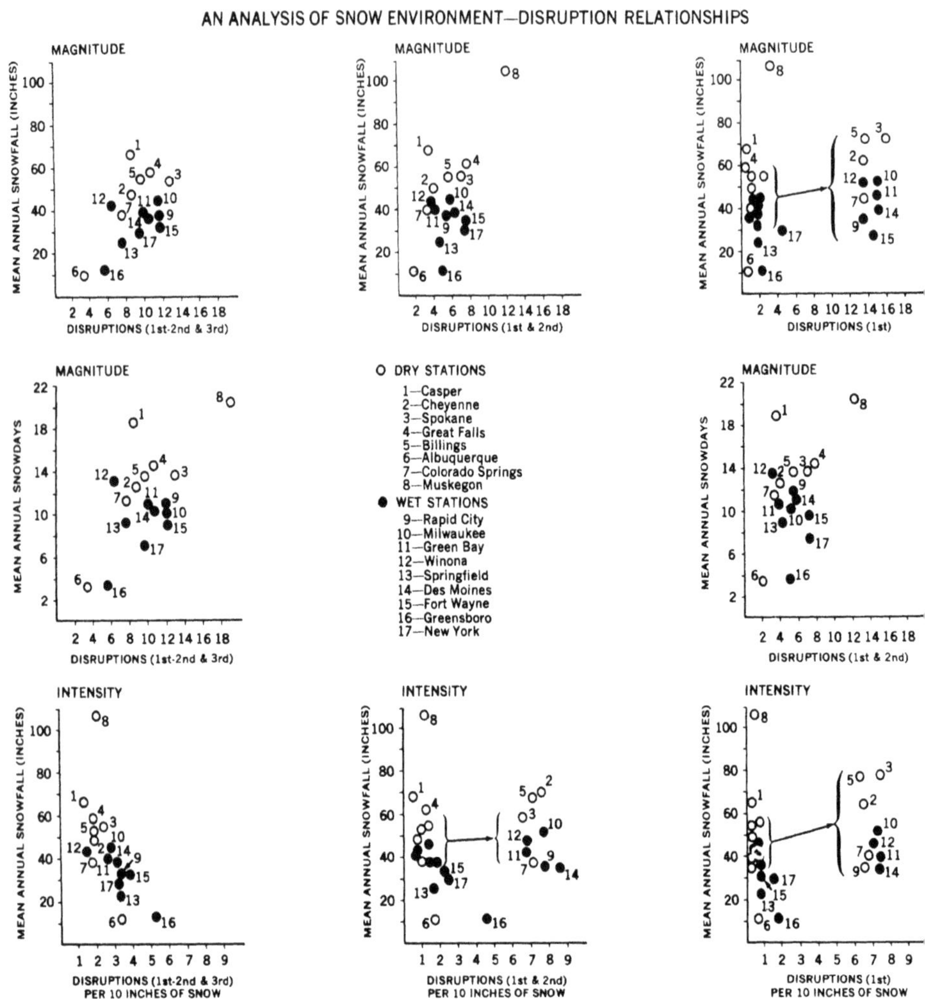

Figure 4.

so at a diminishing rate. Inversely, the intensity of disruption decreases with increasing annual accumulations. The matched pairs of Alburquerque-Greensboro and Cheyenne-Milwaukee are clearly demonstrative of such a pattern. A larger sample of north-south cities would probably further validate this relationship.

SPECIFIC CAUSES OF DISRUPTION

The relationship between disruption a snow depth, water content, and wind in inc vidual storms at the seven original sites presented in Table III. The data illustrate strong correlation between depth of snow a curtailment of human activity. Storms th resulted in first-order situations almost inva

TABLE IV

Effect of Wind in Promoting Disruption

Sites	Wind Velocity	Order of Disruption			
		1st	2nd and 3rd	4th and 5th	Total
ıskegon	Data incomplete				
pid City	≥ 15	9	48	20	77
	< 15	0	62	54	116
lwaukee	≥ 15	13	58	21	92
	< 15	3	44	38	85
een Bay	≥ 15	14	32	15	61
	< 15	1	54	75	130
sper	≥ 15	4	26	43	73
	< 15	2	52	87	141
eyenne	≥ 15	13	35	39	87
	< 15	1	38	67	106
nona	Data unavailable				

.y registered accumulations of five inches more. Only on rare occasions — for in-nce, a sleet storm, drifting of snow on the und from a previous fall, extremely high ıds — did lesser accumulations produce lespread havoc.

Most of the snowfalls that caused first-ler disruption accumulated during twenty-ır-hour periods. Such rapid falls generally ulted in conditions that were extremely dif-ılt to deal with. Only in Muskegon did a ıificant number of paralysis situations de-op over a period of two or more days; on eral occasions it snowed continuously for long as nine days, creating conditions of erity unknown at any of the other sites. wever, even during some of these pro-ged falls, crews were able to prevent com-te curtailment by maintaining operations und the clock.

Although deep snow and severe disruption strongly correlated, deep snow by itself ɜn tends to produce only moderate, and sometimes even minimal, disruption. Most accumulations exceeding five inches were accompanied by winds of more than fifteen miles per hour. However, on many occasions when winds were light and the snow was dry, the problems were substantially reduced.

Wind. A significant statistical difference exists between the impact of snow in association with winds of fifteen miles per hour or more, and that accompanied by winds of lesser velocity.[12] Nearly all first-order occurrences developed with wind velocities in excess of fifteen miles per hour (Table IV). On the other hand, lower-order disruptions char-

[12] For the purposes of this study a statistically significant association is one (measured by chi-square, representing the sum of the differences of the observed distribution and one that might be expected if there were no association between the given variables) that had only one chance in twenty or less of arising purely by sampling variability or by chance (.05 level of significance).

acteristically were associated with lower wind velocities.[13]

A reexamination of Table III reveals average wind velocities to be substantially higher in first-order situations than in lower-order disruptions. Winds of more than fifteen miles per hour are common in first-order disruptions at all sites, and exceed twenty-three miles per hour at Milwaukee, Rapid City, and Cheyenne. In addition, a precipitous drop in wind velocities is associated with second- and third-order cases at all sites, and particularly in these three cities.

Further evidence in support of the impact of high winds was obtained through personal interviews at four of the seven sites. Respondents were asked to select from a list of six types of snow conditions the two they felt presented the most serious hazard in their area. The responses are summarized in Table V. Snow in association with high winds was ranked as the most severe type of hazard in Cheyenne, Rapid City, and Casper, while deep snow in itself was given low priority. The respect afforded ground blizzards in those locations also demonstrates the role of wind. Such evidence, though not conclusive, does substantiate the data procured from newspapers and public records. It also indicates an awareness on the part of the respondents of what proved to be the most significant cause of disruption.

Water Content. Snow made heavier and stickier owing to relatively greater moisture content often tends to present a more formidable hazard, as we have seen. In Cheyenne it was found that wetter snow resulted in a considerably greater amount of disruption.[14] Although a statistically significant differenc between the impact of "wet" and "dry" snov was not characteristic of the other sites, i appears, again, that moisture content of snov is responsible for areal differences in disrup tion. The Midwestern and Eastern cities where higher-order disruptions are consider ably more frequent, experience snowfal which on the average is about 50 percent wet ter than that recorded at the Western sites

Temperature. The role of air temperatur is difficult to measure.[15] An analysis of high low, and mean temperatures provides littl evidence concerning the function of tempera ture in maximizing or minimizing disruptior Mean temperatures associated with all order of disruption did not vary more than four de grees at any of the sites.

Temporal Variation. Another element tha tends to complicate and obscure the relation ship between disruption and the snow envi ronment is the time of occurrence. Tempora variation as used here refers to the rate o fall and the time of day when it occurs. Th rate of fall governs the ability of snow-contrc operations to keep pace; the time of fall ofte means the difference between a fourth-orde and a second-order disruption. For example three inches immediately preceding the morn ing rush hour can cause great difficulty whereas the same amount accumulating dur ing the late evening or early morning hour might result in minimal inconvenience. Week

[13] Chi-square tests were significant at the .01 level of probability (that is, there is less than one chance in a hundred that the association is attributable to chance) at all sites except Casper and Muskegon.

[14] The chi-square test was significant at the .0 level of probability.

[15] At temperatures below 15°F salt has little effe on ice, and when the temperature approache 0°F even calcium chloride is ineffective. A ten perature hovering around the freezing mark critical, for variation in either direction ca make the difference between very slippery c merely wet pavements.

TABLE V

Ranking of Severe Hazards by Persons Interviewed

ites	*Number of Persons Interviewed*	*Number Making Estimate*	*Snow and High Winds*	*Snow and Sleet*	*Snow Over Ice*	*Ground Blizzard*	*Extremely Deep Snow*	*Snow and Extreme Cold*
ıeyenne	43	43	1	3	2	5	4	6
.sper	45	45	1	6	3	2	4	5
ıpid City	62	55	1	3	2	4	5	6
inona	21	20	2	1	4	6	3	5

d snows produce less internal disruption, t generally more external difficulties, and reverse is true of snows that occur during week.

IE FACTORS COMBINED

summarize the impact of snow and asso- .ted weather conditions on human activity, obability matrixes have been constructed the Western and Midwestern sites (Table). The data demonstrate the relative con- oution to disruption of snow depth, wind, ter content, and temporal variation.

Fundamentally, the matrix for the Mid- stern sites illustrates that the probability of ere disruption increases with depth. High nds are shown to be an important catalyst promoting the hazard at all snow depths. r example, the odds against a three-inch ow causing a crippling disruption are five one; add wind and they drop to two and a f to one. Wet snows are not consistent in omoting additional havoc in the Midwest, t this is because most snows there are fairly t. Expectably, weekend snows are gener- y less disruptive than their weekday coun- parts. In the West the role of depth and nd are also apparent from the matrix. Fur- rmore, wet snows appear to cause signifi- ntly more difficulty than dry ones do.

THE ROLE OF ADJUSTMENT AND ATTITUDE

Most cities that lie within the "snow belt" (the area north of 35°N, excluding the Pacific states) have some form of snow-control program. Expenditures for this service range from the $22 million spent by New York City in 1963-1964 to the few hundreds of dollars appropriated annually by many of the smaller communities in the southern part of the belt. The total expenditures for snow control in the United States can only be estimated. The American Public Works Association has been using a figure of $100 million a year, but considers this to be much too low, especially since salt costs alone run to $44 million.[16] The following statement indicates the difficulties encountered in compiling figures for snow removal costs: "One of the main things we uncovered in our study was the general lack of basic operating and especially cost data. Surveys we made were generally inconclusive because data received were not comparable. We obviously were comparing apples with pears. The fact that data on snow opera-

16 Written communication from Robert K. Lockwood, assistant director of Technical Services, American Public Works Association, May 4, 1965.

TABLE VI

Probability of Experiencing Disruption from Snow and Selected Associated Weather Conditions

Depth in Inches	No. of Snow Days	Depth Alone					Winds of ≥ 15 mph					Water Equivalents of ≥ .10/inch					Weekend Snows (time factor)				
		1st	2nd	3rd	4th	5th	1st	2nd	3rd	4th	5th	1st	2nd	3rd	4th	5th	1st	2nd	3rd	4th	5th
MIDWESTERN SITES																					
1 – 1.99	341	.00	.10	.25	.48	.17	.00	.20	.28	.44	.08	.00	.11	.30	.54	.05	.00	.08	.22	.48	.22
2 – 2.99	159	.02	.20	.31	.38	.09	.04	.32	.36	.28	.00	.02	.22	.39	.30	.07	.00	.21	.34	.40	.05
3 – 3.99	86	.04	.19	.44	.31	.02	.08	.41	.44	.07	.00	.06	.23	.39	.30	.02	.02	.16	.46	.34	.02
4 – 4.99	51	.04	.26	.39	.31	.00	.17	.58	.17	.08	.00	.08	.25	.37	.30	.00	.04	.29	.35	.32	.00
5 – 5.99	45	.07	.44	.44	.05	.00	.10	.70	.20	.00	.00	.12	.55	.30	.03	.00	.04	.45	.42	.09	.00
6 – 6.99	30	.20	.43	.23	.14	.00	.30	.40	.30	.00	.00	.27	.43	.20	.10	.00	.14	.47	.19	.20	.00
7 – 7.99	30	.27	.40	.30	.03	.00	.44	.38	.18	.00	.00	.38	.38	.21	.03	.00	.20	.42	.35	.03	.00
8 – 8.99	8	.13	.50	.37	.00	.00	.33	.67	.00	.00	.00	.25	.50	.25	.00	.00	—	—	—	—	—
9 – 9.99	12	.50	.33	.17	.00	.00	1.00	.00	.00	.00	.00	.60	.20	.20	.00	.00	.50	.25	.25	.00	.00
10 & over	65	.52	.42	.06	.00	.00	.64	.36	.00	.00	.00	.54	.42	.04	.00	.00	.46	.48	.06	.00	.00
WESTERN SITES																					
1 – 1.99	151	.00	.06	.11	.33	.50	.00	.08	.06	.44	.42	.00	.07	.21	.33	.39	.00	.05	.10	.40	.45
2 – 2.99	73	.03	.15	.22	.38	.22	.07	.11	.29	.32	.21	.05	.10	.15	.45	.25	.02	.08	.24	.34	.32
3 – 3.99	45	.02	.09	.39	.35	.15	.06	.18	.35	.35	.06	.00	.20	.20	.30	.30	.00	.15	.35	.40	.10
4 – 4.99	26	.04	.27	.19	.50	.00	.08	.42	.08	.42	.00	.00	.25	.50	.25	.00	.04	.18	.24	.54	.00
5 – 5.99	15	.20	.33	.27	.20	.00	.00	.38	.38	.24	.00	.25	.00	.50	.25	.00	.17	.33	.25	.25	.00
6 – 6.99	13	.00	.38	.38	.24	.00	.00	.20	.60	.20	.00	.00	.50	.00	.50	.00	.00	.20	.40	.40	.00
7 – 7.99	17	.18	.30	.30	.22	.00	.50	.33	.17	.00	.00	.38	.38	.24	.00	.00	.25	.38	.12	.25	.00
8 – 8.99	1	1.00	.00	.00	.00	.00	1.00	.00	.00	.00	.00	1.00	.00	.00	.00	.00	—	—	—	—	—
9 – 9.99	1	.00	1.00	.00	.00	.00	.00	1.00	.00	.00	.00	.00	1.00	.00	.00	.00	—	—	—	—	—
10 & over	13	.39	.39	.15	.07	.00	.67	.33	.00	.00	.00	.67	.33	.00	.00	.00	.25	.75	.00	.00	.00

s are limited is understandable. Snow is ted as an emergency and record keeping a secondary priority."[17] But costs are one measure of man's attempt to counct the snow hazard. Other important ts of control comprise organization, comnication with the public, coordination ng city agencies, and among state, city, suburban agencies.

t is impossible within the scope of the sent investigation to assess precisely the tive effects of public adjustment to the w hazard. However, by analyzing and paring the snow-control programs and disruption patterns of the cities studied, a nber of insights can be gained.

now- and ice-control operations at the western sites run the gamut from very d to marginal.[18] By present technological dards the programs of Milwaukee, Muson, and Winona rank high. The Green program is better than average; that of id City is definitely marginal. By Midtern or Eastern standards, neither Casper Cheyenne is unduly concerned with snow ice control.

ince the snow environments at Green Bay Rapid City are basically similar, it is sible to gauge the effect that adjustment on the impact of the hazard. Green Bay nds more money per capita, has an effit alert system, and has a well-organized se of action that calls for operations to n *during* the storm when accumulations h two inches or more. On the other hand, id City has no formal plan, and generally postpones action until after the snowfall has ceased. Moreover, the city's police department is apparently unwilling to enforce snow-emergency regulations.

The more efficient snow-control operation at Green Bay has resulted in holding down higher-order disruptions to 2.59 per ten inches of snow, as compared with the figure of 3.21 for Rapid City (Table II). In percentage terms, Green Bay experiences less than 80 percent as much disruption. Whether or not this reduction can be attributed solely to snow-control differences is a question that can be answered only by the investigation of snow problems at many additional sites, or perhaps by a concentrated economic analysis of one or two cities. It is interesting that nearly half of those interviewed in Rapid City considered the city program to be woefully ineffective. The majority were in favor of immediate improvement, another indication of the general dissatisfaction with the current quality of snow-control service.

Newspaper coverage of storms in Milwaukee and Muskegon revealed that their alert and well-organized public works departments minimized the impact of snow on numerous occasions. Many first-order situations failed to materialize in Muskegon owing to the diligence of the city and county officials.

The Winona program presents an opportunity to measure community attitudes toward the disruptive impact of snow. Snow control in that city was viewed as "excellent" to "very good" by more than 80 percent of those queried. In addition, fewer than 10 percent felt improvements in the program were warranted. If these views can be accepted as factual, and not simply as statements of community pride, it would seem that Winonans are satisfied with the present level of snow-caused inconvenience which their city experiences.

id.

he quality of a snow-control program can be aluated on the basis of expenditures, alert stems, deployment of men and equipment, ganization, public relations, and the degree of operation that exists among the various agenes necessary to its success.

The Winona case suggests that the "satisficer" notion is of considerable value in accounting for the curious relationship that exists between man and the snow hazard.[19] Perhaps the esthetic values of snow make us somehow reluctant to wage all-out war against it. If the "optimizer" approach were applied to the snow hazard, it would emphasize strategy designed to eliminate completely the negative impact of snow. Further evidence to support the applicability of the satisficer concept comes from the field of snow control. Innovations designed to improve snow control have lagged behind, and those that have emerged are being adopted very slowly. Essentially we are coping with snow in much the same way we did thirty years ago. Our attitudes are simply reflected in our actions. If demand existed for real innovations in snow control, they would be forthcoming. Most of the improvements developed thus far (radiant heat, snow melters, street flushing devices) have been rejected owing to their high cost and to the lack of knowledge concerning the losses attributable to snow.

Individuals spend considerably more to protect themselves and their property from the snow hazard than they spend as members of the public sector. Snow tires alone cost the citizens of Rapid City, Casper, and Cheyenne combined more than $600,000 annually.[20] Added to that are expenditures for tire chains, shovels, snow brooms and scrapers, sand, salt, and numerous ice-melting compounds. Since people are willing to funnel these amounts into personal snow control, it seems reasonable to suppose that they would fav additional public expenditure. In fact, th majority of those queried did support the ide of program improvement. However, althoug an occasional clamor is heard after an u usually severe winter, memories tend to b short when the time for increased appropri tions is at hand. This pattern is analogous the attitudes that often prevail with respe to other hazards, particularly floods.[21]

Attitudes can promote or reduce disru tion, largely through their effects on adjus ment. Most of the persons queried in th seven cities tended to underestimate the ha ard potential of snow, considering it to b more of a nuisance than a serious problem — as, indeed, it is most of the time. In th case of minor storms, these attitudes probab lessen disruption; that is, people exhibit litt concern for two, three, or four inches of sno and go about their business normally. On th other hand, they are apt to be grossly unpr pared for a severe storm, and thus experien substantially more disruption.

THE ROLE OF PERCEPTION

Man's attitude toward the snow hazard ca be partly explained in terms of his perceptic of the phenomenon, but this perception is di ficult to measure. Awareness of any eleme varies not only among individuals and group but with the same individuals at differe

[19] Herbert A. Simon, *Models of Man* (2nd ed., New York, 1957), pp. 196-200.

[20] This amount is derived from the number of car registrations and the percentage of vehicle owners who say they install snow tires. The estimate allows for a tire life of three years.

[21] See, for example, Henry C. Hart, "Crisis, Com munity, and Consent in Water Politics," *Lo and Contemporary Problems*, Vol. 22 (1957 pp. 510-37.

[22] This statement does not apply to Muskegc Milwaukee, and Green Bay, where extensi interviewing was not conducted. There is reas to believe that people hold the hazard in high esteem in the Midwest, as evidenced by t existence of more sophisticated snow-contr programs in that area.

ints in time and space. "In any society, in-'iduals of similar cultural background, who ;ak the same language, still perceive and derstand the world differently."[23] "The ecialized literature is replete with examples difference in hazard perception" among the perts themselves.[24] Even with these draw-cks, the study of perception can be ex-mely valuable in identifying the geograph-l implications of the milieu, particularly t part of it which contains an uncertain zard element.

In talking with people in the various cities tendency was detected, especially in the ;st, to minimize the potential danger asso-ted with snow. Statements such as these re common: "The legendary blizzard of 49 was some sort of oddity that will prob-ly never happen again."[25] "Snow doesn't ther or interfere with us or our business." view snow and ice as a challenge, some-ng to break the monotony of the everyday tine." Perhaps the slogan emblazoned on façade of the Engineering Building at the iversity of Wyoming is symbolic of pre-ling attitudes. It reads: "Strive On, the ntrol of Nature Is Won, Not Given."

Perception of any hazard is based largely experience. The slower pace of life in the ;stern cities may account in part for the her low priority granted to the snow hazard there. This does not explain the views that prevail with respect to the "blizzard of 1949," or to the other storms since that have resulted in loss of life and widespread disruption. As many as three or four first-order disruptions may occur during the course of any winter season, yet the population remains largely apathetic and ill prepared, at least as a public body.[26] An examination of additional sites, particularly those with only small and highly variable amounts of snow, should provide greater insight concerning snow-hazard perception. "On the spot" interviews should also be useful.

FUTURE NEEDS

The snow hazard demands more attention. Additional research is needed, both in the physical parameters — accumulations, moisture content, wind velocities, and so on — and in the evaluation of local adjustments at the private and public levels.[27]

An even more pressing need is for improved public service in the field of snow control. A single storm can cost any of the cities investigated considerably more than they spend on snow removal each year. Organization, coordination, and public relations are integral parts of more effective snow-control programs. Disruption could be substantially reduced if more funds were available, and if present funds were more efficiently allocated.

)avid Lowenthal, "Geography, Experience, and magination: Towards a Geographical Episte-nology," *Annals Assn. of Amer. Geogrs.*, Vol. i1 (1961), pp. 241-60; reference on p. 255.

an Burton and Robert W. Kates, "The Per-eption of Natural Hazards in Resource Man-gement," *Natural Resources Journ.*, Vol. III 1964). pp. 412-41; reference on p. 424.

For a discussion of the problems associated with the blizzard of 1948-1949, see Wesley Calef, "The Winter of 1948-49 in the Great Plains," *Annals Assn. of Amer. Geogrs.*, Vol. XL (1950), pp. 267-92.

26 The political ideology in Wyoming and western South Dakota that stresses the role of the individual may be reflected in the inefficiency of public snow-removal programs. However, speculation along this line is difficult to substantiate.

27 The writer is currently engaged in further research on this matter at sites in the south-central and southeastern United States. An analysis of disruption patterns in these areas should provide answers concerning the benefits to be expected from the maintenance of various levels of snow-control programs.

An accurate benefit-cost analysis might produce guidelines for intelligent decisions on expenditures.

We are confronted by an interesting challenge. As a technically advanced urban society we have at our disposal the organizationa and inventive abilities to deal successfull with snow, a menace that often severely ham pers activity in our major centers. Why do w not use them?

39

Let Us Make Ou Cities Efficien

Eric Beecrof

It is an astonishing fact that, in a century dominated by the idea of productive efficiency, the uneconomic organization of our cities has caused so little concern. Today, however, there are signs that the cost of blight, congestion, ribbon development and suburban sprawl is understood better than ever before and that the public's courage is being screwed up to tackle urban redevelopment in a very big way.

The time is past when traffic congestio can be viewed as a minor inconvenience – as part of the price we must pay for all ou other conveniences. It remains one of th greatest wasters of man-hours, a major im pediment both to our industry and to ou recreation. Precise estimates of loss in traffi are difficult to make; but in 1953 on the basi of Montreal experience, Mr. C.-E. Campeau now Director of the Planning Department i

Reprinted from the Canadian Imperial Bank of Commerce *Commercial Letter* (October, 1955) by permission.

at City, concluded that loss from delays of cking alone might be as high as $30 million nually. Mr. Campeau drew attention at the me time to the fact that traffic congestion our central cities "results not so much om people living *in* the cities as from more ople living *outside* them."

The basic remedies for congestion, there-re, are not the makeshift control and park-g schemes which account for so great a part the debating time of public bodies and of ace in the press, but imaginative planning the entire physical growth of the metro-lis — in transport, industry and residential velopment.

In this age of high new capital investment, panding economy and productive effici-cy, industrialists try to plan a long way ead in terms of the physical environment their enterprise. They are attracted to cities ich are taking some care to establish indus-al zones, to assure adequate transport and rehousing facities and to provide economi- and attractive home sites, recreation cen-s and other amenities of daily life. Cities ving natural advantages for industry have und that long-term planning helps to ex-oit these advantages more fully and pays man dividends to employers and employees ke in satisfying a deeply-felt need for civic de and good neighborliness.

Those directly concerned with the building lustry are examining the causes of dis-lerly growth. The Canadian Construction sociation has placed itself firmly behind ng-term community planning to achieve d maintain reasonable standards of hous- and community development through the ovision by appropriate governmental auth-ties of adequate planning controls, person- and facilities, especially at the local el . . ."

As the general state of the Canadian economy is examined with a view toward sustaining a high level of activity, the question "How are we to maintain the present rate of building?" is a vital one. Among builders as well as consumers, it is leading to a closer study of the economics of urban expansion. Hitherto this has tended to be a residual problem — a headache left over for local government — while house builders and house-buyers went single-mindedly about their immediate business. An uneasy feeling is now developing that the backlog of utility financing left to the taxpayers as such may compel a slowing up in the extension of municipal services and put the brakes on new building.

Indeed a great part of the new concern for city planning results from the bitter experience of city authorities in the past 15 years of fast growth. Everyone concerned with the housing of a rapidly-increasing population has been compelled to promote better planning of all the utilities upon which housing depends. The rates of growth shown in the accompanying table can be readily translated into a measure of the task imposed on the governments of our towns.

New housing has required not only the expenditure (mainly private) on land, building materials and labour, but a vast public investment in sewers, water supply, drainage, paved streets and power lines — not to mention schools, buses, fire and police protection and a great many other services. Beset by demands for such services from every quarter, municipal authorities have been forced by acute financial stringency to introduce orderliness and foresight into their programming of public works. Out of this experience, especially in the areas of most rapid growth, has come a determination to prevent the development of scattered residential areas which are costly to equip.

Let us look at the causes of "fringe development." In the absence of controlled growth, and faced by a grave shortage of housing, thousands of people have settled on unserviced land outside the built-up areas — in many cases outside the organized municipalities. Some just want to get away from the crowded city. Others of low income are attracted by cheaper land; many of them can build their own homes without conflicting with building codes. For many people, the lack of sewers and public water supply may seem, for the time being at least, to be minor handicaps compared to the high costs of urban life.

But, at time goes on, with the pressure of population, the sprawling suburbs become a trying problem for the public authorities; some of them eventually have to be serviced for more intensive use. In some areas, the land may be of such a character as to be unsuitable for underground utilities except at excessive cost. Others, suitable for industrial use, cannot be made available for that purpose without the removal of scattered homes already built. In still others, where a replotting is desirable for intensive residential development, the way may be blocked by existing uses. Still more expense is added by requirements for schools, fire and police protection, health services and other collective facilities. Providing such services for scattered communities represents a great financial waste; and, to make matters worse, under our antiquated system of municipal boundaries, the problem of providing the services often falls between two or more autonomous local governments, rural and urban.

It is not surprising that hundreds of municipalities are finally taking or considering measures to ensure that new developments should be compactly planned as successive extensions of the main urban area and that such areas of development should be selected for their most appropriate use. It follows tha other less appropriate uses are forbidden, bu this kind of negative control comes to be accepted where it is seen as a necessary mean to promote a sound positive program of development. Effective planning controls ar now preventing sporadic uneconomic development on the fringes of some of our cities A great deal of wasteful spending on publi services is saved, and to this extent, it becomes possible to finance a greater volume o residential and other development.

A science of land subdivision is finall being developed to enable us as property owners and tax-payers to save financially b better planning. A well-planned neighborhood, for example, has a street system i which an important distinction is made between arteries, circulation streets, and loca access streets. The less travelled streets require less heavy construction. The traditiona "grid" system is expensive because ever street is a "through" street and must be constructed as such. Economies can be effecte because local access streets may need neithe curbs nor sidewalks. In terms of a typica single-family house, such a saving is aroun $250-$300.

Costly street widenings are plaguing almos all our growing towns. If this is to be avoide in future, the major arteries must be laid ou well in advance of urban growth. Plannin must also provide that *traffic arteries* do no become used as *commercial* streets, cluttere with local shoppers and parking. The commerce of the country as a whole is gravel impeded by this confusion of local marke traffic with through traffic flow.

Modern town planning, with judicious us of a master plan, is a prerequisite for the ne form of shopping center, as it is for an efficient network of roads. Shopping centers require careful planning to spot the strategi

tions of these little nuclei of trade. This racteristic of modern cities is probably to stay — immensely preferable to the ribbons of commercial frontage which one of the most ugly and inefficient manifestations of unplanned urban growth. Ribdevelopment throttles commerce. Yet, in absence of planning, our main highways inevitably lined with stores, filling stas, hot dog stands and restaurants. Comtion forces this kind of development, even ıgh it is detrimental to our economy as a le. Setback regulations are a partial rem; by widening the ribbons, they accelerate fic. But a more basic remedy is the modern nned neighborhood and shopping center, oved from the main traffic arteries.

)ne of the advantages of neighborhood nning then is that it frees "through" traffic enables us to go to church and school and nuch of our shopping without using arthighways. Good neighborhood planning at Don Mills near Toronto, Wildwood in ater Winnipeg, Fraserview in Vancouver, ıor Park in Ottawa, or the many new hborhoods in Edmonton) attracts higher ıoderate income families and holds them e, to the benefit of local business people residents alike. For low income areas, d neighborhood planning is not only a rable objective for the residents themes; it is the best security against heavy lic costs of protection for health, welfare, cing and fire. From the taxpayer's viewt, one of the most uneconomic features ıe city is the blighted slum. Assessments low and service costs are high.

Iany of the older American cities, aided 'ederal legislation, are developing urban wal programs of great magnitude to ree old, inadequate residential and comcial areas with new. Slum clearance is a of a still larger program to provide not only decent low-rental housing but industrial and commercial facilities suited to modern needs, and civic centres, parks, recreation areas and throughways. In our own older and still-expanding cities, particularly Toronto, Montreal, Halifax, St. John's and Vancouver, a great need and opportunity is seen to make more efficient use of central space. Large projects are being conceived by both private investors and public authorities. Indeed it seems safe to predict that in the next 15 years: (1) urban redevelopment will be one of the largest fields of investment for private and public funds; and (2) it will be accomplished by bold and ingenious combinations of public and private initiative.

At first there may be painful struggles over the respective roles of public and private enterprise. Both may be hampered, not only by lack of precedents for dealing with such large and complex situations involving difficult issues regarding property rights and welfare problems, but by the absence of satisfactory enabling legislation. A great deal of patience and imagination will be required to devise appropriate legislation and administrative methods to reconcile the many technical, financial and social factors.

The city planning department can play a role that is very useful to all concerned. Its object should be to look at the scene as a whole in day-to-day teamwork with all the operating and financing agencies and, by so doing, to help city authorities to see the full significance of the separate projects of streets, housing, industries, commerce and transportation.

An immediate practical problem in redevelopment is the assembly of central area land. Such assembly of land has to be done on a large scale and with very careful planning. Otherwise the new development, by failing to generate new growth, may result only

TREND OF URBAN GROWTH IN TERMS OF HOUSING STOCK 1951-1955 *

Centre	Housing Stock in Units 1951	Estimate 1955	Annual Rate of Increase 1951-1955 %	No. of Year Required tc Double at this Rate
Metropolitan Areas				
Calgary	40,235	50,671	5.77	12.0
Edmonton	46,395	59,614	6.27	11.0
Halifax	29,640	33,891	3.35	20.7
Hamilton	68,640	77,567	3.05	22.7
London	32,835	37,793	3.52	19.7
Montreal	334,705	394,138	4.09	17.0
Ottawa	66,265	74,282	2.85	24.3
Quebec	54,930	60,993	2.47	28.1
Saint John, N.B.	19,735	20,574	1.04	66.7
St. John's, Nfld.	12,995	14,598	2.91	23.8
Toronto	273,200	322,053	4.11	16.9
Vancouver	153,975	175,257	3.24	21.4
Victoria	31,620	35,405	2.83	24.5
Windsor	41,595	45,674	2.34	29.6
Winnipeg	95,955	106,906	2.64	26.3
Sub-Total	1,302,720	1,509,416	3.68	18.8
Other Major Cities				
Brantford	14,960	15,904	1.53	45.3
Fort William / Port Arthur	19,550	21,005	1.79	38.7
Guelph	8,093	9,238	3.31	20.9
Kingston	13,349	14,121	1.40	49.5
Kitchener	16,700	19,050	3.29	21.1
Moncton	10,989	11,603	1.36	51.0
Oshawa	13,426	15,112	2.96	23.4
Peterborough	10,785	11,787	2.22	31.2
St. Catharines	17,485	18,173	0.96	72.2
Sarnia	11,100	12,920	3.80	18.2
Sault Ste. Marie	8,838	10,604	2.83	24.5
Shawinigan Falls	10,628	11,193	1.30	53.3
Sherbrooke, P.Q.	13,077	14,176	2.02	34.3
Sudbury	15,864	16,943	1.64	42.3
Sydney / Glace Bay	21,418	22,037	0.71	97.6
Trois Rivières	14,079	15,178	1.88	37.9
Sub-Total	220,341	239,044	2.04	34.0
Total	1,523,061	1,748,460	3.45	20.1

* Based on 1951 Census and estimates of Central Mortgage and Housing Corporation.

a waste of investment, public and private. is principle — *that the investment risk is nimized by the scale of the new projects d by carefully envisaging them in their g-term relationship to one another and to entire environment* — is generally ac- ted. But the actual working out of a land embly procedure may require some revi- ns in our traditional thinking. In practice, vate investors are finding it difficult to quire all of the land needed for a well- nned redevelopment project. They tend to k then to the public power to acquire land.
Since, in some Canadian jurisdictions, a nicipality cannot acquire land except for public purpose (a park or city hall, for mple), it cannot assist an otherwise sound ject by acquiring the land and leasing or elling it to private developers. There ap- rs also to be a limitation in present federal islation. The National Housing Act, which horizes the Government to share the costs a city in buying up blighted areas, limits s authority to the assembling of blighted d for low-rental housing or for public pur- ses. This is good as far as it goes, but it is dent that, from the viewpoint of sound nning, low-rental housing is only one of objects — however important — in urban evelopment. And indeed low-rental hous- projects may go forward much faster if they are viewed as a phase of the larger schemes of redevelopment.*

A challenge to investors and governments to tackle redevelopment on a large scale was thrown down by the Hon. H. R. Winters in a recent speech at Montreal at a large joint conference of the Community Planning Association of Canada and the American Society of Planning Officials. This problem, he said, could "never be solved lot by lot and piece by piece." The interest of private investors he thought would be enlisted by "the prospect of a scale of operation that will change the character of large areas." "I am not speaking," he added, "about streets or blocks, much less parts of them. This is not minor surgery but major operations. . . . Planning and the acquisition of land on the scale I have suggested may call for public action. But if redevelopment is more than a salvage operation, if the older parts of our cities have a vital role to play in the whole urban structure, then their redevelopment can properly be made the object of private investment too."

In the same speech, and in connection with new housing, Mr. Winters estimated that in the next 20 years, $25 billion would be in-

* *Editor's Note.* The *National Housing Act* has been amended substantially since to permit "the larger schemes of redevelopment".

---

he table opposite should not be taken as a prediction of future growth. The two right-hand columns, vever, suggest the magnitude of the task which our municipal tax-payers undertook in servicing v housing. Even if the rate of housing construction were to drop considerably, the problem would nain acute. For besides the servicing of new areas, our municipalities are still burdened with the ts of much haphazard, uneconomic real estate development which is already completed or lerway.

Vo one wants the rate of housing construction to diminish as a means of solving the economic ght of our tax-payer. It must be maintained if we are to have an adequate supply of decent hous- . The remedy lies in seeing to it that our future residential sites are selected and subdivided not y with an eye to making them more attractive but on the basis of sound economic principles.

vested in raw land and residential structures and at least $2 billion in residential water and sewer mains and frontage roads (to say nothing of the trunk services required and the additions to central water and sewage disposal plant). This estimate included nothing for schools, churches, hospitals and other institutions. The impact of this investment on Canada over the next two decades would be enormous and would call for vital well-thought-out decisions as to the location of industry within cities, the location and layout of housing areas, the location and design of schools, the routing of roads, the extension of utilities and indeed all those matters which together determine the efficiency of our cities.

Mr. Winters then appealed to city planners to direct their thought to the problem of making the city a more efficient economic unit. There was a danger, he said, in breaking the problem down into parts (traffic, suburban sprawl, slum clearance). Unless the whole unit of the city was envisaged, the parts could not be dealt with soundly. There was also a danger that, by emphasizing beautification and social improvement, we would divert attention from the economic processes which form the basis of the city's existence.

Can we expect a conflict between the concept of the city as an economic unit and the concept of its beautification and social improvement? It seems rather academic to debate such a question at the present stage when many of our cities are both inefficient and ugly and, in some of their quarters, debasing to human character. If our ideals of economic efficiency, welfare and aesthetics are conceived in the broadest terms, we may expect no basic conflict.

It must be admitted that the City Beautiful idea which came into vogue a generation or two ago might have been more successful in practice if it had been accompanied by a de termined attack of the city's rehabilitation a an economic unit. Similarly, we can probabl expect that faster progress will be made to ward elimination of slum-fostered evils if w enlarge our vision of the whole city as a fiel for enlightened investment.

Our cities have greatly increased their tech nical staffs to cope with their planning prob lems. In few places now are there any fals conceptions of the professionatl planner as destroyer of individual freedom. The deman for the planner is in fact outrunning the sup ply; and Canadian cities, once they hav established planning departments, tend to re tain or enlarge them. In most of our cities where planning departments have bee established for some time, they have bee geared intimately into the technical team work of the administration so that they serv and are served by all departments in develop ing a coherent program for the physical devel opment of the city. As part of the administra tive system, the City planning staff acts as kind of intelligence branch. Detailed plans c the whole city are made available to guid executives and elected officials in making im portant decisions. A master plan of develop ment outlines a desirable program of publi works to be carried out and makes possibl systematic capital budgeting.

City planning also helps to bring into bal ance the various purposes for which citie exist. The inhabitants are both producers an consumers. As producers we are concerne with the total efficiency and productivity c the operating plant: public services and trans portation must be economic and free c wasteful interruptions. As consumers, we a want a city that is spacious, beautiful an attractive to live in; we want as much luxur as possible for our home-buyer's or tax payer's dollar. The ultimate decisions as t at we get in urban physical development, oan beauty and urban welfare are made by r elected bodies on the basis of the de- ands made by their constituents and the technical advice they receive from their staffs. We are making gradual progress in both these respects as urban growth is directly examined and better understood.

Sherbrooke District, Edmonton

As might be expected in a city having the ost rapid rate of growth in Canada in the st five years, Edmonton is experimenting dly in the planning of its physical develop- ent. It is becoming widely celebrated for its anned neighborhoods. Thirty of these ighborhoods have been completed since 50 and more are planned.

One of these completed neighborhoods, erbrooke, is briefly described herein. In 49, it was proposed to "develop" this area the "grid-iron" basis. But this plan was ncelled and replaced by a new scheme. It s not easy to incorporate into the new sub- ision 50 scattered houses which had been ilt in the previous 10 years; but, with 5 ex- ptions, this was accomplished.

A comparison of the design features of the l and the new plans shows some of the dif- ences from the viewpoint of economics, pearance, conveience and safety. Mr. Dant nts out that: "a considerable reduction in lengths of both local streets and service es was possible, saving home-owners of the y many dollars of unnecessary street con- uction and maintenance cost. The land so ed was put into more fruitful use, adding total investments and providing the City h increased real estate tax year by year. ere were some slight increases in utility costs, as compared with the 'unplanned' scheme, and some added costs in curved curb- ing, but these were far outweighed by the above savings."

Perhaps the greatest saving is in the diffi- culties and expense to home-owners and tax- payers which, in the absence of such a sub- division plan, arise over the years as a result of failure to provide suitable sites for schools, churches, shops and play space, as well as safe traffic conditions.

The aesthetic and operating advantages of irregular street patterns are still being debated in Edmonton, but it seems unlikely that there will ever be a reversion to the "grid". In fact, it seems more likely that, for reasons of eco- nomy and better neighborhood living, the grid will lose its strong traditional appeal in all of the growing cities of the West.

There may be many imperfections in the new type of subdivision, and the future may witness many changes in the theory and prac- tice of neighborhood planning. But the elected officials of the City of Edmonton and Mr. Dant and his associates in the City ad- ministration are to be congratulated for the bold and extensive pioneer work they have done in a field related so closely to the hearts and pocketbooks of millions of Canadians.

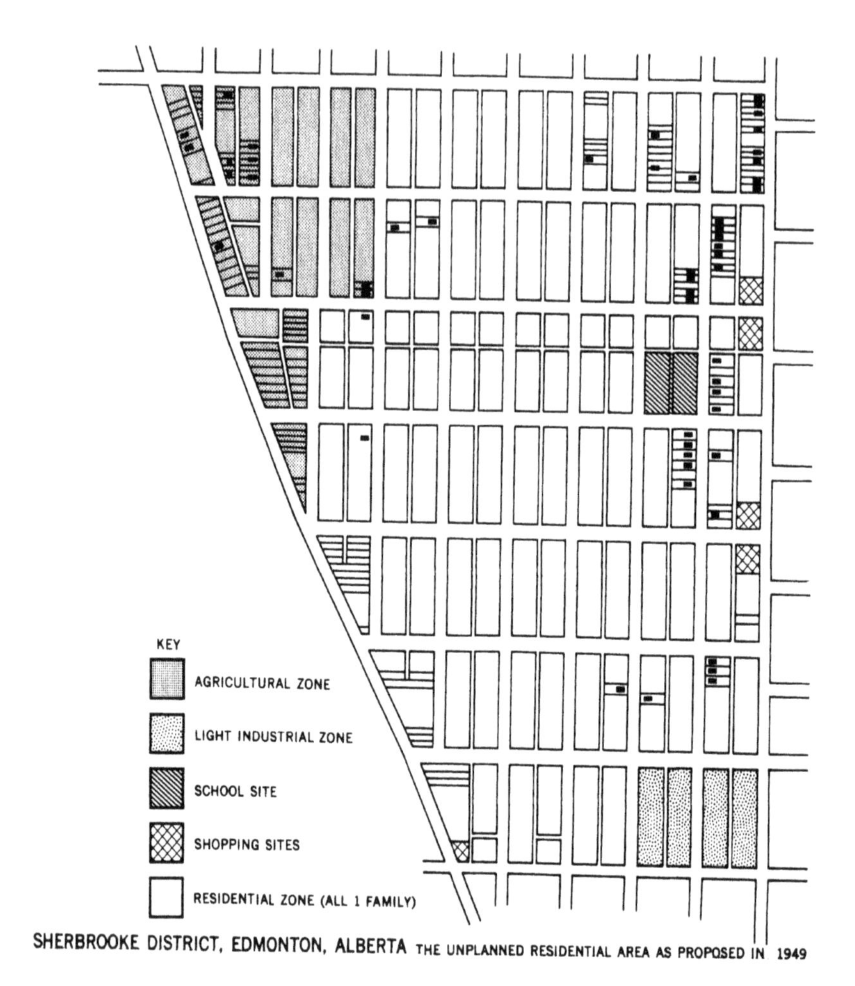

SHERBROOKE DISTRICT, EDMONTON, ALBERTA THE UNPLANNED RESIDENTIAL AREA AS PROPOSED IN 1949

Unplanned Scheme

1. There is a complete absence of parks and recreation areas.
2. No church sites are provided.
3. Shopping areas are small and located on the perimeter instead of being central. No off-street parking is provided.
4. Only one school site is provided, and it is not centrally located.
5. All roads are of equal width and are potential "through" streets.
6. There is no variety in types of residential zones.
7. Some very long, narrow house lots are shown. These are uneconomical. Others are poorly shaped.
8. There is an unnecessary duplication of utility lanes in some places.
9. There are variations in the width of the main road. Half-jogs in the roads in some places are unsafe, and there are some dangerous junctions.
10. The light industrial zone has no place in a residential area.
11. Likewise the agricultural zone seems out of place, especially with houses allowed in it in excess of a ratio appropriate to such a zone.
12. Residential zones are all 1-family.

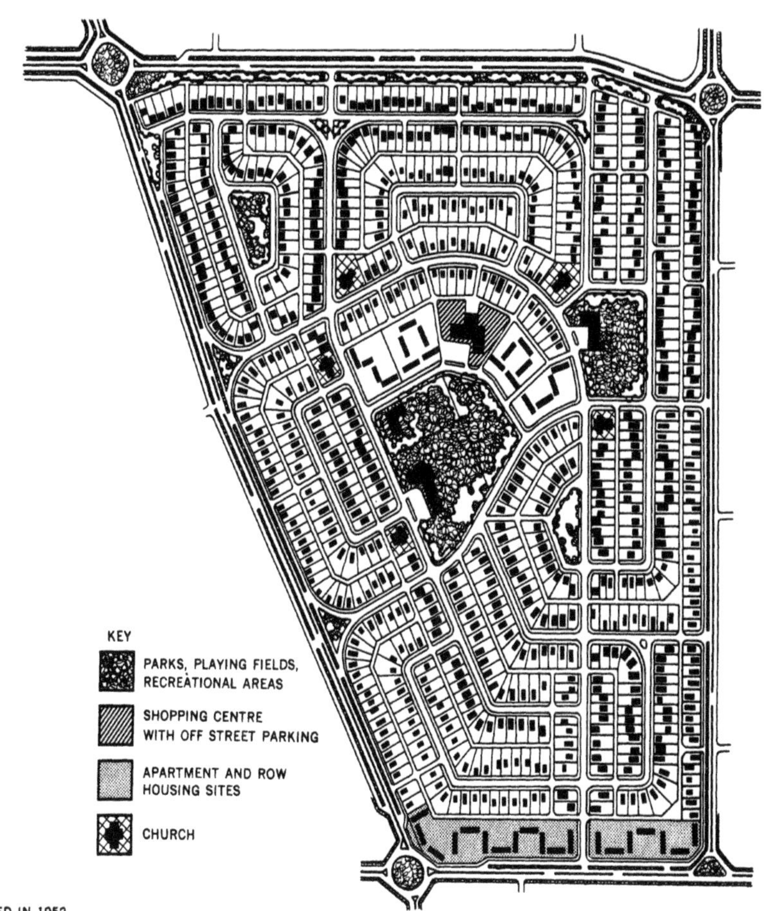

S FINALLY DEVELOPED IN 1952

nned Scheme

Sites are provided for parks, playing fields and recreational areas.

Church sites are provided in convenient ocations.

There is a central shopping area, with off-treet parking.

Two school sites are provided, reasonably ccessible from all parts of the neighbour-100d. The Catholic school is off-center because it also serves adjacent neighbour-100ds.

"Through" arterial highways, of adequate vidth, are separated from local service oads by limited access planted strips. Thus both "local" and "through" traffic re safeguarded. There are feeder roads for bus routes. Local residential streets are designed in such a way as to discourage "through" driving, yet remain adequate for local purposes.

6. At the corners of the area, there are intersections designed to keep "through" traffic moving.
7. One-family housing is created in an aesthetic as well as a functional setting. Setbacks are arranged to allow for a "rhythmic variation". A buffer strip separates housing from an adjacent industrial zone.
8. There are also apartments and row housing in a variety of types.
9. A neighborhood "focus" of larger buildings and open space is included as an essential ingredient of a well-designed residential area.

COMPARATIVE STATISTICS

	UNPLANNED SCHEME		PLANNED SCHEME	
Internal Roads	37,960 lin. ft.	57.2 acres	27,600 lin. ft.	41.7 acre
Utility and Service Lanes	27,460 lin. ft.	12.6 acres	23,400 lin. ft.	10.7 acre
Public School Space		2.14 acres		6.0 acre
Catholic School Space		—		4.0 acre
Recreation League Area		—		2.0 acre
Parks		—		4.0 acre
Churches		—		1.5 acre
Shopping Area(s)	(scattered)	1.9 acres		1.5 acre
Agricultural Zone		19.0 acres		—
Light Industry		6.56 acres		—
Miscellaneous (planted islands, etc.)		—		11.6 acre
Residential				
1-Family Type	1,068 units	122.6 acres	896 units	112.0 acr
	3,400 people	—	3,584 people	
Row Housing	—	—	245 units	17.5 acre
			980 people	
Apartment Type	—	—	190 units	9.5 acre
			760 people	
Total	1,068 units	122.6 acres	1,331 units	139.0 acr

	UNPLANNED SCHEME	PLANNED SCHEME
TOTAL ACREAGE OF SCHEME	222 ACRES	222 ACRES
TOTAL POPULATION (residential)	3,400 PEOPLE	5,324 PEOPLE
GROSS DENSITY (i.e. total no. of dwelling units divided by total acreage of scheme)	4.8 dwelling units/acre	5.7 dwelling units/ac
NET RESIDENTIAL DENSITY (i.e. total no. of dwelling units or persons divided by total acreage of residential property and the roads and lanes which serve them)	26 persons/acre 6.27 units/acre	27 persons/acre 7 units/acre

umbernauld

ew Town

40

H. Wilson

e building of new towns in post-war
tain has been a remarkable enterprise. At
present time work is in progress at various
ges on fifteen towns and six further pro-
ts have recently been announced.

The towns are each under the control of a
velopment Corporation appointed under
New Towns Act 1946 and most of the
ital is loaned by the Government although
re are many opportunities for private
erprise to take part in the erection of some
the factories, houses and shops and such
ldings as schools and churches are pro-
ed by the appropriate authorities. The
rporations are responsible for the planning
d development of the towns to create self
ntained communities where people can live
d work and spend their leisure in pleasant
roundings.

The towns fall into five functional groups.
ght of the towns, Basildon, Bracknell,
awley, Harlow, Hemel Hempstead, Steven-
, Hatfield and Welwyn Garden City are
ked to the London overspill problem and
l provide homes and workplaces for over
f a million people. The sites of these towns are some twenty to thirty miles from London. Three of the new towns in Scotland, East Kilbride, Cumbernauld and now Livingston are similarly related to the problems of Glasgow although in the case of the first two towns the sites are rather nearer to the city. Two towns, Skelmersdale and Runcorn are to be built to take overspill population from the Liverpool conurbation and the towns at Dawley and Redditch will have a similar relation to Birmingham. The remaining five towns are being built to serve the needs of local industry — Newton Aycliffe and Peterlee in County Durham, Corby in Northamptonshire, Cwmbran in Monmouthshire and Glenrothes in Scotland in the vicinity of a new coalfield in Fife. It has since been announced that this coalfield is to be closed and Glenrothes will, therefore, be developed to help in solving the acute overspill problem in Glasgow.

The first fourteen new towns were designated between 1947 and 1950 and the Development Corporations were given population targets, varying from 15,000 to 80,000 although many of these have since been increased. These towns have been planned on

printed from *Plan,* Vol. 4, no. 2 (1963), pp. 70-85 by permission.

the basis of a number of residential neighbourhood units each complete with its own local shopping centre, schools and community and other public buildings.

Cumbernauld is the last new town on which construction work is in progress having been designated in 1956. It was one of the towns proposed in the Clyde Valley Regional Plan prepared in 1946 by Sir Patrick Abercrombie and Mr. (now Sir) Robert Matthew. It is being built to assist in the relief of congestion in Glasgow and the majority of the proposed total population of 70,000 will come from that city. In its siting it differs to a marked extent from all the other new towns so far developed and in its planning a number of principles have been introduced which are either new to or represent developments of recent practice. In their work the planners of Cumbernauld have had the advantage of being able to learn a great deal from the experiences of the other new towns.

Planning Aims

The main planning aim at Cumbernauld has been to achieve a unified coherent structure for the town with an urban character arising out of compact planning at a somewhat higher average density of development than in the other new towns. It should be stressed that this policy has been implemented on the basis of good standards of open space and compactness within the urban area has resulted from the placing of the major elements of open space on the periphery of the town. The neighbourhood unit concept of planning has been abandoned at Cumbernauld and instead of having a series of local centres in the main urban area there will be one town centre which will require to be developed at a relatively early stage. It is felt that the neighbourhood unit has little social significance, social contacts generally being con fined to the residents in small groups o houses or being enjoyed by people of par ticular interests on a town scale.

Towns are essentially meeting places; a Cumbernauld it is considered that this func tion can only be fulfilled by recognising th rights of the pedestrian. Whilst planning fo the motor car, therefore, freedom for th pedestrian has been achieved on the basis tha the conflicting needs of vehicles and pedes trians can be satisfied only by maximun separation.

Much has been learnt from the other nev towns particularly in terms of population an employment balance. New towns tend to at tract a relatively young cross-section of th population with the result that children are i excess of the national average and olde people represent a considerably smaller pro portion. Efforts are being made to redress thi state of unbalance by taking older peopl from Glasgow whenever possible. The popu lation structure will also be helped if a wid range of employment can be achieved wit particular emphasis on service industry in cluding office employment and rather les manufacturing industry than has been ap parent in the other new towns. The Cumbe nauld proposals are for 45 percent of employ ment in manufacturing industry and 48 per cent in offices and services, the remainin balance being taken up by construction mining and agricultural employment.

Site

The site occupies a central position in the low land industrial belt of Scotland between th deep inlets of the Firths of Clyde and Fort and contained to the north and south by up land areas. The trunk road from Glasgow t the north of Scotland passes the site of th

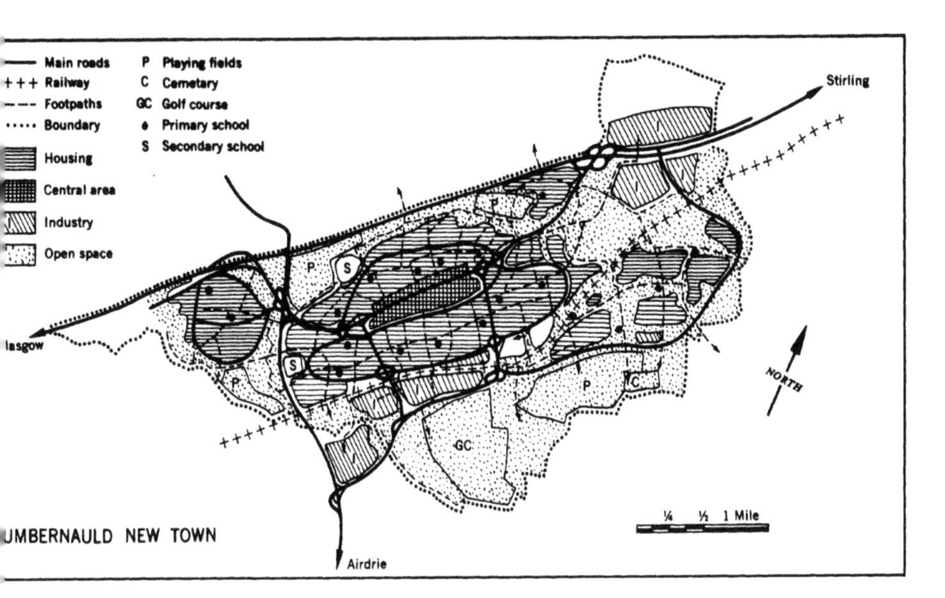

w town where it is joined by the trunk road m England. The railway from Glasgow to north runs through Cumbernauld Station d there are also rail connections to the in line to England. The area designated by Secretary of State for Scotland comprises ne 4,150 acres and is triangular in shape, e miles long from north east to south west d two miles across from the north west to th east at the widest point. The site is erally hilly and the area of land suitable development is restricted by the presence fireclay and coal workings, pockets of peat d steep slopes and part of the site is isolated two deep glens. The main element on the is a hill in the form of a broad hogsback ween the valleys on the lines of the Glas- w trunk road and the Glasgow railway with readth of one mile and a length of two and alf miles. The altitude varies from about 0 feet above sea level in the valley to 485 t at the highest point on the hilltop. There steep and very broken slopes on the north st side of this hill and longer and more ntle slopes on the south east.

The watershed of Scotland crosses the site, the southern part draining to the Clyde and the northern part to the Forth. The whole site is exposed to the prevailing south west winds which blow down the valley and along the length of the site and many tree belts have been planted across the hill in the past to give shelter from these winds. The surface deposits are principally boulder clays.

The conditions and general characteristics of the site determine the areas suitable for development which, apart from the main hilltop, consist of a smaller hill to the south west, three areas to the north east and a series of sites to the south of the main hilltop on level land in the valley including one site, particularly suitable for industrial development, just outside the designated area.

Plan

It soon became clear that the main hilltop, with a gross area of about 930 acres for building, would provide the most suitable site for the major development of the new town since it is the only section of the area capable of being used for a large scale comprehensive scheme. The form of the hilltop, from which there are fine views in all directions and which can be seen from many points outside the area, with its clearly defined limits lent itself to the general conception of this part of the town as a compact urban centre containing a population of 50,000 with surrounding recreation areas, the whole set against the background of open hilly country. The town defines its own urban limits and its own green girdle of open space. The linear town form has many advantages, particularly in association with the linear central area. Apart from ensuring a more efficient and economical road and footpath system, the linear town allows the great majority of the population to live within a ten minutes walk of the town centre or of the open space. At Cumbernauld every part of the town can be in close touch visually and physically with the surrounding countryside.

The ultimate population of 70,000 will be achieved by the development of a series of satellite villages around the main hilltop each with its own local centre but relying on the central area of the town for the main shopping and recreational facilities. The overall pattern for the town can, therefore, be regarded as that of a cluster city in which there is one major urban centre surrounded by but separated from a number of smaller compact urban units of varying size the whole being linked together by an efficient road and footpath system. This process might well be extended in the future providing a very flexible approach to the ultimate size and detailed pattern of the town.

Housing

Housing design in Cumbernauld has been strongly conditioned by a number of factor including:

(a) attempt to achieve integration o dwelling design and layout
(b) need to achieve within ruling densitie the maximum number of houses o the ground
(c) site conditions with slopes varyin from almost level to very steep an with fine views out of the site
(d) the need to secure adequate daylight ing and sunlighting standards
(e) privacy for main rooms and gardens
(f) flexibility in the use of space withi dwellings
(g) maximum separation of pedestrian and vehicles
(h) the provision of one car storage spac for every house plus parking spac for visitors' cars

Net densities average about 80 persons pe acre and vary on particular sites between 6 and 120 persons per acre with some sma areas at a rather lower density. The solution which have been evolved in response to thes conditions have produced housing scheme far removed in form and character from mos of the post war designs of local authoritie and private enterprise developers in thi country. The concept of streets lined wit houses has been dropped in favour of design in which the movement of vehicles and pedes trians and the planning of the individua dwellings to achieve good living condition are highly integrated. The result is a strongl developed pattern of buildings and publi and private spaces carefully related to th site conditions and with considerable atten tion being given to all the details of change of level, pavings, planting, fencing, etc.

In higher density schemes it is possible t achieve separation of vehicles and pedes ns vertically but at the lower densities izontal separation is being used at Cum- nauld. The general pattern is for the car nove outwards (in contrast to the pedes- n who moves inwards) to join a collector d forming part of the traffic distribution em. The effectiveness of a road pattern ving housing where high speeds are not controlling factor can be measured by the ay or waiting time experienced at any nt. At the start of his journey near the se where the pedestrian and the car meet car should move slowly and fairly long ts are permissible. As the car moves ough the hierarchy of roads, delays at k flows are decreased until there is no ay at all at the town throughway junctions. estrian movement varies according to the ition in the town; on the main roads there omplete separation, at the house, pedes- n and car meet. Within housing areas the estrian is focussed inwards to the main n footpaths with access on foot to the p, pub, primary schools and play areas. object is to provide not only safe routes safe areas for social life to develop. With egation of traffic there are opportunities increased social relationships within the sing group while the mobility of the car eases the range of social selection outside group.

At Cumbernauld the housing layouts are erally based on a meshed system of roads footpaths. Houses are approached by r footpaths and garages are grouped in ks alongside the roads at the ends of ter- s. In some areas road access is provided ne side of the house and footpath access he other. This involves quite separate pat- s of roads and footpaths with complete aration. This form of layout has many antages and it has been used at Cumber- nauld with a wide frontage house with a through hall with an entrance door at either end. On another site the area is divided into units each containing about 200 houses sur- rounded by peripheral roads giving access to garaging areas and to a number of tree planted culs-de-sac driven into the housing areas to provide facilities for service traffic and visitors' cars only. The centres of the units are kept clear of traffic although of necessity there has to be a greater average distance between house and garage.

Car storage on housing sites is being dealt with in various ways including blocks of in- dividual lock-ups in straight or curved ter- races, garages attached to houses or at the ends of the gardens, car ports consisting of a roof and back wall, garages or car ports under the houses or blocks of flats and paved parking space reserved for particular cars. All these types are being used in various parts of the town according to site conditions and layout.

Privacy in houses and private gardens is being achieved not by the amount of space between buildings but by the design of the house and the space involving in particular the use of single aspect houses to prevent overlooking. These are of wide frontage type of houses which are single storey on the entrance side and two storey on the garden and sunny side. On steep sites split level de- signs are being used with living rooms on the first floor, particularly where there are fine views out of the site.

Great care is being taken with the land- scaping of the areas with considerable use of pavings of various materials and textures to- gether with the planting of trees and shrubs. Toddlers' play spaces occur at frequent in- tervals and provision is also being made for sites for playgrounds for older children.

Central Area

With a compact plan the central area of the town can be within easy distance of most of the houses and at Cumbernauld the main shopping provision will be concentrated in the centre. Local shopping needs are being satisfied by the provision of individual shops throughout the residential areas on the basis of one shop to about 300 houses; these shops will sell a wide variety of merchandise. Local centres will be provided in the satellite villages. In order to prepare a programme on which to base the design for the central area a study of shopping provision was made and particular notice was taken of shopping trends on the North American Continent. It is felt that the successful centre should be capable of adapting itself to unforeseen patterns of consumer goods distribution if it is to avoid obsolescence. Indeed the need for flexibility in the use of space and for protection for the shopper from weather leads to the design of very much larger building units of a more comprehensive nature each accommodating varying uses at different levels with the necessary circulation space contained within. In assessing the amount of shopping space required in the town, probable sales for each trade category were calculated and converted into the floor area required by assessing the sales per square foot for each trade. For this purpose figures were obtained from a sample group of one hundred towns with broad similarities to Cumbernauld. A total floor area of 620,000 sq. ft. for the town was calculated, divided to give 415,000 sq. ft. in the central area and the remainder for the local shops and village centres.

The amount of car parking space was calculated at some 3,000 car spaces with a possible rise in the future by another 2,000 spaces. Separation of pedestrians and vehicles is, of course, an essential requirement for major centre and at Cumbernauld this wi be achieved vertically by the provision of multi deck centre with roads and car park underneath. The site is south of and just be low the ridge of the hill with road access fro the major junctions on the radial roads an with pedestrian access from the main foo path system on the northern and souther slopes of the hill. In this way it is possible fo the centre to take a linear form and to b planned on various levels. Road access wi be by a dual carriageway running along th ridge of the hill and under the centre. Th roads give access to car parks and also areas where goods vehicles can load and un load. After leaving the car the driver an passengers are able to move to the pedestria decks above by escalator or lift. Goods a taken directly from the loading docks int the storage areas of the shops or are co veyed by hoists to the shopping or storag levels above. A 'bus station is provided with the centre from which access can be gaine to the upper floors. All the floor levels abov the road and car parks are for the exclusiv use of pedestrians and these decks are linke by steps and ramps and bridges to give eas communication throughout the centre. Th upper floors are laid out in squares and pro menades, some covered and enclosed to pro vide shopping halls, others open. All th facilities and amenities provided in a larg town centre will be available in a compa yet spacious form and will include nearly te acres of shopping floor space together wi churches, hotel, restaurants, cinema, danc hall, public houses, business and profession offices, local and central government office town hall, technical college, library, club etc. Above the centre there are terraces flats and maisonettes where the residents ca enjoy the advantages of being in the centr thout the bustle and noise. Around the ntre there is space for open terraces, tree anting, an arena for sports events, exhibins and fairs and there are also sites for fire ation and hospital.

dustry

vo main industrial areas have been proded, to the north and to the south of the ain hilltop site and as well as providing es for purpose built factories to be erected the Corporation or by individual industrists, the Corporation are building advance ndard factories in units of 20,000 sq. ft. ese have been specially designed to ensure aximum flexibility in the use of space and n be divided to form two smaller units or n be enlarged to provide a wide range of or areas up to 60,000 sq. ft. The Corporan are also erecting three storey flatted facies within the housing areas to provide ace for the small industrialist or as a stag; post for the larger industrialist who wishes train staff while his factory is being built. e units of accommodation in these build;s vary from 300 sq. ft to 3,300 sq. ft. on y one floor.

mmunity Services

mary school sites are being provided in ation to the housing areas with access from footpath system and sites are being reved for secondary schools on the fringe of hilltop where they can be situated in close oximity to a main road and 'bus service d their playing fields can be in the recreanal area surrounding the town. Churches the various denominations are being sited convenient locations throughout the town hough some of the denominations with ly one church will have sites in the central a. Residents meeting rooms are being provided throughout the housing areas and there will also be provision for larger community halls although there will be a concentration of such facilities in the town centre.

Open Space and Recreation

A comprehensive scheme for the provision of open space of various kinds has been prepared on the basis that the whole of the peripheral areas of the town should form a continuous recreational and amenity system which will vary in use from place to place according to the demands of a particular development or to the requirements of the town as a whole; thus there are playing fields, woodland areas, parks and walks of various kinds and also reservation for a golf course. The open space system runs between the satellite villages and the main hilltop site and these are linked together by main pedestrian routes taken through the open spaces well away from all vehicular traffic. The aim has been that just as nobody on the main hilltop will live within much more than ten minutes walk from the town centre so no one will live more than ten minutes walk from the girdle of open space.

Landscaping has been considered as much more than being concerned with the design of the green spaces within the town. Everything within the town is landscape or, more properly, townscape. There must be as much shelter planting as possible on this exposed site with bold masses of trees on a scale large enough to contrast effectively with the bold mass of buildings and it is hoped that this interplay of trees and buildings may be one of the chief means of attaining ultimately a clear coherent pattern for the town as a whole. Except where openness is functionally required space within the town is regarded mainly as a site for trees contributing to shelter but also reinforcing a sense of enclosure that is one of the underlying qualities of urbanity.

Communications

The organisation of an efficient communications pattern in the town has been considered as a basic principle of the plan. The conflicting needs of vehicles and pedestrians can, of course, only be satisfied by maximum separation; this should be achieved to cause both the minimum unnecessary inconvenience but in general the car can be made to take the longer journey. The principle adopted at Cumbernauld is that if the car is to be accepted in the town then its reasonable requirements must be met; on the other hand it must be subjected to proper and adequate control.

A completely separate system of pedestrian paths is being provided throughout the town while conversely footpaths will not be placed alongside main roads. Local shops, primary schools, churches, pubs and other elements meeting the communal needs of the population are associated with these paths, many of which will pass through or terminate in the town centre. Where a path crosses a road other than a housing development road a bridge or underpass is being provided.

The town road system has been worked out after the preparation of traffic projections forecasting vehicular movements some twenty years ahead with an assumed rate of car ownership of one vehicle per family. Any study of traffic in Great Britain has to be made against a background of inadequate research and a great many of the assumptions in the Cumbernauld study had to be based on experiences in America. Desire line diagrams indicating ideal paths for traffic moving about the town were prepared and from these diagrams an ideal road pattern was evolved. After a study of various alternatives based o this pattern, one eventually emerged as pro viding the best solution when considered i detail in relation to the comprehensive de velopment of the whole hilltop area.

To ensure an easily recognisable communi cations pattern the multi purpose road wa discarded and a traffic flow hierachy worke out with local development roads, collecto roads, main town radial roads and trun roads. The road plan for the main town are incorporates three radial roads running fron the trunk roads up on to the hilltop wher they are connected by radial link roads wit junctions giving access to the central are and by a ring road and a spine road to th south of the centre serving the various hous ing areas. The major junctions have been de signed to be grade separated although it wi be possible to carry out the work in stages.

The pattern of roads and paths has bee evolved as part of the general planning pro posals for the town; it has not been impose on the town to the possible detriment of de tailed development proposals. The large scal of the road works will present considerabl visual problems and it is felt that the onl reasonable solution lies in the very carefu treatment of cuttings and embankments o viaducts together with the planting of tre belts alongside the major roads.

Rail provision exists although the presen station will have to be considerably enlarged The station is well situated in relation to th town and there will be direct vehicular acces to the ring road as well as a pedestrian rout to the town centre.

Engineering Services

Various services are being provided, by th County Council in the case of main drainag and water and by the statutory bodies in re ect of electricity, gas and telephones. The stence of the watershed requires the proion of two sewage disposal works with in trunk sewers running on either side of watershed, along the foot of the main hill. separate surface water drainage system is ng provided with balancing ponds at either d of the main hill to deal with discharges der abnormal storm conditions. These nds are being designed to be of amenity ue in the town open space system. The t that the site is situated on the watershed ans that the existing streams are very all and, therefore, care has to be taken to oid flooding in the lower reaches.

Agreement has been reached with the eleccity and the telephone authorities to place derground all cables within the town area; cisions which have a marked effect on the al environment. It is also worth noting that mbernauld is close to the local television tion and, therefore, no external aerials are quired.

nning Methods

om the beginning every effort has been de at Cumbernauld to achieve unity of sign in the town with the integration of ldings, planning and townscape. One of principal methods adopted to this end by Chief Architect and Planning Officer was t of designing by teams drawn from the ious professions involved. The whole of technical staff of the Corporation has erated in his department and at all levels office is organised in teams consisting of hitects, planners, engineers, quantity surors and landscape architects. Projects are cussed at all stages and at all levels, each ofession making its own contribution. It is t possible for any one profession to work isolation on these problems and it is also felt that solutions arrived at as a result of team working are more likely to fit in to the overall conception of the town than by other methods.

The planning proposals for the new town are set out in a document called the 'basic plan' accompanied by various reports. It is intended that this should be a plan to guide but not to stifle the growth of the town and quite deliberately there has been no attempt to produce a master plan to fix future development once and for all. Most important of all there is an overriding need for a comprehensive approach to the detailed design of the whole town within the framework of the plan. With compact planning complete integration of all the space-consuming elements in the town is important and this involves a special technique in terms of planning procedure. Instead of reserving sites for individual buildings or uses it becomes necessary to settle the design form of the development at an early stage and to fix the site boundaries after this has been done. In this process the various architects engaged in the design of the individual buildings are briefed in terms of the requirements and problems to be dealt with and also on the character of the site in relation to the surroundings. All this involves great flexibility in the legal and technical processes involved. It is appreciated that these matters are comparatively simple in the context of a new town development where there is complete ownership of the land by the Development Corporation. On the other hand it does seem to be a problem which must be faced in the redevelopment of existing towns.

Conclusion

By March, 1963 some 1,800 houses had been completed and another 2,000 were under contract. Detailed designs have been produced for almost half of the ultimate number of houses required in the town and the design of the first stage of the central area has been prepared. Work is also complete on over half of the ring road and the first of the major interchanges, that at the junction of the northern radial and the ring road is now under construction.

Cumbernauld has been called the first of the Mark II new towns. In its design an attempt has been made to face up to the problem of the motor car but at the same time t create a town which will provide a good setting for the people who have to live and wor there. British town planning has moved fa from the garden city days and Cumbernaul represents, perhaps, a return to an earlie phase of town design, when towns achieve an urban quality which has so often bee lacking in recent examples. Design, however must look to the future and here is the crucia test of the success or otherwise of the Cumbernauld approach.

www.ingramcontent.com/pod-product-compliance
Ingram Content Group UK Ltd.
Pitfield, Milton Keynes, MK11 3LW, UK
UKHW020410010325
455677UK00029B/830